SCIENCE
ANNUAL

A Modern Science Anthology for the Family

1985

This edition of the YEARBOOK is published for use as a supplement to
THE FUNK & WAGNALLS NEW ENCYCLOPEDIA
THE NEW ILLUSTRATED ENCYCLOPEDIA

DISTRIBUTED BY FUNK & WAGNALLS, INC.

Funk & Wagnalls offers a 2-Volume Record
Collection, "Great Music of Our Time."
For details on ordering, please write:
Funk & Wagnalls, Inc.
Dept. BPM
P.O. Box 391
Radio City Station
NY, NY 10101

ACKNOWLEDGMENTS

Sources of articles appear below, including those reprinted with the kind permission of publications and organizations.

UNVEILING THE INFRARED UNIVERSE, Page 6: Dennis Overbye, © 1984 *Discover* Magazine, Time Inc. Article originally titled "The Secret Universe of IRAS."

SPACE CAMP, Page 13: First appeared in *Science Digest*, © 1983 by the Hearst Corporation.

ASTROARTISTS, Page 24: First appeared in *Science Digest*, © 1983 by The Hearst Corporation.

OUT ON THE STREETS, Page 34: Reprinted with permission from *GEO* Magazine. Article reprinted from the April 1983 issue.

MIND HUNTERS, Page 40: Reprinted from *Psychology Today* magazine. Copyright © 1983 American Psychological Association.

EIGHT WOMEN IN THE WILD, Page 48: Reprinted with permission of the author; article first appeared in the January-February issue of *International Wildlife*.

MOTHERESE, Page 56: Copyright © 1983 Parents Magazine Enterprises. Reprinted from PARENTS by permission.

PSYCHOBIOLOGY, Page 66: Reprinted from *Fortune* Magazine. © 1983 Time Inc. All rights reserved.

THE ASSASSIN BUG, Page 74: With permission from *Natural History*, Vol. 92, No. 5; copyright the American Museum of Natural History, 1983.

COCONUTS, Page 78: Reprinted with permission of the author; article first appeared in *Oceans* magazine, September-October 1983.

ENGINEERING THE BIRTH OF CATTLE, Page 86: Copyright © 1983 by The New York Times Company. Reprinted by permission.

FRACTAL IMAGERY, Page 100: Reprinted with permission of the author; article first appeared in *Smithsonian*, December 1983.

THE MICROCHIPPED DUGOUT, Page 108: Reprinted from *Psychology Today* magazine. Copyright © 1983 American Psychological Association.

THE ORGANIC COMPUTER, Page 114: *The Sciences*, May/June 1983 © by The New York Academy of Sciences.

SEAFLOOR PANORAMA, Page 130: Reprinted from *Popular Science* with permission © 1984, Times Mirror Magazine, Inc.

THE OTHER COLD WAR, Page 137: Reprinted with permission, *Technology Illustrated* magazine, May 1983. Copyright © 1983 by TECHNOLOGY ILLUSTRATED Publishing Company, 38 Commercial Wharf, Boston, MA 02110.

A NEW EYE ON THE WEATHER, Page 142: Reprinted from *NOAA*, volume 14, number 1, winter 1984.

COLUMBIA: A GLACIER IN RETREAT, Page 148: Reprinted with permission of the author; article first appeared in *Smithsonian*, January 1983.

PEACEFUL NUCLEAR EXPLOSIVES, Page 160: Malcolm Browne, © 1983 *Discover* Magazine, Time Inc. Article originally titled "A Soviet Nuclear Boom."

A NEW WAY TO BURN, Page 166: Reprinted by permission of *Science 83* Magazine, copyright the American Association for the Advancement of Science.

ICE PONDS, Page 178: Reprinted with permission of the author; article first appeared in *Smithsonian*, August 1983.

TROPICAL FORESTS IN TROUBLE, Page 190: Reprinted with permission of the author; article first appeared in the May-June 1983 issue of *International Wildlife* Magazine.

PLASTICS AT SEA, Page 196: With permission from *Natural History*, Vol. 92, No. 2; Copyright the American Museum of Natural History, 1983.

MEETING GROUND FOR THE MASSES, Page 204: Reprinted with permission of the author; article first appeared in the December-January 1983 issue of *National Wildlife* Magazine.

STAFF

EDITORIAL

Editorial Director
Bernard S. Cayne

Executive Editor
Lynn Giroux Blum

Managing Editor
Doris E. Lechner

Art Director
Suzanne Shaw

Contributing Editor
Barbara Tchabovsky

Copy Editor
Grace F. Buonocore

Chief Indexer
Jill Schuler

Production Editor
Cynthia L. Pearson

Indexer
Pauline Sholtys

Proofreader
Stephen Romanoff

Manuscript Typist
Joan M. Calley

Staff Assistant
Jennifer Drake

Director, Central Picture Services
John Schultz

Chief, Photo Research
Ann Eriksen

Manager, Picture Library
Jane H. Carruth

Photo Researcher
Jane DiMenna

Picture Library Assistant
Mickey Austin

MANUFACTURING

Director of Manufacturing
Joseph J. Corlett

Production Manager
Alan Phelps

Production Assistant
Marilyn Smith

CONTENTS

ASTRONOMY AND

SPACE SCIENCE

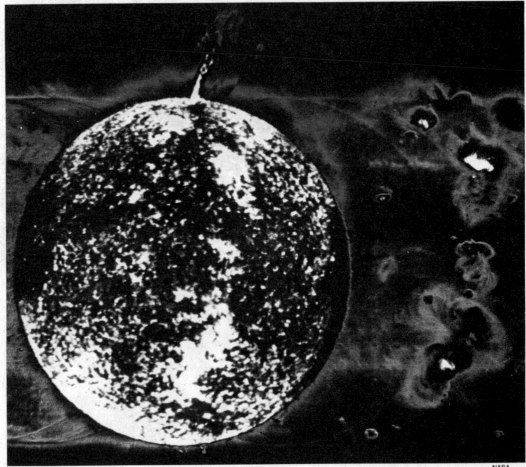

REVIEW OF THE YEAR

ASTRONOMY AND SPACE SCIENCE

Astronomers' ideas of the solar system, the Milky Way galaxy, and the universe continued to change as new data became available during the year. In space activities the ongoing manned and unmanned projects of the United States and the Soviet Union were joined by increasing—and successful—efforts by the European Space Agency, Japan, and China.

THE SOLAR SYSTEM

There was continued revision of our ideas about the Sun, the heart of our solar system. Observations of solar flares show that these flares produce neutrons that arrive on Earth before charged particles emitted at the same time. Not detoured by spiraling down Earth's magnetic field as charged particles do, the neutrons may provide an early warning of the arrival of charged particles, which can be dangerous to astronauts and disrupt radio communications on Earth. ■ Astronomers have detected a great dust sphere encircling the Sun. They suggest the dust originated in the outer solar system and is attracted to the Sun by gravity.

The unveiling of Venus continued. Observations by the U.S. Pioneer Venus orbiter, confirmed by the Soviet Venera 13 and 14 probes, found sulfur dioxide on Venus, most likely the result of volcanic eruptions. ■ Venera 15 and 16, launched this year, provided pictures of the Venusian surface and observed impact craters, fractures, mountain ridges, and objects as small as 1 to 2 kilometers (0.6 to 1.2 miles). ■

Meanwhile, Earth-based observations using the Arecibo Radio Telescope in Puerto Rico are being used to map one large continent—Ishtar Terra—and its towering peak Maxwell Montes, the highest point on Venus.

The Earth is slowing down. The U.S. Naval Observatory on June 30, 1983, added one more second to that day to keep clocks synchronized with the Earth's rotation. The Earth is slowed by friction in the atmosphere, the oceans, and in its own core. ■ The Moon—did it break off from Earth? Was it captured by Earth's gravitation? New tidal calculations demonstrating that the Earth and the Moon were never closer than 225,000 kilometers (140,000 miles) suggest that the two bodies probably were formed at the same time out of the primordial solar nebula. ■ The origin of meteorites has also long been puzzling. Continuing analysis of the meteorite Allan Hills 81005 found in Antarctica seemed to confirm earlier speculation that it is from the Moon—most likely ejected from the young lunar crater Giordano Bruno. Another meteorite found on Earth may be from Mars, according to analyses of the abundances of rare gases and nitrogen isotopes it contained.

That part of space around Jupiter was also studied during the year. The Infrared Astronomical Satellite (IRAS) observed three vast rings of dust between Mars and Jupiter. These rings may be part of a single thicker ring tilted 9 degrees to the plane of the solar system. ■ Analysis of data provided by the U.S. space probes Voyager 1 and 2 suggests that Jupiter's own ring is doughnut-shaped and surrounded by a thin gossamer ring that extends out to the orbit of the satellite Amalthea. ■ Jupiter's magnetotail also stretches as far as Saturn, nearly 650 million kilometers (about 405 million miles) away. ■ Ground-based infrared observations of Io— Jupiter's most famous and active moon— revealed that Io's huge volcano named Loki may be ejecting sulfur dioxide, causing a kind of snow. Meanwhile, Jupiter's recently discovered satellites were officially named by the International Astronomical Union. Jupiter XIV has been named Thebe; Jupiter XV, Andrastea; and Jupiter XVI, Metis.

Analysis of Voyager observations continued to reveal a complex and varied system of rings and satellites around Saturn. A gossamer ring between the narrow braided F ring and the outer edge of the A ring was discovered. Meanwhile, the outer edge of the A ring itself was found to be scalloped, suggesting the possible presence of more moonlets. An eccentric ringlet, 31 to 107

kilometers (20 to 67 miles) wide and not more than 1 kilometer (0.6 mile) thick, was found in a gap in the C ring. And scientists now believe that the mysterious spokes in the B ring that generally last no more than two hours may be due to electrostatically charged particles levitated out of the ring plane. ■ Hyperion, one of Saturn's outermost satellites, was found to have a very chaotic rotation and an irregular shape. ■ Titan, Saturn's largest moon and long the object of speculation about possible prebiotic conditions, was found to contain carbon monoxide in its atmosphere. Carbon monoxide is one of the first chemical steps toward life as we know it. ■ And Saturn itself was observed to have a ten-month-long gargantuan lightning storm.

The "first direct evidence of an ocean on an extraterrestrial body" was found on Triton, one of Neptune's satellites. Scientists detected nitrogen, presumed cold enough to be in liquid form, on the moon. The surface temperature of Triton is no higher than $-218°$ C ($-360°$ F).

Models of the development of the solar system and of life within it were revised during the year. Data on the nature and abundances of elements present in the early solar system were gathered through an analysis of meteorites. And, perhaps most significant, the finding of prebiotic material in the Murchison meteorite (which fell to Australia in 1969) fostered ideas that the formation of chemicals necessary for life is probably not very difficult nor unique to Earth.

OUR GALAXY—THE MILKY WAY

An exciting new view of our galaxy is emerging. No longer is the old view of a thin round disk about 30,000 parsecs across with the Sun two-thirds of the way from the center considered correct. (One parsec equals approximately 30.9 trillion kilometers, or 19.2 trillion miles). The new picture is of a four-armed disk about 60,000 parsecs across with a large transparent and nonluminous spherical halo made up of objects larger than dust and smaller than the smallest star, and containing most of the mass of the galaxy.

The first direct evidence of solid material— believed to be tiny particles—around stars other than the Sun was found by IRAS observations of the stars Vega and Fomalhaut. ■ Meanwhile, ground-based observations of the variable star T Tauri have led some astronomers to propose that it has a companion that is a planet in formation (the possibility that the companion could be another star has also been suggested). ■ Other astronomers have suggested that planet-sized objects—which they termed milli-suns—may orbit the stars VB8 and VB10.

Several astronomers now think that some planetary nebulas have close binary systems at their centers, not single stars. They suggest that one star of the pair eventually becomes engulfed by its companion, at which time its atmosphere is expelled to form the spherical halo of material characteristic of planetary nebulas. ■ In separate investigations, four stars of a new class, a kind that has just given off the spherical halo of a planetary nebula, were discovered. These new stars appear to be in a transitory, unstable state of stellar evolution never before observed.

The amount of cosmic rays has changed in the past several million years, with variations about every 200 years—such is the conclusion of scientists who have analyzed meteorites and moon rocks for evidence of cosmic rays. And for the first time, cosmic rays were noted to come from a particular direction—specifically, from a region in the constellation Cygnus or from the north pole of the galaxy.

Pure stardust—interstellar grains from which planets and new stars are thought to form—was identified for the first time in a meteorite. This discovery may provide important clues to how the galaxy evolved.

THE UNIVERSE

A massive cloud of rapidly rotating hydrogen gas has been found in intergalactic space. Some scientists believe it may be a galaxy in formation; others think it may have been created by tidal streamers from nearby galaxies. ■ A black hole has been found in the galaxy nearest us, the Large Magellanic Cloud. LMC X-3 containing the black hole is a binary system with an egg-shaped primary star around which revolves a massive dark companion, identified as a black hole.

KATHERINE HARAMUNDANIS

THE U.S. SPACE SHUTTLE

Four flights of the U.S. space shuttles in 1983 were successful, but not without problems. The newest shuttle, Challenger, was launched April 4 on a five-day mission. It was commanded by Paul J. Weitz and Karol J. Bobko. Two mission specialists, Story Musgrave and Donald K. Peterson, spent more than four hours outside Challenger, testing new space suits in the first space walk by Americans in nine years. The crew deployed the Tracking and Data Relay Satellite from the shuttle, but a rocket malfunction prevented it from reaching its intended orbit. However, by the end of June, multiple firings of the satellite's small thruster jets had nudged the $100 million vehicle into its proper orbit, where it began operating as a communications link between shuttles and the ground.

On June 18, Sally K. Ride became the first U.S. woman to travel in space. She and four men—Robert L. Crippen, Frederick H. Hauck, Norman E. Thagard, and John M. Fabian—deployed two commercial communications satellites from Challenger's cargo bay and practiced releasing and retrieving a satellite with the shuttle's mechanical arm. Bad weather forced the crew to cancel plans to make the first landing at Cape Canaveral; they came down instead at Edwards Air Force Base in California on June 24.

The first nighttime shuttle launching occurred on August 30 when Guion S. Bluford, Jr., the first black American to fly in space, and four others—Richard H. Truly, Daniel C. Brandenstein, Dale A. Gardner, and William E. Thornton—rode Challenger into orbit. They deployed a communications satellite for India, conducted experiments, and landed September 5 in the dark at Edwards. They had a close call, it was discovered later; the nozzle of one booster rocket had almost burned through during lift-off and could have thrown the spaceship out of control.

The final mission of 1983—the ninth of the shuttle program—was a joint venture with the Europeans in which the shuttle Columbia carried a scientific laboratory into orbit. Spacelab, built by the European Space Agency, remained in the cargo bay and was used as a base for dozens of scientific and engineering experiments. The mission, launched November 28, had the largest crew; the six were John W. Young, Brewster H. Shaw, Jr., Owen K. Garriott, Robert A. R. Parker, and two scientists—Byron Lictenberg and West German Ulf Merbold—who were the first space travelers not trained as astronauts. A cascade of malfunctions marred the flight's ending on December 8. Two computers failed, one navigation unit shut down, and just before touchdown, a minor fire broke out in a rear compartment.

OTHER U.S. SPACE ACTIVITIES

Troubles plagued the National Aeronautics and Space Administration's (NASA's) most ambitious scientific satellite project, the Hubble Space Telescope, which is being designed to "see" seven times farther than any ground-based observatory. Problems with installing a mirror and with guidance sensors caused a delay in the scheduled launch until 1986.

More carefree was the hardy Pioneer 10 space probe. After a voyage of more than 11 years, Pioneer 10 on June 13 became the first man-made object "to leave the solar system," crossing the orbit of Neptune, at present the outermost of the known major planets. It is still radioing data.

The International Sun-Earth Explorer 3 (ISEE 3), a spacecraft launched in 1978 to study the Sun, and shifted in June 1982 to study the Earth's magnetotail, was given yet another duty— it will make the first-ever visit with a comet. The ISEE 3 was directed to fly around the Moon in

Astronauts Story Musgrave and Donald Peterson float about in the cargo bay of the Challenger during their mission.

December 1983 so that the Moon's gravity could then send it on to rendezvous with comet Giacobini-Zinner in 1985, preceding by half a year the international examination of Halley's comet planned for March 1986. (See "The Return of Halley's Comet" on page 18.)

The most productive satellite of the year was the Infrared Astronomical Satellite (IRAS) launched January 25. For a report of its achievements, see "Unveiling the Infrared Universe" on page 6.

SOVIET SPACE ACTIVITIES

The Soviet Union's continued operation of the Salyut 7 space station was beset with problems. In March the Salyut was almost doubled in size when an unmanned Cosmos 1443 ship docked with it. However, the year's first attempt to send a crew to the enlarged station failed in April. Three cosmonauts—Vladimir G. Titov, Gennadi M. Strekalov, and Aleksandr A. Serebrov—were launched April 20 in Soyuz T-8, but the mission was terminated in two days because they were unable to link up with the Salyut. On June 28, two cosmonauts, Vladimir Lyakhov and Aleksandr Aleksandrov, succeeded in Soyuz T-9; they spent 149 days aboard Salyut 7 and activated the attached Cosmos 1443 module. Then trouble set in: a massive fuel leak left half of Salyut's steering jets unusable, and an explosion on the

launching pad on September 27 prevented a new crew from visiting the Salyut. Lyakhov and Aleksandrov finally returned safely to Earth on November 23.

The Soviet Union continued its unmanned exploration of Venus with two craft, Veneras 15 and 16, that transmitted radar images of the cloud-shrouded surface. ■ Cosmos 1514, launched December 14, carried two monkeys and several rats into orbit for biological studies. ■ In all there were 98 Soviet launchings in 1983 (compared with 22 U.S. launchings); most of the Soviet launches were short-duration military reconnaissance satellites.

OTHER SPACE ACTIVITIES

For the 11-nation European Space Agency, the high point may have been the flight of Spacelab, but there was also good news in the two successful launches of Ariane rockets. ■ The European Exosat satellite, launched in May by a U.S. rocket, is charting some 2,000 known X-ray-emitting astronomy sources.

China launched its 13th satellite in 1983. ■ Japan launched domestic communications satellites in February and August and announced plans to develop its own powerful launching rockets and orbit more than 75 advanced satellites by the year 2000.

JOHN NOBLE WILFORD

Unveiling the Infrared Universe

by Dennis Overbye

To astronomers confined to Earth, a planet in the suburbs of the Milky Way galaxy, it was like seeing the lights of downtown for the first time. Amid blobs of dust and gas lit by unseen suns, the heart of the galaxy blazed like Times Square, New York, on New Year's Eve with the glow of infrared radiation. The images compiled from a satellite called IRAS (InfraRed Astronomical Satellite) are revealing to astronomers a universe that, until now, has been inaccessible and invisible to them. It is a cold universe painted not with stars but mostly with dust—trails and bands of dust strewn by comets rushing pell-mell through the crowded solar system; streams of dust blown like autumn leaves from hot stars across light-years of space; great clumps of dust decorating the Milky Way like neon signs, forging new stars in their dense folds; fleecy shreds of dust curled and tufted in the chill void between the stars. (A light-year equals approximately 9,500,000,000,000 kilometers or 5,878,000-000,000 miles.)

Revelations Abound

This is the picture of the universe that was slowly transmitted, bit by bit, by IRAS during 1983. The satellite's telescope was chilled to just above absolute zero ($-273.15°$ C or $-459.67°$ F) in order to sense faint emanations of heat, or infrared radiation, from the universe. Every day, during the two passes that IRAS made over Chilton, England, 350 million bits of information poured down from the satellite into a 3-meter (10-foot) radio antenna. From there they were fed next door into the computers and minds of the harried occupants of a cluster of mobile offices huddled like serfs' cottages

Clouds of interstellar dust and gas heated by nearby stars glow brightly in this IRAS image of the Milky Way.

NASA

select wavelengths of celestial infrared radiation.

That is unfortunate for astronomers and casual skygazers alike, because everything in the universe, every dust mote warmed only a few degrees above absolute zero by distant starlight, every planet, star, and even the universe itself emits some infrared (heat) radiation. The hotter the object, the more intense—and the shorter the wavelength of—the radiation it emits. Moreover, infrared waves pass virtually unimpeded through interstellar dust that obscures starlight (which is why IRAS got such a good view of the galactic center). By perching infrared telescopes on high, dry mountaintops, or in jet aircraft, scientists managed a few peeks at the infrared universe. But until now, says California Institute of Technology's (Caltech's) Gerry Neugebauer, leader of the IRAS science team, "we have been limited to studying objects discovered at other wavelengths."

An observatory in space was clearly the answer, but until the mid-1970's, the best infrared detectors were classified by the Defense Department. About the time that the National Aeronautics and Space Administration (NASA) started thinking about IRAS in 1974, the Netherlands Agency for Aerospace Programs (NIVR), flush with oil money, was fishing about for a space project. NIVR and NASA joined forces, and were followed shortly by Britain. The Dutch built the spacecraft, and the British put up the ground tracking station and preliminary data center in Chilton, while U.S. scientists crafted and launched the telescope, and are analyzing the final data.

A Mighty Sensor

The heart of the satellite is a telescope 56.9 centimeters (22.4 inches) in diameter, surrounded by a giant thermos bottle of liquid helium to keep its own heat from swamping that of the sky. The telescope focuses the incoming infrared light on an array of 62 tiny solid-state chips that produce an electrical signal in response to heat. The chips are sensitive to four infrared wavelengths, corresponding to temperatures between the freezing point of water (0° C or 32° F) and about 20° C (68° F) above absolute zero. The temperatures of planets, asteroids, and interstellar dust clouds where stars form all lie

around the Rutherford Appleton Laboratory.

IRAS has unveiled a new infrared universe to the world. Astronomers hardly know what to talk about first. With discoveries ranging from dust bands and comets in the solar system to mysterious galaxies that seem to radiate almost all their energy as heat, IRAS's cold eye has spied a dustier, but in a way friendlier, universe that is busily birthing stars and perhaps planetary systems, nestling whole galaxies in dusty cocoons.

A Good Look at Long Last

Earthbound astronomers are doubly blind to infrared, which is a form of light or electromagnetic radiation with wavelengths longer than those of radio frequencies. Not only are human eyes insensitive to the infrared, but the water vapor in Earth's atmosphere blocks all but a few

During its 11-month life, the Infrared Astronomical Satellite made many fascinating discoveries and revealed that the universe is much dustier than astronomers ever imagined it was.

NASA

within this range. IRAS, says Neugebauer, can see the heat of a dust mote 3.2 kilometers (2 miles) away, or of a baseball across the continent.

As IRAS circled Earth in a polar orbit, the field of view of the telescope moved along a swath of sky half a degree wide that changed slightly with each orbit until the whole sky had been "painted." Drifting through the telescope's view, the image of a star or heat source crossed a row of detectors, producing signals that were stored and finally recorded as a long, bumpy strip chart at Chilton each time IRAS passed overhead. After being sampled and checked, the data were transmitted to the Jet Propulsion Laboratory (JPL) in Pasadena, California.

Vega Discovery Brings Mixed Reactions

For the first three months after IRAS's nighttime launch from Vandenberg Air Force Base in California on January 25, 1983, data piled up at JPL while the science team tested their computer programs. As a result of that, and of communication problems between Pasadena and Chilton, the first major find of the mission—Comet IRAS-Araki-Alcock, which passed within 4.8 million kilometers (3 million miles) of Earth in May 1983—leaked out from the Chilton group before JPL knew about it. "JPL was supposed to get it first, then the press," says George Aumann, who is stationed at Chilton. "In the end, well, people got their noses bloodied on it."

It was Aumann, along with his office-mate Fred Gillette, from Kitt Peak National Observatory in Arizona, who made one of the most exciting of the IRAS discoveries. The pair noticed during one sighting that the star Vega was unexpectedly bright in the infrared, and that the bump its radiation was producing on the strip chart was too wide for a mere star. The extra heat, they concluded, was coming from a ring of pebbly material 24 billion kilometers (15 billion miles) across that was orbiting Vega. Although several solar-system experts opined that the debris circling Vega was unlikely ever to condense into planets if it had not done so by now, the discovery excited hopes that other planetary systems exist in the galaxy. (In December 1983, IRAS astronomers announced that a band of solid material was also discovered orbiting the star Fomalhaut, also known as Alpha PsA.)

Vega inspired a different reaction from NASA officials in Washington, D.C., who found out about it from the newspapers. Embarrassed that the discovery had not been announced through proper channels, NASA prohibited the release of future IRAS results—even to the rest of the astronomical community—until a previously scheduled November 9 press conference. Scientists are naturally cautious about discussing ongoing work, both out of a desire to safeguard their "goodies" and to make sure their work is sound, but the NASA censorship made it even more uncomfortable for the IRAS team. Says JPL astronomer Charles Beichman: "I had a lot of people buying me

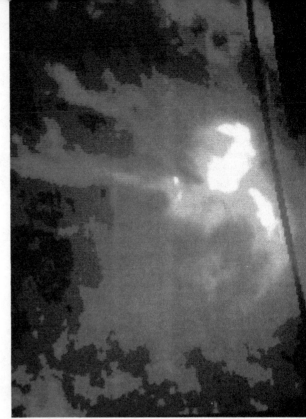

Photos: NASA

Above: The white areas in this false-color image of the region around the star Rho Ophiuchus are recognized as sites of active star formation. Left: Dust clouds swirl about in a region of the sky around the constellation Orion, another active stellar nursery.

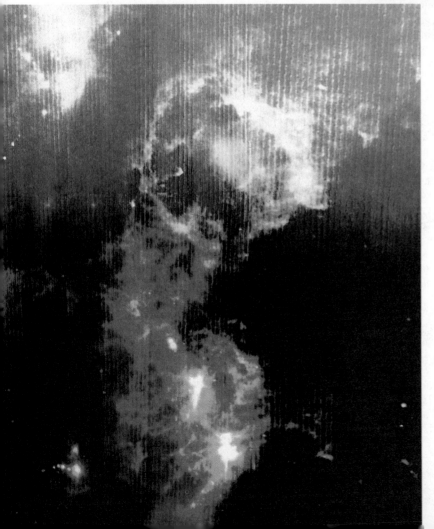

drinks, trying to get information about IRAS.''

The absence of news added to the drama of November 9. The wait was finally worth it; IRAS had already accumulated a variety of wonders, most of them in a preliminary ''mini-survey'' that covered only 1 percent of the sky. Says Nancy Boggess, a NASA project scientist: ''A lot of astronomy books will have to be rewritten when all the results are in.''

A Dead Comet and Stellar Nurseries

The rewriting starts right in the solar system, where in addition to discovering five live comets, IRAS may have found a dead one. It is an asteroid, 1.9 million kilometers (1.2 million miles) in diameter, with the unglamorous name of 1983 TB. It travels in an elliptical orbit that takes it within 14.5 million kilometers (9 million miles) of the sun and out to the asteroid belt. Its path coincides with a stream of dust and debris. Earth crosses that stream every December 14, causing the Geminid meteor shower. Because other meteor showers are associated with the dust left behind by comets, Russ Walker, a scientist from Jamieson Engineering, a NASA subcontractor, suggests that 1983 TB ''may actually be the remains of a comet nucleus that passed close to the sun and lost its ice and volatiles.''

IRAS also found intriguing hints that planets and stars, like the sun, may be common in this galaxy. Spurred by the Vega discovery, Gillette and Aumann are searching for other bright stars that seem too fat and bright in infrared light—stars that may be surrounded, in other words, by rubble in the process of becoming worlds. Out of 9,000 stars examined, Gillette has found about 50 candidates for deeper study.

Stars are the ultimate source of life, and dust is the source of stars. Early infrared observers found that thick clouds such as the Orion nebula were heated up like lampshades by the fires of invisible, massive new stars deep in their dusty hearts. But IRAS is finding that the best-known stellar nurseries are only the most intense parts of vast webs of gas and dust curling and streaming across the sky. ''The streamers aren't understood at all,'' says Beichman, who speculates that they may be dust blown by radiation pressure from luminous young stars. Even in the small dust clouds called Bok globules, which dot the Milky Way and were previously believed to be cool and inactive, too small

to make stars, IRAS has detected the telltale heat from baby suns.

In a small knot of a cloud known as Barnard 5 nestle four new sunlike stars, one of which may not have actually started to burn yet. All told, Beichman says, IRAS's results show that in the Milky Way, sunlike stars form more frequently—about one a year—than astronomers had thought.

Many Mysteries

Old stars also made an infrared appearance. Dutch astronomers perusing IRAS data found three partial shells of material around the bright star Betelgeuse, which is nearing the end of its life as a puffed-up red giant. (At this late stage in a star's evolution, it expands greatly; becomes more luminous; and has a relatively low temperature, which causes its red color.) According to Harm Habing, from Huygens Laboratory in the Netherlands, the partial shells were blown off during the last 100,000 years as the star rehearsed its final blowout and collapse to a tiny white dwarf. The missing pieces of the shells, he thinks, could have been lost to the drag of interstellar gas and dust as Betelgeuse cruised through the void.

Space itself looks different through the cold eye of IRAS. Between the stars and the roiling nursery clouds, the satellite discovered, the sky is laced with cold, fleecy dust clouds. Says Neugebauer: ''The interstellar cirrus clouds are a new component in the galaxy, not previously suspected.'' Some of the cirrus clouds seem to be identified with interstellar gas clouds that have previously been mapped by radio astronomers, but others are a mystery. One possibility: they could be part of the so-called Oort cloud of comets that orbit in a perpetual deep freeze at the outer edge of the solar system. Almost as surprising as the existence of the cirrus clouds was their temperature, warmed by starlight to 35° C (1.6° F) above absolute zero. ''You have to take my word for it that 35 degrees is warm,'' says University of Arizona astronomer Frank Low. He concludes that the cirrus clouds are made of graphite dust, which absorbs heat more readily than the silicate dust of which the inner planets are made.

An artist envisions the formation of new planets from the whirling disk of debris that orbits the bright star Vega.

David Egge

Another mystery is the source of a faint infrared background radiation that appears to permeate the galaxy. According to Low, astronomers will have to analyze more data before they can say whether there is enough known dust and other matter in the galaxy to account for this extra radiation, or whether estimates of the mass of the Milky Way will have to be increased.

A New View of Galaxies

Astronomers were particularly eager to see what other galaxies look like in infrared, and many of the blips on IRAS's data charts coincided with distant spiral galaxies. Most of the infrared emission from a galaxy is thought to come from dust clouds heated by new stars, and the intensity of the emission is thus considered a good measure of how fast stars are being created in a galaxy. IRAS found that galaxies are surprisingly diverse in their productiveness. The Milky Way, for example, emits about half its energy as infrared, indicating a healthy rate of star formation. But its nearby twin, the Andromeda galaxy, puts out only 3 percent of its energy as infrared rays.

Other galaxies that appear insignificant in visible light are pouring out 10 or even 50 times as much infrared as optical energy. This may suggest that they are enshrouded in dust and spawning stars at an enormous rate. Many of these strong emitters, reports Tom Soifer, a Caltech astronomer, appear distorted or have peculiar shapes in optical pictures, as if they had been disturbed by neighboring galaxies. Theorists have suggested that the gravitational forces from a close galactic encounter could trigger the collapse of dust clouds in a galaxy, setting off a burst of star formation.

When compared with optical photographs, an obscure handful of the IRAS sources studied so far seem to correspond to nothing at all. They apparently have the temperatures of galaxies, according to Cornell astronomer Jim Houck. But if they are galaxies, he says, they must be radiating more than 100 times as much infrared as optical energy. One problem is that IRAS has no way of determining how far beyond the solar system the unidentified objects are. Says Low: "They could be common little dwarf galaxies, very close, that wouldn't amount to a hill of beans, or they could be very distant galaxies."

Success Supports Future Projects

Now that the word is out, help in identifying these mysterious features will come mainly from other wavelength bands, while IRAS astronomers continue to plow through the overwhelming data the satellite has sent down. But NASA is beginning to plan a more powerful infrared telescope, the Shuttle Infrared Telescope Facility (SIRTF), to be flown on the shuttle during the 1990's. It is a mark of IRAS's success that astronomers are now arguing that SIRTF should be a separate satellite that can fly on its own, instead of being limited to two-week stints in the shuttle's cargo bay.

Even while IRAS was being celebrated, its life was already over. On November 22, 1983, the last of the liquid helium that cooled its telescope wafted into space, and the detectors began to warm and go blind. The helium had already lasted far beyond its original projected lifetime of six months. Engineers on the ground had no warning of the demise, because as the helium boiled off, the remaining liquid spread into a thinner and thinner coating that still cooled the telescope. IRAS was good to the very last drop □

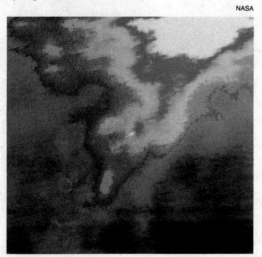

A star about 100,000 years old (arrow) believed to resemble our early Sun appears in a dark cloud of molecular hydrogen known as Barnard 5.

NASA

SELECTED READINGS

"The planet hunters" by Marcia Bartusiak. *Science Digest,* January 1984.
"Throwing wide the IR window" by Jonathan Eberhart. *Science News,* November 19, 1983.
"IRAS: mission invisible" by Frithjof Sterrenburg. *Astronomy,* April 1983.

© Flip and Debra Schulke/Black Star

SPACE CAMP

by Jeffrey Kluger

The exchange between spacecraft cockpit and flight control remains unruffled even in the face of potential disaster.

"Shuttle, we show a glitch in section A, number 8. Do you copy?" the systems director asks briskly.

"Copy. A8 shows no-go. Standing by for analysis," a crew member calmly replies.

In voices that are sober, well drilled, almost flat, the pair keeps up a staccato dialogue that grows ever more technical.

Suddenly, a payload specialist giggles. It's a tiny incongruity, lasting only an instant—but it's enough to jar the senses.

It's then we remember who's running this flight. Despite all the clipped, ship-to-shore chatter, this is something less than the real thing: no astronauts are calling down from orbit, no flight controllers are laboring in Houston, Texas. What we're watching is simply a group of children conducting a simulation. But what a simulation!

The youngsters have gathered, from 41 states and 3 countries, at the Huntsville, Alabama, Space and Rocket Center. They'll be here for only 6 days, but what will happen in that time will be impressive indeed: a hundred 12- to 14-year-old children—having been strapped into simulators, bounced in near-zero gravity, tutored in rocketry, and drilled at computers—will absorb vast amounts of astronautical and astronomical schooling. Over 1,000 children participate in the U.S. Space Camp's sessions each summer.

Both Recreational and Realistic

The Space Camp program—like the Mercury, Gemini, and Apollo programs—originated in large measure with rocket designer Wernher von Braun. As early as the late 1960's, at the first signs that the public was taking manned moon flight seriously, von Braun turned his attention to advertising and popularizing the sciences. Cajoling, encouraging, and lobbying, he helped persuade the state of Alabama to appro-

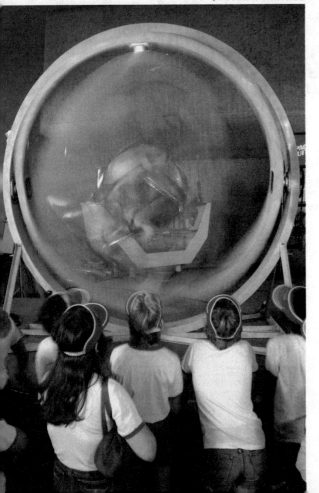

A camper spins around in a multi-axis machine that simulates the tumbling of free fall in space.

© Flip and Debra Schulke/Black Star

priate funds for the Alabama Space and Rocket Center, a 14-hectare (35-acre) museum complex that opened in 1970. Twelve years later, with gate receipts garnered from more than 4 million visitors, the museum gave rise to the Space Camp.

Developed with the encouragement of the National Aeronautics and Space Administration (NASA), the museum and camp—which currently share the same facilities—bristle with simulators, spacecraft capsules, and other hardware donated from the agency's stockpiles of surplus. On the grounds surrounding the main building is a picket fence of Atlas, Redstone, and Saturn rocket boosters. Within the museum are hands-on displays and climb-in mock-ups.

While such machinery both amuses and educates the museum's guests, it also comprises a laboratory for the space campers. Lodged at a nearby University of Alabama campus, the children—who are chosen from among thousands of applicants—are brought to the museum each morning. There they are put through rigorous paces by counselors who are college students majoring in aerospace or engineering.

The campers learn to hop and bound in a counterweighted sling Apollo astronauts used to prepare for the moon's reduced gravity. They familiarize themselves with a functioning reproduction of the shuttle's cockpit and conduct simulated flights presided over by camper ground controllers. They learn to feel comfortable in a chair that rolls, pitches, and yaws to duplicate the gyrations of a craft maneuvering in space. They construct miniature rockets and launch them with insects aboard, quickly learning—from the occasional crew fatality—the importance of fastidious booster design.

Despite the recreational aspect of the facilities, Space Camp involves more than gamboling through an elaborate amusement center. The program includes extensive study of substantive science, much of it taught by computer. After arrival and orientation, the campers are divided into groups and shown to a bank of terminals equipped with keyboards and glowing display screens. The computers offer a chatty, gently instructive program dubbed PLATO, which explains the day's schedule, as well as teaching, drilling, and quizzing the children on carefully selected lesson material.

Though PLATO is conversant with such idioms as "kinda" and "see ya' around," it appears that the campers would be comfortable even if the computer weren't so colloquial.

They are at ease with these machines and need only a moment to familiarize themselves with the keyboard before plunging in.

"I guess I first toyed around with computers when I was eight or nine," says Stephen Gregg, now 13. "They're not really hard. You just have to learn their languages." How many languages are there? "Oh, plenty. There's BASIC, FORTRAN, LOGO, APL. . . ." Fifty years ago, these kids would have been memorizing batting averages. Now it's an electronic litany.

Exacting Taskmaster

First on the campers' agenda each morning is a session with PLATO, an exacting taskmaster. The computer's lesson plan includes spacecraft technology, astronaut training, lasers, electronics, robotics, the physics of rocketry, and more.

Under ordinary circumstances, children in their teens—who are apt to have a low tolerance for academics in July—would approach such a study load with apathy, at best. But the Space Camp program assures that the kids' enthusiasm doesn't wilt. After each morning's lesson, they participate in experiments designed to make the concepts PLATO describes become real to them. Thus, for example, the campers are not expected to take the principles of rocket propulsion on faith. Rather, they will see the theories tested when they try to operate a miniature rocket and then a propeller in a vacuum that has been produced in a bell jar. Firsthand experience shows them why a shuttle can maneuver in space, but an airplane cannot. Even PLATO's history lessons are graphic: in an air-conditioned Plexiglas cage in the museum lives Miss Baker, an aging, well-tended squirrel monkey. In 1959 Miss Baker and a companion, Able, rocketed aloft aboard a relative popgun of a booster, becoming the first of a long string of space voyagers. The campers are told of the space program's past and can then touch a bit of it.

One of the Space Camp's most valuable assets is its location, just a short drive from the Marshall Space Flight Center. It is this NASA installation—where rocket engines are tested and astronauts frequently trained—that earned Huntsville a space-age notoriety shared only by Houston and Cape Canaveral. Many of the Space Camp's activities are conducted at Marshall. The children explore the yawning maw of the shuttle's central fuel tank, examine a one-third-scale wind tunnel model of the orbiter itself, and study a space-station mock-up designed by von Braun over a decade ago. Most significantly, however, they are given the opportunity to see NASA's massive neutral-buoyancy tank. With a capacity of over 5.3 million liters (1.4 million gallons) of water, the tank has been used for years to prepare astronauts for weightlessness and to test orbital equipment.

"We get a lot of cooperation from the folks at Marshall," smiles Lee Sentell, Space Camp spokesman. "They usually know when we're bringing the kids over, and somehow always seem to be conducting an interesting tank experiment on just that day."

With scripts in hand, the mission-control team prepares to swing into action to oversee a staged shuttle flight.

Splashdown Rehearsal and Simulated Flight

Though the campers do not use the tank themselves, they do conduct underwater work in a university pool. Practicing splashdown-rescue techniques, they learn to inflate and clamber aboard rubber life rafts and to use an evacuation cage that a helicopter would lower if the splashdown were the real thing. In other exercises, they submerge in shallow water, slip their feet into brackets, and practice operating a piece of spacecraft machinery while suspended in watery weightlessness. The same training techniques were used in the Skylab program.

But by far the most significant activity of Space Camp week is the simulated shuttle flight that concludes the campers' stay. The panels in the camp's mock-up cockpit and control room are authentic to the point of being almost incomprehensible. That the children can approach these blinking hieroglyphs with any degree of understanding is a testament to the thoroughness of their training.

As they prepare for the brief flight, the campers are each given a mission script that they will use from lift-off to landing. "The kids are surprised to find that the real shuttle pilots also use scripts," Sentell says. "It's the only way to assure that all of the elements on the flight-plan checklist are attended to."

With verisimilitude thus established, the campers go about rehearsing their parts in the flight. On the front of each mimeographed flight plan is a warmly parental reminder: "Not everyone will be able to claim the title of shuttle commander or pilot. But keep in mind that all people must work together as a team before any flight can be a success." Though the campers are playing adult roles, they are still indisputably children. And occasionally the child must be bolstered.

Nevertheless, the flights are conducted with silky professionalism. The campers know their jobs, perform them flawlessly, and even cope confidently with surprise malfunctions the ground controllers cook up to test their preparedness. When the glitch in section A, number 8 turns up, there's no question that they'll be ready to roll right into action.

© Flip and Debra Schulke/Black Star

Campers enjoy practicing splashdown-rescue techniques with a NASA Bird Cage that was used in the ocean recovery of Apollo astronauts.

© Flip and Debra Schulke/Black Star

Konrad Dannenberg, an engineer for the V-2 rocket, gives a fascinating explanation of rocket design and fuel systems.

Farsighted Enthusiasts

Later, as the kids describe their experiences, their vocabularies are studded with casually dropped—if baffling—nuggets of jargon. "I was surprised at how few mistakes we could afford to make," says Bart Bedford, age 12. "At one point, we went EVA, took too much radiation, and lost the TPS." Other campers listening laugh heartily at the anecdote. Nearby, uninitiated adults chuckle uncertainly, concluding that something funny has been said, but dubious about what it was. (Actually, the story is amusing enough once it's explained that EVA means ExtraVehicular Activity, which means space walk, and TPS stand for Technical Payload Specialist, a member of the crew.)

As the campers grow more nonchalant about what they're being taught, it becomes clear that if the Space Camp curriculum is unique, the Space Camp kids are extraordinary. They're a beguiling group, really, not easy to account for. Their parents, who cut their technological teeth in the 1950's and 1960's, came of age in an era in which science and space were unhappily tangled up with militarism and cold war politics. "Our counselor told us about the way everybody reacted when Sputnik was launched," says David Cole, age 13. "People were so afraid because of a little beeping globe." Cole and his peers have no such fear of the scientific unknown. They understand technology's potential, while remaining aware of its shortcomings. "It's science that got us into space, but it's also science that's gotten us into trouble, especially with warfare," he says. "There are too many missiles, too many weapons. It's way too easy to light a match." The kids nearby nod in assent.

These kids, whose career plans range from neurosurgery to space-computer design, from engineering to acting, are simultaneously dreamers and realists, idealists and pragmatists. "I really would like to join the space program," says Dacia Jessick, 13. "Maybe one day we'll go to Jupiter or Saturn. Probably not in my lifetime. But I could certainly help things get under way."

What a strange comment from a 13-year-old: "probably not in my lifetime." It's the remark of someone decades older, a thought simultaneously practical and visionary. But it typifies these children. Too young to remember the giddiness of the Apollo moon-program era, they understand both the limits and the promise of space. They know that exploring such a frontier takes time, and they're prepared to allow the effort to move slowly—as long as they're permitted to help it along□

The Return of Halley's Comet

by Benedict A. Leerburger

From late 1985 to the spring of 1986, people around the world will have a unique opportunity to observe a spectacular celestial object that few have ever seen before. That's because Halley's comet is finally coming back! The famous apparition has influenced people throughout the world for thousands of years. It has not been seen since 1910.

Halley's comet travels in a long elliptical orbit around the Sun. It is observable from Earth for a brief period every 75 to 77 years. Although historical records indicate that Halley's comet was sighted by the ancients, its behavior was not predicted until 1705, when English astronomer Edmund Halley published his observations.

The Comet Mystique

Throughout history, the appearance of a comet has been received with astonishment and fear. They have been considered the source of natural disasters, famines, and wars. When an earthquake and subsequent tidal wave destroyed the cities of Helice and Bura in the Peloponnesus, the Greek philosopher Aristotle was quick to note that the tragedy occurred when a comet appeared in the heavens. Even Shakespeare noted in *Julius Caesar*: "When beggars die, there are no comets seen; The heavens themselves blaze forth the death of princes." Shakespeare was aware that Caesar's death, as well as that of Atilla the Hun and the Roman emperor Valerian, had coincided with the appearance of a comet.

During the Middle Ages, comets were considered to be the cause of every plague and epidemic. Many people have even believed that their appearance heralded the end of the world.

Not everyone, however, has cited the appearance of a comet as a bad omen. When Napoleon—whose own birth was marked by a comet—invaded Russia in 1811, he credited the comet seen that year with his success in battle. The American writer Mark Twain credited Halley's comet, which appeared the year he was born, with his success in life. Ironically, when Halley's comet next appeared in 1907, it also marked the year that the great author died.

Apparition Inspires Artists

The periodic appearance of Halley's comet has inspired some interesting depictions by artists through the ages. More than 800 years after the comet was sighted in A.D. 684, an illustration of it appeared in the book *Nuremberg Chronicles*,

which was printed in Germany in 1493. A crude woodcut created by German artist Michel Wohlgemuth of what had to have been Halley's comet accompanies the text that describes the year A.D. 684.

An artist's lighthearted depiction shows people flocking to watch Halley's comet blaze a fiery path across the sky.

When the comet was observed in the spring of 1066, it inspired another work of art. Queen Matilda, the wife of William the Conqueror, wanted to honor her husband following his victory at the Battle of Hastings. She commissioned an embroiderer to design and create a tapestry commemorating the events of 1066. One panel of the now-famous Bayeux Tapestry, which hangs in the town hall in Bayeux, France, shows a crowd of Englishmen pointing to the comet. A legend above them reads, "They are in awe of the star."

Perhaps the most famous artist's rendition of Halley's comet was created by the great Florentine painter Giotto di Bondone. Giotto was commissioned by Enrico Scrovegni, a wealthy businessman, to design a private chapel in Padua to memorialize Scrovegni's father. Giotto erected the renowned Arena Chapel between 1303 and 1304. He decorated the interior with paintings including a realistic depiction of Halley's comet, which he had observed during its 1301 appearance.

Frozen Two-tailed Objects

Comets appear in the heavens to be much larger objects than they actually are. The nucleus, or main body, of a comet is an extremely dense mass of frozen matter consisting of ice, carbon dioxide and other frozen gases, and dust.

Though usually just several kilometers in diameter, the compact nucleus can weigh billions of tons. As a comet approaches the Sun, radiant heat causes the nucleus' ice particles to evaporate without actually melting. These icy vapors flow out in all directions from the nucleus, forming a gaseous halo called the coma, which appears to the earthbound observer as a fluorescent glow. (The nucleus and coma make up the head of the comet.)

As the comet draws still closer to the Sun, the powerful solar wind forces dust particles to flow out of the coma and form a tail of dust. Meanwhile, solar ultraviolet radiation ionizes the comet's gaseous molecules to form a plasma. (A plasma is a collection of charged particles—positive ions and electrons—that conducts electricity well and is affected by a magnetic field.) The magnetic field of the solar wind catches the ions in the plasma and pulls them away from the comet's head, creating an ion tail. Since this tail is fluorescent, it appears as a long, glowing stream. It can extend for millions of kilometers behind the coma. From our line of sight on Earth, the two tails usually look like one.

Mysterious Origins Still Debated

Although comets have been an observable celestial phenomenon for centuries, their origin is still a mystery. For years, comets were thought

One panel of the commemorative Bayeux Tapestry illustrates the appearance of Halley's comet in the spring of 1066.

to have originated outside our solar system. By the 1700's, however, most astronomers assumed that comets were part of our solar system. Eighteenth-century astronomer George Adams wrote that approximately 450 comets were in elliptical orbits around the Sun. By the 19th century, astronomers realized that Adams' numerical estimate was far too low. They believed that hundreds more existed; most of them unseen.

It wasn't until 1950 that our current understanding of comets and their origins was formulated. In that year, Dutch astronomer Jan Oort concluded that a vast reservoir, or cloud, of comets existed in the outermost reaches of the solar system. He wrote that comets in the cloud are "pristine remnants of the solar nebula from which the sun and the planets formed." This huge spherical cloud—now known as the Oort Cloud—extends out more than 500 times the distance between the Sun and Pluto.

Oort stated that the gravitational influence from passing stars may cause a few cloud-bound comets to alter their orbits so that they pass close enough to the Sun to be heated and thus visible. Oort believes Comet Kohoutek, which was observed in 1973–74, is one of these "new comets."

Astrophysicist A. G. W. Cameron differs with Oort on the origin of comets. He believes that when the solar nebula contracted, it assumed the form of a giant spinning disk. This disk spun off satellite disks that orbit the solar nebula at great distances from the center. It is these satellite disks, according to Cameron, that serve as the birthplace of comets. Paul R. Weissman, a cometary astronomer at the Jet Propulsion Laboratory in California, performed numerical calculations on tens of thousands of hypothetical comets with the aid of a computer. He believes that there are more than a trillion comets in the Oort Cloud. These comets, he stated, are not as likely to be ejected from their present orbit by passing stars as by the gravitational pull of the planets Jupiter and Saturn.

Although astronomers differ on the number of comets in the Oort Cloud and what happens to the unobservable mass when it comes under the influence of stars and planets, they do agree on one point. They do not know the exact origin of the Oort Cloud or how comets originate.

Edmund Halley Commemorated

Historically, a comet carries the name of the individual who first sights and reports it. Having the skilled eye of a professional astronomer is not required to make such a discovery. In 1861 a sheep rancher in Australia discovered a comet. In 1910 a group of railway workers on the way home from their late-night shift observed a "new" comet. In 1961 a pilot, an airline stewardess, and an astronomer spied a new comet simultaneously. Kaoru Ikeya, a former Japanese lathe operator, became a national hero for his first comet discovery. He reported his first "new" comet in 1963 and later discovered two more.

Ironically, the comet we know today as Halley's comet was not first sighted by Halley. It had been observed in the heavens many centuries before Halley identified it and predicted its movements. In 1684 Edmund Halley, a young, talented astronomer, visited the famed physicist Isaac Newton. Newton had observed the movements of a comet four years earlier and had attempted to describe general cometary movement mathematically. Newton believed that all celestial bodies were affected by gravitational attraction. Halley understood that if Newton's gravitational theory was correct, it would demonstrate that "bodies mutually attracted would tend to move in elliptical, parabolic, or hyperbolic orbits." Since neither Newton nor the Royal Society of London had the funds to publish Newton's findings, Halley financed the printing of what was to become one of the most influential works of modern science—*Principia*.

After Newton's *Principia* was published, Halley set aside his work with comets to devote his energies to the study of navigational theory and compass variations in the North Atlantic. It

This 1910 photo of Halley's comet shows its glowing ion tail streaming behind the nucleus and a second tail of dust.

Yerkes Observatory

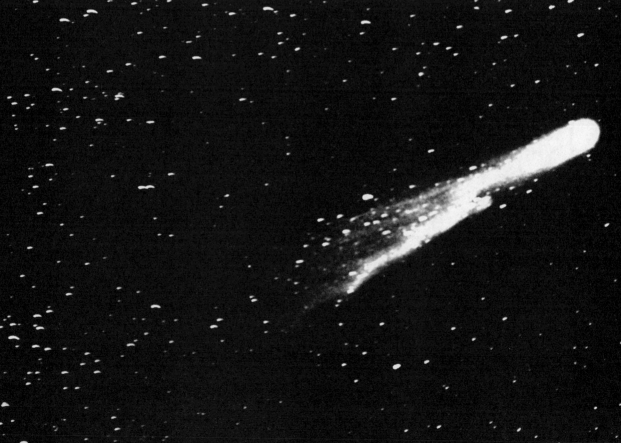

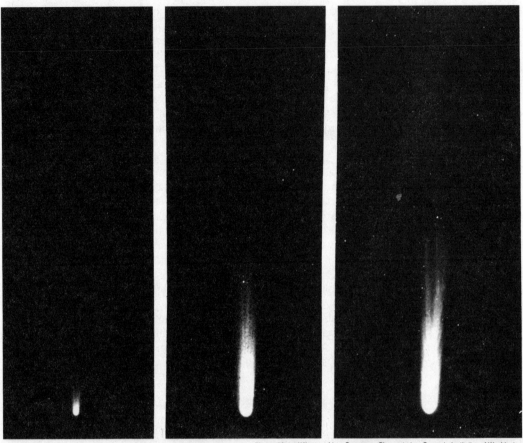

Photos: Mount Wilson and Las Campanas Observatories, Carnegie Institution of Washington

Above: This series of photos shows three stages in the growth of the comet's long tail of dust particles. These images of the famous celestial phenomenon, from left to right, were made on April 26, April 29, and May 3, 1910. Below: The head of the comet is made up of a nucleus of frozen gases and dust, and a gaseous halo called a coma.

wasn't until he was appointed to a position at Oxford University two decades later that he returned to studying comets.

By applying Newton's theories, Halley computed the orbits of 24 comets, including the one observed in 1682. He was surprised to discover that the comet of 1682 had an elliptical orbit and thus, conceivably, had appeared in the skies before and probably would again. By checking astronomical records, he noticed that the movement of the comet of 1682 closely resembled those of comets recorded in 1531 and 1607. Based on his data, Halley predicted that the same comet would appear again in late December 1758.

Unfortunately, Halley died 16 years before the comet was due to reappear. Astronomers throughout Europe understood that if the comet appeared as Halley predicted, it would tend to confirm Newton's theory of gravitation. To the surprise of many, it wasn't an astronomer who first spotted the expected comet. On Christmas night a German farmer named Palitzch observed the comet through a homemade, 2-meter (7-foot) telescope. However, to perpetuate Halley's name and to proclaim his achievement, scientists named the comet after him.

Probes Plan Close Encounter with Comet

In March 1986, several space probes are expected to encounter Halley's comet to gather scientific data that will greatly improve our understanding of cometary phenomena. The unmanned, flyby probes are planned by the European Space Agency (ESA), Japan, and the Soviet Union. A Halley's mission was planned by the United States but was abandoned because of high costs. The U.S., however, is planning to observe the comet with the Space Telescope, and is coordinating the International Halley's Watch program. The Soviet mission will carry experiments developed in cooperation with scientists from France, West Germany, Austria, Czechoslovakia, Hungary, Bulgaria, and Poland.

The ESA probe, known as Project Giotto (after the famous Italian artist who painted Halley's comet in the early 1300's), is scheduled to be launched in July 1985. Its goals include determining the coma's composition and studying the processes that occur within it, investigating the interaction between the coma and the solar wind, analyzing the nature and size of dust particles in the comet, and photographing the nucleus.

Japan plans to launch a scientific probe to Halley called Planet A. Planet A will carry a solar wind analyzer and an ultraviolet camera, which will take detailed photographs of the coma.

Unlike the ESA and Japanese probes, the twin Soviet craft—designated Venera-Halley—have a twofold purpose. In June 1985, the spacecraft will deliver landing probes to explore Venus before continuing on toward Halley's comet. The probes are expected to encounter the comet about a week apart. They will collect data and conduct a wide range of imaging and plasma experiments. The instrumentation on board the Soviet probes include: an integrated-imaging system that will rapidly transmit to Earth detailed images of the comet's nucleus and coma, a spectrograph to study the makeup and behavior of cometary gases, a dust-detecting system, plasma analyzers, and a magnetometer to measure magnetic fields.

Observing Halley's Comet

It is difficult to predict exactly where and how bright Halley's comet will appear. At best astronomers can make reasonable estimates based on the comet's past behavior. During December 1985, Halley's comet can be seen with the aid of a telescope or a good pair of binoculars shortly after sunset low over the western horizon. By January 1986, as the comet brightens and moves closer to the Sun, it should be a terrific sight with viewing aids, and may be visible to the naked eye. In early March, as the comet's tail grows large, Halley may be seen quite well with the naked eye an hour before sunrise. The comet will be closest to the earth in mid-April, and can be seen high overhead in the Southern Hemisphere. In the United States it will appear very low on the southern horizon. Astronomers expect Halley will still be visible through June. But eventually the comet will make its way back toward the outer solar system, thus ending one of the rarest of astronomical shows □

SELECTED READINGS

"Brighter prospects for Halley's Comet" by John E. Bortte and Charles S. Morris. *Sky and Telescope*, January 1984.

"Four probes to Comet Halley" by Jesse Eichenlaub. *Astronomy*, September 1983.

"Halley watch '86" by Stephen J. Edberg. *Astronomy*, March 1983.

The Comet is Coming! The Feverish Legacy of Mr. Halley by Nigel Calder. Penguin Books, 1982.

ASTROARTISTS

by Andrew Chaikin

What is it like to float within the rings of Saturn? How does Mars' Olympus Mons, the largest known volcano in the solar system, look in the midst of an eruption? How would Jupiter have looked if you had been standing on Europa, one of its moons, over 3 billion years ago? These questions seem most appropriate for astronomers and planetary scientists. Yet artists are tackling them, too, and providing realistic answers that combine imagination and scientific facts. Such mental excursions are the stock-in-trade of space artists, who began touring the solar system a century before artificial explorers. Unmanned probes have flown past, circled, or landed on every planet known to the ancients, radioing to Earth thousands of remarkable images. Far from upstaging the efforts of space artists, these images have fueled their imaginations. Scenes beyond the reach of the Voyagers and other space probes, which are limited in their photographic coverage by their trajectories and fuel supplies, appear on canvas with vivid realism, as if painted from an explorer's sketchbook.

Visions of Alien Beauty

As the flood of new data has revolutionized astronomy in the past two decades, space artists have kept pace, producing visions that convey the stark beauty of other worlds and contribute to our understanding of them. Ultimately, it is the desire to know not only the appearance but the true nature of worlds in space that motivates these craftspeople.

For no one is this dual purpose truer than William Hartmann of Arizona. Hartmann holds an uncommon position among space artists, for he is also a professional astronomer, and is often on the receiving end of inquiries from his fellow artists. "As an astronomer," he says, "I'm able to go to many of the scientific meetings at which the very newest results are often announced. I can sit and listen to the papers with part of my focus on visual content that the scientists at the

Mr. Kim Poor's Green Piece *blends fact and fantasy.*

Mr. Kim Poor

meeting may not be paying too much attention to. It's nice to be able to go up to someone who talked about the clouds on Titan and ask, 'How much light actually gets down to the surface?' I often come back with more ideas for paintings than I'm likely to be able to do in months."

In many ways Hartmann is a modern counterpart to the acknowledged grandfather of space art, Lucien Rudaux. The keen perception of this French astronomer-artist is evident in his renderings from the 1920's to the 1940's, which stand up remarkably well against lunar photographs and images of Mars from the Viking space probe.

But the greatest influence on space artists came from a North American painter who was trained as an architect, named Chesley Bonestell. By the 1940's, Bonestell was producing planetscapes of unparalleled realism, spread over the pages of *Life* magazine. At a time when scientists were thinking seriously about space travel, Bonestell was one step ahead of them—he painted astronauts walking on his worlds. "We're following in his footsteps," says Hartmann, "but the environment has changed—there is new knowledge about the places we're painting."

Hartmann himself provides inspiration to others. "His were the first paintings of other worlds I had ever seen," recalls illustrator Pamela Lee. "I would not have attempted such painting myself if Bill had not urged me to."

Computer Sketches

Lee and other space artists make up for a lack of formal scientific training with remarkable thoroughness. At first glance, a space artist researching a painting could be mistaken for an astronomy student preparing a term paper. Detailed calculations, such as those necessary to determine the apparent size of a planet as seen from one of its moons, often find their way into the efforts. Joel Hagen, another Californian, took this approach one step further by writing a computer program to do the figuring for him. Hagen need only supply the name of the planet or moon and the distance it is viewed from to obtain a sketch showing it, precisely scaled and

complete with longitude and latitude lines, hanging in the sky.

Even orbits are of importance. As Hartmann points out, Jupiter's four major satellites circle in such a way that they never line up on one side of the planet, however pleasing that might be to an artist. Such detailed figuring, far from being nit-picking, enhances the feeling of being there, so valued by the artists.

For Kentucky-based James Hervat, perfectionism extends to precise renderings of the robot planetary explorers. Hervat, who sees his role as a kind of space historian, explains, "To me the exciting thing is depicting the machines in these very alien but beautiful environments."

He will stop at almost nothing for accuracy's sake. He has even gone as far as obtaining spacecraft blueprints from the National Aeronautics and Space Administration (NASA) and from aerospace companies. For a view of the Galileo probe descending through the lightning-streaked clouds of Jupiter, he kept up with design changes even for the parachute lines. "I had nightmares about that one," he recalls.

Space artists must keep up to date on the constantly changing array of new theories. *Primordial Europa,* a work by California artist Michael Carroll, exemplifies this blend of aesthetics and scientific curiosity. Carroll was inspired to do this icy moon of Jupiter after seeing high-altitude photographs of the Earth's Arctic regions. Today the surface of Europa is a smooth expanse made up largely of water ice, but scientists believe that for a short time after the planet formed, an ocean of water may have covered the moon. Eventually, it froze and produced the network of hairline cracks we see today.

Not long after Carroll first thought of the Europa project, he noticed an article discussing the possibility that its early ocean could have harbored life, and his vision took a different direction. "I thought it would be interesting to try and depict what Europa might have looked like during the brief period when it may have been somewhat Earthlike."

He could imagine the basic scene, set roughly 4 billion years ago: A huge wave-swept ocean stretches to the horizon beneath a tenuous

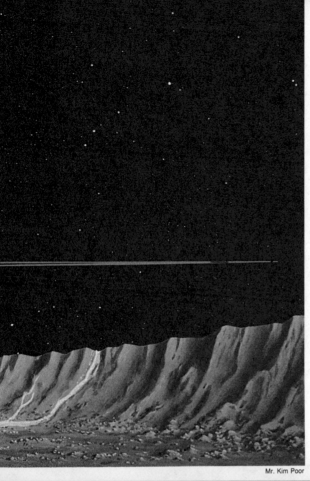

atmosphere (essentially composed of boil-off from the sea of water), while a youthful Jupiter hangs imposingly in the sky. For the particulars, however, Carroll had to make some phone calls. He learned in conversations with Hartmann and other planetary scientists that 4 billion years ago Jupiter would have looked very different from the way it looks today. Early in that period, the gaseous giant would have been half again or even twice its present size. Furthermore, the heat generated by the planet's continued contraction would have made it glow like an ember. At that stage, Jupiter would have looked something like a star as well as something like a planet. For his portrayal, Carroll relied on both present-day images of the planet and photographs of the Sun made from the Skylab space station.

For the rocky outcrops dotting the primeval ocean, Carroll photographed the wave-beaten coast near Carmel, California, realizing, however, that on the infant Europa, rocks might not have appeared quite so eroded. After determin-

Left: Saturn looms behind two of its moons as seen from a third in this interpretation by Mr. Kim Poor. Below: Pamela Lee painted this fiery view of Io.

ing how large Jupiter and its inner moon, Io, would look in Europa's sky, he was ready to paint. The result serves not only as a striking image of another world but as a representation of a scientific theory.

Studying Nature's Subtleties

Largely through Hartmann's efforts, Carroll and other artists have been enjoying some other-worldly scenery here on Earth. "An important aspect of capturing realism," Hartmann says, "is going to places that have interesting geological phenomena that may be represented on other planets." The old lava flows of Hawaii hold a particularly strong attraction for Hartmann, who makes frequent trips to the extinct volcano Mauna Kea.

In 1982, Hartmann brought some of his fellow painters to nearby Kilauea, the largest active volcano in the world, so they could experience the kind of landscapes they were depicting on other worlds. Lava flows abound on many planets and moons so far observed.

To Marsha Morrison, whose Volcano Art

Center served as a base for the Hawaii workshop, the combination of aesthetics and theory that infuses space art is reminiscent of the "edge effect" noted by naturalists. "At the boundary between two different environments," she told the artists one night, "biological activity is more intense and more varied than in the surrounding areas. Space art is at such an edge; it is at the frontier between art and science."

The duality inherent in space art produces some unusual challenges. On an airless planet, for example, there is no atmospheric haze to provide clues to distance. Mountains on the horizon appear just as sharply defined as rocks a few meters away, a fact that gave moonwalking astronauts trouble. To prevent the same ambiguity in a painting, Carroll says, "You have to fudge it a little, fool the eye by putting slightly less detail in background features. It gets the point across."

It isn't coincidental that Carroll and the other artists take an astronaut's-eye view of their subject—any one of them would jump at the chance to make a trip. "I've always wanted

William Hartmann's conception of the bright star Antares with a smaller companion star from an imaginary planet.

William K. Hartmann

Mr. Kim Poor's depiction of Jupiter, the solar system's largest planet. Io can be seen in the foreground.

to be an astronaut,'' confides Canadian artist Maralyn Vicary Diddams. To her and the other workshop participants, wandering over the bleak lava flows at Kilauea was the next best thing to setting foot on Venus or Mars.

Such experiences, Carroll explains, are invaluable to the artist. ''The subtleties of nature that are so important in space art are very difficult to derive solely by imagination,'' he explains, ''so it's very beneficial to come to a place like Hawaii.''

Resurgence of Realism

Carroll believes space art is on the verge of being accepted as a school of painting, aside from its visionary aspects, as part of a resurgence of interest in realistic art. Indeed, the views are reminiscent of landscape art that flourished about 100 years ago, epitomized by the works of Thomas Moran.

Hartmann feels that space artists could play just as influential a role. ''Moran and his con-

temporaries were seeking out exotic scenes on the frontier of their day, and that's what we're doing now. Spacecraft have shown us much, but there are even more places we haven't seen—that's the terrain of space artists.'' Showing the destinations of the U.S. Voyagers and the Soviet Veneras as places you can stand on, he adds, could increase the public's receptiveness to space exploration.

The Second Annual Space Art Workshop convened in the rugged wastes of California's Death Valley in November 1983. There, on the artists' canvases, the rocks took on a reddish hue, the sky was replaced by a salmon-colored haze, and the sun-tortured Mojave Desert was hard to distinguish from the surface of Mars. In ''touring'' the Red Planet, the artists foreshadowed explorations the astronauts will one day make. In this way, space artists show how far the human reach can extend. Before too long, these visionaries know, we'll be able to see for ourselves what they have been showing us □

BEHAVIORAL
SCIENCES

REVIEW
OF THE
YEAR

BEHAVIORAL SCIENCES

During 1983 there were advances in the understanding of human memory, learning, and dreams as well as of schizophrenia and another ailment, termed "ascetic disorder."

MEMORY AND LEARNING

Scientists studying amnesia in humans and in monkeys are providing clues about how the brain handles memory. They may ultimately settle a long-standing dispute about the fundamental nature of human learning. Larry R. Squire, a neurologist at the University of California at San Diego, has compared the memory deficits caused by chronic alcoholism (known as Korsakoff syndrome) and by electroconvulsive therapy (ECT) with the amnesia associated with well-known brain lesions. He has found that

Korsakoff patients and a man known as NA have one form of amnesia—a problem with the initial encoding of information—while ECT patients and another subject known as HM have a distinctly different form—namely, difficulty with the consolidation and elaboration of memories during the weeks and months following an experience. This distinction is important, Squire says, because the two research subjects—NA and HM—are known to have very different brain lesions. NA (like Korsakoff patients) has a damaged thalamus, that part of the brain thought to give the original "print" order for a memory; in contrast, HM (like ECT patients) has a damaged hippocampus, suggesting that this brain region plays a role in fixing memories.

Interestingly, all of the amnesic patients studied remain capable of a certain kind of learning. While they are totally unable to lay down memories of new information, they can acquire new skills. This suggests that there may be yet another brain region involved in skill learning, or what scientists call procedural memory. Indeed, independent research on monkeys lends support to this theory. Mortimer Mishkin, a psychologist with the National Institute of Mental Health, has found that monkeys with experimentally made incisions in the known memory structures of the

A new theory suggests that dreaming is the brain's way of ''debugging'' its overloaded cerebral cortex.

brain (the thalamus and hippocampus) are still capable of mastering new procedures. The brain appears to make a fundamental distinction between knowledge and skills, or habits.

One of the most divisive issues in modern psychology has been human learning. Do humans acquire and store knowledge, and then act on the basis of ideas? Or are they merely conditioned by interactions with their world, forming numerous stimulus-response bonds that then guide future behavior? The answer, it now seems, is that both kinds of learning—memories and habits—are involved in laying down the effects of an experience. These findings may also shed light on another phenomenon that has long puzzled memory researchers—so-called childhood amnesia. Infants are obviously learning, yet adults rarely have memories from early infancy. Mishkin has shown that very young monkeys are capable of learning habits, but they cannot memorize information until much later. This suggests that the (as-yet-unknown) brain structure involved in habit formation develops much earlier than the memory system.

DREAMS

Nobel laureate Francis Crick, co-discoverer of the structure of DNA (the basic hereditary material), has proposed a theory about dreams that psychoanalysts are certain to find unsettling. The Salk Institute biologist suggests the vivid imagery and mysterious symbolism of dreams, rather than bearing messages from the unconscious, may be nothing more than the brain's way of ''debugging'' an overloaded cerebral cortex (that part of the brain concerned with thinking and other higher mental processes). And what takes place in the mind during sleep, he adds, is probably best forgotten. Writing with Graeme Mitchison, a biologist at the Medical Research Council in England, Crick theorized that dream sleep (also called rapid eye movement, or REM, sleep) functions as a mechanism for ''reverse learning.'' As a result of normal brain development and experience, accidental and meaningless connections in the brain occur— connections that must be eliminated if the cortex is to remain an efficient thinking system. Everything that is known about dream sleep indicates that it functions to dampen these ''parasitic'' connections. The leading theory about REM sleep is that it is somehow involved in memory storage and consolidation. According to Crick, this idea is consistent with his theory. By cleaning the cortex nightly of its spurious ''memories,'' REM sleep in effect enhances the memory system. The implication for dreamers is that remembering one's dreams—the cornerstone of Freudian psychoanalysis—may be unhealthy; dreams are parasitic connections that, for the health of the organism, should be unlearned. ''We dream,'' Crick concludes, ''in order to forget.''

SCHIZOPHRENIA

A group of scientists reported evidence that the predisposition for schizophrenia, a debilitating thought disorder, is inborn. Psychiatrist Arnold Scheibel and anatomist Joyce Kovelman of the University of California at Los Angeles did post-mortem analyses of the brain tissue of people with schizophrenia. They found that the pyramid-shaped cells of the hippocampus were dramatically misaligned. Normally these cells are arranged in very precise layers, but in the brains of schizophrenics they were rotated as much as 180 degrees. Furthermore, the scientists reported, the degree of misalignment appeared to reflect the seriousness of the illness. Because there is no room for these hippocampal cells to shift position once the brain begins developing and making connections after birth, Scheibel and Kovelman interpret their findings as evidence that the disorder develops prenatally (before birth).

The most dramatic cellular disarray was found in a brain area called the subiculum, which is a kind of gateway from the hippocampus to other brain regions. It could be, the researchers speculate, that normal precise cell layering is necessary for processing incoming information and that misalignment causes distorted messages. Distorted messages to the nearby amygdala, the brain region where memories and emotions are linked, could explain the emotional flatness characteristic of schizophrenia, while distorted messages to the higher thought centers of the brain might explain such symptoms as hallucinations and paranoia.

AN ASCETIC DISORDER

Scientists studying compulsive runners suggest that extreme commitment to running should be recognized as a destructive and pathological behavior. Compulsive running, they say, is a primarily male manifestation of what they call an ''ascetic disorder,'' which, in women, tends to show up as anorexia nervosa—self-imposed starvation. Psychiatrist Alayne Yates and her colleagues at the University of Arizona found that compulsive runners are very similar psychologically to anorexics, and that both abnormal behavior patterns tend to start at times of stress and may be reinforced by mood changes associated with increased levels of endorphins (natural morphinelike brain chemicals) found in anorexics and implicated in the emotional ''high'' some runners experience.

WRAY HERBERT

Douglas Kirkland/Sygma

Out on the Streets

by Ira Mothner

There are homeless women in almost every city in the United States now, and not all fit the "bag lady" stereotype. Many do indeed wear layers of filthy clothing and camp in doorways or storefronts. But others are harder to spot, pulling neatly loaded carts along the street or mingling with luggage-laden travelers in bus depots and train stations. Only torn footwear or matted hair or a vacant look in their eyes gives them away.

These are women who have lost jobs or

homes or just lost their way—widows who cannot cope, the mentally ill who cannot make it on their own. They are the socially, emotionally, or economically fragile.

The New Homeless

Little more than a decade ago, only a handful of women were on the streets. Since then the numbers have swelled. Today women are the most visible segment of a homeless population estimated at between half a million and 2 million.

To the traditional hoboes and derelicts have been added discharged mental patients and the victim's of today's hard times and harsh fiscal policies. Widespread unemployment has put whole families on the road. In the public shelters, men are younger, better educated, and more frustrated. Skid Row has become more dangerous. There are more drugs and more violence, for many of the new homeless do not know how to survive without preying on the weak and vulnerable.

To the rest of the nation, the homeless seem a frightening mix. In Phoenix, Arizona, emergency shelters were shut down in the summer of 1982. "The Bums Are Back," reads a leaflet distributed by merchants near a shelter in Chicago, Illinois. This past winter in New York City, community groups protested plans for housing for the homeless. Least frightening but most disturbing are the women. Many of us would like to believe they are on the streets by choice. We relish the myths of bag ladies with bankbooks and wads of cash in their satchels, for the reality of women forced to live without shelter or privacy or the comforts we take for granted says that something is terribly wrong.

Free-roaming Mentally Ill Add to Problem

The country's recent past is littered with bright ideas that didn't quite work out. A few of them have even helped to create homelessness. That has been the downside of "deinstitutionalization," the policy of hustling patients out of mental hospitals. It is also a price of "gentrification," the move of the well-to-do back to run-down areas.

Deinstitutionalization seemed a logical step once psychotropic (mind-altering) drugs were available to control the behavior of the emotionally disturbed. If patients no longer "acted crazy," far better to care for them in the community. So public mental hospitals were able to cut patient loads by nearly 75 percent between 1955 and 1979. Today estimates of the chronically mentally ill outside institutions range from 800,000 to twice that number. Between 300,000 and 400,000 are in boarding-houses and adult homes, many of which provide only bare subsistence.

Joan, who showed up broke, with torn and infected feet, at a women's crisis shelter in Phoenix, had walked out of one such place. Although she had signed over her disability check and left before the month was out, she still owed $88 for tea and toast.

Jean-Louis Atlan

The rapid release of mental patients from overcrowded institutions has added to the problem of homelessness.

Recently released from the county hospital, Joan, a middle-aged woman, carries medication but rarely uses it. "It takes away my feelings," she says; but it also stills the voices she hears. The night before, the voices had told her to take off her shoes and run barefoot for kilometers. Police found her by an irrigation ditch and brought her to the Salvation Army. There was no room. "I tried to crawl into a dumpster and finally got some sleep in an alley."

Cynthia, a pleasant-faced, white-haired woman over 50, showed up at the crisis shelter, Sojourner Center. She'd gone to the police and asked for a cell. ("My mother said that was the proper thing for a respectable person to do.") They ordered her out, and she slept on the ground one night, in the rest room at the bus station the next.

It's hard for the homeless in Phoenix. There are few places for single women. Sojourner can take a few, and the Salvation Army sometimes will. For men, there are even fewer places. Just about all that's available for most of the homeless is the midday meal at the Saint Vincent de Paul Charity Dining Room. More than 1,000 can show up. "The ladies are afraid," says Judy Knight of the Salvation Army. "The street has changed in the past nine or ten months. So many young people resort to violence. They abuse the bag ladies, so the ladies have found hidey-holes.''

At the Phoenix South Community Mental Health Center, Carl Brown points out that few of the homeless are in touch with the center. Working on a survey with anthropologists Louisa Stark and Ron Paredes, Brown was ''amazed at the number of young homeless who are really isolated, the ones who refused to participate.'' Even so, the survey showed that 18 percent of all the homeless had serious mental problems, and 30 percent of the women did.

In New York City, about one-third of the men and women in public shelters have some history of psychiatric hospitalization. About half of the homeless on the street do. But there are no figures to show how many are being refused hospitalization under today's rigid admission criteria, and none show the impact of homelessness itself on stability. What happens to a woman who has no history of mental illness when she loses her home?

Survival on the Streets

Each year 2.5 million people in the United States are pushed out, priced out, or forced out of where they live. Each year half a million units of low-rent housing disappear. They are burned down, torn down, or converted for new young city dwellers who are renting, buying, and renovating old apartments and buildings—gentrifying marginal neighborhoods. Cheap hotels and SRO (single-room-occupancy) buildings are hard to find, and the displaced must compete for space with new immigrant families coming to the cities. What they can find is often more frightening than the streets. ''They're not rooms where you can put a decent woman,'' says Joyce, who has been looking for a room in New York for many months. ''What you get are hotels where men are pounding on your door at 2:00 A.M. to find out if you want to make a couple of dollars.''

In the fall of 1982, when a Los Angeles, California, street character died, she was memorialized in the *Los Angeles Times* by a poet

Rosa, a street-savvy survivor, roams around Los Angeles, California, with all her life's possessions in a shopping cart.

Homeless women care more about safety than comfort when it comes to finding places to sleep. In New York, subways are popular resting places.

F.B. Grunzweig/Photo Researchers

and novelist: "The bag lady of 6th and Western was someone who could not or would not fit into the bureaucracy, the categories on government check-yes-or-no boxes, the nursing homes and institutions, the packaging our society increasingly insists upon. For reasons we will probably never know . . . she said no."

That romantic view of bag ladies is both comforting and common. But if bag ladies choose the streets—and some indeed do—they choose it over a limited number of less attractive alternatives. Often there is no choice at all, according to Marsha Martin, who runs New York's Manhattan Bowery Corporation's Midtown Outreach Program. She wrote her doctoral dissertation on homeless women, "because resources for women are so scarce."

Until quite recently, services for the homeless were virtually only for men. In Los Angeles there are plenty of mission beds for men but still very few places for women. There are even jobs for men on the street. They can wash cars or dishes, deliver takeout lunches, or pass out handbills—jobs that are rarely open to women. "When women go to the streets," says Martin, "they are opting for survival. They are creating systems for themselves," using their ingenuity to meet the main challenges of street life—

sleeping, keeping clean, and eating.

Women sleep where they feel safe. Security, not comfort, is the criterion. They steer clear of parks and the greenery along highways and freeways, stay out from under bridges and overpasses. Bus and train stations are top choices. All-night movies also rank high, as do the subways in New York.

Pauline has spent nights at theaters in downtown Los Angeles. "You find an awful lot of bums there. I got hit over the head with a wine bottle once." Still, she prefers movies to a doorway, where "it's cold and ugly." But doorways will do if the street is well lighted and people are about.

Some Prefer Homelessness

"The women I met did everything they could to keep clean," says Martin. "They took little bottles of water from fire hydrants and washed their hair in the street." In Phoenix, Michele will pay a dollar to shower at a motel. When she doesn't have the money, "I use the rest room in the park to keep my body clean. Some women strip right there."

Hygiene is a problem homeless women sometimes cannot solve. Many have lice. Literally living on the street, they are prey to infec-

tion. The hard life shows in the swollen and bandaged legs, the result of too many nights spent upright. Over the years, fluid accumulates in the legs. Circulation is reduced, tissue breaks down, sores develop, and infections spread easily.

While there are soup kitchens and charity dining rooms in most cities, many homeless women are likely to scorn public shelters. Martin calls them "unconnected." The tendency toward isolation is strong among these women. While men group together, women most often go their own way, not only homeless but alone as well.

"To eat," Martin explains, "they will hang out near a McDonald's or a Burger King. They will get to know fruit-stand owners." They neither steal nor beg. "They are just there, and people give them things." Far from being coldhearted, New Yorkers are a benevolent bunch, according to Martin. "New York pedestrians have kept an incredible number of people on the streets alive, giving them food and money and bringing them clothing.

"But when people drop a quarter or buy the woman on the corner a sandwich, it keeps

Left: Sleeping in an upright position night after night is taking its toll on this woman's legs. Fluid buildup may lead to tissue damage. Below: A woman's novel way of washing meets with disapproval from a passing policeman.

Photos: UPI

that woman on that corner. She is getting reinforcement. What we have to do is break through that survival system.'' The purpose of the Midtown Outreach Program is to move people off the streets, help them get the assistance they need and a place to live.

Some of these people can be frightened of living indoors. ''What frightens them,'' says Martin, ''are the responsibilities. On the street, they can avoid responsibility. You either give them a quarter or you don't. They don't have to act a certain way for you. But once they move indoors, they have to pay rent, respond to neighbors, talk to the landlord. They have to interact.''

Temporary Shelters Only Buffer the Plight

It is not easy to work with homeless women. It can be unnerving. ''Other women look at them and shudder at what they themselves might become,'' says one church official. Yet plenty of volunteers are eager to help at the centers where homeless women can get off the streets and back in touch, places like the Downtown Women's Center in Los Angeles and the Olivieri Center for Homeless Women in New York.

Jill Halverson opened the Downtown Women's Center in 1978. She runs it with only one part-time assistant, raises a $30,000-a-year budget from private contributors, and keeps it open from 9:00 A.M. to 5:00 P.M., seven days a week. There is a row of beds for naps, and a shower that runs pretty much full time in the two long and narrow storefronts that house the center. Yet it is a bright and homelike place, with a full schedule of activities—bingo, art classes, group therapy—a pot of coffee always on the stove, lots of snacks. There is a light breakfast and a substantial lunch, often brought by volunteers. As many as 100 women use the center, but rarely more than 40 at one time. It offers little in the way of formal social services, but Halverson will go to bat for the women and has gotten several their Social Security benefits or welfare. She runs the center firmly and will bar women for fighting and, in one case, for refusing to be deloused.

The Olivieri Center, near Pennsylvania Station in New York, is something quite different. It is larger, less intimate, more institutional, and its clients appear to be more disturbed. Opened in 1981, it is supported by the city and runs 24 hours a day with a staff of 24. There are no beds, and women sleep sprawled across the landing outside the office. Three meals a day are served. Many women have their mail delivered there. A medical team comes by three times a week, and staff counselors are aggressive welfare advocates.

Dorothy, a slight woman with short white hair, has been at the center for six months. A former mental patient, she complains: ''There's nothing to do. In the hospital you could earn privileges, go to the movies. Here it's so boring, and you sit up all night.''

Helene Fishman, who started the center and was its first director, has complaints of her own. ''I'm an angry woman,'' she says, ''because there is no provision for released mental patients, and lots of these people cannot maintain themselves. Many have been off medication for years.'' It's a complaint that can be heard across the country, and so can her objection to ''the misguided protection of individual rights'' that makes it ''so difficult to get someone hospitalized.'' She has her share of horror stories. One woman was at the center four months because no psychiatrist found that she was a danger to herself. ''She was totally incontinent,'' says Fishman. ''Her legs were infested with maggots. She refused food, wouldn't sign her Social Security checks, and talked about 'a Communist conspiracy.' ''

Paving the Way for Permanent Housing

Shelters and centers are no real answer to homelessness. Both Fishman and Halverson know this, and both would like to see permanent housing for their clients. Fishman wants to start what she describes as ''someplace where people want to stay for the rest of their lives.''

Currently, New York City is contracting with several groups to run transitional housing for the homeless, a first step toward permanent homes. The city is out in front on this, in part because the issue surfaced there but most likely because a suit on behalf of homeless men in New York established their right to decent shelter. Kim Hopper of the New York Coalition for the Homeless told a congressional panel in the winter of 1982 that the mentally disabled on the streets are not ''the victims of their own unwillingness to seek assistance or to accept it when offered.''

He said, ''Where decent, humane shelter has been made available, it has never lacked willing recipients. . . . As the range of options offered the homeless poor has increased, so has their demonstrated willingness to come in from the cold'' □

THE MIND HUNTERS

Terry Arthur

by Bruce Porter

When the New York City police finally
called the Federal Bureau of Investigation (FBI) into the case in October
of 1979, the investigation was beginning to
stall. The nude body of a 26-year-old special-
education teacher had been found on the roof of
the Bronx public housing project where she
lived. She had been badly beaten about the face
and strangled with the strap of her purse. Her
breasts had been mutilated, and scrawled on the
inside of her thigh in ink was: "___ you. You
can't stop me."

Zeroing in on a Killer

"We get a lot of murders, but not this type of
mutilation," says homicide detective Thomas

Foley, who was in a quandary as to what kind of
suspect to look for. "Frankly, I didn't see where
the FBI could tell us anything, but I figured
there was no harm in trying." A few days after
delivering pictures of the murder scene and a
copy of the autopsy report to the FBI Academy
in Quantico, Virginia, Foley got back a description of the probable killer. He would be a white
man, probably 25 to 35 years old, who knew the
victim and either lived or worked nearby, possibly in her apartment building. He would be a
high school dropout, would live by himself or
with a single parent, and own an extensive collection of pornography. What's more, in all
probability he would already have been interviewed by the police.

It took Foley and other detectives 10 or more months of digging before they were ready to turn the case over to the district attorney for prosecution. By then they had found out that the murderer was indeed 32 years old and a high school dropout who knew the victim and lived on the fourth floor of her building. The police had already questioned the young man's father—with whom he shared both an apartment and a pornography collection. Detectives had lost interest in the youth after they were told that he had been in a mental hospital at the time of the killing. But because he fit the sketch so closely, detectives checked the hospital and discovered that security was lax enough to permit patients to come and go more or less at will. This led them to concentrate on the young man, and eventually they built up enough evidence to convict him of the murder and send him off to prison for 25 years to life. "What the FBI description did," says Foley, a 10-year veteran of the force, "was to keep me on course."

The sketch not only provided the key that broke the case, but also, in its uncanny accuracy, gave New York detectives an eye-opening illustration of the latest weapon in the FBI arsenal: psychology profiling. What's more, it persuaded Foley and his colleagues to join a growing number of police officers across the country who think that when it comes to solving certain kinds of crime, profiling can provide crucial help. "They had him so right that I asked the FBI why they hadn't given us his telephone number, too," says Lieutenant Joseph D'Amico, Foley's boss and head of the homicide squad for the New York City Housing Police.

Mental Disorders Motivate Bizarre Murders

The FBI agrees to provide profiles in only a narrow selection of crimes—mainly multiple rape or child molesting, or so-called "motiveless" murders, in which the nature of the killing points to major psychological abnormality in the killer. One reason that such cases are chosen for profiling is that deviant crimes lend themselves much more readily to the technique than do the mundane varieties. "The more bizarre the crime scene," says agent John Douglas, who helped work up the Bronx profile, "the easier it is to tell what kind of person did it."

Another reason is that bizarre crimes have increased significantly in recent years. About 20 years ago, according to the FBI, the rule of thumb was that in more than 80 percent of all murder cases, the killer had some kind of previous relationship with the victim. The motive was passionate anger or a desire for revenge, and a quick canvass of the neighborhood usually turned up a list of likely suspects.

Recently, though, the 80 percent figure has been dropping precipitously. Out of the 22,516 killings in the country in 1981, some 45 percent were either "stranger murders," in which killer and victim had never seen each other before the crime, or murders in which the killer was listed as "unknown." Many such cases are felony murders—killings that occur as outgrowths of robbery or some other crime, and are motivated by the need to escape. But in an increasingly large number of "stranger" homicides, the killer seems driven to murder not by some "rational" reason or easily understood emotion, but by a serious psychological disorder.

Overwhelmingly, the victims of bizarre murders are women or children; the killers are almost invariably men. They are usually intraracial—blacks killing blacks and whites killing whites. And the list of victims killed by a single murderer often runs into double figures.

Success with a Bomber; Failure with a Strangler

Using psychology to catch criminals hardly qualifies as a recent discovery. Its literary origin goes back at least to 1841 and the publication of "The Murders in the Rue Morgue" by Edgar Allan Poe. His detective, C. Auguste Dupin, demonstrated the ability to follow the thought pattern of a companion while the pair strolled through Paris, France, for 15 minutes without uttering a word.

In real life, undoubtedly the most ingenious piece of profiling was performed in the late 1950's, by a Greenwich Village psychiatrist named James A. Brussel to help the police catch the Mad Bomber of New York. The bomber turned out to be George Metesky, a disgruntled ex-employee of the local utility company. Metesky had set off 32 devices over an eight-year period. After poring over letters written by the bomber and looking at pictures of the bomb scenes, Brussel theorized that the criminal was an Eastern European man, 40 to 50 years old, who lived with a maiden aunt or sister in a Connecticut city. He hated his father but loved his

Just as a profiler had predicted, the Mad Bomber was a middle-aged man with a penchant for double-breasted suits.

mother, something Brussel divined from the way Metesky rounded out the sharp points in his *W*'s so that they resembled cartoon versions of a woman's breasts. Brussel diagnosed him as a paranoiac who was meticulous in his personal habits. When he was found, Brussel said, he would be wearing a double-breasted suit—buttoned.

When Metesky was captured, shortly thereafter in Waterbury, Connecticut, the portrait turned out to be an extraordinary likeness, right down to the suit. (Metesky actually lived with two maiden sisters.) But other profiles by independent psychiatrists and psychologists—in contrast to those by FBI agents—have not always proved helpful. Some are so vague as to point to practically anyone. In the worst cases, psychological profiles can severely hamper an investigation by sending the police off in the wrong direction. During the 1960's, a committee of psychiatrists and psychologists set up to help catch the Boston Strangler portrayed him as not one man but two, each of whom lived alone and probably worked as a schoolteacher. One of them, the committee said, was a homo-

sexual. The person who later confessed to being the strangler was one man, Albert DeSalvo. He lived with his wife and two children, was employed as a construction worker, and would never have been found by any of the police assigned to search the city's homosexual community.

"I don't think psychiatrists or psychologists have any business pretending to be experts in profiling criminal suspects," says Park Elliott Dietz, an associate professor at the University of Virginia Schools of Law and Medicine, and chief psychiatrist for the prosecution during the trial of John W. Hinckley Jr., who attempted to assassinate President Reagan. "What's different about the FBI's effort is that they process a crime scene through an experienced investigative brain."

Minute Details Yield Valuable Clues

In working up its profiles, the FBI pays microscopic attention to autopsy reports and to maps and photographs of the crime scene. How the victim was treated reveals a lot about the killer. "A person who covers up the body with clothing, or hides it, is saying that he feels pretty bad about what he's done,'' says FBI agent Douglas, who has a master's degree in educational psychology from the University of Wisconsin and has profiled over 450 murders. "If he moves the body so it will easily be found, this may show that he has some feeling for the person. He doesn't want them exposed to the elements. He wants them to have a funeral and decent burial.''

Profilers pay particular attention to the manner in which a person was killed, the kind of weapon that was used, and something the bureau calls "post-offense behavior," or what the killer did to the victim after he or she was dead. Sex murders typically are stabbings, strangulations, or beatings, rather than shootings. If the killer brought along his own weapon, it points to a stalker, someone fairly well organized, even cunning, who came from another part of town and probably drove a car. If the killer used whatever weapon was available—a knife from the kitchen or a lamp cord—it points to a more impulsive act, a more disorganized personality.

It also means that the person probably came on foot and lives nearby.

Was there a lot of beating about the face? The general rule is that a brutal facial attack, as in the case of the murder in the Bronx, means that the killer knew the victim; the more brutal the attack, the closer the relationship. Was the victim killed immediately in a blitz style of assault? This usually indicates a younger killer, someone in his teens or early 20's, who feels threatened by his victims and needs to render them harmless right away. On the other hand, if the killer showed mastery of the situation, if he killed slowly and methodically, it points to a more sadistic personality, a man in his late 20's or 30's.

Equally significant is what the killer does right after the murder. Does he seem to be hanging around the scene, enjoying himself, going through the victim's things, setting the body up in a ritualistic position? Or does he kill and run? And has he taken something? Killers often carry away an artifact, such as a bracelet or compact, to use afterward as a way to re-create the expe-

The committee that created a psychological profile of the Boston Strangler (middle) portrayed him incorrectly.

Wide World Photos

rience in memory. Certain kinds of killers also keep diaries and scrapbooks about their crimes.

Even evidence that at first seems contradictory and confusing can be exceedingly helpful in pointing to a culprit. In 1981 a 22-year-old woman was abducted one night from a baby-sitting job in a small town in Pennsylvania. Her body was found several days later at the local garbage dump. When Douglas eventually got the photographs and autopsy report, he saw evidence that pointed to two totally different personalities. On the one hand, the fact that the victim was murdered in a fierce, blitz-style attack, and had been mutilated after death, pointed to a disorganized, frightened killer who was able to carve her up only after she was no longer alive. On the other hand, the girl was murdered at one location, then taken by car to the dump, something that Douglas felt could be done only by an organized, calculating killer. There was also evidence of postmortem rape. Again, judging from similar cases, Douglas reasoned that this act was probably not done by a disorganized man.

In the end, Douglas told the nonplussed chief of police that he should be looking for two killers rather than one. And that's the way it turned out. One of them—the man responsible for the frenzied attack—was the girl's live-in boyfriend; the other was his brother, who organized the transportation of the body from the original scene of the crime to the dump.

Teten's Uncanny Ability

How the FBI gets each piece of the puzzle seems understandable enough. But a completed profile can astound even the most seasoned investigator. The oldest and one of the best-known profilers at the agency is Howard Teten, a 20-year veteran of the FBI and now director of its Institutional Research and Development Unit. Teten is famous for his ability to come up with detailed descriptions of killers on the scantiest of information.

Teten began doing profiles in 1970. At that time, he taught a course at the FBI Academy in applied criminology, and students from various police departments would bring him their cases. On one occasion, a California policeman telephoned about a baffling case involving the multiple stabbing of a young woman. After hearing just a quick description of the murder, Teten told him he should be looking for a teenager who lived nearby. He would be a skinny kid with acne, a social isolate, who had killed the

girl as an impulsive act, had never killed before, and felt tremendous guilt. ''If you walk around the neighborhood knocking on doors, you'll probably run into him,'' Teten said. ''And when you do, just stand there looking at him and say, 'You know why I'm here.' '' Two days later, the policeman called back to say he had found the teenager as Teten said he would. But before the officer could open his mouth, the boy blurted out: ''You got me.''

Insight into the Criminal Mind

Along with poring over crime-scene photographs, the FBI uses data from basic research among murderers themselves. In 1982 the Behavioral Science Unit received a $128,000 grant from the National Institute of Justice for the purpose of building a file of taped interviews with at least 100 notorious mass murderers and assassins, and computerizing the similarities in their cases. This is the first methodical study ever made of so many killers. The bureau has talked with such infamous murderers as Charles Manson, Richard Speck, David Berkowitz, Sirhan Sirhan, and Arthur Bremer, the man who tried to kill George Wallace. Agents even visited the criminally insane ward of the Mendota Mental Health Institute in Madison, Wisconsin, to see the elderly Ed Gein, the ''Ghoul of Plainfield,'' whose nocturnal excavations in the graveyard of a small Wisconsin town supposedly provided Alfred Hitchcock with his inspiration for the movie *Psycho*.

What the agency hopes to gain from the interviews is insight into how criminals actually work—something that, for all the academic research in crime, remains largely uncharted territory. How do killers approach their victims? What do they talk to them about before killing them? How do they react immediately after the murder?

''When we went to New York to talk to the 'Son of Sam,' David Berkowitz,'' says Robert K. Ressler, the agent in charge of the project, ''he told us that on the nights when he couldn't find a victim to kill, he would go back to the scene of an old crime to relive the crime and fantasize about it. Now that's a heck of a piece of information to store somewhere to see whether other offenders do the same thing.''

In writing up its profiles, the FBI steers clear of psychiatric terminology and couches everything in plain English. For one thing, local police officers tend to regard the mental-health professions, and their language, with consider-

David Berkowitz, the ruthless Son of Sam killer, was one of many murderers interviewed by the FBI in the agency's effort to gain insight into how killers really work.

able suspicion. For another, psychiatric language is not terribly helpful in catching criminals. "We don't get hung up on why the killer does the thing he does," says FBI agent Roy Hazelwood. "What we're interested in is that he does it, and that he does it in a way that leads us to him."

A Dead-Wrong Profile

The agency warns local police officers not to take any profile too literally—not to limit their investigation to people who exhibit the characteristics in the sketch. A profile is supposed to describe a general type of person, not point to a certain individual. And there is always the possibility that an FBI profile could be dead wrong. Hazelwood, for instance, holds the dubious distinction in the agency of having drawn perhaps the most inaccurate profile on record. He did it in a Georgia case in which a stranger showed up at a woman's door one day and for no apparent reason punched her in the face and shot her little girl—not fatally, as it turned out—in the stomach. Hazelwood told the local police to look for a man who came from a broken home, had dropped out of high school, held a low-skilled

job, hung out in honky-tonk bars, and lived far from the crime scene. When the culprit was finally caught, he turned out to have been raised by both of his parents, who had stayed married for 40 years. He had a college degree and had earned above-average grades. He held an executive job in a large bank, taught Sunday School and regularly attended church, never touched a drop of alcohol, and lived in a neighborhood close to the scene of the crime. "I keep that profile around," says Hazelwood, "as a reminder that we're still in the stage where profiling is an art rather than a science."

Pro-active Strategy

Along with helping the police narrow down an investigation, profiles are also frequently used to lure killers into the open. The police refer to this as a "pro-active," as opposed to "reactive," technique—and it often requires getting cooperation from the local press. When the FBI determines that a killer is jumpy or under a great deal of stress as a result of his crimes, it encourages the local police to promote newspaper stories saying that the investigation is getting closer and closer to a solution—even if it's get-

Alon Reininger/Contact Press Images Inc.

Police often use newspaper stories to psychologically lure killers into the open. This is called pro-active strategy.

ting nowhere at all. "You never want to let the guy off the hook psychologically," Douglas says. "You put enough stress on him and it can cause a change in his behavior so he'll give you something to go on."

When the profile suggests that the killer is experiencing strong feelings of guilt over his crime, the pro-active tactic might be to encourage a newspaper story on the anniversary of the crime, perhaps a sympathetic piece about the victim's family. "What you're trying to do here is to draw the person psychologically back to the victim," Douglas says. The police might also be told to watch the cemetery where the victim

is buried, or even place a listening device on the tombstone. Several killers have been caught when they went back to put flowers on the grave.

Some professionals in the mental-health field get a little queasy over the use of psychological approaches to create or compound stress rather than relieve it. Dietz, for instance, occasionally consults with the FBI on profiling, and has joined with agents in publishing papers on crime in professional journals. But when it comes to planning pro-active strategy, he begs off on the ground that the tactic could lead a suspect to kill himself. "It is generally unethical," he says, "for a physician to apply medical knowledge that can result in direct harm to a human being."

The general police reaction to the possibility of a suspect's suicide is "good riddance." And the police argue that whatever the moral complications, pro-active techniques are justified by the deaths they may prevent. If the police aren't careful, however, such techniques can prove to be dangerous. In one Western state, the FBI profile of the killer who stabbed a girl 84 times was presented over a local "crime-stopper" television program, together with vivid pictures from the crime scene. The police, however, had not thought to stake out the girl's home. Shortly after the program, the killer returned to the house and smeared blood over the wall of the girl's room. Had her mother not been out at the time, she might well have met the same fate as her daughter.

A Growing Role for Profiling

The FBI has recently been using psychological profiles in the later as well as the earlier stages of investigations. The agency instructs local police officers in the best techniques for interrogating different kinds of suspects—whether, for instance, to take them to the station in the daytime and use bright lights and hard grilling, or whether to see them at night and act gentler.

The agency has also begun advising prosecutors during trials on how to cross-examine the accused on the stand. During the prosecution of Wayne Williams for 2 of the 27 Atlanta, Georgia, murders he is believed to have committed, Douglas sat next to Assistant District Attorney Jack Mallard during most of the trial. The problem, from the prosecutor's point of view, was that in the beginning, Williams appeared too cool and composed—hardly the picture of a man capable of murderous outbursts. "We were concerned," Douglas recalls, "that the jury was seeing him as a creditable type of person." Douglas therefore advised Mallard to keep Williams on the stand as long as possible and to rattle him with detailed questions about the killings. "It's difficult for him here in the courtroom where he's not in control and where all the flaws in his personality are coming out. The longer you can keep him up there, the more he'll become agitated and the greater the chance that he'll create an outburst," Douglas explained.

That's exactly what happened. As Mallard bored into how the victims were strangled— "What did it feel like, Wayne, when you

UPI

During Wayne Williams' prosecution for two murders, FBI profiling helped to break his composure while testifying.

wrapped your hands around their throats?" he asked. "Did you panic, Wayne?"—Williams became increasingly uneasy. Suddenly, he interrupted the cross-examination, pointed a finger at the district attorney, and called him a "fool." Then he went off into a rambling tirade during which his manner of talking changed from calm, educated English into street slang. "You could see the jury suddenly look up in astonishment," says Douglas. "They were shocked; here was a completely different side of his personality coming out."

As for the future of profiling, the FBI says the system will soon be computerized, so that a police officer anywhere in the country can punch the characteristics of a bizarre murder into a terminal and get back an educated guess as to who did it. The killer will be categorized on the basis of whether he fits into an "organized," "disorganized," or "mixed" personality type. And his characteristics, such as age, race, and how near he lived to the crime scene, will be given numerical weight, depending on how frequently they showed up in similar crimes of the past. No one, of course, is willing to say how the computer can factor intuition and instinct into its analysis. But the day does not seem far off when the police will be able to identify a criminal by the psychic loops and whorls he left at the scene, just as surely as if he had covered the wall with fingerprints □

Eight Women in the Wild

by Anne LaBastille

When women poke their underwater cameras into the faces of great white sharks, track rhino through South African thickets, and round up bands of deer poachers, many people still react with astonishment. Why do they take such risks? How do they handle those rigors? Why are these women working in the wilds? Are they oddballs or renegades?

Women Flourish in Male-dominated Jobs

In the past 15 to 20 years, a social phenomenon has taken place across the United States and in many foreign lands. Women are entering traditionally male-dominated outdoor professions. They are living and working in ever-increasing numbers as marine and wildlife biologists, federal and state game wardens, park rangers and supervisors, field zoologists, and environmentalists.

Sometimes they work alone, sometimes with families. Many are U.S. scientists conducting research abroad. Others are virtually pioneers among professional women in developing countries. They are simply people doing jobs they like. And in the process, they are providing valuable data and conservation efforts for wildlife and wildlands, while proving that women do belong in the wilds.

Of course, such professional women face the same kinds of conflicts as other career women. Changing roles produce problems among marriages, families, work, and lifestyles. Some manage to balance it out; others don't. But these dilemmas are largely symptomatic of the times.

Females Fully Participate at Long Last

There were a few earlier outdoorswomen, to be sure. Names such as Elinor Stewart, Calamity Jane, Isabella Bird, Sacajawea, Martha Maxwell, Annie Peck, Elizabeth Agassiz, and Delia Ackley come to mind from the 1700's and 1800's. They were intrepid, unconventional, and observant. Yet it's safe to say that only since the middle of the 20th century have females been participating in every aspect of the natural sciences, including the full field experience. Even as late as the 1950's, there were practically no women foresters, ecologists, wildlife managers, ethologists, fire fighters, or conservation-enforcement officers.

The main thrust came in the 1960's and 1970's when a combination of circumstances finally gave women the opportunity to engage in outdoor field research. What were those circumstances? I would list: the early feminists' efforts, followed by the current women's liberation movement; increased industrialization and mechanization, which have freed women from housework and given them incomes; improved education and birth control methods; enhanced legal status for both single and married females; and a revolution in attire and equipment that allows anyone to function comfortably and efficiently outside.

Taken in concert, these factors have fostered profound changes in the degree of independence and professional involvement of women. The Wildlife Society, the only professional organization for wildlife workers, recently published results of a 1977–78 survey in which it found that women make up almost 25 percent of the total college enrollment in undergraduate wildlife sciences. Women make up 18 percent of master's students, and 10 percent of Ph.D. candidates. I estimate that 10 to 15 percent of all professional environmental or natural sciences scientists in the United States today are female.

Actually, good physiological and psychological reasons exist why women can excel at—and even outperform men in—outdoor work. Research tends to show that women have greater patience and a less violent nature than men; better manual dexterity, memory, and perception; greater constitutional strength; keener auditory, olfactory, and taste senses; better capacity to withstand adverse conditions; and more flexibility. Clearly, they can use these psychobiological factors to excel in new professional fields with wildlife and wildlands. The

David Doubilet

Dr. Eugenie Clark, a world-traveled marine scientist, examines a deep-sea shark aboard a Japanese fishing vessel.

eight women in the following profiles have done just that. They are fascinating people in their own right, but their lives illustrate the possibilities for all women in the wild.

A Passion for Sharks

A large black-tipped shark cruised almost within touching distance of the doe-eyed woman diving on a reef in Micronesia. Carefully, it seemed to look her over. Rather than cringing in fright, she turned to admire the creature before it swam off into the blue Pacific. From that moment in 1949, Eugenie Clark has been "turned on" to sharks.

At the time, Clark was virtually the only female marine scientist working in the field. In 1952 she completed her Ph.D., and since then she's gone on to rack up over 30 years of diving and research, much of it on sharks, in remote oceans. Today appropriately enough, Clark is

nicknamed the "Shark Lady," which is also the title of her children's biography.

As a youngster, this small, graceful only child of a hard-working Japanese mother would spend hours at a time at the New York Aquarium. Years later, as a full-fledged scientist, she was able to make her own aquarium. She helped found and direct the Cape Haze Biological Laboratory (now the Mote Marine Laboratory) near Sarasota, Florida. There she collected hundreds of live sharks to study their behavior.

But her most compelling interactions with the fish have been in faraway seas. Clark has traveled to Mexico's Yucatán and to Japan to investigate "sleeping" sharks. Often she is accompanied by her children—she has four and has been married four times—or by students. Other trips take her to the Red Sea, which she has visited more than 30 times to collect the Moses sole, a shark-repelling fish.

The Red Sea, near Ras Muhammed, is her favorite underwater diving area. She and other conservationists are working to have it declared a marine park. Meanwhile, she continues to champion the conservation of sharks everywhere, pointing out that the vast majority of the world's nearly 350 species are unaggressive and rarely attack human beings. "No shark normally feeds on man," she states. "Mankind should not persecute sharks; rather protect them."

So avid a professional is Clark that she continued to dive through all four of her pregnancies. What plans does this vivacious, soft-spoken scientist have now? "I plan to keep diving until I'm at least 90," she says.

Protector of Lemurs

The U.S. zoologist was picnicking in southern Madagascar. Suddenly, three Antandroy tribesmen emerged from the spiny forest, dragging a half-choked sifaka by a vine around its neck. The animal, a kind of lemur, had been smashed in the muzzle by a stone. A jagged bone protruded from its arm. One of the tribesmen held out his hand. Would the foreign woman give them a few francs for such an unusual prize?

The zoologist flew at the tribesman and pretended to choke him. He and his comrades had stumbled on a world authority on lemurs, a university teacher and author who knew that the victim—a distant relative of monkeys, apes, and humans—was one of the earth's rarest animals. Yet even while her anger cooled, she could not help thinking that the tents in her Land-Rover cost more than all the houses of their village. For Alison Jolly, that moment sums up a world dilemma: How can we weigh short-term economic need against saving our priceless natural heritage?

Born in Ithaca, New York, Jolly lived in England with her husband and their four children, and taught at Sussex University before returning to New York. Yet her heart is in Madagascar. The 1,600-kilometer (1,000-mile)-long island is a separate experiment in evolution. Ninety percent of its forest-plant species are found only there, as are nearly all its reptiles, half its bats and birds, and all its land mammals, including 40 races of lemurs.

Jolly's extensive research there has resulted in several scholarly books and a popular account of her experiences: *A World Like Our Own: Man and Nature in Madagascar*, pub-

Zoologist Alison Jolly attends a village ceremony on Madagascar, where local people believe lemurs are sacred.

Maria Tereza Jorge Pádua was formerly director of Brazil's national parks system and one of the world's top female conservationists.

lished by Yale University, her alma mater. This tells the story of the wounded sifaka and also more cheerful tales, such as Jolly's visit to an island of sacred lemurs, where local people have saved a patch of forest because they believe lemurs guard the tombs of dead kings.

Asks Jolly: "What do you do if you actually like people as well as animals?" She partially answers the question in her Madagascar book: "Saving the wilderness is saving what we do not yet know: the drug untested, the genetic stock untapped, the species undescribed. It is saving what we do not yet love: the white sifaka leaping in beauty against the desert sky."

A Brazilian Park Chief

Dressed in a chic, flowing white suit, she looks like an elegant society lady at a government gathering. Yet watching Brazil's chiefs of state and ministers shake her hand deferentially, one senses that there is more to this attractive mother of three than that. Indeed, there is. Until 1983, Maria Tereza Jorge Pádua directed both Brazil's national parks system and its wildlife service. She was probably the most highly placed woman of conservation worldwide. She left the park service for political reasons and is now head of environmental affairs for a huge Brazil-ian electrical company serving São Paulo, one of the largest cities in the world.

Mrs. Jorge Pádua, who is married to an ecologist, began her outdoor interest on her father's farm near São Paulo. She went on to study agronomy, earn an M.S. degree in ecology, and take on agricultural engineering in postgraduate work. In 1968 she began her career as a parks administrator, and in 1979 was appointed assistant director, then director, of her country's National Parks and Equivalent Reserves and Wildlife Department. As a woman, she had to prove herself along the way. "I had to show the men in the field that women can ride horses like them, drive heavy machines, and walk for hours in the jungle," she says.

Her accomplishments speak for themselves. In the past few years, the Brazilian government has officially approved a National System for Conservation Units and legally established more than 7 million hectares (17 million acres) of protected areas. This amounts to roughly 1.2 percent of the nation, or a total area the size of Cuba. Two of these parks are in the rich Amazonian rain forest and will encompass more than 2 million hectares (5 million acres). It's a grand breakthrough that in 1982 Jorge Pádua and a colleague won the prestigious Getty

Zoologist Kes Hillman is waging a vigorous battle against poachers who illegally kill endangered African rhino just for their valuable horns. Here, Hillman poses with a collection of confiscated horns.

Esmond Bradley Martin

Wildlife Conservation Prize, a $50,000 award from the World Wildlife Fund.

Looking back at her productive career, Jorge Pádua says reflectively, "It was a good work so far." Then, with her usual verve, she adds, "But much, much, much more needs to be done."

Advocate for Rhino

The small blue and white plane swooped low over bushy terrain surrounding a water hole in an African game reserve. The pilot, a lean, curly-haired woman in her 30's, was searching for rhinoceroses. She was eminently qualified to do just that. Not only does Kes Hillman have a pilot's license and her own plane, but she also has a Ph.D. in zoology and a deep interest in African cultures. For several years, she has chaired the African Rhino Group, which keeps tabs on the animals for the International Union for Conservation of Nature and Natural Resources.

The decline of the two African rhino species—along with elephants—is probably the chief wildlife conservation problem in the continent. "It is," Hillman reports, "an intensely desperate situation." Indeed, according to Hillman's survey, only 13,500 to 19,000 rhino remain in Africa.

Economics is the problem. The wholesale value of rhino horn has risen 20 times since 1970, mainly in Asia and Yemen. A pound of horn may bring more than $350. Yet, asks Hillman, "how can you blame an African who might earn a mere $400 in one year, for being tempted to kill a rhino and make $400 in one swoop?"

Before she began her rhino work, Hillman, an Englishwoman with Kenyan registration, had been keenly interested in Masai handicrafts, technology, and settlement problems. She and her husband built a tiny stone-and-wood house in Masailand with their own hands. To live so simply was "wonderful but difficult," recalls Hillman. In fact, it became impossible when she started her survey work, and finally the marriage ended.

Now, working with rhino is all-consuming, and Hillman and her colleagues are fighting a battle against time. What's needed, she emphasizes, is antipoaching action, stiff fines, control of illegal international trade, and better parks and reserves. "At the present rate of devastation, rhino could be all gone by 1991," she warns.

Undercover Game Cop

"Yes, sir, I'll try anything!" That was the confident response a petite young woman gave to a senior supervisor interrogating her almost 10 years ago. She had just received a degree in criminology and had flown—at her own expense—to the employment interview in hopes of being hired as a federal game agent. She looked unblinkingly at the burly man twice her age—and got the job.

Since 1974, despite dealings with poachers, bad weather, and all types of outdoor equipment, Federal Agent X (her identity and location are protected because she often works undercover) has proved that a female federal game agent can handle anything. Early in her career, she discovered an illegally baited mourning-dove field in North Carolina where several prominent businessmen were shooting. Calmly, she walked up, collected photographic evidence, and advised the hunters of their court appearance. Later, disguised as a tourist with big sunglasses and a camera, she worked with a fellow game agent in a covert operation to catch game poachers. While her "boyfriend" chatted with people suspected of taking hundreds of pounds of bluegills or dozens of deer to sell, she wandered around getting evidence. The result: 40 arrests.

Currently there are about 180 special federal game agents—7 of them female. "I found in the beginning that a female wildlife officer has to be twice as good at her job to be considered half as good," she says.

Agent X recently married a wildlife biologist and moved out to the western U.S. "It takes a very special, open-minded man to be married to a female agent," she muses. "Mine's very supportive—even if I have to leave home at 4:00 A.M."

Agent X chose her lifework because, she says, "in wildlife enforcement, I feel I'm on the front line fighting for the very survival of wild creatures and the places they live." It is a job that keeps her on her toes. Even when she dresses in plainclothes, she carries a gun.

Preserver of Giant Otters

A choir of tropical birds woke a Doberman named Pegs, and the big dog gently nudged her mistress. In a flash, a tanned, blonde zoologist dressed and climbed into a small canoe and paddled quietly down a creek in search of rare giant river otters. Suddenly, a sleek head with whiskers broke the surface just ahead. Nicole Duplaix smiled to herself. She had come to wildest Suriname on a two-year expedition, and after months of travail, she had finally found some of the 2-meter (7-foot) creatures to study. This work in the jungles of northern South America was to earn her the undying respect of her native

Top: Federal Agent X holds a dead golden eagle. Below: Nicole Duplaix seeks giant otters in Suriname.

guides and a doctoral thesis.

Duplaix's fascination with otters began when she was a volunteer worker at New York's Bronx Zoo. It continued into graduate class- and fieldwork at the University of Paris. Then, over the next ten years, Duplaix researched captive otters in several U.S. and European zoos, at the same time editing the International Zoo Yearbook for the Zoological Society of London.

After a brief marriage, Duplaix moved back to the United States, becoming scientific assistant to the director of the Bronx Zoo, and, at age 32, prepared for her otter expedition to Suriname. The giant river otter is one of the 200 most endangered mammals on Earth, and largest of the otter family. It hangs on only in remotest Guyana and Suriname. After finding a viable population of them, Duplaix was able to start a nature preserve in Suriname—a key step in saving the species.

Concerned about endangered wildlife, Duplaix returned to a desk job, directing TRAFFIC, an organization that monitors international trade in wildlife and plants. On one occasion, she joined a "sting operation," seizing hundreds of pounds of illegal walrus ivory.

She now lives in Washington, D.C.—writing about her pet otters and various conservation matters. One of her latest assignments was on fleas and plague for *National Geographic* magazine, which she researched for 18 months. From time to time, she jets back to Suriname to check up on her study animals there. "A lot of development has come to that country," she reports, "but my families of giant otters are holding their own."

Underwater Film Maker

Tropical reef fish hovered among purple sea fans and yellow sponges. Bizarre corals grew every which way. To a 19-year-old commercial artist making her first dive, this was a fantastic peek into a different world. From this beginning, on the Great Barrier Reef off Australia, Valerie Taylor went on to win several Australian diving and spearfishing championships. At one of these events, she met a diver named Ron. Never did she imagine that one day they'd get married and form a team that is the quintessence of underwater adventuring and film making.

Blue-eyed and blonde, Taylor makes a stunning appearance in her orange neoprene wet suit, fins, and face mask. She is the underwater star in many of her husband's films, and she made her first major movie appearance in Peter Gimbel's *Blue Water, White Death*, about great white sharks. Yet Taylor is also a deft and artistic underwater photographer, with work featured in *National Geographic* and Time-Life books, among others.

Daring Valerie Taylor tests the durability of a special mesh suit against the razor-sharp teeth of an interested shark.

Jeremiah S. Sullivan

Taylor loves getting close to marine animals. She vividly recounts swimming with a huge leatherback turtle and hand-feeding it jellyfish—probably a first in diving history. Or spending a month with a mother right whale as it suckled and cared for a calf.

These days, Taylor—who has a special interest in studying sharks—has given up spearfishing and any type of underwater hunting altogether, and focuses her energy on conserving the Great Barrier Reef. Thanks in part to her work—and her husband's—two new marine reserves are being developed. Says Taylor wistfully: "I would like to live forever. There is so much to conserve in the sea."

Hers is not a life without danger, however. In May 1982, she was accidentally bitten in the leg by a blue shark off the California coast. It took a wild helicopter ride, many stitches, and three weeks to recover. Despite this harrowing event, the spunky woman was diving within the month and insisting, "I sought out the shark—not the shark me. I bear it no malice or fear."

Costa Rican Warden

The sturdy young woman on horseback was following jaguar tracks in moist, black sand. Strings of pelicans flew overhead against the sunset sky. Maria Elena Mora and two of her rangers forded a river, fighting against strong currents and the outgoing tide. Continuing along the rugged coast, they checked for lights or signs of illegal hunters, smugglers, or shipwrecked people.

Then chief administrator of Costa Rica's most spectacular national park, Ms. Mora was on a routine inspection trip. Her park, Corcovado, had 36,000 hectares (89,000 acres) of prime, virgin, lowland rain forest filled with peccaries, monkeys, jaguars, manatees, snakes, parrots, and brilliant butterflies. It had been under her supervision for more than four years, and recently it had been especially vulnerable to intrusion from gold miners.

Overseeing a park like this would be a challenge for anybody. Before taking her first job as assistant administrator of a smaller, more popular park, Mora studied natural resources management at the University of Costa Rica in San José. None of this really prepared her for Corcovado, which is huge and totally isolated. She saw no more than 200 to 300 visitors a year, since the only safe and easy access to the park is by lightplane.

Anne LaBastille

Maria Elena Mora, now manager of several Costa Rican national parks, patrols lush forestland in Corcovado.

Still, Mora met the challenge. There were always orders to be made up and reports to write in her thatched-roof office. But she spent many days in the field supervising her staff of 25, and she was not above picking up a machete to clear trails, or driving a small tractor to repair washouts on the tiny grass airstrip when extra help was needed. "At first my all-male staff was skeptical," she reminisces. "They really 'baptized' me. But once they saw I am not a desk person and can do a little of everything, they changed attitudes. Also, I'm very firm. That helps establish respect."

Once a month, Mora flew out to her home in the capital, donned a dress, and visited family, friends, and headquarters. Weather permitting, that is. In October 1981, it rained for 26 days nonstop. No planes could land. That didn't seem to bother Mora. "I love Corcovado," she said at the time. "Even if I got married, I'd live and work in a park somewhere."

Mora's dedication has obviously paid off. She is now regional manager of several Costa Rican national parks in the northwestern part of the country□

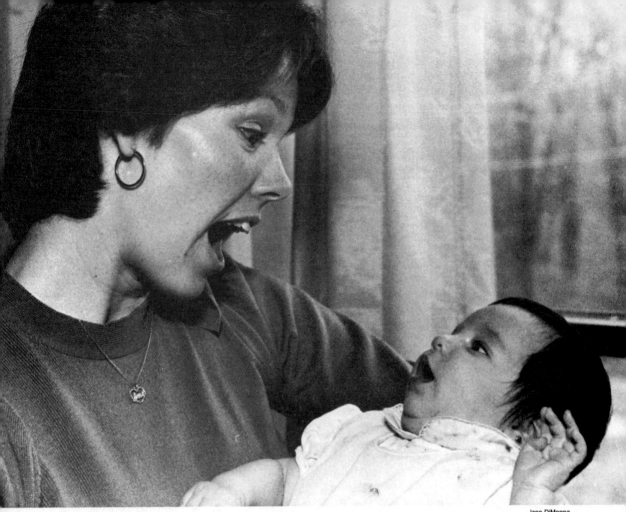

Motherese by Beth Birdsong

L ike all parents, Mike and Laura Daniels brought their baby home from the hospital resolved to do everything right. High on the list of priorities was helping their son, David, to speak well. "We didn't know much about the theories of how children learn to speak," says Laura, "but it seemed reasonable to us that David would grow up speaking in the same way he was spoken to." With this in mind, both parents agreed that they would always talk to David as though he were an adult.

A Special Kind of Speech

As time passed, however, Laura noticed that when she talked to David, she often changed her tone of voice and speech patterns. Despite her intentions, the way she spoke to David was not the way she spoke to Mike. To her dismay, when speaking to David, she even found herself referring to the family pet as the "doggie." Without realizing it, Laura had begun to use motherese.

Motherese is the term sometimes given by psychologists and linguists to the speech that adults—especially mothers—use with babies and small children. Current research on the subject is focusing on the idea that the changes mothers make in their speech when talking to their young children actually play a significant role in encouraging their children's language acquisition.

In the past, some linguists thought that children learned to speak merely by imitating the adults around them (hence, baby talk was to be avoided so that children would learn to talk correctly and clearly). Other linguists pointed out, however, that if children merely imitated adults, they would not make the charming mistakes that nearly all young children make as they learn to speak: using "foots" for feet, for example, or "runned" for ran. These linguists suggested that the ability to acquire speech is something innate in every human being that develops with maturity.

But observation also tells us that when children begin to speak, it will be only in the languages they hear spoken. Those who are deprived of hearing language may not learn to speak at all. Thus, it appears that environment—in this case, the speech that young children hear from their parents, older children, and other adults—must play some role in how they learn to speak.

Working with this concept, researchers in both psychology and linguistics began paying more attention to the samples of speech that young children hear in their early stages of development. Conversations between adults—particularly mothers—and children were recorded and analyzed. Again and again, similar speech traits were noted as mothers from different social and cultural backgrounds modified their speech in predictable and consistent ways. Studies of languages other than English—such as Greek, Arabic, Berber, Comanche, Japanese, Latvian, Romanian, and Spanish—came to the same conclusion: motherese seems to be nearly universal.

Distinct Speech Characteristics

Of all the traits that have been identified in mothers' speech to infants, perhaps the most obvious is the change to a higher pitch. And the younger the child, the higher the pitch, researchers report. Dr. Jean Berko Gleason, professor of psychology at Boston University in Massachusetts, describes one typical mother who spoke in a normal voice to her husband, a slightly raised voice to her eight-year-old, and a high voice to her four-year-old. When she talked to her baby, she nearly squeaked.

There appears to be a good reason for this. Experiments show that infants only a few days old can discriminate between differences in pitch and that they seem to respond to certain frequencies more readily. Indeed, infants favor the higher frequency that corresponds to the speech their mothers direct to them. Consequently, the use of a higher pitch aids a mother

in getting her child's attention and establishing eye contact, an essential step in the bonding process between mother and child.

Besides being in a higher pitch, motherese in many languages is characterized by other sound differences. These include a deemphasis of consonants and a corresponding stress on vowels to produce an exaggerated effect. Also, certain sounds or words may be stressed, and some words may be simplified or shortened. Syllables are commonly duplicated, as in "choo-choo" for train or "bye-bye" for goodbye. Linguists note that the syllables chosen for duplication are often those that appear in the infant's early babbling vocabulary. Words with difficult-to-pronounce consonant clusters such as "stomach" are often replaced with simpler words such as "tummy."

An interesting exception to the simplification rule is the addition of the -ie ending to many words such as "doggie" and "horsie." The latter seems especially hard to resist. One university professor who was certain she had always addressed her daughter as an adult admitted upon reflection to having used "horsie." Languages other than English share this trait. For example, -ko is often added to Japanese words addressed to small children. Some linguists say that focusing attention on the endings of words in this way helps prepare the child for the -es and -ed endings he or she will eventually learn to add to words.

A third characteristic of motherese is that its sentences are generally short, simple, and complete, with distinct pauses between words or phrases. Motherese contains few of the fragmented sentences that tend to make up conversation between adults, and the pauses between motherese sentences are longer than the pauses in adult conversation. Also, the subject matter of motherese is likely to be concrete rather than abstract—there are few references to past or future events.

Motherese contains many questions and commands. In fact, questions may make up more than a third of what mothers say to young children. For mothers of younger children, linguists observe, questions are used more to test comprehension—or simply to get the child's attention ("What is that?" a mother asks as a child picks up a toy)—than to request actual information. Still later in the child's development, mothers begin to use directive questions ("Can you put the toy on the shelf?") to help the child learn how to do things himself.

"What's that?" a mother asks her toddler. Questions are used to test a child's comprehension and give directions.

Repetition is another important characteristic of motherese: almost a third of what mothers say to their children will be repetitions (or partial repetitions) of earlier sentences.

Psychiatrist Daniel N. Stern of the Cornell University Medical College in New York points out another important characteristic of motherese: that it varies significantly with the age of the child. According to Stern: "Mothers exhibit a greater range of pitches from high to low during any particular speech segment when their infant is between the ages of 2 and 6 months than when their infant is either a newborn or is between 12 and 24 months, when language comprehension is well under way."

First Birthday Marks a Change

Along with this change in what experts call "pitch contouring," the language content of a mother also changes with the age of her infant.

About the time a mother senses that her child is ready to say his first words—somewhere around the first birthday—she begins to simplify her language even more than she did when the child was younger. This trend toward simplification coincides with the "label-the-environment" game mothers begin to play with their babies as they seek to teach them the names of animals, parts of the body, and names of family members. Verbs are also introduced around this time, but with the purpose of naming specific actions—*smile*, *run*, and so on.

One fundamental way in which mothers simplify speech at this stage is by nearly eliminating the pronouns "he," "she," and "it" and replacing them with nouns. There is also a strong tendency to avoid "I" and "you." Following this unwritten rule, mothers typically speak to their one-year-olds in sentences like: "Give Mommy a kiss. Mommy is going bye-bye. Sara will stay with Grandma. Sara will have fun with Grandma. Then Mommy will come back and Sara will say, 'Hi, Mommy.' " Of all the basic pronouns, only *they* seems to be used in motherese the same way as it is used in adult speech.

Expansions and Paraphrases

As a child becomes able to combine words to make two-word sentences (usually when he or she is around 18 months old), sentences in motherese gradually become longer and more complex. However, a mother will continue to use a level of complexity only a few months ahead of that of her child.

Some of these longer sentences may take the form of expansions. If, for example, a child forms the two-word sentence, "Doggie eat," the mother is likely to reply, "Yes, the dog is eating his dinner," demonstrating the correct word order for a complete sentence. She may also expand the scope of the child's meaning by paraphrasing or elaborating on the original statement with sentences such as "Yes, the dog is hungry," or "What is the dog eating?" Expansions probably serve mainly as a check for comprehension. Paraphrases serve as an introduction to the richness and variety of language by offering the child new words and new concepts. At the same time, the child is being provided with the opportunity to observe how different sentences are constructed.

The use of expansion and paraphrases illustrates another significant element in motherese—the role of the child. Motherese depends upon the interaction between mother and child. The mother who believes her baby to be advanced and able to understand will simplify and adjust her speech more than a mother who does not have this perception of her child. Later, as a child begins to speak, a mother adjusts her speech in response to each new language skill the child acquires.

The importance of the child's role is illustrated in one study in which it was observed that mothers modified their speech less when talking to children whose response they could not see.

How Long Does Motherese Last?

"By the time a child is around four, the usefulness of motherese has passed," says Thomas Woodell, professor of linguistics at the University of Houston in Texas. Dr. Woodell explains that although language learning is by no means complete at this age, a substantial amount has been accomplished. Parents will continue to use a vocabulary appropriate for the child's age, but by four years old, most children have a fairly clear grasp of sentence structure as well as a larger vocabulary, and this encourages parents to converse with them on a more adult level.

Mothers, of course, are not the only ones who use motherese. Fathers have been observed to make similar changes in talking to their young children. However, it appears that they show less sensitivity to the child's level of speech development than mothers. This probably results from the fact that most fathers simply spend less time with young children than do their mothers.

Older brothers and sisters may also use motherese. As Dr. Woodell notes, "Even children as young as three or four modify their speech when addressing younger children, just as adults do." Eight-year-olds have been shown to make definite changes in their speech to children under the age of four by consistently using

By the time children are four years old, parents begin to converse with them on a more adult level.

Older sisters and brothers often use motherese when addressing their younger siblings.

Jane DiMenna

short, repetitive sentences, a singsong style, and special intonation.

Motherese Expresses Affection

Both the use of motherese by siblings and its occurrence in other languages support the view that motherese is instinctive. And perhaps it is. But to say that mothers make changes in their speech for the purpose of teaching language is more than most psychologists and linguists are willing to assert. Language is a complex phenomenon serving many purposes at once.

For example, regardless of its strength as a tool for teaching language, motherese serves the very important purpose of creating intimacy and expressing affection between mother and child. And the mother who says to her child, "Put the cars in the red box. That's right—the cars. In the box. The red box," is probably more concerned at the moment about getting the clutter on the floor cleared away than she is in encouraging her child's fluency in his native language. Similarly, the intent of the mother who changes her voice when she talks to her infant is to attract and hold his attention, to establish eye contact, but the effect may also be to teach him language skills at the same time.

Built-in Language Lessons

This is supported by studies done by David Furrow and Katherine Nelson at Yale University and by Helen Benedict at Michigan State University. Their work shows that even if motherese is used primarily to serve other functions, it still has a strong positive influence on a child's language growth. One might say that motherese results in a set of built-in language lessons. The mother who wants her child to pick up his toys simplifies her request by isolating her sentence into components that he can understand; at the same time, he is receiving instruction on how words are arranged to make such requests.

Without such a set of language lessons, acquiring a working knowledge of his language would be an almost insurmountable task for a child. Ordinary adult conversation is full of rambling, complex sentences as well as broken sentences and other interruptions. It may deal with abstract subjects and be dependent on a prior knowledge of the subject—not a very promising model for a child who must cope not only with learning what a word means, but where it is to be placed in order to make sense.

The language lessons provided by motherese aid a child in absorbing the language he hears and in internalizing its grammar. This process—and the child's ever-expanding vocabulary as he matures—allows the child to become the possessor of language.

Decoding Speech, Deciphering Emotions

Dr. Stern explains that "even before the baby begins to produce speech, the mother's special way of speaking—in particular, her use of exaggerated pitch changes and musical phraseology in her speech—serves two very important functions.

"The first is that it helps infants to decode what are meaningful units of speech. For instance, in a foreign language that an adult has never heard, it's hard to know where a word or a phrase begins and ends. It's easier from listening to the musical features of the language to begin to get an idea. This is how motherese functions for babies—it teaches them things they have to know before they can even begin to understand the language.

"Also," Dr. Stern says, "because motherese communicates emotions, it teaches about the emotional content of speech. From the beginning of life, how you say things matters just as much as what you say. Motherese helps teach the deciphering of emotional signals. Just as there are facial characteristics of emotion, there are qualities of voice, and these are made up mainly of pitch contours, or changes, intensity, and tension. Babies are learning this early on, long before they start to talk."

What Should Parents Do?

Knowing the important role parents play in their child's language acquisition, as well as in his or her emotional development, causes many parents to wonder if there are special "rules" or practices to follow when speaking to their infants. Fortunately, the advice of both linguists and psychologists is easy and enjoyable: just talk.

Dr. Woodell explains that there is a gap between a child's comprehension and his production of language. Like a sponge, a child must absorb a fair amount before getting anything out. The richer the linguistic environment provided by parents, the greater the language growth of the child. Primarily, this means direct conversation and—for an older child—reading to him. Overhearing speech on television or in adult conversation has little, if any, learning

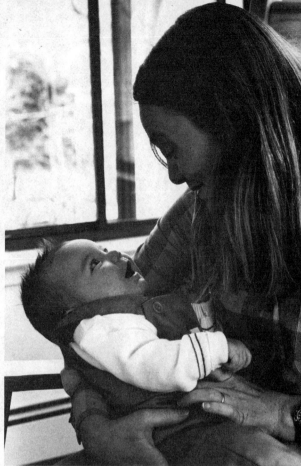

Jane DiMenna

Psychologists encourage parents not to worry about how they talk to their children but to do what comes naturally.

value for very young children and only limited value for older children.

It is the parent's and child's desire to communicate that is most important, emphasizes psychologist Roger Brown. Parents who seek to communicate will find themselves making changes in their speech whether they think about it or not. Doing what comes naturally is the key.

Dr. Stern agrees, adding that "to teach babies the other part of motherese—the emotional message—the parent should remember that you are most expressive when you are being spontaneous and having fun. You don't do all these special things with your voice when you're trying to be superconscious of what you're saying." As Dr. Brown explains: "There is no set of rules on how to talk to a child that can even approach what you as a parent unconsciously know. If you concentrate on communicating, all else will follow" □

BIOLOGY

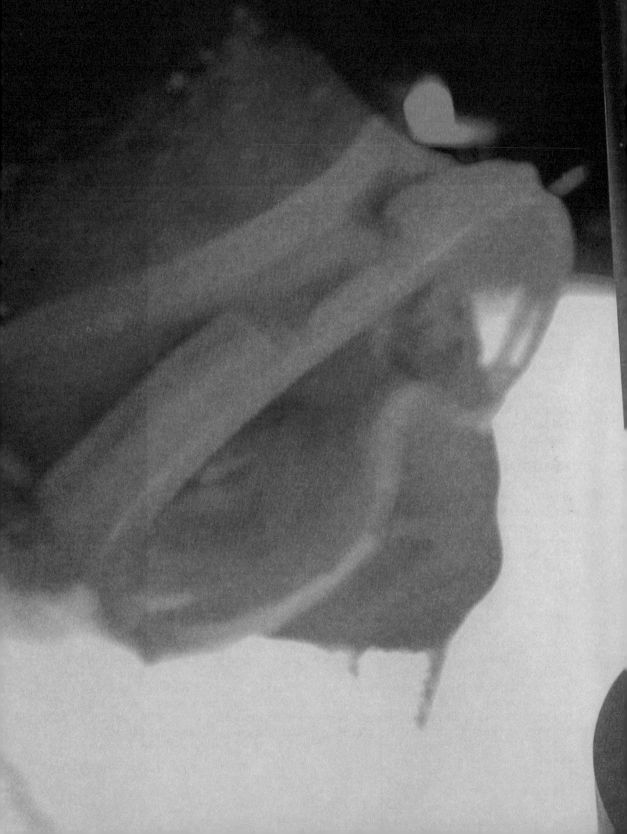

A scientist examines a test tube containing glowing bands of DNA (deoxyribonucleic acid), the molecular basis of heredity in all living cells.

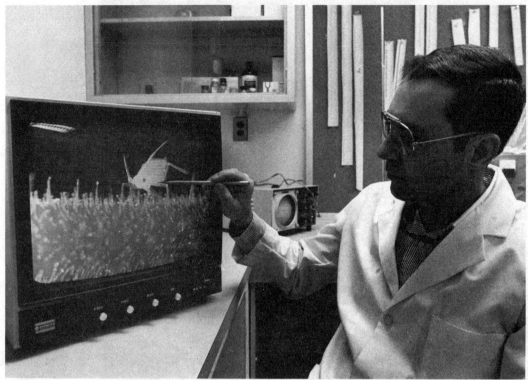

A Cornell University entomologist points to a magnified view of an aphid trapped by the sticky hairs of a potato plant.

REVIEW
OF THE
YEAR

BIOLOGY

Advances in biology during 1983 included new understandings of genes and viruses, the discovery of some new and unusual creatures, and observations that continued to reveal the diversity and surprising behavior of life-forms.

GENES AND VIRUSES

Viruses were used during the year to correct a genetic defect in human cells. Lesch-Nyhan syndrome is a severe brain disorder caused by the inability of certain cells to produce an enzyme known as HPRT. Researchers from the Salk Institute and the University of California at San Diego loaded viruses with the HPRT gene and used the viruses to induce production of the enzyme in human cells unable to produce it. The researchers say it will be four or five years before the procedure is ready to be tried on humans.

A genetic finding with more immediate application was the discovery of a genetic marker for Huntington's disease, a progressive neurological disorder that does not manifest itself until the afflicted person is about 40 years old. The identification of the marker will enable people in families prone to Huntington's disease to learn before that age if they have the disease and thus the potential of passing it on to offspring.

Scientists also made progress in understanding the role of cancer genes. It takes two steps for a cell to become cancerous—so say researchers in the U.S. and Britain. A chemical carcinogen or virus first makes the cell immortal, and a cancer gene causes rapid cell division and an alteration of the cell's surface.

Just how viruses get into cells has long puzzled biologists. Now the "doorway" may have been identified. Viruses, report researchers, latch onto specific proteins on cell surfaces. This finding may help explain why some people get fewer colds and other virus-caused illnesses than others—they may have fewer receptor "doors." It may also explain why viruses tend to be species-specific and why certain tissues are more prone to viral infection than others. The researchers hope to learn how to block the receptors as a way of combating viral infection.

GENETICALLY ENGINEERED ORGANISMS

The first officially approved release of a genetically engineered organism into the environment was postponed during the year in the face of court action by certain environmental groups. Scientists had sought to use genetically modified bacteria to protect plants from frost—which causes more than $1 billion in damages each year in the United States alone. Ice needs a "seed" around which to form; on plant surfaces that role is played by two species of bacteria. University of California at San Diego researchers altered the genetic makeup of one of these bacteria by recombining DNA segments so that the "new" bacteria wouldn't promote ice formation until air temperatures dropped below −5° C (23° F). The lawsuit cited a specific risk of the "new" bacteria possibly escaping to the upper atmosphere and affecting climate, and a general risk of introducing "exotic" organisms into new environments where, the suit charged, they may pose a potential threat to plant, animal, and human health.

CLUES TO THE ORIGIN AND LIMITS OF LIFE

How the first proteins were assembled in the primordial soup in which life presumably originated without the aid of other proteins to act as catalysts for assembly reactions has long puzzled scientists. Now researchers from Yale University and the University of Colorado believe they may have found the answer—RNA (ribonucleic acid). They say that RNA, a compound that plays a major role in building proteins in cells, has another role: it may act as an enzyme (a biological catalyst), snipping other pieces of RNA into active forms.

The wide range of conditions under which life can thrive was also demonstrated during the year. Researchers from Oregon State University and Johns Hopkins University subjected bacteria collected from submarine hot springs in the Pacific Ocean to conditions of high temperature (250° C [482° F]) and pressure (11,400 kilograms per square centimeter [3,900 pounds per square inch]), and found that they thrived. The bacteria may shed light on the origin of life, and suggest that life could exist in the harsh conditions that prevail deep within the Earth as well as on many other planets.

NEW CREATURES

A new creature has been discovered that is not going to win any beauty contests but has been awarded its own phylum, a high-ranking division in the scientific classification of living things. Reinhardt Kristensen of the University of Copenhagen in Denmark had found a strange and tiny larval (immature) animal but was unable to find an adult of the species. Then one day while rushing to extract microscopic marine life from gravel dredged from 27 meters (90 feet) of water, he used water instead of the customary magnesium chloride solution, and found some adults. The freshwater evidently shocked the minute creatures into releasing their hold on the gravel. Kristensen named the creature Loricifera, Latin for "girdle-wearer," since it wears a ring of tiny platelike scales around its midsection. Loricifera adults are less than 0.025 centimeter (0.01 inch) long. The creature rates its own phylum, Kristensen feels, because of its peculiar characteristics, including a telescoping mouth tube and clawlike and club-shaped spines protruding from its head.

Elsewhere on the new-creature scene, Norman Platnick of the American Museum of Natural History found a tiny spider with its own peculiar characteristics—two eyes instead of a spider's usual six or eight, no lungs (it gets its oxygen by absorption), and a minute body—barely the size of a pinhead.

PREDATION—AND ITS AVOIDANCE

Several curious forms of predation and predator avoidance came to light during the year. An alert cameraman shooting a television science program in the Arizona desert noticed something strange in the mud. When the show's scientists took a close look, they found a reversal of the status quo between toads and flies. Instead of toads eating flies, larval flies buried in mud were grabbing small young toads and sucking them dry. ■ An unusual form of predator avoidance was observed in the orb-weaving spider: it drops off an appendage to fool a predator, much as lizards do.

With the help of humans, potatoes may get their own novel defense. Cornell researchers have had some early success getting potato plants to grow their own flypaper. It seems a wild potato from Bolivia has sticky hairs on its stem and leaves, which trap insects. The scientists have gotten a cross between the two potatoes; now they have to ensure that the hybrids have the marketable qualities of the domestic brand.

If a tree falls in a forest, does it scream? Researchers from the University of Washington and from Dartmouth College report that when willow trees and sugar maple and poplar seedlings were attacked by tent caterpillars, they made their own leaves unpalatable—and so did their uninfected neighbors. The discoverers suggest that the infected trees somehow emit a chemical warning that tells the other trees to prepare for attack.

JOANNE SILBERNER

PSYCHOBIOLOGY

by Tom Alexander

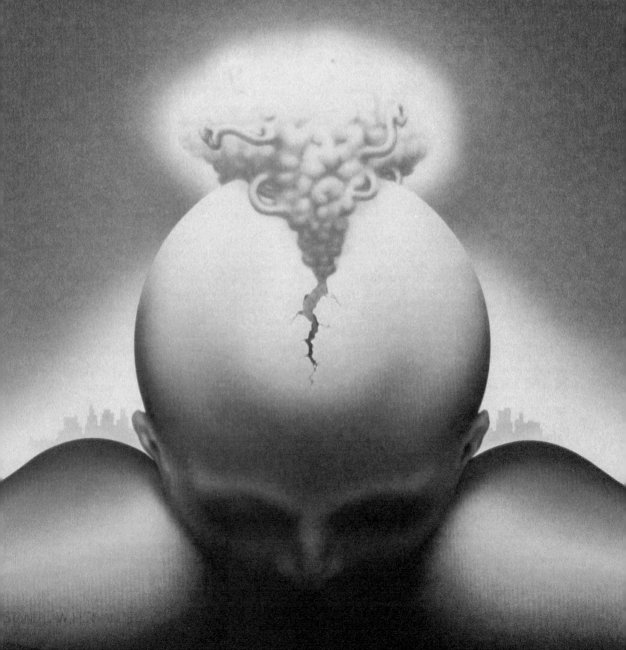

O f all the mysteries it is science's business to resolve, the two that remain most awesome are the ultimate nature of matter and the ultimate nature of the mind. The first of the two began to yield early in the 20th century, with discoveries about atoms, elementary particles, and forces. Within the past few years scientists of the mind may have reached their takeoff point, too. New research tools and discoveries have provided the kind of critical mass of principles and concepts that often explodes into a scientific revolution. Already they have furnished insights into the nature of learning, thinking, and consciousness—and some practical payoffs.

A Merging of Disciplines

Says Maxwell Cowan, a brain scientist at the Salk Institute: "Until quite recently, most of us felt the problem of the brain was simply impenetrable; we were lost in a dense fog. Now the fog is starting to lift. We can't see the whole thing, but we can see the glimmers of cats' eyes in our headlights. It's a terribly exciting time for a scientist."

So many common threads have emerged recently in the numerous neurosciences—including psychology, neuroanatomy, neurophysiology, and neurochemistry—that concepts are beginning to flow together into one interdisciplinary pool known as psychobiology. If there's a central dogma in psychobiology, it's that all behavior and all operations of what's referred to as the mind will ultimately be explainable in terms of immensely complex cellular and molecular interactions in the brain.

As happened in physics, these explorations into the nature of the mind are already yielding important practical applications, primarily drugs for treating mental illnesses such as depression, schizophrenia, and memory loss, as well as physical illnesses such as hypertension, ulcers, and parkinsonism. One of the revelations of psychobiology, in fact, is the degree to which the brain governs the scores of delicately tuned organs and systems elsewhere that keep us warm, nourished, intact, and procreating. It exerts this control through scores, perhaps hundreds, of distinctly shaped chemical molecules that a decade ago weren't even known to exist. Whatever the ailment or the therapy, psychological or physiological, it probably involves changes in the number or effectiveness of these molecules.

Opposite: Stanislaw Fernandes

These revelations have set off a pell-mell rush among pharmaceutical companies to develop central nervous system drugs to cure both disorders of the mind and disorders of the body that originate in the mind. The commercial implications are stupendous. The classical mental disorders, ranging from anxiety or sleeplessness to full-scale diseases such as schizophrenia or senile dementia, afflict at least 35 million Americans. The number of bodily ailments that are wholly or partially under the mind's control, from high blood pressure to perhaps cancer, can't even be guessed at. Finally, there are prospects of drugs to improve or simply alter the functioning of the nominally healthy, such as memory- or concentration-improvers or recreational mood-changers that duplicate nature's own way of changing a person's outlook on life.

The conceptual merger of body and mind also implies that, at bottom, psychology is not all that different from physiology. Life's experiences, it turns out, alter the cellular and chemical arrangements of the brain. Among the implications of this are nonpharmaceutical therapies not too unlike Eastern mysticism.

Amazingly Complex

Unlike the science of matter, whose very quest and beauty is simplicity, the science of the mind must seemingly wrestle forever with complexity. The human brain is far and away the most elaborately constructed entity known in the universe. Scientists don't even agree on how many cells it contains. Various experts estimate the brain has anywhere from 10 billion to a trillion neurons—cells that do the actual information processing—plus many times that many "glia," which provide skeletal support, nourishment, and perhaps other services.

But what really distinguishes the brain from other organs is its intricate interconnectedness. Each of those neurons has somewhere between 1,000 and 10,000 infinitesimal tentacles enabling it to communicate with other neurons or other cells. According to some estimates, every neuron in a brain may be no more than a step or two removed from direct communication with every other neuron.

While under a microscope all this appears a hopelessly matted tangle, these interconnections are not random. Each region of the brain has its own functions to perform. This first became known years ago during brain operations. Since the brain has no pain sensors, surgeons could open a fully conscious patient's skull under local anesthesia and locate various sites simply by probing them electrically and asking patients what they saw or felt. Sometimes patients reported that entire scenes, conversations, or melodies from their past were recalled with startling clarity and immediacy. In more recent years, new research techniques for displaying activity throughout the brain graphically reveal how different sites are activated depending upon the task at hand.

One instance of this principle of localization that has captured public attention is the specialization of the left and right halves of the brain. In most right-handed people, the left half not only controls the right side of the body, but is mainly responsible for sequential tasks such as language comprehension and speech, logical deduction, and mathematics. The right hemisphere, by contrast, seems to specialize in nonsequential processes such as judging spatial relationships and aesthetic appreciation.

Localization extends all the way down to the microscopic level. Most of the investigations have been made in the cerebral cortex—the wrinkled, one-half-centimeter- (one-fifth-inch)-thick shell of "gray matter" that envelops almost the entire outer surface of the brain. It carries out high-level information-processing tasks like sensory analysis, control of bodily movements, and thought. The brain breaks each of these tasks into myriad subtasks, assigns them to separate circuits for processing, then reintegrates them.

One area of the left brain's cortex, for instance, interprets nerve sensations arriving from the skin covering the right side of the body. Each bodily part—each half-lip, arm, each finger on that side—is represented by its own spot within that area. What's more, adjacent nerve endings in the skin are represented by adjacent spots in the brain. The result is that the cortex is mapped with several homunculi—grotesque caricatures of half a body, with features such as lips or thumbs comically exaggerated because of the large numbers of nerves servicing these areas.

Among the more recent discoveries is that these homunculi and all the other areas of the cortex are subdivided into myriad minute circuits. These take the form of columns of a few hundred interconnected neurons, extending from the outer to the inner surface of the thin shell of cortex. Large numbers of these columns are arranged in hypercolumns. Apparently, these columns and hypercolumns specialize in processing small fractions of the reality the brain deals with.

Each column in a hypercolumn has a specialty. In the visual system some concentrate on perceiving movement, some on color, some on straight lines slanted at specific angles. Thus, a given hypercolumn might report, in effect: "I have a moving line in my sector, slanted at a 45° angle, with red on the left side of the line and blue on the right." Presumably, all the hypercolumns pool their information somewhere else in the cortex. There, by some process still unknown, it all gets reassembled into the image in our consciousness. That image is then compared with previously stored models from past experience. The person can then identify the object as, say, a low-flying red airplane.

One intriguing question about the still-mysterious regions where these final operations take place is whether the columns there maintain the same spatial relationships that existed in the eye. If so, it might be possible to stimulate them electrically and thus permit the blind to see with the aid of a video camera. Another question is whether these columns are activated when one simply imagines a visual image. If so, it might be possible to scan the area and convert the activity into video signals—in other words, telepathy.

Similar subdivisions of tasks among regions, hypercolumns, and columns apparently exist for all the functions in the cortex. In summary, the brain as a whole can be regarded as somewhat like an immense pyramidal hierarchy of interconnected computers, each layer reporting to a higher layer, each higher layer dealing with an ever more abstracted version of reality. The most primitive of these computers is the individual neuron.

Sea Snail Yields Important Clue

A long-standing puzzle is how a structure so elaborate and highly organized gets wired together. Elementary calculations suggest that there simply can't be enough information encoded in the DNA (deoxyribonucleic acid) molecules that constitute the body's genetic blueprint to specify how two neurons are connected. Presumably, then, the developing brain gets additional guidance from elsewhere—either from some automatic molecular mechanism within the body or from the outside world.

Most likely, both are involved. Prior to birth, neurons begin sending out axons. Like

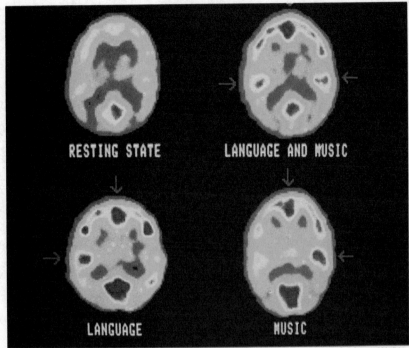

A computer created these four images of the brain from data obtained by a PET (positron emission tomography) scanner. The pictures show that the left half of the brain is more active when the person hears speech and the right half more active when responding to music (regions of highest activity in red).

RESTING STATE   LANGUAGE AND MUSIC

LANGUAGE   MUSIC

Courtesy of Drs. Michael E. Phelps and John Mazziotta/UCLA School of Medicine

plant shoots seeking sunlight, the axons worm toward the sources of certain chemicals diffusing outward from target regions.

Though our brain circuits are apparently all wired together by the time we are born, they perfect their operations only through practice. Even the visual system's clever "feature detector" circuits must learn their job during a specific period in infancy. For example, kittens raised in a chamber where no horizontal lines exist are never able to perceive them. In later life they will fall off a table because they can't see its edge. And babies with uncorrected astigmatism, in which some lines don't focus properly on the retina, never acquire perfect vision, even when fitted with glasses.

It now appears that the principles by which these visual circuits learn their job may not be all that different from everyday psychological events such as memory, learning, and stimulus-response conditioning. In 1982 Eric Kandel and James Schwartz of Columbia University announced an apparent explanation of habituation and sensitization in one admittedly simple animal: the sea snail. The secret, it appears, involves synapses, the sites where chemical messengers, called neurotransmitters, pass between neurons to activate receptor molecules.

The processes by which physical and mental skills and long-term memories are stored are probably not all that different. The major differences seem to involve changes at synapses that turn on certain genes in the neuron's DNA-RNA (ribonucleic acid) protein-manufacturing machinery. A likely guess is that this mechanism increases the amount of neurotransmitter released, thereby strengthening connections and associations between different neuron circuits. Drugs that suppress protein manufacture can destroy the ability of experimental animals to develop long-term memories.

Neurotransmitters Play Vital Role

Synapses, neurotransmitters, and receptors are the bases of the psychobiological revolution. Until about a decade ago, neuroscientists believed that only about two neurotransmitters were needed or employed—one to excite neurons and one to inhibit them. But just as physicists were embarrassed in the 1950's and 1960's when their oversimplified ideas about the nature of matter were overthrown by discovery of hundreds of unsuspected subatomic particles, psychobiologists have had their ideas thrown into fruitful disarray by a rapid proliferation of substances that qualify as neurotransmitters.

In recent years the number of known or suspected neurotransmitters has mushroomed

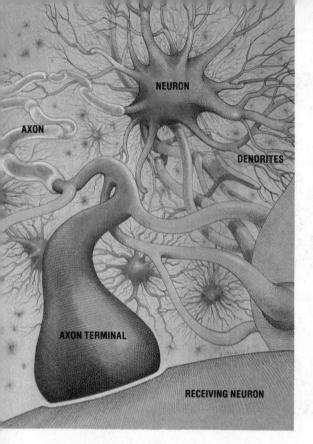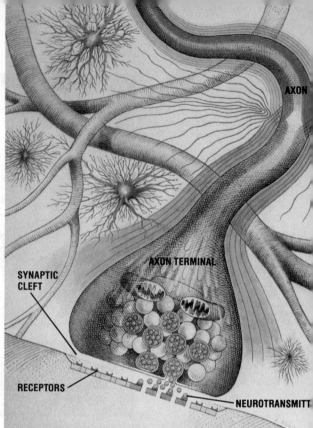

Left: In the intricate network of cells in the brain, neurons receive signals from each other through a dense array of dendrites and send out signals through their axons. Right: A schematic shows the synapse—the site where chemical messengers called neurotransmitters pass from one neuron to another to activate receptor molecules.

from four or five to between 30 and 40. Solomon Snyder of Johns Hopkins University estimates that the total number of neurotransmitters could range as high as 300. Many of these substances are identical to such hormones as adrenaline and testosterone, which are produced by glands and which activate organs in the body. Within the brain, however, they are produced by neurons instead of glands. The absorbing question is: Why does the brain need so many?

The answers probably lie in the number and complexity of tasks the brain performs. Virtually all the organs of the body are ultimately regulated by the brain; the glands, such as the pituitary, are not autonomous, as was once thought. With all its responsibilities, the brain needs complex reporting and control mechanisms to cope with changing internal and external environments. It needs neurotransmitters to address specific control centers and organs, to obtain feedback about their readiness or effectiveness, and to vary the strength of synaptic signals. Some overall control mechanisms apparently manifest themselves in our consciousness in the form of emotion or mood. Thus, there may be specific neurotransmitters for every state of mind—a vindication of the ancient Greeks' theory of humors, which ascribed emotions to preponderances of different fluids, or "humors," in the body.

Most of the substances controlling emotion and the functions of the body originate in neurons in the hypothalamic and limbic regions at the bottom of the brain. Those regions are often called the reptile and animal brains because human beings have inherited them almost intact from primordial ancestors. More advanced regions of the brain, such as our elaborate cortex, seem largely designed to moderate the reptile brain's fiercely instinctive responses. Scientists have identified hypothalamic neurotransmitters that promote feelings of hunger, thirst, satiety, sleepiness, alertness, anger, anxiety, and well-being, as well as ones that stimulate sexual interest, physical growth, and memory.

Some of these substances activate only a specific gland, such as the nearby pituitary. Others are shipped out along nerve pathways to distant parts of the brain or the body. One group of a few thousand cells may send a neurotransmitter to the cortex, making it more alert to sensory signals. Drugs have been developed that seem to affect this region, such as pramiracetam, which in tests has reportedly produced improvements in 100 percent of an experimental group of patients suffering from Alzheimer's disease, a type of senile dementia. Another substance, MSH/ACTH 4-10, being tested at Boston University Medical Center, produces brief but substantial improvements in concentration span in almost anyone.

A fundamental assumption of present-day drug research is that most mental disturbances or incapacities arise from deficiencies or excesses of neurotransmitters or receptors. "Every effective mind drug works by affecting a neurotransmitter," declares Snyder.

Drug Research Reflects New Understanding

The breakthrough in the treatment of mental illness with drugs began in the mid-1950's after the development of tranquilizers and antidepressant and antischizophrenic drugs. Among other things, the new medicines have helped reduce the population of U.S. mental institutions from 550,000 to 150,000. No one knows for sure how all these psychopharmaceuticals work. Apparently, though, some of them make up for faulty synaptic operation because of their fortuitous resemblance in molecular shape to some neurotransmitter.

While many drug companies still employ a trial-and-error approach to neuropharmacology research, universities and a few companies are beginning to pursue a more theoretical route based on the new understanding of brain mechanisms and molecules. Their hope is to develop a second generation of neuropharmaceuticals, tailor-made for desired effects, with minimal side effects. A number of companies, notably Pfizer, Merck, and Warner-Lambert, are building new central nervous system research labs.

One of the most dedicated corporate believers in the basic-science approach to drug development is Hoffman-La Roche. A major project at the Roche Institute in New Jersey is the study of the characteristics of brain receptors. The Roche researchers are trying to find the natural neurotransmitter that binds to the same receptors as do current drugs. Having already identified a "Valium receptor" in the brain, scientists can now search for the natural neurotransmitter that performs Valium's tranquilizing role. If it can be found, the pharma-ceutical possibilities include duplicating it in a drug, stimulating its natural production in the brain, or simply improving the efficiency with which it acts. Such an approach has been made possible only by the advances of the past few years.

Along with broader understanding of brain anatomy and chemistry, researchers have acquired sophisticated knowledge about how receptors work and how to tailor-make molecules with the correct shape to fit them. For instance, they are learning to predict the shape that any given sequence of atoms will assume. "The point is," says Sydney Spector, head of pharmacological research at the Roche Institute, "we're not going about this search in a random way anymore. Now we know something about molecules and how to manipulate them. We have computer graphics that can give us three-dimensional views of a molecule and how it interacts with a receptor. All this is helping us design new drugs. We can put a formula into a computer and compute what the molecule will look like."

Once a receptor is isolated, researchers then have more rapid and precise techniques for screening candidate brain drugs than the animal-feeding tests that are normally used at this stage. In principle, at least, the sheer economics of this increase in experimental productivity should make it feasible to develop more precisely tailored drugs.

Agonists and Antagonists

Basically, neurotransmitter-like drugs seem to take two forms. These are what scientists call "agonists" or "antagonists." An agonist not only fits a receptor molecule but also activates it to initiate some operation in a cell. An antagonist, by contrast, plugs into the receptor but doesn't activate it. The antagonist just sits there, jamming the receptor so that the natural neurotransmitter can't activate it either. Some drugs may combine the roles of both agonists and antagonists.

Drug designers may be able to construct more discriminating substances than nature employs. Different parts of the brain and body sometimes have slightly different receptors for the same neurotransmitter. Knowing this, drug developers can design agonists or antagonists that will affect one and not the other. A prime example is the beta-blocker drug Tenormin, which blocks receptors for the natural substances adrenaline and noradrenaline in the

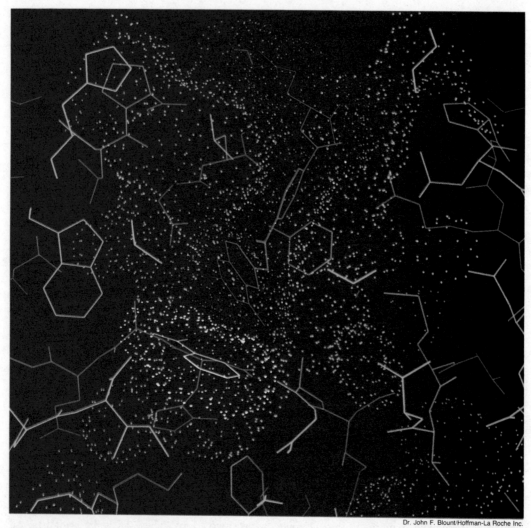

Dr. John F. Blount/Hoffman-La Roche Inc.

Computer-generated images showing molecules interact with receptors help researchers design new drugs.

heart but not the receptors for those substances in the bronchial tubes. The effect is to lower blood pressure without promoting bronchial spasms.

Stress-related Illnesses

Blurring the distinction between mind and body also blurs the distinction between illnesses formerly classified as physiological—involving some chemical maladjustment or damaged part—and those classified as psychological—involving some mysterious psychic maladjustment to everyday life. One factor in both kinds of malfunctions, for example, is the collection of loose concepts that go under the term "stress."

In benign forms, stress—the reaction of an organism to its inherently hostile environment—is the shaping force of mental and physical growth. It toughens the muscles, immunizes the body, and drives the synaptic alterations that constitute development and learning. But an increasing number of scientists believe that the psychological strains between the ancient human organism and its modern alien environment lie near the heart of many, perhaps most, nervous-system and non-nervous-system illnesses.

The most extensively studied psychosomatic diseases initiated or aggravated by mental stress include peptic ulcers, hypertension, migraine, and asthma. But considerable evidence

72 BIOLOGY

suggests that resistance to cancer and other diseases is affected by psychological factors. One likely explanation lies in the immune system, the bodily mechanisms that fight germs, viruses, or malignant growths. Some experiments carried out by Robert Ader and his colleagues at the University of Rochester, for example, have revealed that the immune systems of rats can be switched on and off by conditioned stimuli, just as dogs' salivary glands were turned on by a bell in Pavlov's famous experiments.

Stress is also often the trigger for psychic disorders such as chronic depression, which may afflict as many as 20 million people in the U.S. alone. One pharmaceutical company makes use of a stress phenomenon in animal tests to screen substances for antidepression effects. The phenomenon is the so-called learned-helplessness reaction in rats, a reaction some believe to resemble certain forms of depression in humans. At Pfizer Pharmaceuticals, researchers dunk rats into tubs of water. At first most of the animals swim around, trying to find a way out of their predicament. After a short time, though, some quit struggling and simply float, which they do quite well. When dunked again, these same animals don't even attempt to swim; they immediately float. Researchers, though, have discovered that when they give antidepression drugs to the rats, they resume swimming and struggling.

The work at Pfizer is part of an industry-wide search for a more effective antidepressant. It appears that at least some forms of depression originate in a synaptic deficiency. Postmortem examinations of the brains of suicide victims, for example, often reveal abnormally few receptors that bind with the antidepressant drug imipramine.

Standard Professions May Be Enhanced

Since most of the emphasis in the psychobiological revolution has been on drug development, conventional "talking" psychiatrists and even neurosurgeons sometimes refer to themselves as members of dying professions. But that reaction is probably premature in view of studies that reveal how experience shapes the brain and its chemistry. More knowledge about the control of the body by the mind may even strengthen the case for nondrug treatments.

Psychobiology may also have uncovered a new role for neurosurgery as well. The brain, it turns out, is one of the few organs in the body with little immune response. It doesn't reject

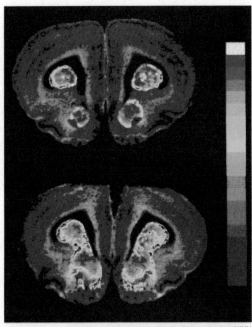

Candace Pert and Miles Herkenham/NIMH

The red areas in these images of rat brains reveal those regions richest in opiate receptors, which filter pain.

tissue transplants from other individuals, sometimes not even from other species. Numerous experimental grafts of fetal brain tissues into animals demonstrate that the grafted immature neurons somehow perceive their surroundings, wire themselves into contact with their surrounding neurons, and begin producing the appropriate neurotransmitters. One brain graft in a human patient in Sweden with a deficiency of the neurotransmitter dopamine may have improved the condition. For all the recent progress, no one in psychobiology really believes that the field is near the verge of a comprehensive theory. Still, just as the physicists' unfinished journey into the heart of matter left in its wake such things as nuclear energy, lasers, and semiconductors, the psychobiologists' newer voyage may one day enable the retarded to learn, the disturbed to function, and the blind to see □

SELECTED READINGS

"Mind over body" by Dianne Hales. *Medical World News,* October 10, 1983.
"How the brain works" by Sharon Begley et al. *Newsweek,* February 7, 1983.
"The educated nervous system" by Robert Pollie. *Science News,* January 29, 1983.

THE ASSASSIN BUG

by Elizabeth A. McMahan

Humans have long held the conceit that only they use tools. Today we know of many other tool-using animals, not only vertebrates but also invertebrates. ("Tool" in this instance follows the definition given by specialists in the field—an object separate from its own body that an organism manipulates to extend its innate capabilities.) Chimpanzees thrust twigs into termite mounds and then lick off the clinging termites. Galápagos finches flush hidden insects from crevices by probing with sticks held in the beak. Ants transport fluids by means of absorbent leaves. Solitary wasps close the entrance holes to their newly provisioned burrows by tamping the earth with a pebble.

In a lowland rain forest in Costa Rica, I recently discovered an unusually complex use of tools by an insect. The user was an assassin bug, and, as with the chimpanzee, the tool served as a termite extractor. The bug *Salyavata veriegata,* a member of the order Hemiptera and family Reduviidae, is found from Mexico to the southern Amazon region. Apart from its general morphology and distribution, little has been known to date of this bug's biology.

Keepers of the Nest

I first observed the assassin bug's unusual behavior serendipitously in the summer of 1980 while I was investigating termite behavior at Finca La Selva, a field station of the Organization for Tropical Studies located in northeast Costa Rica. Among the termite species prominent in the area are several belonging to the genus *Nasutitermes*. These termites build egg-shaped nests of a brownish gray material called carton, produced by the termites themselves out of anal fluid and undigested debris. The nests are usually arboreal (in trees) and average about 46 centimeters (18 inches) in height and 36 centimeters (14 inches) in width. The interiors are molded into a mass of galleries and chambers that connect up with foraging tunnels leading to cellulose food sources in the vicinity. The continuous outer surface of carton, though relatively thin, is usually hard enough to resist penetration by small predators. It also protects the myriad termites within from downpours and other hazards.

The typical *Nasutitermes* colony consists of tens or hundreds of thousands of individuals. All of them are offspring of a single pair of reproductives, the king and queen. Colony chores are carried out by wingless workers, colony defense by wingless soldiers. Winged reproductive forms called alates arise seasonally and swarm to found new colonies as kings and queens. Young are produced continuously, and when the nest becomes too crowded, the workers, using their hardened jaws, remove portions of the nest's covering skin in order to extend the exterior wall. During nest expansion the colony is more vulnerable to predators. Soldiers line the surface holes in a defensive stance, ready to prevent the entry of ants and other small carnivores, while the workers busily proceed with expansion and repair activities. The soldiers of *Nasutitermes* are called nasutes because they have a prolonged snout. Through it a noxious substance can be ejected as a spray from a gland inside their heads. The substance, which is both

Opposite: The assassin bug is a master of disguise. It camouflages itself by gluing tiny bits of termite-nest material onto its body. Right: A typical termite nest may house hundreds of thousands of individuals, all produced by the same king and queen. Below: After capturing a termite without alarming nearby soldiers, an assassin bug liquefies its victim's innards for its meal.

Dangling the empty carcass of its first victim into a hole in the nest, the tricky bug tries to fish out a second termite.

sticky and toxic, is a strong deterrent to most would-be predators.

A Perfect Disguise

Although the termites live in a protective, fortresslike nest and have a soldier caste capable of chemical warfare, the assassin bug has evolved several means to circumvent these defenses. Two adaptations are especially important. In one, the bug camouflages itself by covering its body surface with crumbs of carton scraped from the nest surface with its hind legs and stuck onto glue-secreting setae, or hairs. In the other, it uses the carcasses of previously captured termites as bait to fish out additional termites from within the nest.

The assassin bug has five nymphal (immature) stages before it reaches maturity. The youngest nymph is about one-third centimeter (one-eighth inch) long—approximately the size of a termite worker—and the oldest nymph is about 1.3 centimeters (0.5 inch) long. All of the nymphs camouflage themselves with carton crumbs, a strategy that is probably very effective in preventing visual detection by such natural predators as birds, lizards, and spiders. Adults are patterned naturally in a mixture of concealing grays and browns, and therefore

have no need for artificial camouflage. I observed the termite-baiting technique only among nymphs in stages three, four, and five, but representatives of all stages of the bug, including the adult, killed and ate termites presented to them in the laboratory.

Before baiting with a carcass can proceed, the bug must catch its first victim. When an assassin bug discovers a hole in the nest surface, it creeps slowly up to the edge and waits patiently. Its antennae seem to touch lightly the heads of the workers, which are busy making repairs just inside the hole. Suddenly the bug leans forward, snatches a worker from the rim of the hole, and retreats. That capture often alarms the attending soldiers, which run out over the adjacent exterior nest surface. Termite workers and soldiers are blind, so the bug's camouflage plays no visual role. If the soldiers encounter the bug, they usually climb over its back and tap it with their antennae and palps—fingerlike projections on their mouthparts. They are capable of taking chemical action that would probably repel the bug, yet they seem not to recognize it as an intruder. To an investigating soldier, the camouflaged bug probably feels, smells, and tastes like a protrusion of the nest surface. Its pasted-on camouflage, plus the

As each termite is eaten, the assassin bug uses the body to lure out another victim until it is finally satiated.

bug's tendency to flatten itself into a surface depression on the nest, make it seem almost a part of the nest. Eventually the soldiers abandon the search and reenter the breach. The assassin bug is then free to feed on termites at a leisurely pace.

Bugs Use Bait to Fish for Prey

The bug inserts its long, sharp sucking mouthparts into the captured termite's body (usually in the neck region first). Then it injects salivary enzymes that liquefy the body contents, and sucks out the fluid. After three or four minutes, only an empty carcass, consisting chiefly of the exoskeleton, remains. Now comes the fishing procedure. Holding the termite carcass between its front legs, the bug returns to the hole. Using the carcass as a tool, the bug pushes it over the hole's edge and jiggles it as bait among the worker termites engaged in nest repair. Since termites consume their own dead and dying nest mates as a means of conserving nutrients, a worker usually seizes the dangling carcass with its mandibles and attempts to pull it into the interior of the nest. But the bug holds on tightly and draws slowly backward, gradually pulling the tenacious worker out of the hole. As soon as the worker is fully exposed, the bug drops the dead bait, seizes the new victim, and sucks it dry. If it is not disturbed, the bug repeats the process of baiting, capturing, and feeding without pause until it is satiated. An average meal may consist of seven or eight termites, but I saw one bug fish out and eat 31 termite workers in a single feast that lasted more than three hours.

During the fishing process, the carcass bait serves to insulate the bug from the termites. The dangled carcass arouses no antagonism among the soldiers guarding the nest hole, and the workers inside the nest are not alarmed by the slowly withdrawn bait. During the fishing process, the concealed predator is often not even detected by the soldiers. And the about-to-be-eaten worker, which has been lured from the safety of the nest, discovers its peril too late to save itself from becoming the next victim.

The complexity and specificity of the assassin bug's fishing behavior, coordinated as it is with the habits of the termite prey, imply that this bug (at least in its nymphal stages) feeds mostly, perhaps exclusively, on nest-dwelling termites. As is often the case with predator-prey relationships, a process of coevolution seems to be at work, in which capture techniques and defensive strategies evolve in response to each other in a never-ending catch-up game □

COCONUTS

by James C. Simmons

A h, the coconut! It contains everything necessary to support a man from the cradle to the grave,'' eulogized Robert Dean Frisbie, a vagabond and writer from the U.S. who spent 20 years drifting through the Pacific islands in the first half of this century. And a piece of Filipino folk wisdom insists, ''He who plants a coconut tree plants vessels and clothing, food and drink, a habitation for himself, and a heritage for his children.''

A Vital and Versatile Plant

Most people in the United States would probably dismiss such sentiments as ridiculous exaggerations. For most of us the coconut is little more than a large brown nut found in the produce section of the local supermarket or a minor ingredient in our favorite candy or cake. Few of us appreciate the vital importance of the coconut palm to tens of millions of people throughout the tropical regions of the world—or know that most of the smaller Pacific islands would never have been permanently settled had it not been for an abundance of coconut palms on their shores.

The tall, straight, branchless trunk topped by spreading fronds makes the coconut palm the most striking and easily distinguished plant in the tropical flora—and an indispensible part of the popular fantasy of the South Pacific island paradise. If there were such a thing as a *Guinness Book of Records for Plants*, then the coconut palm would take first place in numerous categories: the largest seed, leaf, and flower cluster, as well as the world's most extensively cultivated nut. And the world's most useful plant—one recent study put the number of ways contemporary people utilize the coconut palm at an astounding 360.

Ninety percent of the world's total acreage of coconut palms lies in a belt along the equator extending to 20-degrees north and 20 degrees south latitudes. The trees thrive best in areas with a temperature range of 24° C to 29° C (75° F to 85° F) and an annual rainfall of 152 to 203

Opposite: A coconut sprouts on a craggy beach in the Cook Islands. Coconuts will still germinate after spending about four months at sea. Left: The dwarf coconut offers delicious fruit that can be plucked off the tree from the ground.

centimeters (60 to 80 inches) evenly distributed. An old Sinhalese proverb states, ''The coconut will not grow out of the sound of the sea or a human voice, nor will it thrive unless you walk and talk amongst the trees.''

Scientists have disagreed for more than a century on the origins of the coconut palm. Fossilized coconuts have been found in New Zealand. Coconut palms have been cultivated for over 4,000 years in India. Because of this, some botanists argue that the ancestor of the modern coconut palm originated somewhere in the Indian Ocean. But this is simply an educated guess.

Myths for Mysterious Origins

The fact is that the coconut's origins are as much a mystery today as they were to plant scientists a century ago—a mystery to everyone, that is, except the Pacific islanders, whose very survival depends upon an abundance of coconut palms. They know precisely where their coconuts came from, although, of course, each island culture has a somewhat different myth. An ancient Hawaiian legend tells us that long ago in the dawn of history, Piti-ri, his wife, and

three children suffered terribly in a time of famine. Piti-ri's wife selflessly gave her own food to her children and soon died. In desperation Piti-ri set off for the foothills to search for more food. Almost a week passed before he returned with an armful of bananas. He found his three children huddled together outside their hut, dead from hunger. But wonder of wonders, Piti-ri saw that from each of their heads grew a small tree. In time those trees became coconut palms. From that day on the islanders never suffered from famine, for there were plenty of coconuts. And if you doubt the story, then look at the end of a coconut and you will see there the face of one of Piti-ri's children.

The islanders on Mangaia in the Cook Islands tell a different story. A beautiful young girl named 'Ina bathed every day in a stream near a clump of trees. One day an eel startled her, as it brushed against her leg. A moment later it had changed its form and assumed the appearance of a handsome youth. ''Oh, lovely 'Ina,'' he said to her, ''I am Tuna, the god and protector of all freshwater eels. Your beauty induced me to leave my gloomy home to win your love.'' In time the couple became lovers,

Tuna always returning to his eel shape after each encounter. Then one day he told the young girl, "We must part, but in memory of our love I will bestow upon you a gift. Tomorrow there will be a great rain, flooding all the valley; but do not fear it. It will enable me to approach your house on the high ground in my eel form. I shall lay my head upon your wooden threshold. At once cut off my head and bury it. Then visit the spot daily and see what will happen."

'Ina never saw her lover in human form again. But that night torrential rains started to fall. The next day she kept watch as the floodwaters swirled around her hut. Suddenly, a giant eel appeared and put its head on her doorstep. 'Ina cut off its head with her ax and buried it on the hillside behind. She visited the spot faithfully every day. Soon she saw a green shoot breaking forth. In time it became a coconut tree. And in the nut even today one can see the eyes and mouth of the lover of 'Ina.

Widespread Distribution: Many Varieties

Wherever the coconut originated, the tree spread quickly from island to island. The coconut is especially well adapted for dispersal by sea. Studies have shown that coconuts will sprout after spending 120 days at sea, time enough for a nut to travel 4,830 kilometers (3,000 miles) under ideal conditions of wind and currents. As any beachcomber on tropical shores knows, coconuts are common flotsam at the high-water mark. (In 1957 an underwater volcano threw up a brand new little island in western Indonesia. When it finally cooled off enough for scientists to safely land, they discovered on the rubble shores 41 coconuts already sprouting.)

However, much of the coconut's distribution in the past 1,500 years would appear to be the work of humans. The early Polynesian voyagers carried many coconuts in their canoes and planted them on distant islands visited during far-flung expeditions throughout the central and South Pacific. The Portuguese apparently introduced the coconut palm to the coast of West Africa sometime after the 1500's and soon afterward to Brazil. The Spanish carried the plant throughout the Caribbean and also along the Pacific coast of Mexico, where coconuts were unknown as late as 1539. (The word *coco*, by the way, is Portuguese for "hobgoblin" and dates back to the 1500's when Portuguese seamen used it to refer to the monkeylike "face" on the end of the nut.)

There is a bewildering variety of coconut palms, varying in denseness of their crowns, the length of their leaves, and the size and quality of their nuts. Some send up slender trunks 27 meters (90 feet) into the air, twisting themselves into strange shapes and curious angles against the sky, while others, thicker and stockier, produce huge, delicious coconuts that can be easily plucked from the ground.

Another Hawaiian legend tells us that in the beginning there were only dwarf palms that possessed magical powers and, unlike other trees, were loyal to their owners. The chief Kane was the first to bring one of these wondrous trees to Hawaii from Tahiti. One day another chief ordered his servant to fetch him coconuts from Kane's tree. However, when the man tried to pluck the nuts, the palm tree quickly grew out of reach. The servant then tried to climb the tree's trunk. But the palm tree grew higher and higher, keeping the nuts always out of reach. In time the coconut tree lost its magical powers, but it compromised by growing very tall, making it difficult for thieves to steal the nuts.

Everblooming and Everbearing

Botanists tell us that the coconut palm is not a true tree but rather a close relative to such plants as orchids, irises, lilies, and grasses. Unlike the familiar tree, the coconut palm has no hard central core of wood surrounded by a spongy ring of bark through which courses the vital sap. Nor does it have growth rings. Rather, its truck is spongy on the inside and exceedingly hard on the outside with sap veins throughout. You can tell the age of a coconut palm by counting the spiraling row of scars left by leaves that have dropped away—one scar per month.

The common tall coconut palm tree (*Cocos mucifera*) is a true tropical plant that knows no seasons. The bloom is produced within a woody spathe (a sheathing bract), resembling a huge ear of green corn, 0.9 to 1.2 meters (3 to 4 feet) long. Tightly wrapped inside are the male and female flowers. A single spathe can produce a cluster of up to 20 nuts. A tree will always have blooms, baby nuts, and mature nuts at all times.

The nut of the coconut palm attains full size after 160 days and is fully matured after 330 days. People in the U.S. who have shopped for coconuts only in supermarkets know them strictly in terms of the sweet white meat and "coconut milk." However, in the tropics the

A thirsty boy climbs a coconut palm to retrieve an immature, or "drinking," nut. The nutritious liquid found in the nut is often the only source of water available to the Pacific islanders for weeks.

James C. Simmons

word "milk" applies only to the creamy white liquid that can be pressed from the meat of the ripe nut. The fluid inside the nut is always called the "water." A fully mature coconut has a dark husk and contains only a small amount of water. But when the nut is unripe, it has a green husk and holds up to a quart of clear, cool liquid, rich in vitamin C, minerals, and amino acids. People who have experienced only the unpalatable cloudy water of the mature nut are usually not prepared for the refreshing, slightly sweet water of the immature, or "drinking" nut. On many tropical atolls unripe coconuts are often the islanders' only source of fresh water for weeks on end. (An atoll is a coral island made up of a reef surrounding a lagoon.)

Coconuts have also saved many a distressed mariner from a slow death by thirst.

Captain Bligh's epic voyage after his famous mutiny might not have been successful had it not been for the chance presence of coconuts on board his small boat. A typical entry in his journal reads: "May 6, 1789—our allowance for the day was a quarter of a pint (0.1 liter) of coconut milk, and the meat which did not exceed two ounces (57 grams) to each person." Thor Heyerdahl carried on board the *Kon-Tiki* 200 coconuts. "They gave us exercise for our teeth and refreshing drinks," he noted later.

An Integral Part of Island Life

The coconut palm lies at the heart of the Pacific island culture, permeating all aspects of the daily lives of its people. "What a tree is this palm!" enthused one longtime observer of the Pacific atoll scene. "Picture just one day in the life of a Polynesian: a morning coconut oil massage; rations of coconut mash for the domestic animals; a full-course dinner with coconut milk entrée, coconut-fattened crabs in a salad with fresh coconut buds, served on platters fashioned from coconut shell, followed by a dessert of coconut pudding. After dinner comes conversation in the flicker of coconut oil lamps, which illuminate the faces of a household reclining on coconut mats in their shelter fashioned from coconut wood and thatched with coconut leaves. Outside, serenely nodding over all, what should one find? Why, of course, a coconut palm!''

It would be hard to imagine a more versatile plant than the ubiquitous coconut palm. This ''sky-farm'' poised on its pedestal 15 to 27 meters (50 to 90 feet) above the ground sends down almost everything an islander needs. The smaller leaves, singed over a fire to make them more pliant, are made into baskets, durable hats, and fans. They can also be shaped into kiltlike lavalavas for the men and skirts for the women. The wirelike midrib of the green leaflets makes an excellent broom. The fiber of the husk makes a sturdy rope. (The husk is soaked in water until the fibers separate; then they are dried and woven.) This same fiber will furnish a fisherman with his nets while the dry spathe provides him with a torch for night fishing on the reef. And if he decides to fish in the lagoon, then he will probably do so in an outrigger canoe fashioned from the trunk of the coconut palm. He may rub coconut butter (made from the milk) regularly into his hands and face to prevent chapping and windburn while handling his nets in the salty spray. The shell of the coconut, highly polished and elaborately carved, makes a handsome cup, holding up to 1.1 liters (1 quart) of liquid. And should a dreaded hurricane sweep across the atoll, then the coconut palm may provide the islanders with their only secure refuge. Tightly bound to the swaying top of a coconut tree, many a Polynesian man, woman, and child has survived when all else below has swept out to sea.

The coconut palm is also the islanders' pharmacy. They use coconut water to kill intestinal worms, subdue the effects of cholera, check vomiting, and soothe stomach problems.

Left: An islander slices open some ripe nuts to expose the tasty meat inside. Right: Coconut leaves are collected to make baskets and clothing; fiber from the husks is made into rope and fishing nets.

Split coconuts are placed on drying racks for several weeks before the meat is extracted and bagged as copra.

During World War II in the Pacific, both U.S. and Japanese military doctors often relied upon coconut water (which comes sterile from the nut) as a substitute in an emergency for glucose solution, dripping it directly into the patient's veins. And on many atolls the coconut palm suckles the human young, who are fed almost from the day of birth with nourishing "milk" collected from the flower stalk.

Copra for Cash

The coconut also provides islanders with what is often their only industry, the production of copra—or the dried meat of the mature nut that, when pressed, yields coconut oil. Copra is the islanders' link with the outside world. It provides a source of revenue with which the islanders can buy those things they need to make their lives a little easier. And the interisland freighters, or copraboats, are often the sole means of travel among the other islands far beyond the horizon.

Copra production is a communal affair. Johnny Frisbie, the daughter of Robert Dean, has given us in her book, *The Frisbies of the South Seas*, an excellent account of a coconut harvest on Pukapuka, an atoll in the northern Cook Islands, where she lived as a child for many years. Every six months the islanders would leave their village on the main island to take up residence on one of the smaller *motus*, or islets, on the other side of the lagoon. "The first and most important preparation for the copra festival was the rebuilding of the old racks and the making of new racks for the drying of the coconuts. Then, when this was done, the chiefs of the village ordered the coconut gathering to commence. Exuberant families ran to the coconut groves, yelling at the tops of their voices and clapping their hands. Each family staked out a small section of the coconut trees. The competition was keen, for although the coconut trees belonged to the people, the gatherers were paid only for the amount of nuts they actually collected and turned into copra.

"It was a wonderful sight to watch: bodies bending, hands speedily reaching for the dried nuts and hastily throwing them toward family piles. Sometimes all one could see was the nuts flying through the air. As long as one had lifted a coconut off the ground and thrown it toward the *putunga* (pile), no one else had the right to touch that coconut. But as long as there were untouched coconuts, anybody could fight for them, and there were many conflicts and long discussions over a single coconut."

Most commercial coconuts are grown in natural groves of wild trees that are managed by family operations.

Soames Summerhayes/Photo Researchers

Then the men husked the coconuts and passed them on to the women, who with quick swings of their machetes cut the nuts in half. The shells were soaked in the sea to toughen the meat so that it would be less likely to spoil while being dried. After two days the coconut shells were arranged on the racks to dry in the hot sun. After several weeks the islanders scooped out the meat, cut it into small pieces, and stored it in large bags to await the next copraboat.

The Emergence of Commercial Production

The coconut palm is of enormous commercial importance to many Third World countries. The worldwide coconut crop is estimated at 17 billion nuts grown on approximately 3.5 million hectares (8.7 million acres) distributed primarily over the tropical coasts of Asia and the Pacific. The Philippine Islands rank first in the world production of coconuts, with nearly 5 billion nuts grown annually on 1.07 million hectares (2.65 million acres), a cash crop that brings in over $150 million in foreign exchange and employs 8 million people, one-third of the country's entire population. India is the second-largest producer of coconuts, with 650,000 hectares (1.6 million acres) under cultivation.

The favored tree for commercial production is the tall palm, a long-lived (70 to 90 years), high-yielding (100 to 200 nuts annually), and pest-resistant plant. Most of the world's commercial coconuts are still grown from wild trees in natural groves, harvested and processed in crude, laborious fashion. There are large commercial coconut plantations, to be sure, but the vast majority of the world's coconuts come from small family operations involving little more than the planter, his family, and several hundred nut-bearing trees scattered about a palm-thatched home. Virtually the entire time of the family is taken up with the monthly harvest of the coconuts. Long poles are used to knock down the ripe nuts while sweepers gather the fallen coconuts into piles.

Much of the world production of coconuts goes toward copra, which ranks second to soybeans as a world source for oil. The trade began a century ago when coconut came into demand as an ingredient in soap. This led to the first expansion of commercial production. The early missionaries pioneered the copra industry throughout much of the Pacific. They were quick to see the advantages of a profitable trade in copra, both for the betterment of the islanders and as a means of financial support for their churches. Large acreages in the Pacific were planted in palms on church-owned estates. Even today travelers to the remote atolls find that most leading churches have plantations from which they derive regular incomes. And their parishioners often donate a day every week to the gathering and processing of coconuts from the church's property. European planters followed the missionaries, sometimes buying or leasing entire islands on which to establish their plantations.

Many Industrial Uses

The industrial world, too, has a myriad of uses for the coconut palm and its products. The United States consumes over 33 million kilograms (72 million pounds) of desiccated coconut annually, chiefly in candies and confectioneries. Coconut shells yield a superb charcoal that, when powdered, finds widespread use as air filters in gas masks, submarines, and factory smokestacks for the removal and abatement of industrial smells. It is also used in the natural gas industry for the recovery of gasoline from natural gas and the purification of gases contaminated by sulfur. In India, Sri Lanka, and the Philippines, smoke-free coconut shells are the major fuel for tens of thousands of kilns, bakeries, brickyards, iron foundries, smithies, and other local industries.

The nut itself comes packed in a thick mass of fibers called coir, which yields mats, rugs, carpets, bags, chair and cushion stuffing, and twine. There is a good chance that your car seats are made of coir fiber covered with rubber latex. India and Sri Lanka together produce over 350 million tons of coir products a year.

Even more important to the world market is coconut oil, whose high degree of saturation and long stability render it one of the world's most desirable natural oils for confections, bakery goods, deep-fat frying, and candles. But it is not only used in food. Apart from suntan lotion the oil can be found in plastics, quick-lathering soaps, synthetic rubber, hydraulic brake fluids for airplanes, toothpaste, and insecticides.

Many U.S. servicemen and servicewomen stationed in the Philippines know from firsthand experience another product of the coconut palm—a highly potent and fiery liquor called

James C. Simmons

Copra is a vital source of revenue to many Pacific islanders. Here, workers ready sacks of it for shipment.

"arrack" that is formed when the sweet juices of the palm's spathe are allowed to ferment. On Guam the GIs (members of the U.S. Armed Forces) call it "sudden death" because of the fierce speed with which the alcohol is absorbed into their systems.

James Norman Hall of *The Mutiny on the Bounty* fame loved to tell stories of Mama Tu, an ancient, wizened Tahitian woman, famous on the island for her profound knowledge of medicinal plants. One day she told him a Tahitian variation on the story of 'Ina and the eel. "And the eyes of the coconuts always look down before the nuts fall; for the coconut palm is man's friend and will not allow him to be injured by falling fruit," she concluded.

Hall was skeptical. "A year or two ago, I heard of a man in the Hitia district who had his arm broken by a falling coconut," he insisted.

"I know of him," said Mama Tu. "It was his right arm that was broken. He was always beating his wife. He was justly punished!"

Ah, the ubiquitous coconut palm! □

John Messineo

Engineering the Birth of Cattle

by Harris Brotman

Peering through a microscope at the embryo impaled on the blade of his micromanipulator—a $25,000 machine that scales down surgical dexterity to a fraction of a millimeter (0.04 inch)—Timothy J. Williams announced quite casually: "This is my 618th egg whacking."

Away from these instruments and Colorado State University's Animal Reproduction Laboratory, 89 kilometers (55 miles) north of Denver, Williams might be taken for a ranch hand. He wears a tall cowboy hat, jeans, denim jacket, and a plaid shirt. Mud and manure are caked onto his boots. But Williams is a scientist. In 1981 he was the first in North America to successfully use microsurgical methods to turn single cattle embryos into identical twins, although the feat had been accomplished in England a year earlier.

A Technological Boom in Animal Breeding

The commercialization of genetically engineered cattle embryos, which includes such highly advanced technologies as twinning, sex identification of six-day-old embryos, and the freezing of embryos for short- and long-term preservation, portends a revolution in the animal-breeding business. Such technology could endow male calves with the ability to convert feed to beef more quickly and efficiently, and make dairy cows produce more milk on the same amount of feed. The large-scale production and reproduction of superior strains of cattle for breeding means perhaps the biggest leap in the beef and dairy industries since the domestication of cattle 10,000 years ago. "We're going to do more to improve every breed of cattle in a shorter period of time than has ever been done in history," says rancher Harry Stafford.

Already cattle breeders can order male or female embryos from laboratory breeders as easily as their grandparents ordered overalls and mackinaws from the Sears, Roebuck catalog. The approximately 140 companies dealing in cattle-embryo transfers in the United States as of 1983 were making anywhere from $300 to $3,000 per embryo. So breeders, whose championship breeding cows can go for hundreds of thousands of dollars, are currently their only customers. One cow that grew from a transferred embryo was sold for $825,000. The high price had nothing to do with the transfer; it was simply a function of breeding. Breeders can learn of available embryos from advertisements in breed journals, from catalogs, or from the International Embryo Transfer Association based in Denver. Currently only about 1 percent of all cattle in the U.S. are involved in embryo transfers. But as prices come down, the new technologies will begin to benefit the average dairy farmer and beef rancher as well.

The number of live births through twinning has increased reproduction in the participating cattle by 50 to 100 percent. (Only 50 percent of natural pregnancies among cattle result in live calves.) Forty percent of all frozen embryos develop into healthy cattle. And the ability to select cattle embryos by sex is 92 to 94 percent accurate. Sex selection will be a particular help to dairy farmers, since bull calves are essentially useless to them—and they bring in only $50 to $100 apiece. Ranchers, who prefer bull calves because they grow faster than heifers (young females), will also have a more efficient operation when they can decide in advance the sex of their calves.

Embryo Transfer

At the Animal Research Laboratory, at least 25 calves were born from the spring of 1982 to the spring of 1983. Each of these animals began its life as an embryo that, on the sixth day of gestation, was gently flushed from its mother's uterus with a saline solution piped through a catheter (a tubular medical device). Through a second channel in the same catheter, the embryos floated out to awaiting collection cylinders to be cut in half by Tim Williams' blade. Each half-embryo was then repackaged in a separate but natural shell, called a zona pellucida.

This shell is a tough but pliable covering that normally surrounds the developing ball of cells for the first eight days of its existence. Extra zonae pellucidae are easily supplied from other nonviable (nondeveloping) eggs. Finally, the twinned embryos were placed, generally one at a time, in the uteri of recipient mongrel cows to gestate for the next nine months.

The greatest agricultural benefit of embryo engineering can be found in the field of embryo transfer—the development of a simple, commercially viable, virtually "goof-proof" method of freezing, thawing, and implanting bovine embryos. Breeders can now not only choose when and where they want their calves to be born, they can also ship frozen embryos anywhere in the world. Foreign markets from China to Saudi Arabia, from Latin America to Eastern Europe, are clamoring for the sophisticated products of U.S. genetic science packaged and shipped as frozen, unborn livestock. U.S. producers and marketers of frozen embryos will profit enormously. According to the best predictions, the exportation of embryos—of goats, sheep, and swine, as well as of cattle—is likely to shape up as a billion-dollar-a-year business within the decade.

A cattle embryo is cut in half under a microscope to create identical twins, a feat having great consequences.

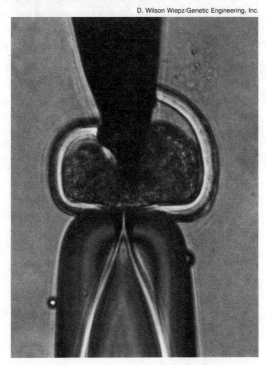

These new techniques may, in a short time, produce populations of efficient dairy- and protein-producing animals in countries where the need for milk and meat is severe. Because these exported embryos will gestate in wombs and suckle the milk of native livestock, they will develop immunity to the infections that so often strike down imported animals. But they will grow up to produce more milk and meat on the same amount of feed as local animals. Finally, these new methods will make it possible for Third World countries to begin establishing their own high-quality food-producing breeds as soon as the embryos are born, mature to adulthood, and give birth to the next generation—a matter of no more than a year or two.

In the United States the embryo-transfer business is already grossing $25 million a year. Reproductive engineers have begun using hormone treatments to transform the finest cows—many of which are valued at up to a quarter of a million dollars—from once-a-year calf producers to machines that can produce embryos every

Photomicrographs show the early stages of a calf's development: unfertilized oocyte (A); embryo after fertilization on days 2, 3, 4, 7, and 8 of the pregnancy (B–F).

Colorado State University

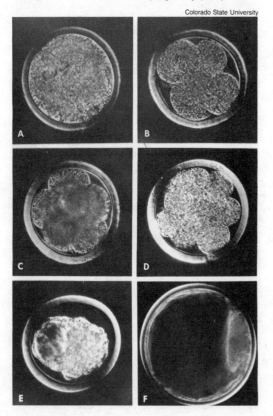

two months, superovulators that generate 30 to 40 or more calves a year without ever having to give birth. Embryo transfer also allows the finest cows to extend their reproductive lives long beyond what nature would normally allow. And, because only the best cows and bulls are given the opportunity to have offspring, the new reproductive technologies make it possible to increase a herd's value vastly in just a few generations.

Genetic Manipulation for Superiority

Experts believe that within a decade, they will be breeding cattle of a quality far superior to any that exist today. The commercial emphasis in the field of genetic engineering has focused on pharmaceuticals, on human health, and, to a certain extent, on plants, says Thomas E. Wagner, a molecular biologist who was the scientific director of Genetic Engineering Incorporated (G.E.I.) in Denver during the 1981–82 academic year, while he was on sabbatical from Ohio University in Athens. "But almost nothing has been applied to the improvement of livestock."

This potent area of life science formed by the merger of applied livestock embryology and recombinant DNA (deoxyribonucleic acid) technology is G.E.I.'s domain. Its long-term goal is the genetic manipulation of livestock through injection of cloned genes from other animals. Such an extra endowment of genes can give an animal physiological properties never before possessed by members of its species. The gene for growth hormone for more efficient and quicker use of feed is one example. Another is the gene for interferon, a naturally produced antiviral agent that may help the animal resist viral infections.

Molecular geneticists have succeeded at transferring genes from the cells of one animal species to the cells of another. "But it's not simply putting more of a gene product into an animal," says Wagner, who is also a family farmer. The trick is to get these transferred genes to express themselves properly—that is, to work effectively—in their new locations. Typically, these genes artificially transferred between animals either have not produced at all, or they have overproduced uncontrollably—they "overexpress" themselves.

In 1982 biochemists got their breakthrough. They caused a mouse to double its size by giving it, in an embryonic stage, a growth-hormone gene already attached to a receptive stretch of DNA. The goal now is to engineer cattle genes designated for relocation—to make them receptive to the chemical language of their new host by attaching them to regions of the host's DNA known to have this property of receptivity.

More specifically, scientists hope to engineer a gene that has a more timely expression from a livestock farmer's point of view. A dairy cow that overproduces growth hormone as an adult would generate more milk, but in a beef animal, the excess hormone would be more valuable during puberty, when muscles are in their growth stage.

In order to achieve commercial success in the embryo-engineering business, several technologies will have to be successfully combined. "The future in the embryo business should almost be like an assembly line," says Charles Srebnik, chairman of the board of G.E.I. "If we can create an animal that would have greater milking capabilities or greater capabilities of converting feed to beef, or an animal that will have twins consistently, that really is where the embryo business is going, and this is G.E.I.'s direction."

On the Brink

According to many experts, embryo engineering today could be compared to the semiconductor business in the early 1960's. The key to success at that time was getting more chips out of one wafer. Now, says Srebnik, the key question in the embryo business is: How do you get more cattle out of one embryo?

"Here we sit on the brink," says Wagner, explaining that scientists will soon be able to manage and utilize more completely the first six or seven days of a calf's embryonic existence. "The bits have all been done and, just now, we're at the point of stringing them together."

In 1980, working in tandem, Wagner and Peter C. Hoppe, a reproductive biologist and the senior staff scientist at the Jackson Laboratory in Bar Harbor, Maine, reported the transfer of genes from rabbits to mice, an experiment many consider a milestone and one that demonstrated the possibility of applying the technique to cattle. The feat illustrated an important principle: That while still a one-celled embryo, an animal's physiology can be heritably altered with an injection of DNA.

In the Wagner-Hoppe gene-transfer experiment, a fertilized mouse egg was injected with a rabbit gene. The gene, which directs the synthesis of the protein beta-globin (a component of hemoglobin), found its way into the chromosomes of the developing mouse. (Hemoglobin is the protein found in red blood cells that carries oxygen from the lungs to the tissues of the body.) After the mouse was born, blood samples showed traces of the rabbit protein. More significantly, the mouse delivered the gene for beta-globin protein to some of its offspring, showing that the rabbit gene became heritable in mice.

For the gene-insertion work to come to fruition in cattle, however, the more complex embryology of the animal has to be understood. Already Wagner's group and others are getting unfertilized eggs out of the cow, fertilizing them in dishes (in vitro), and keeping them alive. They are now perfecting and practicing how to make the one-celled embryo survive in the lab long enough to divide into 16, 32, or 64 cells, at which time its sex can be determined. At this point the embryo can be microsurgically dissected to create twins or even quadruplets and then reimplanted in the uterus of a recipient cow. Freezing the embryo in liquid nitrogen for storage or shipment completes the process.

Egg Banks

Another group at G.E.I., under the direction of Jonathan Van Blerkom, is working on the development of "egg banks." Late in 1982, Sabine 2A, a quarter-million-dollar championship cow, and the calf she was birthing died in the midst of a Cesarean procedure. Sabine's embryos had been fetching $10,000 or more on the embryo-transfer market. One of her embryos developed into a bull calf, which was later sold for $100,000.

Genetically, however, Sabine is still alive. G.E.I. embryologists removed her ovaries and salvaged 36 eggs. This egg bank now lies in deep freeze. The plan to thaw and mix them with sperm from prize bulls awaits improvements of the in-vitro fertilization technique. Breeders are likely to pay as much for embryos from the postmortem Sabine as they did for the ones she produced when she was alive.

In order for egg and embryo banking to work in an animal-reproduction program, the recipient animals must be at the proper stage in their estrous cycle. Formerly, the reproductive rhythms of a recipient herd would have had to be tediously synchronized with the donor cows; otherwise, the recipient would not be in the cor-

A researcher removes eggs from a championship cow; they will be frozen and stored until surrogate mothers are available.

rect hormonal stage to start a pregnancy the day the donors provided embryos ready for implantation. Although egg banks are slated for future use, warehouses of frozen embryos are already available, making it possible for breeders to maintain smaller, more cost-effective recipient herds.

Under the new system, the breeder merely pulls the required number of frozen embryos out of storage and transfers them to the surrogate mother cows as they come into the receptive stage for pregnancy.

Artificial Insemination Inspired Technique

"It's the key to the industry," says Harry Stafford, a partner in the S Bar W Ranch, pointing to a full-blood Simmental bull calf that developed from an embryo-on-ice. As an embryo, the bull calf was successfully transferred from Rio Vista International, a genetics lab and ranch in San Antonio, Texas, and thawed on Stafford's ranch in Woodville, 160 kilometers (100 miles) northeast of Houston, Texas.

Rio Vista is one of the most advanced of the several companies involved in developing these techniques. In January 1982, with its own ranch and 1,500 to 2,000 cows on hand for research and for its commercial embryo-transfer business, Rio Vista began to market a technique called the Rio Vista One-Step Straw, a simplified method for freezing, thawing, and implanting cattle embryos. Such a development, says Albert West 3d, the chairman of Rio Vista's board, "will have about the same effect on the embryo-transfer business as the ability to freeze, store, and ship semen did in the artificial-insemination area."

Artificial insemination was one of the earliest areas in the improvement of cattle breeding. Before techniques were developed for freezing bull semen in 1949, artificial insemination was more of a lab curiosity than a practical advantage for breeding. Bull sperm would lose its potency in two to three days. Special arrangements had to be made to get the bulls in the right place at the right time, that is, when the cows were in heat. "But all that changed when you could freeze sperm," says West. "Then you always had the bull available in the frozen state, and whenever the cow was available, you could breed her. We could ship the bull's semen without shipping the bull. It became immensely more usable."

Equipped with that insight, West began searching for a cryobiologist (a scientist who studies the effects of cold temperatures on living things) to bring into his embryo-transfer business. In January 1981, he hired Stanley J.

Right: A scientist prepares to remove a thawed embryo from a plastic straw filled with cryoprotectant-rinsing agents, according to Rio Vista's One-Step technique. Below: After this straw is filled with an embryo and the necessary fluids, it will be frozen in liquid nitrogen for storage.

Leibo, a cryobiologist who was working under a grant from the Atomic Energy Commission at the Oak Ridge National Laboratory in Tennessee.

In 1972 Leibo and two co-workers had discovered a technique that could reliably freeze and thaw mouse embryos. Their report, published that same year in the journal *Science*, forecast that the application of the freezing technique to large domestic animals would promote worldwide dissemination of stock.

Solving the Rinsing Problem

Before working with embryos, Leibo studied the effects of cold temperature on bacteria, viruses, and yeast. Then, in 1971, a report came from England that a British scientist, David Whittingham, had successfully frozen and thawed a mouse embryo. That observation could not be repeated, however, not even by Whittingham himself. The British researcher was invited to spend several months at the Oak Ridge Laboratory. Within about a month, he, Leibo, and Peter Mazur, the head of the cryobiology group, solved the problem of freezing and thawing. They found that the warming rate was just as important as how quickly the embryos were frozen.

Leibo immediately began working on the freezing of cow embryos. But he ran out of research funds, and the following year, the first live calf from a frozen embryo was reported by still another group of British researchers. "Here was a case where we clearly had a lead but couldn't find funds to translate fundamental biology to the solution of a practical problem," says Leibo. "The Cambridge group solved the problem first."

Leibo continued to study embryo freezing, and by 1977, it was clear that bovine embryos could be reliably frozen and thawed. However, the cumbersome technique involved an hour's worth of detailed rinsing to rid the embryo of the chemicals that protected it from ice-crystal damage but would poison its growth after implantation.

The rinsing problem now appears to be nearly solved. With the commercial introduction of Rio Vista's One-Step in 1982, Leibo is urgently working out methods to improve em-

Frozen-egg banks may someday become as common as cryostorage facilities for bull semen, such as this one.

bryo survival. Currently 40 percent of the embryos survive removal from the mother cow, the quick-rinse One-Step technique, and gestation.

Basically, Leibo's One-Step technique involves freezing the embryo in what looks like a cocktail straw. It is essentially the same type of straw that is used for freezing bull semen. Variations on this technique have been developed by other scientists, such as those in the lab where Tim Williams does his embryo twinning. Williams works with George E. Seidel Jr., a professor of physiology and biophysics, and the research head of the embryo-transfer unit in the College of Veterinary Medicine at Colorado State University.

The embryo is situated in a freeze-protection fluid in the middle of the narrow plastic straw with various fluids for diluting out the cryoprotectant molecules arrayed on either side of it. The straw is then frozen in liquid nitrogen. When it is removed from liquid nitrogen, it is tipped or shaken so that the embryo floats through the other fluids, which rinse away the cryoprotectant. This streamlined technique takes the embryo from liquid nitrogen to uterus in 10 minutes.

Stopping Biological Time

To upgrade their herds, many U.S. dairy farmers and a few beef ranchers are purchasing sexed embryos frozen in a straw. Arriving on the ranch with a stainless-steel storage tank filled with liquid nitrogen, the technician from Genetic Engineering Incorporated or Rio Vista or Colorado State University can, in a matter of minutes, transfer the bovine embryo to its surrogate womb. More and more, in fact, the local veterinarians are beginning to be able to perform the transfer process. Around the world, frozen embryos will become the foundation of new food-producing herds.

By amplifying the production of the best animals and enhancing their inherited traits, the embryo engineers have wrought for the livestock ranchers a brave new world: frozen banks of male embryos, female embryos, and cloned embryos; ordinary sperm banks divided into male-determining and female-determining sperm; and, perhaps a few years down the road, banks of frozen ova or ovaries. Insertion of genes for hormones, antiviral agents, and digestive enzymes portends the reinvention of livestock that grow faster, yield more milk and beef on less or cheaper feed, and survive harsh environments.

The traditional goals of livestock breeding—raising animals better than the best we have now and matching them for a particular environment—are now subject to cellular and molecular genetic techniques. The test tube in which all this is carried out—the embryo, carrying all the genetic information of livestock in the 1980's—can be frozen for a century or a millennium. By stopping biological time, it has now become possible to transmit intact to the farmers of the 30th century the genetic information and embryo experiments of the 20th□

THE 1983 NOBEL PRIZE
Physiology or Medicine

by Barbara Tchabovsky

The 1983 Nobel Prize in Physiology or Medicine was awarded to U.S. biologist Barbara McClintock of the Cold Spring Harbor Laboratory on Long Island, New York, for her discovery that genes can move from one site to another on the chromosomes of plants and can change the characteristics of future generations of plants. This work, done in the 1940's and early 1950's, proved to be a seminal contribution to an understanding of genetic processes. Until her discovery—and for some time afterward—it was generally believed that genes, the basic units of heredity, were lined up immovably on chromosomes much like beads on a necklace.

Calling her discovery "the second-greatest discovery of our time" in genetics, the Nobel Committee compared her to Gregor Mendel, the 19th-century Austrian monk who established the basic principles of classical genetics. Like Mendel, McClintock worked alone and her findings were largely ignored by her fellow scientists. In announcing the award, the Karolinska Institute in Stockholm, Sweden, said, "The discovery of mobile genetic elements by McClintock is of profound importance for our understanding of the organization and function of genes. She carried out this research alone and at a time when her contemporaries were not yet able to realize the generality and significance of her findings." Commenting that she was "far ahead of" the enormous advances in genetics that have occurred in recent decades, the Nobel Committee cited her for "experiments carried out with great ingenuity and intellectual stringency . . . [that] reveal a whole world of previously unknown genetic phenomena."

An Unusual Award

The award to Dr. McClintock was unusual in several respects. She is the first woman to win an unshared prize in the Nobel category of Physiology or Medicine. She is only the third woman to win in that category after Marie Curie and Dorothy Hodgkin, the third to win any unshared Nobel science prize, and the seventh woman to win a Nobel in science.

The award was also unusual in that it came so long after the prizewinning research was completed. Her wait was the second-longest in Nobel history. Only the 1966 Physiology or Medicine winner, U.S. biologist Francis Peyton Rous, waited longer: 60 years for recognition of his 1906 discovery of a cancer virus.

Dr. Barbara McClintock's understanding of corn genetics won her the Nobel Prize in Physiology or Medicine.

Diego Goldberg/Sygma

A Loner

Barbara McClintock was born June 16, 1902, in Hartford, Connecticut. Her family moved to Brooklyn, New York, and she attended Erasmus Hall High School. After graduating, she enrolled at the School of Agriculture at Cornell University to major in plant breeding, but the department would not accept a woman as a major, so she switched to botany. Scientists at Cornell were then doing pioneering work in genetics, using corn in their experiments. McClintock quickly became involved in this work, and showed great skill in analyzing the genetic structures of corn. She received her doctorate in botany in 1927.

While a graduate student at Cornell and later while working and doing research at the University of Missouri, McClintock studied the relationship between the hereditary characteristics of plants and animals and the material in the nuclei of their cells. She made many important contributions to this then-burgeoning field of cytogenetics.

In 1941, out of work and considered too much of a maverick for many university teaching-research positions, McClintock obtained a

McClintock at the Cold Spring Harbor Lab in 1947.

Wide World Photos

grant for a research position at the Carnegie Institution's department of genetics and genetics research based at Cold Spring Harbor. This marked the beginning of her long and continuing relationship with the Cold Spring Harbor Laboratory, now a world-renowned independent institution engaged in biological research.

Jumping Genes

By the early 1940's interest in corn genetics had waned, eclipsed by the studies of fruit flies, bacteria, and other organisms, using sophisticated equipment such as ultracentrifuges, radioisotopes, and electron microscopes. McClintock persevered, however, and continued her studies of Indian corn, or maize, using a simple light microscope and tweezers—tools of a high school biologist. One description of her solitary work states that ". . . each spring, she plants her corn, judiciously fertilizing the budding kernels according to a carefully worked out plan of genetic crosses, watches the plants grow over the summer, and spends the long quiet winters analyzing the results."

Dr. McClintock noticed that the colors of the kernels and leaves of the corn plants sometimes changed from one generation to the next in an unexpected manner. Pursuing this observation, she studied the nuclei of the plant cells involved and found that genes on chromosomes normally found in one place in the parental generation were found in another place on the chromosomes of the offspring plants that had changed unexpectedly. The genes had somehow moved, or "jumped," on the chromosomes. This finding challenged the widely held belief that genes were lined up immovably on chromosomes. McClintock further found that genes could move only if another gene—which she called an activator—was present. She proposed that the movable genes—control elements, as she called them—underlie the diversity that arises during an organism's development. A Nobel Committee member, in commenting on McClintock's findings, said that due to the complexity of the work, "only about five geneticists in the world could appreciate them."

McClintock presented a paper summarizing her findings at a symposium at Cold Spring Harbor in 1951. Fellow scientists largely ignored her findings, considering them irrelevant, heretical, or, as one geneticist who had earlier heard of her work said, "Bah." McClintock remembers they called her crazy.

Undaunted and confident that "sooner or

Dr. McClintock's Understanding of Kernels on a Cob

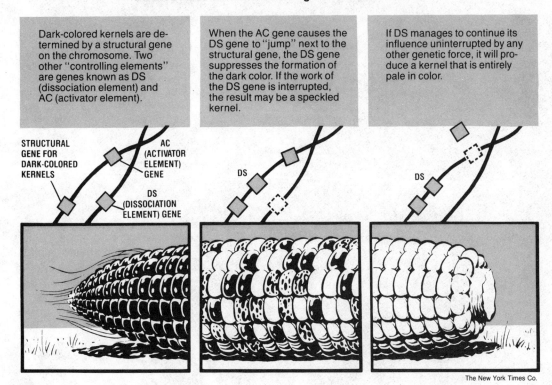

Dark-colored kernels are determined by a structural gene on the chromosome. Two other "controlling elements" are genes known as DS (dissociation element) and AC (activator element).

When the AC gene causes the DS gene to "jump" next to the structural gene, the DS gene suppresses the formation of the dark color. If the work of the DS gene is interrupted, the result may be a speckled kernel.

If DS manages to continue its influence uninterrupted by any other genetic force, it will produce a kernel that is entirely pale in color.

STRUCTURAL GENE FOR DARK-COLORED KERNELS

AC (ACTIVATOR ELEMENT) GENE

DS (DISSOCIATION ELEMENT) GENE

DS

DS

The New York Times Co.

later it will come out in the wash," McClintock continued her work. Vindication came slowly, as jumping genes, also called transposable genes, were discovered in other organisms. "It's really that science caught up with Barbara," remarked James D. Watson, longtime director of the Cold Spring Harbor Laboratory and Nobel laureate for his codiscovery of the structure of DNA, the hereditary material.

Wide-ranging Implications

Transposable genetic elements have now been found in many organisms—yeasts, bacteria, and other microorganisms; and insects and other animals, including humans. They are studied as part of the investigations of many biological phenomena, from cancer to growth and differentiation to immune responses. Jumping genes have already been found to play a role in the transference of antibiotic resistance among bacteria and in the ability of a parasite to avoid the immune defenses of its host. In biotechnology a transposable element is used to alter the genetic makeup of fruit flies, and attempts are being made to use such an element in maize.

Studies with jumping genes may also help

unravel some of the major puzzles of biomedicine—what transforms a normal cell into a cancerous one or why one cell in an embryo becomes a muscle cell while another becomes a bone cell, for example.

Accolades—and Walnuts

In recent years, as the significance of Dr. McClintock's work has gained recognition, she has received many awards and honors. Among these were the National Medal of Science, the Lasker Award, the Lewis S. Rosensteil Award for work in basic medical research, and awards from the MacArthur Foundation and Israel's Wolf Foundation.

When McClintock learned she had been awarded the Nobel Prize, she exclaimed, "Oh, dear." The 81-year-old loner, clad in her usual dungarees, then went out for her morning walk along a wooded path, picking walnuts along the way. In a later statement, she said, "The prize is such an extraordinary honor. It might seem unfair, however, to reward a person for having so much pleasure, over the years, asking the maize plant to solve specific problems and then watching its responses" □

COMPUTERS AND

MATHEMATICS

REVIEW
OF THE
YEAR

COMPUTERS AND MATHEMATICS

The volatile computer industry made deeper inroads into American life in 1983, but intense competition saw two major companies withdraw from the market at the same time more than 100 others entered it. Scores of major advances in both the machines and the programs that run them were announced. In mathematics, a new definition of the meter was adopted, a major number theory problem was solved, and researchers found evidence that may link human math skills and biology.

EASIER THAN EVER, COMPUTERS OPEN WINDOWS ON THE WORLD

Two words not usually associated with data processing—window and mouse—became the most revolutionary new elements in expanding personal-computer capabilities and ease of use.

The combination of the two, first introduced in January by Apple Computer, Incorporated, in its new machine called Lisa, helps convert the computer's display screen into a replica of a typical desktop. Advanced software (the programming that makes a computer do a job) allows a user to divide the screen into many separate "windows," each containing something different. One window, for instance, might show a chart or graph, while another shows a page of text, a financial summary, or other data. The "mouse" is a small hand-held device that controls movement of a marker on the screen. Sliding the mouse around on a hard surface makes the marker move. When the marker is pointing to the desired command on the screen that tells the computer what to do, the user clicks a button on the mouse to activate the command. The next mouse movement can slide an entire window around on the screen, and can even overlap windows just like pieces of paper might be stacked on a desktop. Moving the mouse-marker to menus (lists of options) that appear on the screen, the user also can select any one of dozens of commands. Most computer functions thus are controlled by the mouse, with the standard keyboard needed only to type in text or numbers. ■ Apple researchers found that inexperienced computer users, even those who had never touched a computer before, can learn

The lap-size PowerPad computer serves as an electric paintbrush, a piano keyboard, or a board for games.

word processing or create original graphics displays in less than one hour. By year's end, dozens of new window-and-mouse systems were on the market for other computers, including International Business Machines Corporation's popular personal computer—the IBM PC.

In only its second full year since entering the personal-computer field, IBM swept to the number one position, selling more than 80,000 PC's each month. Despite its relatively high cost and unsophisticated design, the PC became the de facto industry standard. Virtually every major software developer introduced new programs designed for the PC and for the many look-alike computers brought out by competitors. Near year's end, IBM unveiled PCjr, a scaled-down version of the PC aimed at home computer users. Though limited in function, PCjr has an innovative cordless keyboard that communicates with the main computer box from up to 6 meters (20 feet) away via a beam of infrared light.

As computers themselves become more useful, designers and engineers continue to develop add-on devices for special applications. One such device introduced in 1983 is called IntroVoice, which includes a circuit board and a microphone that plugs into Apple computers. Once IntroVoice is installed, the computer listens to verbal instructions to carry out numerous functions. It can even be "trained" to respond only to authorized voices. ■ Chalk Board, Incorporated, of Atlanta, Georgia, developed a lap-size unit called PowerPad that can be connected to Apple, Atari, Commodore, IBM, and other computers. ■ With special software, the PowerPad—intended for young people—becomes an electric paintbrush, a piano keyboard, or a board for hundreds of games. ■ Another device, called Keyport, designed by the Polytel Corporation of Tulsa, Oklahoma, is an oversized membrane keyboard that plugs into Apple computers. With multicolored overlays that convert complex computer instructions into a single keystroke, it simplifies basic programming or serves as a child's gameboard.

TWO DOWN, MANY TO GO

Intense demand for home and personal computers resulted in more than 250 companies competing to sell machines in 1983. IBM, followed by Apple, dominated the field for personal computers, with Commodore taking more than 40 percent of the market for smaller home computers. Industry experts believe that no more than 100 of the companies can survive by 1986. By late 1983, Osborne Computers (the pioneer in transportable machines) and Texas

Instruments withdrew from the home market.

NEW TECHNOLOGY FOR MAINFRAMES

Leapfrogging the technology that packs thousands of microcircuits onto a tiny chip, a Cupertino, California, company is developing a new computer no bigger than a filing cabinet that will outperform the largest IBM-class mainframe computers. Trilogy Systems, Incorporated, is basing its new machine on "wafer-scale integration," a process that packs the equivalent of 100 current computer chips onto a 6-centimeter (2.5-inch) silicon wafer. The small size means that circuits are much closer together, reducing the time it takes for an electronic signal to travel between them.

A MATH-BIOLOGY LINK?

A biological basis for superior mathematical ability may have been found by researchers at Johns Hopkins University. The tentative conclusion came after Camilla Benbow and Julian Stanley studied 300 mathematically gifted children and found that only 20 were female. In addition, the children deviated from the average population in other ways. About 20 percent were left-handed, compared to 8 percent of the general population; 60 percent had allergies, versus 10 percent of the population; and 70 percent were nearsighted, compared to 15 percent of the population.

Their discovery matched research done at Harvard University, which indicated that left-handedness and allergies may be caused by exposure to the male hormone testosterone during fetal development. The Harvard theory said that high levels of the hormone during pregnancy caused the right hemisphere of the fetal brain to become dominant; that hemisphere is believed to control mathematical reasoning and spatial abilities. The findings by Benbow and Stanley indicate a biological reason for male dominance in math skills, since few females are exposed to excess testosterone in the womb.

A NEW METER AND A NEW MATH PROOF

A new definition of the meter, based on a unit of time, was adopted in 1983. The meter is now defined as the distance light travels in a vacuum during 1/299,792,458 of a second. This measurement is 10 times more accurate than that used previously to define the meter.

An outstanding number theory problem that has stumped mathematicians for over 60 years has finally been solved by Gerd Faltings of Germany. Faltings successfully resolved the Mordell conjecture, which postulates that there are only a finite number of rational solutions to polynomial equations in two variables.

JIM SCHEFTER

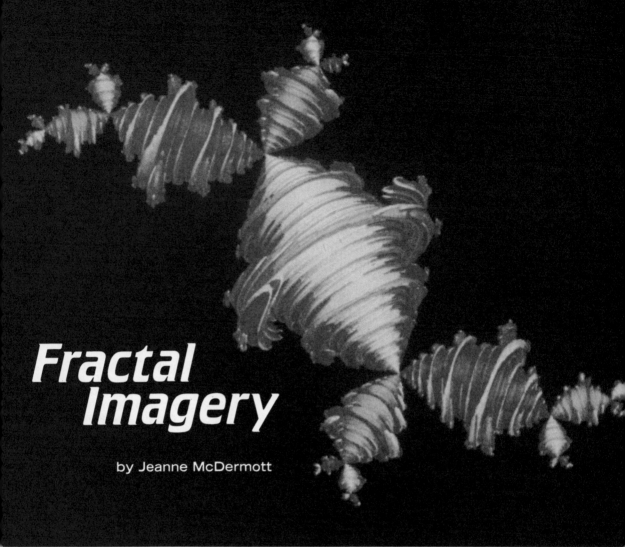

Fractal Imagery

by Jeanne McDermott

Alan Norton/IBM Research

A t ungodly hours of the morning or the evening, when the town's computers are freed from mundane activities, Palo Alto, California, hums. A simple two-story building that looks as though it should belong to a dentist rather than a computer company is no exception. Outside is parked a beat-up car with paper coffee cups strewn across the back seat. Inside, computer scientist Bill Gosper sits in front of a terminal, eyes scanning the screen. He is reed-thin and ashen from the hours he keeps. But with fingers as long as heron wings, he plays the computer and it plays back.

Gosper is searching for an image that he made a few weeks earlier to please a friend. He finally finds it. It appears on the high-resolution color monitor, drawn in line by line. A white lightning bolt darts within an orange lightning bolt, which darts within many more multihued lightning bolts. The jagged shape repeats itself with an eye-pleasing regularity at different scales. "This is a very simple fractal," he nearly apologizes, but he obviously enjoys it.

Simulated Realism

Fractals are not so hard to find across San Francisco Bay, in Marin County, California, where Loren Carpenter makes fractal images for Lucasfilm. The movie company that created the *Star Wars* saga is housed not in high-tech splendor but in an oak-paneled, leaded-glass, arts-and-crafts–style building that immediately feels homey and familiar. Carpenter is a member of

Above: This three-dimensional fractal "dragon" was generated by computer using mathematical equations. Opposite: The overall pattern of this flat dragon is composed of ever-smaller dragons that all have the same fractal shape.

the staff for computer-graphics projects (the first such unit within a movie company). His specialty is coaxing imaginary landscapes out of mammoth computers. "We want the image to look indistinguishable from live action. It should look like someone went in there with a camera and made a movie," he explains.

Tall order. But Carpenter and his colleagues are perfecting the techniques that will bring realistic computer graphics to the movie screen. In a darkened room with several shelves of videotape machines, he demonstrates what is already possible. "Everything that you see in this animation is synthetic," he says with pride. The tape begins with hills of green polygons (figures having three or more sides), cold artificial shapes that you instantly associate with a computer. Suddenly they erupt, forcing up a mountain range as craggy and majestic as the Rocky Mountains. Exhilarated, you fly through them, up and down steep cliffs scattered with ledges, covered with ice, until you land safely on an outcropping. "This is fractal, this is fractal, this is fractal," Carpenter hurriedly explains, pointing to the most realistic images with the same delight that Gosper voiced.

What They're All About

Fractal? Close the dictionary. The word was coined in 1975. But a sense is already building in fields as diverse as physics and ecology, pure mathematics and computer science, that fractals are changing the way we look at the world. "No one will be considered scientifically literate tomorrow who is not familiar with fractals," offers John Wheeler, professor of physics at the University of Texas in Austin. "Fractals delineate a whole new way of thinking about struc-

Lucasfilm Ltd. (LFL) 1983. COURTESY OF LUCASFILM LTD.

Computer programmers at Lucasfilm used fractals to create this realistic scene titled Road to Point Reyes.

ture and form,'' writes Paul Davies, professor of physics at the University of Newcastle-upon-Tyne, England.

Fractals describe a new geometry of nature. They form a field of mathematics that may have a profound impact on how we view the world, not only in art and film but in many branches of science and technology, from astronomy to economics to predicting the weather. Straight lines, planes, and spheres—the pure shapes that many students study in high school geometry—describe the world of built things. Fractals tackle the chancy intricacies of nature—bark patterns on oak trees, mud cracks in a dry riverbed; the profile of a broccoli spear. They are a family of irregular shapes with just enough regularity so that they can be mathematically described.

Fractals wriggle and wrinkle, meander and dawdle, while remaining infinitely rich in detail. Magnify one again and again, and more detail always emerges. Just as a twig resembles a branch and a branch resembles the tree, each part of a fractal is like the whole.

That indeed is the definition of fractal, according to Benoit Mandelbrot, who coined the term. If you look at a circle, he explains, then look at it more and more closely—you will see a smaller and smaller segment of the curve, and it will appear to become straighter and straighter. There is no new structure in a circle at higher magnifications. It simply looks more and more like a straight line. But imagine a shape in which increasing detail is revealed with increasing magnification, and the newly revealed structure looks the same as what you have seen at lower magnifications. This shape is a fractal.

The "Father" of Fractals

Within the family of fractals are two clans. Geometric fractals, like Gosper's, repeat an identical pattern over and over at different scales. Random fractals, like Carpenter's landscapes, introduce some elements of chance. To get to the bottom of fractals, you must meet their "father" and champion, Benoit Mandelbrot. His colleagues describe Mandelbrot as a genius—

eccentric, literate, and contrary. He describes himself as a self-taught nonconformist. Whatever else, he is one mathematician who does not speak in dry, perfectly balanced equations. Born in Poland in 1924, Mandelbrot skipped most of college but did pass the entrance exam to the leading French science school, the École Polytechnique. He then pursued a master's degree in aeronautics at the California Institute of Technology, and finally returned to the University of Paris, where he got a doctorate in mathematics. While at the École Polytechnique as a professor of mathematics, Mandelbrot accepted a position at International Business Machines' (IBM's) Thomas J. Watson Research Center in Yorktown Heights, New York, where he is now an IBM Fellow and manager of a group working on fractals.

Mandelbrot traces his work on fractals back 25 to 30 years, to a time when "science looked at things that were regular and smooth." He was intrigued by what are called chaotic phenomena. At IBM he turned his attention to a chaotic problem in data transmission by telephone. Every electrical signal is subject to random perturbations called noise. Usually, the noise is not so overpowering that it interferes with the signal's message. But under certain conditions the noise does interfere with the signal in destructive ways. Mandelbrot found a way to describe the chance fluctuations.

Defanging Mathematical Monsters

Mandelbrot's methods built on the ideas of several mathematicians who worked between 1875 and 1925 on shapes so strange that their colleagues labeled them "pathological" and "monsters." Scientists of the day were convinced that such shapes were mathematical abstractions with no relation to nature. Helge van Koch added ever smaller triangles to the side of a large triangle to create an infinitely intricate snowflake curve. Giuseppe Peano's curve writhed in contortions until it nearly filled a plane. Georg Cantor's shapes evaporated into mere dust particles while they repeated a pattern into infinity. Mandelbrot "defanged" some of the monsters. He saw their similarities and called them "fractals" (from the Latin *fractus*: broken or fragmented).

This stunning fractal image resembling our planet and the lunar surface was created on a computer by implementing Mandelbrot's fractal theories of the shapes of topographic features of Earth and the Moon.

While the turn-of-the-century mathematicians never thought their monsters bore any relation to reality, Mandelbrot believed that, on the contrary, they described nature much better than any ideal shapes. He set out to prove it. He scavenged problems that scientists had swept under the rugs of their disciplines, problems that did not fit conventional thinking. He characterized price jumps in the stock market, turbulence in the weather, the distribution of galaxies, the flooding of the Nile, even the length of coastlines. "The construction of this theory of chaotic behavior was itself chaotic," he observes.

In the late 1960's, a wider scientific community noticed Mandelbrot's work in a now classic paper titled "How Long Is the Coast of Britain?" Mandelbrot answered: It depends. As the crow flies, the coast is one length. As the person walks, it stretches even longer. As the spider crawls, it stretches still longer. In essence, a coastline with all of its microscopic points and inlets is infinitely long. Mandelbrot suggested that it makes more sense to treat the coastline as a random fractal than as an approximation of a straight line.

Mandelbrot's first book on fractals appeared in 1975 in French. The turning point came when he started using computer graphics to illustrate fractals. His book dazzled the eye

A two-dimensional fractal dragon is encased in a three-dimensional exoskeleton in this intricate design. It came as a great surprise to fractal researchers that simple mathematical rules could generate such complicated shapes as this.

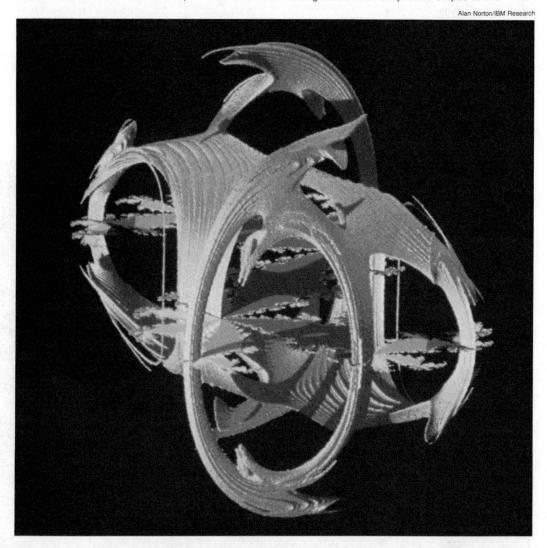

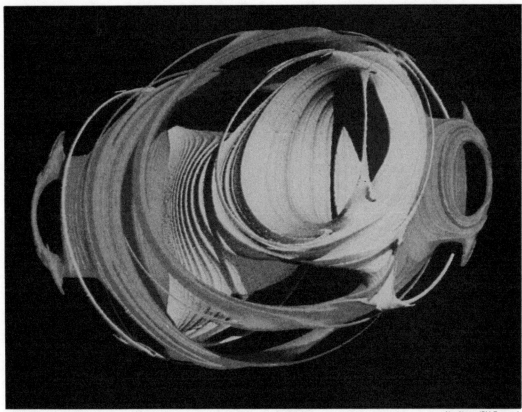

Alan Norton/IBM Research

Two intertwined fractals are distinguishable in this vivid example of fractal geometry. The aesthetic nature of these mathematical designs has contributed to broadening interest in the field.

with state-of-the-art computer images unlike any ever seen. Richard Voss, a physicist at IBM, created stunning landscapes, earthly and otherworldly, from Mandelbrot's ideas. "Without computer graphics, this work could have been completely disregarded," Mandelbrot acknowledges.

Masters of Math Designs

These images helped produce a change in the scientific world view. Until recently, scientists believed that the only shapes that were useful in science were those simple figures, lines, planes, and spheres; all else was chaos. There was order and there was disorder. Now there is order (simple shapes), manageable chaos (fractals), and unmanageable chaos.

"The esthetic beauty came as a total surprise," Mandelbrot adds. "A premium. When people first come to work on this project, their first reaction to making the images is invariably a kind of intoxication."

For Doug McKenna, one of those drawn to

Mandelbrot's circle at IBM, fractals crystallized a lifetime in mathematical designs. As a child, he painstakingly drew them by hand on graph paper with colored pens. In school he got interested in computers, in part because they meant automating the drawing process. But practical considerations, like pursuing a master's degree in electrical engineering at Yale, superseded his artistic inclinations.

After programming fractals at IBM for one year, McKenna struck out on his own as a computer artist. With his reddish beard and eyes made large by his glasses, a habit of precision, and a quietly comic manner, McKenna probably comes as close as anyone to looking the part. Yet he has his doubts about the label. "What I am doing is not quite art and it is not new mathematics, but a synthesis. Computer art is a bad term. Most artists are afraid of computers except for a few who are discovering how wonderful they are," he says. "That's why I call it Mathemesthetics, I guess." Well aware of the sheer pleasure fractals evoke, he hopes to turn them into a variety of designs, from fabrics to architectural ornamentation to fine art.

McKenna has a library of "seed" shapes stored on the computer. He can take one out, glance at it on the monitor, and then instruct the computer to repeat the pattern, with mechanically controlled ink pens that dance across a piece of paper. The results are pure abstract images that attract your gaze and attention. Some are "replicating tiles," or "reptiles" for short. Others are boldly colored mazes. "I should have been a hedge designer," McKenna laughs, "but I was born in the wrong century." He also works with ornate spiraling forms, creating artful "dragons."

Loren Carpenter was working in computer graphics at the Boeing Company in Washington when Mandelbrot's images inspired him to action. "I saw the picture of the mountain range and said, 'Hey, I've got to do this!' But the methods Mandelbrot uses are totally unsuitable for animation, for making a picture where you stand in the landscape," Carpenter says. So

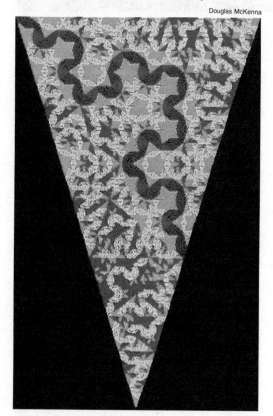

The tiling pattern of this fractal "reptile," which its developer calls Golden Quilt, was derived by subdividing a triangle into smaller and smaller related triangles.

Douglas McKenna

Carpenter used his own techniques that allow the viewer to move around in the landscape, using mathematical shortcuts to produce many images quickly. Carpenter made an animated film to demonstrate his technique. It won him a job at Lucasfilm—and the irritation of Mandelbrot, who disdained these shortcuts.

Practical Uses Exploited

Shortcuts or not, random fractals capture the texture of reality. In another few years, Carpenter expects computer graphics to replace some movie sets and models. The lure is making the impossible look real. "You can re-create the pyramids. Or a civilization from another planet. You could change colors, twist, or deform shapes, do things that are completely fantastic," he says.

The true beauty of fractals is that they describe the pedestrian as well as the fantastic. But the practical uses of fractals have developed at a slower pace. "A lot of people say, 'Okay, so you have a new name for these things, you call them fractals. But what can you do with them?' " says Alan Norton, a former associate of Mandelbrot's who is now working on computer architectures at IBM. It is a question currently being asked by physicists and other scientists at many professional meetings. For a young idea still being translated into the dialects of each scientific discipline, the answer, Norton says, is: Quite a bit. The fractal dimension (a number expressing the complexity of a particular fractal form) may give scientists a way to describe a complex phenomenon with a single number.

Harold Hastings, professor of mathematics at Hofstra University on Long Island, is enthusiastic about modeling the Okefenokee Swamp in Georgia with fractals. From aerial photographs, he has studied vegetation patterns and found that some key tree groups, like cypress, are patchier and show a larger fractal dimension than others. In analyzing shapes that are hard to describe with any exactness—"patchy" ones, for example—using fractals may provide a more precise measure. Eventually, he hopes that slight and unexpected changes in the fractal dimension of key species can be used as an early-warning system for harmful disturbances like pollutants and acid rain.

Shaun Lovejoy, a meteorologist who works at Météorologie Nationale, the French national weather service in Paris, confirmed that clouds follow fractal patterns. Again, by ana-

Richard F. Voss/IBM Research

The basic structure of this awesome fractal landscape was created by an equation that scientists consider simple.

lyzing satellite photographs, he found similarities in the shape of many cloud types that formed over the Indian Ocean. From tiny puff-like clouds to an enormous mass that extended from central Africa to southern India, all exhibited the same fractal dimension. Prior to Mandelbrot's discovery of fractals, cloud shapes had not been candidates for mathematical analysis, and meteorologists who theorize about the origin of weather ignored them. Lovejoy's work suggests that the atmosphere on a small-scale weather pattern near the Earth's surface resembles that on a large-scale pattern extending many kilometers away, an idea that runs counter to current theories.

A New View of Mathematics

The occurrence of earthquakes. The surfaces of metal fractures. The path a computer program takes when it scurries through its memory. The way our own neurons fire when we go searching through our memories. The wish list for a fractal description grows. Time will tell if the fractal dimension becomes invaluable to scientists interested in building mathematical models of the world's workings.

Whatever the purpose, fractals touch the imagination in a way that no other computer-generated image has. "They produce unprecedented visual effects. They are pure art, pure playthings," says Bill Gosper. But their images have changed the world of mathematics as well. Mandelbrot explains: "Imagine 100 years ago that singing was outlawed and a great science of analyzing scores arose. Now think that 100 years later someone looked at these scores and found that they were really much more beautiful and accessible when sung. Beautiful opera scores were appreciated by only a few but beautiful music was appreciated by everyone. I have done that for branches of mathematics." Who would disagree? □

SELECTED READINGS

The Fractal Geometry of Nature by Benoit Mandelbrot. W. H. Freeman, 1982.
"A place in the sun for fractals" by Dietrick E. Thomsen. *Science News*, January 9, 1982.
"Mathematical games" by Martin Gardner. *Smithsonian*, April 1978.

COMPUTERS AND MATHEMATICS   **107**

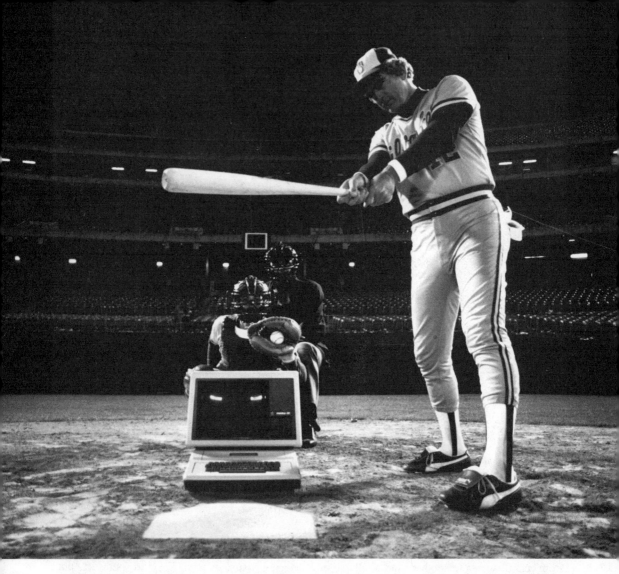

The Microchipped Dugout

by Sy Weissman

chilly spring evening in 1983 at Boston, Massachusetts', Fenway Park: one hour before game time, the fans are beginning to fill the seats; some are wearing zippered jackets, others carry ponchos or umbrellas against the possibility of rain. On the playing field, the visiting baseball team from California, the Oakland A's, are grouped behind the batting practice cage, waiting for their turns at the plate. They gaze longingly at Fenway's Green Monster, the tempting left-field fence, only 96 meters (315 feet) away and one of the shortest outfield barriers in any major-league ball park.

An Electronic Consultant

In the Oakland dugout, manager Steve Boros, his nose deep in the accordion folds of computer-printout paper, is studying the data, looking for a replacement for his regular designated hitter, Jeff Burroughs, who is sidelined with a painful muscle pull. Ignoring the crackling line drives off batting-practice pitches, Boros slowly turns the folded pages, and then stops. He runs a finger across a line of type and makes a note on his batting-order card: Mitch Page will substitute for the injured Burroughs tonight. According to the computer, Page, a .267 hitter and part-time

Steve Boros, manager of the Oakland A's, picks a starting lineup from a computer print-out during the 1983 season.

Above and opposite: Tom Zimberoff/Gamma Liaison

bench warmer, has a love affair with Fenway Park, where his batting average soars to .330. (A player's batting average is determined by dividing his number of hits by his number of at-bats. For example, a player with 150 hits/500 at-bats has a .300 batting average.) Boros again checks the data for Page's record against Boston starting pitcher Dennis Eckersley.

Satisfied with the choice, Boros moves back to the printout. Hidden in this dense forest of numbers are precise measures of every member of the Oakland team: individual batting performance at Fenway Park, in night games, against Eckersley and every other member of the Red Sox pitching staff; performance with runners in scoring position, with the count 0 and 2 (balls and strikes), or 3 and 2, in any inning from the first to the last; stolen-base records against any Boston pitcher-catcher combina-

tion; double-play figures on Fenway's slow grass infield.

The numbers are packed together, but they seem to make sense to Steve Boros, who will refer to them constantly during the course of the game. He will use them to make judgments on when to bunt, or steal, or send in a pinch hitter (a substitute hitter), or in a thousand other tactical situations. The computer is the 10th man in Oakland's starting lineup.

Some Like It; Others Don't

Steve Boros is a pioneer. In a game that has seen few pioneers, Boros belongs to an elite group of managers who have only recently discovered the mathematical wizardry of the microcomputer.

As an enterprise heavy with record keeping and the divining of statistical probabilities, baseball is decades late in accepting the very tool that could help it the most. We have only just entered the age in which computers are being utilized for the on-field management of major-league baseball. As of 1983, only three clubs had adopted them: Oakland, the Chicago White Sox, and the New York Yankees, none of which finished the season higher than third in their respective divisions. According to statistician Jay Alves, the man who runs the Oakland computer, "It's only a matter of a few years before every team has one, if only to prevent their opponents from gaining an unfair advantage."

The computer is still disparaged as a toy by many players, veteran and rookie alike. But in a game dominated by folklore and fetish, that's hardly surprising. It still leaves a lot of thoughtful concern among those who fear that baseball will lose its zest to the thunk of disk drives (devices that use magnetic disks for high-speed transfer and storage of data) and the clatter of printers to become a game run by actuaries.

Elizabeth Durbin, an economist by training and a baseball watcher by habit, would like to put these fears to rest. "The whole point of the game is to beat the odds," says Durbin, an associate professor who teaches at the Graduate School of Public Administration at New York University. "Thus, while computers can provide fancier ways to figure out the statistical probabilities, the fun and drama will always come from managers who outwit the other team and from players who outdo themselves and undo the opposition."

Will Machines Displace Scouts?

"The common manifestations of computerphobia are no different on the ball field than they are elsewhere," notes Matt Levine, chairman of Pacific Select Corporation, the company that developed the first computer program specifically for baseball. "Managers' pride is the biggest problem. They think they know it all, and like to keep everything in their heads. They wouldn't dream of turning game decisions over to a machine."

The computer is seen by some players and managers as an unwelcome insinuation in the dugout. This is particularly true after the widely reported incident involving manager Don Zimmer of the Texas Rangers, baseball's first known computer-age casualty. Zimmer was fired shortly after statistician Craig Wright submitted a computer-generated report detailing Zimmer's poor personnel-development strategy. The story reverberated through league clubhouses, adding injury to the insult of the computer's presence, and giving credibility to the threat of technological unemployment, especially among veteran talent scouts.

Dick Cramer works with the computer program he designed called EDGE 1.000 that produces visual coaching advice.

Sal DiMarco © 1983 Discover Magazine, Time Inc.

These venerable judges of talent—usually ex-ballplayers, for whom scouting is a well-earned reward—are paid to hang around the minor leagues and sandlots in search of promising youngsters. For the most part, they're not trained to collect the kinds of precise data the computer can use: pitch-by-pitch records of a hurler's velocity, his stamina—his stuff—or the trajectory and placement of a batter's drives. Floppy disks are replacing handwritten reports; dead-accurate radar-gun readouts have retired the somewhat vague estimates of the past, such as that lyrical report from sportswriter Bugs Baer on fastballer Lefty Grove's stupefying delivery: "He's got an arm that can throw a lambchop past a wolf."

A Major Asset

The computer is already having a noticeable impact on the fortunes of ballplayers. More than 3,000 men are being tracked by computer, simplifying the tedious and error-prone process of separating the truly brilliant performers from the merely outstanding. Gifted minor-leaguers can be spotted quickly and nurtured with greater understanding of their strengths and weaknesses, thanks to the computer's analyzing and teaching capabilities.

The computer is also moving into the coach's domain. Pacific Select Corporation's computer program, called EDGE 1.000, can produce visual coaching advice for batters, in the form of computer-generated graphics. By looking at the hit zone as a nine-cell grid on the computer's display screen, hitters are able to see where and when they're swinging in relation to the oncoming pitch. They can follow the ball's trajectory and where the ball passes through the strike zone. The "honeyspot" is shown as a red ball, the teaser zone by a blue area.

"Players catch on quickly to these simple picture lessons," says Matt Levine. "Even the older stars, who are sometimes deaf to the nagging of their batting coaches, understand what the computer is telling them."

But the computer is best at what it was originally designed to do: make sense out of statistics. The very limited menu of data categories found in the heavyweight baseball almanacs are merely pencil-and-paper attempts to give form, structure, and relevance to a game that defies easy quantification. Baseball is filled with random events influenced not only by the inequalities of human physiology and psychology, but by the laws of physics, which are tested by the

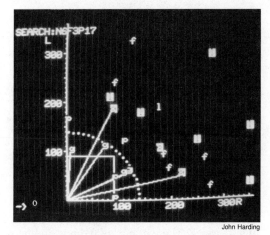

John Harding

A computer screen shows the distribution of all the fly balls hit in one season by Oakland A's Rickey Henderson.

irregularities of weather, ball-field architecture, artificial illumination, horticulture, and synthetic carpet weaving, to name a few.

The New-style Oddsmakers

Despite this apparent disorder, or perhaps because of it, no other sport has been so painstakingly quantified and analyzed as baseball. Its history has been reduced to simple numerical abstractions representing averages, home-run production, stolen bases, fielding efficiency, strikeouts, shutouts, putouts, and so on, adding very little to the understanding of the game.

You can look it up—as the late Casey Stengel was fond of repeating. Yet smart managers like Stengel held the statistical tables in low esteem and doubted whatever they implied. They knew when and why to stop "playing the percentages" in favor of "psyching out" the opposition with high-risk tactical maneuvers, nerve-wracking delays, beanballs, and other threats of bodily harm. (Ty Cobb's record of stolen bases was as much a function of his deadly spikes as his speed on the basepaths.) As manager of the original miserable New York Mets, Stengel learned all about the relative usefulness of percentage baseball as he watched his team collapse under the weight of accidents and negative psychology, a tangle of interrelated factors as dense and confusing as Stengel's own version of the English language. The percentages, as he knew them, made no sense at all.

There must be something wrong with the way statisticians have been rating the players and figuring the odds. Such is the conclusion drawn by the new breed of baseball savants, who call themselves sabermetricians (SABER is an acronym for Society for American Baseball Researchers). Exploiting the computer's prodigious mathematical capabilities, the new-style oddsmakers have produced a revised catalog of data categories to describe the phenomenology of the game—giving managers a more reliable fix on statistical probabilities for a wide variety of game situations—and entirely new systems for quantifying individual player performance.

Using new categories of offensive and defensive data, computers are simplifying the task of calculating the risks—the percentages—of any strategic play. The computer is able to digest, weigh, and extrapolate the probabilities for any combination of several dozen specific game factors, including inning, score, the batter-pitcher match-up, number of outs, and number of men on base, as influenced by such physical circumstances as temperature, humidity, wind velocity and direction, park dimensions, playing surface, and total attendance.

Dramatic Applications

The last two items were shown to have significant bearing on player performance by Bill James, the dean of sabermetricians. He uncovered the so-called "Birthday Effect," in which batters outperform their season batting averages by some 50 points in games played on their birthdays. He also offers the information that Reggie Jackson, when playing before crowds larger than 40,000, hits .368. In front of smaller crowds, Jackson's batting average falls to .255. James himself minimizes the importance of these data (they represent the 1982 season only), and yet they indicate the computer's ability to quantify the stress-related factors that influence performance. Data collectors are searching for more meaningful indicators of performance under stress hidden in the remotest thickets of the statistical forest.

It is now theoretically possible to make instant computations of the most complex strategic event and produce reliable odds favoring or not favoring any number of optional moves such as a bunt, attempted steal, hit-and-run, or whether the batter should swing on a 3-0 pitch. No mere manager could possibly keep all the relevant data in his head, not even the brilliant Earl Weaver, the Orioles' legendary professor of baseball, who claimed he could do just that; and not Billy Martin, who, during his several tenures as manager of the Yankees, amazed

sabermetrician Jay Alves by coming up with the correct tactical decisions a few milliseconds ahead of the computer.

The controversy over whether computer-generated percentages are more reliable than the old seat-of-the-pants variety will take more than a few years to resolve—if, in fact, it ever gets resolved. For the present, at least, evidence that the computer abets winning baseball is scant.

Far from scant, however, is the computer's clout when it's time for salary negotiations. Sentiment is no longer the tie-breaker at the bargaining table. The drawing power of a popular star—an easy calculation with pencil and paper—has diminished in importance compared with the player's real value to the team. Oakland's outstanding starting pitchers, who boasted the lowest American League Earned Run Average (ERA) in 1980, were dismayed to learn at contract time that their wonderful statistic was actually the product of the strength of the Oakland defense. Given a computer-generated "average" defensive outfield, the same pitchers would have had the worst ERA's in the league, said EDGE 1.000.

Stored in the Apple II computer's memory bank is a hit-by-hit record of each batter's drives, showing the distance and placement of each batted ball. This information can be superimposed over a graph representing the contoured dimensions of every major-league ball park. Armed with such unimpeachable data, the Oakland management decided to trade slugger Tony Armas to the rival Boston Red Sox, who looked forward with great relish to the acquisition of this long-ball hitter. The Oakland computer, however, showed that Armas's towering flies are clustered in deep left center field, a ways short of Fenway's distant no-man's-land outer fence.

All Aim to Preserve Game's Randomness

Will the computer change the game for the better? Current opinion is deeply divided. On one side stand the advocates, who cite the computer's capacity for processing performance data. One result, they say, will be to cleanse the system of its less able players and speed up their replacement by talented minor-leaguers, who can now be swiftly identified, quantified, and developed.

In addition, they believe that strategy making will soon be elevated to a high science. Jay Alves, for example, is concerned with the refinement of the statistical probabilities of swing-

Statistician Jay Alves uses an Apple computer to update information on the Oakland A's and on opposing teams.

ing on a 3-balls, 0-strikes pitch, heretofore one of baseball's most sacred prohibitions. The result of adding such a technological twist to the standard batter-pitcher guessing game may add a new dimension of conjecture. Alves is understandably silent on the subject.

From the other side we hear the voices of well-informed observers—not to be confused with the protests of computerphobes—who question the computer's ability to combine and process the relevant situational and circumstantial data and produce a truly reliable table of statistical probabilities. The events in a baseball game are simply too random to give off data more useful than a smart manager's experience-generated instincts, they say. To do otherwise would require a kind of calculus of infinite differentials, the Indian Rope Trick of reckless mathematical genius. While the rigors of training and competition will produce increasingly better players, they say, games will always be won by smart managers and good talent.

The most relevant statements of all come from the bleachers, from the fans themselves, who see the game as a spectacle made up of unexpected thrills and random brilliance. They perceive the potential irony lodged in the imposition of the risk-minimizing power of the computer on baseball's drama of ritualized randomness. No two baseball games are ever exactly the same.

Even the club owners, the people who are responsible for bringing the computer to baseball in the first place, agree that the aim of the game is to create more thrills, not fewer. No one on or off the playing field believes in the remotest possibility that the computer will turn baseball into chess, not even manager Steve Boros, a confirmed believer in the miracle of the computer, his trustworthy "predicting machine."

The game at Fenway Park is over. The new Red Sox slugger, Tony Armas, failed to lift one over the left-field wall. Designated hitter Mitch Page lived up to his readout by going one for four with a stolen base. But the Oakland A's lost the game.

So much for the statistical sciences in the game of baseball as it is played today. Casey Stengel, a man of no great faith in rational analysis, had what is probably the last word on the subject: "Baseball ain't nothing more or less than the science of getting 27 outs" □

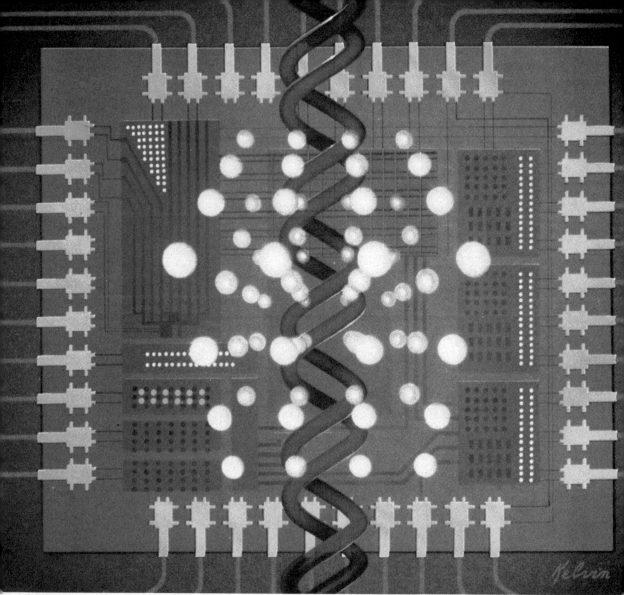

The Organic Computer

by Robert Haddon and Angelo Lamola

In the design of computer hardware, small is beautiful and smaller is even more beautiful. No longer satisfied with tiny silicon chips that hold thousands of bits of information, and computers delicately wired chip to chip, some engineers now dream that the molecule itself will someday function as an electronic device—that computers will be built not merely chip by chip, but molecule by molecule. The workings of these revolutionary thinking machines, they suggest, might even be manufactured by millions of bacteria, all set to the task by genetic engineering.

Computer engineers enthusiastic about this dream believe that the technology can eventually be made to work, perhaps in a matter of decades. They are convinced that the components of electronic memory can in principle be reduced to molecular scale. If the human body can be built up from molecules, the argument runs, and if we can already harness bacteria to manufacture such complex hormones as insulin molecule by molecule, why can't we someday make a biochip, or even a biocomputer, from individual molecule components?

Even Smaller and More Sophisticated

Today's computer technology has already presented us with wondrously small machines, of course. When ENIAC (the Electronic Numerical Integrator and Computer), the world's first electronic digital computer, went into operation in 1946, it weighed 30 tons. Its 18,000 vacuum tubes, which pulled information-transmitting electrons here and there, consumed 140 kilowatts of power—enough to light a small town. Only a year later, the transistor was invented at Bell Laboratories. This device directed the flow of electrons much more cheaply and reliably than vacuum tubes, and on a vastly smaller scale. Since then computers have been shrinking relentlessly. Computer engineers have competed to pack more and more transistors into a single chip, to achieve what is known as very large scale integration, or VLSI. These days a sophisticated chip weighs only a few grams, holds 200,000 transistors, and is 30,000 times cheaper to build than ENIAC. Even a sliver of

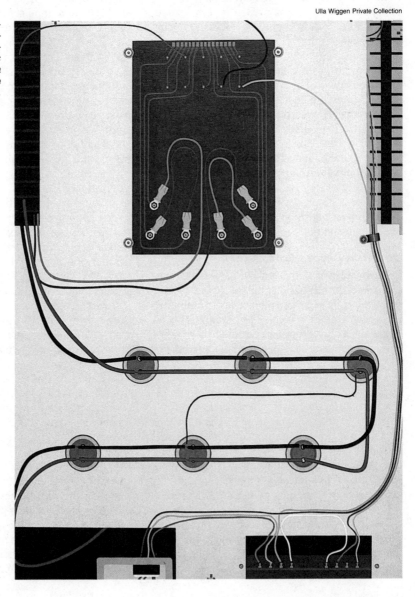

Opposite: An artist's conception of an organic computer combining microelectronics and genetic engineering. Right: Ulla Wiggen, Kanalväljare (Channel Selector), 1967. Artwork turned sideways.

silicon the size of a baby's thumbnail can hold as many as 256,000 binary digits, or bits, of information. Electrons now fly through circuitry patterned onto these silicon chips by ultraviolet light. And advances in new lithographic techniques, which use X rays or electron or ion beams to engrave circuitry in silicon, promise chips many times more densely packed, with memory capacities in the megabit (million-bit) range.

Since such tiny chips seem well within the grasp of our present technology, why has the dream of building a molecular device engendered such excitement? Some insist that the reason is pragmatic, that we are approaching the limit of our ability to miniaturize silicon circuitry. We may someday reach a point where even electron or ion beams cannot engrave circuits any more finely. Then, some engineers believe, the only way to proceed will be from small upward, rather than large downward, to construct molecular components into microscopic arrays, so that the molecules may act individually. At present, a single unit, or bit, of information in a silicon chip involves many billions of atoms. But in a molecular electronic device, a single molecule would store a single bit of information. Such technology would lead to an extraordinary reduction in the size of the electronic computer, a reduction perhaps comparable to the transition from ENIAC to today's pocket calculator.

Lure of the Unprecedented Biochip

In many respects, the appeal of a molecular electronic device lies not in the realm of the practical, in fitting more memory on a chip, but in the domain of the ideal. The glamour of biomolecular engineering has captured the imagination of computer engineers and tempted them to dream that computer components could be produced and assembled like proteins. This visionary idea is still at a stage where it would fit more comfortably in a science-fiction novel than in a working laboratory. But though the biochip may prove to be impracticable, the obstacles to its manufacture will force us to delineate the limits of our knowledge, both of computers and of biomolecular engineering, and to identify the next hurdles in these fields.

Nothing found in either nature or computer technology is like a molecular electronic device. Electronic computers are not organic, and biological processes do not involve the conduction of electrons. The Nobelist Albert Szent-Györgyi, currently at the Marine Biological Laboratory, in Woods Hole, Massachusetts, did propose an electronic theory of biology in 1941. But it is now understood that electron movement is highly restricted in biological systems. Although processes such as respiration and photosynthesis rely on electron transport, each electron must pass from carrier to carrier in a complex series of chemical reactions, and the transport is considerably slower in these than it is in a computer's circuitry.

Even the computer we carry inside our heads does not make use of electronic carriers. Nerve impulses are not transmitted along axons and dendrites by electrons, but by sodium and potassium ions. These ions are far heavier than electrons, and impulses move along much more slowly than electrons through a wire. Indeed, they move only about as fast as a car on a highway—about 25 meters (85 feet) per second, or 90 kilometers (56 miles) per hour. But while the brain perhaps speeds up its functions by parallel processing—handling many impulses at once—most computers can do only one thing at a time. Until we learn exactly how to duplicate the architecture of the brain in computers, electrons, which travel tens of millions of meters each second, make far better messengers than sluggish ions do.

There are similarities between the electron switches in present-day computers, which control the flow of information, and the gated ion channels in neural membranes, which control the flow of nervous impulses. But even this analogy is stretched. While electronic switches are simple and are either on or off, ion channels are neither simple nor totally on or off; rather, their opening and closing are random processes with overall membrane conductance reflecting the activities of many channels. And while electrons carry signals in computer switches, chemicals called neurotransmitters carry the signals from nerve cell to nerve cell. In short, there does not appear to be in nature machinery that operates anything like the electronic switches in a computer.

First, a Switch

In order for the molecular electronic device to share all the advantages of electronics—conducting velocities near the speed of light and switching speeds on the order of picoseconds—this hypothetical molecule could not be anything like the garden variety proteins that occur in nature. (A picosecond is one trillionth of a

Ernest Trova, Falling Man/Perspective Shadow Man, *1972*

second.) Altogether new molecules would have to be employed, for natural proteins, which are the building blocks of biology, simply do not conduct electrons. And unfortunately, no precedent suggests that any protein that genetic engineers can coax a bacterium to produce would conduct electrons much better than natural proteins. Despite the fact that nature has not invented the molecular electronic device, many engineers, pushing aside the theoretical demons, have decided to tackle problems of production: constructing electronic molecular switches and molecular wires that could connect the switches together.

The first component a computer needs is a device that stores and then surrenders information—a switch. A computer's information is broken down into a set of binary digits, or bits, a series of yes-or-no answers to a series of questions, as in a game of twenty questions. So the components that store information must be able to switch back and forth between two different states (on and off, for instance). And the person using the computer must be able to interrogate the components, to find out which state the switch is in. In modern electronic computers, transistors act as switches, allowing current to flow when the answer is yes and stopping current when it is no. And the computer's memory can be read by seeing whether electricity flows at a particular address. If a molecular switch were developed, the feat would be comparable to the discovery of the transistor, which ultimately gave birth to the so-called Information Age. It would be the second great leap in electronics: from the bulky vacuum tube to the tiny transistor to the submicroscopic molecule.

Any switchable element—an element that is stable in either of two states—can be used to build a memory. Photographic film, for instance, is a switchable element; when it is exposed to light, a chemical change is induced, and information—a picture—is recorded. But unlike photographic film, which makes permanent records, a computer memory must be capable of continuous change; it must be reusable. Thus, it must be an array of molecules that can undergo reversible alterations. No such memory yet exists, synthetic or natural.

Tunneling Poses Major Barrier

It is true that many simple organic compounds undergo reversible switches between two relatively stable forms—changes in geometry, group bonding, or the positions of well-localized electrons or protons. These processes are understood well, and some have even been used in memory functions. Of these compounds, the so-called photochromic materials, which can be switched and probed for information by means of light, have been most successful. In fact, one reversible photochromic system, developed by H. G. Heller of the University of Aberystwyth in the United Kingdom, was used to make an optical tape with greater storage capacity than conventional magnetic recording tape.

But it is not clear that light could be used to switch molecular components on and off, since the ability to "aim" light accurately depends on its wavelength. Even wavelengths of light 2,500 angstroms long—about the shortest wavelength that might be used with photochromics—could not be used to address, interrogate, or switch individual molecules separated by distances of, say, only 100 angstroms. (The angstrom is a unit of length used especially to measure the wavelength of light. One angstrom equals one ten-billionth of a meter.) The only alternative is to separate the molecular components by distances greater than the light's wavelength. But there's no gain there, since such distances would be the same as the distances separating components in today's silicon chips.

Unlike simple organic molecules, hydrogen atoms are small enough to "tunnel" through barriers, so they can change positions quickly in some molecules under certain circumstances (though as yet we have no way to read these changes). For this reason, molecules in which reversible changes involve the motions of hydrogen atoms have been marked as candidates for a role in memory functions. But while hydrogen atoms do indeed tunnel quickly, they also leak back through the barrier, causing inadvertent loss of information.

Indeed, tunneling may be one of the most formidable obstacles to the dream of building a memory from an array of molecular electronic devices. As individual components are made smaller and packed closer together, electrons will begin to tunnel between the components, to cross talk, destroying information that depends on their remaining in one place. In today's electronic devices, crosstalk is eliminated simply by keeping components far enough apart from each other. Ultimately, it may well be the threat of crosstalk, rather than the dimensions of the active components, that sets the limit of miniaturization. If the connections among components must be relatively large, why even bother to make the components out of molecules?

Modeling a Molecular Wire

How one would go about connecting components in a molecular-scale device is, at the moment, anybody's guess. For some 20 years proposals for a "molecular wire," a one-dimensional molecular conductor, have been in the scientific literature, starting with a model for a high-temperature organic superconductor suggested by William A. Little of Stanford University in 1964. This complex polymer has never been constructed, but the idea inspired a burst of interest in the field of organic conductors, and considerable progress has been made.

Current proposals for a molecular wire, such as those advanced by Forrest Carter of the Naval Research Laboratories, are based on polymers that are found in bulk, rather than in single strands. When doped (treated with small amounts of impurities that alter the properties of a pure substance), these polymers, though they are a tangle of many strands, conduct electrons best in the direction of the polymer strands. Thus, they behave roughly as though they were an array of conducting single-molecule wires or individual molecular sheets. However, it is not clear at the moment that such materials could serve as prototypes for a molecular wire, since although most electrons move along the strands, the interactions that occur between the strands are also very important.

Even if we could isolate a single molecular strand from the bulk, it is not clear that it could conduct electrons very well. Materials that act like a single molecular wire are prey to many phenomena that hamper conductivity. In metals, for example, conductivity changes when dimensions are reduced to the dimensions of large biomolecules. Jerry Dolan and Douglas Osheroff at Bell Laboratories recently showed that at low temperatures the conductivity of 30-angstrom-thick, 1,000 angstrom-wide "wires"

of gold-palladium is reduced significantly from that of the bulk material. Apparently, atomic-scale distortions appear in the wire, producing comfortable niches in which the electron carriers lodge. Since conduction speed depends on electrons moving smoothly, this effect, called localization, would drag conduction considerably and limit the usefulness of the molecular wire—assuming we could even find a way of manufacturing such a wire.

Still a Long Way to Go

If we ever design molecular logic and memory elements—switches—that can be probed, and if we ever find an appropriate way to connect

Ulla Wiggen, Vägledare (Micro-Circuit), *1967*

them, we would still have to put the machine together. At present, within conventional chemistry, there appears no obvious way to assemble a structure as complex as the molecular computer. Recently, James H. McAlear and John M. Wehrung of EMV Associates in Rockville, Maryland, and J. S. Hanker of the University of North Carolina have demonstrated that a pattern with molecular-scale features can be created on a thin layer of the synthetic protein polylysine. This has been touted as the first step in the construction of a biomolecular-based electronic device, because the pattern produces a molecular conducting network when treated with silver salts. But it does not represent a breakthrough toward the development of a molecular electronic device, because it is created by conventional photolithographic techniques—reproduced from a model on a metal plate—rather than from individual molecules. In other words, it is still a structure created from large downward rather than from small upward.

Can present-day genetic engineering help us create a biocomputer built up from molecules? With considerable effort and expense, we can direct bacteria to produce many copies of a protein having any desired sequence of amino acids, so long as the protein is not a poison for the microbe. This is a revolutionary achievement, but it does not go very far toward building a biochip. Since proteins do not conduct electrons, they are not very useful either as components or as conductors. However, they may prove suitable as a matrix or a template on which to graft active electronic elements made of other organic molecules. If these components are to be electronic devices, they will have to be "unnatural"—not products of bacteria, but creations from the chemist's flask. Only if we figure out how to create a computing device incorporating molecular processes that do not require electron flow might it be built naturally from proteins.

We are far from engineering proteins that will function as computing devices or even as templates for such devices, since we do not yet know how to specify an amino acid sequence that will yield a synthetic protein with the right three-dimensional configuration to carry out a specific function. We cannot even predict very well the three-dimensional structure of a natural protein from its amino acid sequence, nor the function of a protein from its three-dimensional structure. In fact, the relationship between the structures and functions of natural proteins, where both form and function are known, are far from being fully understood, though they have been actively investigated for three decades.

Even if at last we find the right molecular structures for building a molecular computing device, be it electronic or nonelectric, would it work as well as a silicon chip? The organic nature of the components would certainly affect their function. When complex molecular structures are built, there are bound to be some mistakes within them, and organic materials, which are more complex than the inorganic materials of present-day silicon chips, will be more prone to error. Living systems have solved these problems by devoting a substantial fraction of their free energy to error correction, repair, and replacement. For example, enzymes and other functional proteins degrade at characteristic rates and are removed and replaced at commensurate rates. But if the molecular chip were to include repair functions, several more layers of complexity would have to be added to it. So, rather than include repair functions, the engineer would simply put in extra components with like functions to take over in case of a breakdown. This, of course, would increase the size of the chip.

Clearly, the molecular electronic device does not lie around the corner. There are so many fundamental uncertainties and difficult technologies in the path to the biochip computer that it is possible to say only that one cannot say it cannot be done. Before we design complex functional molecules, we must learn how nature designs them: how protein structure translates into protein function; how faulty enzymes are removed and replaced; how information is processed in the brain; indeed, how human memory works. Because nothing like the hardware envisioned in the biochip computer seems to exist in nature, it is possible that Mother Nature has figured out that the biochip does not fit her needs; it is even possible that she has not yet arrived at the idea. Or perhaps she has already tried the device only to decide that it simply does not work □

SELECTED READINGS

"Organic superconductors" by Klaus Bechgaard and Denis Jérome. Scientific American, July 1982.
"The organic computer" by Natalie Angier. Discover, May 1982.
Extended Linear Chain Compounds by Joel S. Miller. Plenum Press, 1982.

Bill Sanders/
The Milwaukee Journal

Computer Capers

by Benedict Leerburger

I n 1971 a 19-year-old student stole $1 million worth of telephone equipment through the phone company's own computerized ordering system. In 1973 a convicted drug pusher studying computer operations in jail used the prison's computers to commit a $1.1 million payroll fraud. In 1983 a group of Milwaukee, Wisconsin, teenagers was taken into custody by the Federal Bureau of Investigation (FBI). They were suspected of having broken into more than 60 business and government computer systems in the United States and Canada, including computers at the Los Alamos National Laboratory in New Mexico, the Security Pacific National Bank in Los Angeles, California, and New

York's Memorial Sloan-Kettering Cancer Center. In all cases the culprits had two things in common: an extensive knowledge of computer operations and a desire to "beat the system."

Hidden Losses; Poor Laws

The number of crimes committed with computers has risen dramatically over the past several years. However, the exact number of "computer crimes" may never be known, according to Jay BloomBecker, director of the National Center for Computer Crime Data in Los Angeles. He estimates that no more than 15 percent of all the people detected in computer crimes are ever reported to the authorities. Furthermore,

says BloomBecker, "The average loss in a computer-crime case is about $450,000. Those 85 percent who aren't reporting cases are absorbing enormous losses."

To complicate matters, law enforcement representatives complain that the present laws regarding computer crime are inadequate and that many of the so-called crimes do not actually violate existing statutes. For example, it is against the law to break into someone else's house, but in most states there is no law that prohibits breaking into someone else's computer.

A High-tech Bank Robbery

Stanley Mark Rifkin was a 32-year-old computer specialist in 1978 when he conceived what he thought was the perfect crime. He schemed a way to steal a fortune from the Security Pacific National Bank in Los Angeles, where he worked as a consultant.

Rifkin had access to the room where Security Pacific transferred funds to other banks via a nationwide electronic wire. On October 25, he somehow learned the secret security code of the day and was able to plug into the wire.

Late that afternoon, just minutes after the wire went down for the day, Rifkin called the wire room. He identified himself as an authorized international banker, used the correct security code, and ordered the transfer of $10.2 million to a bank in Switzerland via a bank in New York City. The money order was executed first thing the next morning.

Agents from the FBI suspect that Rifkin flew to Switzerland and purchased $8.1 million in diamonds from a Soviet diamond-trading group. He then smuggled the gems and the rest of the stolen cash into the U.S.

At this point the "perfect crime" began to collapse. FBI agents believe Rifkin referred to the heist when speaking with a prospective diamond buyer, who tipped off the FBI. Rifkin was arrested soon after in a friend's apartment. According to some reports, he appeared relieved when taken into custody. Ironically, the bank had no suspicion that it was the victim of the largest wire fraud in history when notified of the theft one week later by the FBI.

The Rifkin case is significant not so much for the potential dollar loss to the bank, but because it alerted many thousands of businesses and government agencies that rely on computers to the fact that neither secrecy nor safety of assets can be assured.

The 414s

Although the number of computer-related criminal activities has risen since Rifkin's arrest, the public heard little about such high-tech crimes until the summer of 1983, when a group of high-school and college students in Milwaukee, Wisconsin, shocked the nation. According to the FBI, about a dozen students calling themselves

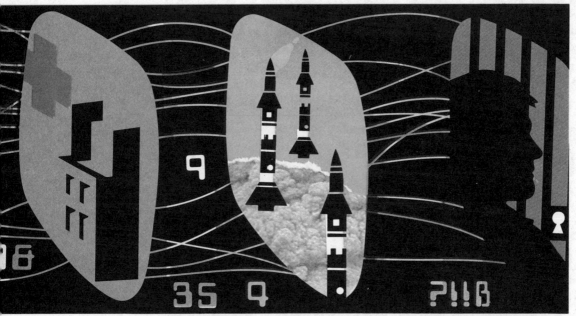

the "414s"—after Milwaukee's telephone area code—found ways to break into various private computer systems around the country. What surprised many people was the ease with which individuals could gain entry into a computer's vast memory, where data can be stored on anything from a person's credit status and medical history to top-secret U.S. Department of Defense documents.

Most computer systems are protected from unauthorized entry. Some form of password is usually required before total access to a system's data banks will be granted. The 414s picked up many secret passwords from nationally operated electronic bulletin boards. An electronic bulletin board, like the message board at a school or supermarket, allows computer buffs to post messages ranging from items for sale to newly discovered computer-related information—including passwords and access telephone numbers.

If a system's password wasn't known, the 414s would play a trial-and-error game until they accidentally hit upon the correct code. When they broke into Sloan-Kettering's computer, for example, they discovered that the word "test" opened all doors.

Neal Patrick, one of the 414s, said that trying to discover the right password was like climbing a mountain: "Once you reach the peak you make a map of the way and give it to everybody else." Patrick used his father's TRS-80

Model II computer and an automatic redial modem. (A modem is a device that converts computer signals into telephone signals and vice versa for data transmission.) Together with a group of fellow 414ers, he used the modem to hook into Telenet—a national network owned by GTE Telenet Corporation that enables its authorized customers to tie into more than 1,200 computers.

Sometimes the 414s didn't know whose data bank they had entered until they were on the inside. "In the case of the Security Pacific National Bank," says Patrick, "we knew it had something to do with world banking because of instructions for a program to set various loan limits on countries around the world." Once inside a private data bank the 414s just "looked around." "It was just a game," said another member of the group. Patrick, however, did admit that a line from "WarGames"—the hit movie about a defense-department computer break-in—inspired them to reprogram one computer to respond: "Would you like to play a nice game of chess, Dr. Falken?"

Hackers and Phreaks

A growing group of computer enthusiasts has developed a passion for cracking computer codes. They are primarily high-school students who devote most of their time and energy to computers. Once they master the ins and outs of their own computers, they branch out and try to

see how much more they can learn about other systems and prospective applications. These computer "hackers" are totally fascinated by technical challenges.

The hackers' goal, according to John Kender, assistant professor of computer sciences at Columbia University, is some "evidence of cleverness. They are showing that they have the technological mastery to do some very surprising and powerful things."

One of the more infamous early hacks was a Massachusetts Institute of Technology (MIT) program known as "Cookie Monster." A hacker would gain entry to a computer and plant the special program in the system. Whenever an unsuspecting programmer was working on the computer, the word "cookie" would suddenly flash on the screen. If the startled user didn't respond, the words "cookie" and "gimme cookie" were written over and over, obliterating whatever the user had on the screen. The only way to stop the destruction was to type in the word "cookie," thus feeding the "monster."

Hackers trace their beginnings to the so-called "telephone phreaks" of the early 1970's. Phreaks gained attention by developing technical devices and methods to gain illegal use of telephone lines. One of the early "phreaks," for example, was known as Cap'n Crunch. He

discovered that a toy whistle given away in boxes of the breakfast cereal Cap'n Crunch produced a tone of exactly 2,600 cycles per second. He further discovered that by dialing a long-distance number and blowing his whistle into the phone, he could open a long-distance trunk line over which he could make unlimited free calls. His claim to fame—he reportedly was the first to place a series of calls all the way around the world and back to himself.

Phreaks and hackers have something else in common. They see nothing wrong with their activities and do not believe in property rights. "Hackers have their own ethics," says Kender. "It is OK to explore another system, or to make the life of another hacker miserable."

A Question of Ethics

For the most part, hackers consider their exploits little more than a challenge—certainly nothing unethical or immoral. One 19-year-old hacker in New York City, who calls himself Stainless Steel Rat, says he breaks into computers "just to see what's around. It's sort of like a challenge to see if you can beat the security system somebody else made." Another hacker, a 15-year-old from Los Angeles who calls himself Red Rum, doesn't believe he is hurting anyone by breaking into computers. He rationalizes, "I don't think the big companies are really concerned. The amount of money they lose because of computer freaks is nothing compared to what they make."

In fact, the implications of computer break-ins involve a great deal more than money. "Some of the systems are very fragile," according to Donn B. Parker, a computer-crime consultant and author of *Fighting Computer Crime*. "If you hit the wrong key, you can wipe out files or cause the whole system to crash. The hackers say they're not malicious, but the victims lose all this computer time and they have some stranger roaming around in their system. I call it electronic vandalism."

The question of computer ethics is not just limited to hackers. Many large companies are issuing codes of conduct governing the use of computers by their employees. At International Business Machines Corporation (IBM), for example, all company computer systems are for business use only. At General Electric, employees can use company computers "for private tasks" but only on their own time. The insurance company Equitable Life goes one step further and encourages employees to take company computers home "to allow their families to use them."

However, the use of company-owned hardware and access to company-owned data are two entirely different situations. R. Vincent Conant, president of the information services division of Carter Hawley Hale, notes that a person with a key to 50 locked file cabinets cannot easily browse through those cabinets. But the same person with the password to 50 computerized files can easily look for and locate specific information. "Does that employee have more responsibility now? I think so," says Conant.

The ethics of computer vandals breaking into corporate computers is overshadowed by the many attempts to snoop in government files. "Some hackers spend 12 hours a day trying to break into computers at the Central Intelligence Agency (CIA) or the Pentagon." According to Larry N. Hurst, an FBI agent who monitors computerized bulletin boards, "They have a keen interest in the systems of the U.S. military. Federal authorities are extremely concerned with the growing expertise of these 'amateur spies.' " Defense Department officials admit that the growing band of young hackers has exposed weak links and points of vulnerability in many military systems. The fear, of course, is that Soviet agents will discover how to break into classified computer files. Needless to say, the Soviets care little about ethics.

New Laws Needed

Despite the increase in computer-related crime, state and federal prosecutors find the current laws either inadequate or lacking entirely for prosecuting those arrested. Only about 20 states have adopted laws governing computer abuse. However, even in those states, the application of computer-crime laws is confusing. For example, prosecutors in New York are unsure which laws may apply in the prosecution of the Milwaukee 414s who violated the Sloan-Kettering computer system. To complicate the matter, as Harvard Law School professor Arthur Miller points out, "It is not clear where the crime was committed. Was it in New York? In Wisconsin? This is why the Federal Government will have to take a major role."

Unfortunately, Congress has debated several proposed Federal statutes since 1977, but as of early 1984, no legislation had been enacted to prevent unauthorized access to computers. Says Arthur Miller, "This is an area in which the kids in blue jeans are often far ahead of the guys in three-piece suits. The law is beginning to catch up, but it is a heck of a track meet" □

SELECTED READINGS

"Armchair outlaws" by Robert R. Rhein. *Micro Discovery*, November 1983.
"Beware: hackers at play" by William D. Marbach et al. *Newsweek*, September 5, 1983.
Fighting Computer Crime by Donn B. Parker. Scribner's, Charles, and Sons, 1983.

EARTH SCIENCES

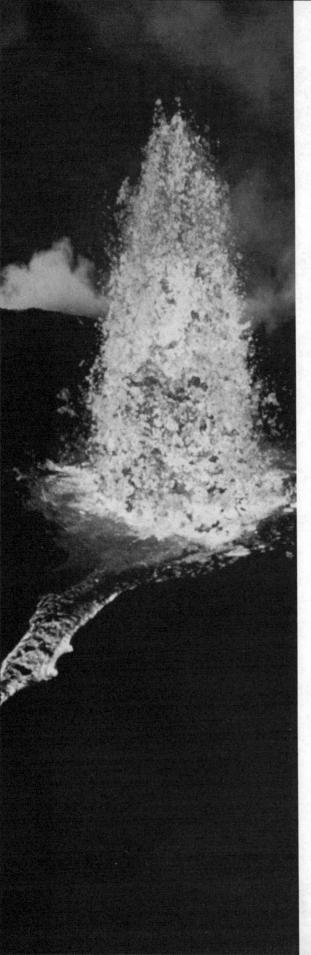

REVIEW
OF THE
YEAR

EARTH SCIENCES

UNUSUAL WEATHER PATTERNS

In many parts of the world the year 1983 was
marked by unusual weather patterns that were
caused by the El Niño of 1982–83—the greatest
oceanic-atmospheric disturbance recorded in over
100 years. El Niño is a periodic (every 3 to 10
years) massive climate change initiated by a
warming of Pacific waters (usually starting off the
coast of Peru) and a change in wind patterns.
The unusually intense and long-lived 1982–83 El
Niño is thought to be responsible for the severe
droughts that hit Indonesia, eastern Australia, and
parts of South America and southern Asia; the
torrential downpours of Ecuador and Peru; the
unusually wet winters and spring flooding in much
of the western United States; and the severe
coastal storms that hit California during 1983. ■
Meanwhile, scientists were also trying to
determine how future climate patterns may be
affected by the April 1982 violent explosion of
Mexico's El Chichón volcano, which spawned
widespread sulfuric acid clouds and atmospheric
changes.

EARTHQUAKES AND SEISMOLOGY

Seismic activity increased during 1983 in the
United States and elsewhere, with several strong
quakes occurring, some very puzzling. On May 2
an earthquake of 6.5 on the Richter scale
occurred near Coalinga, California, injuring at
least 47 people, demolishing 300 buildings, and
causing $31 million in property damage. (The
Richter scale is a widely used indicator of an
earthquake's severity, with a magnitude of 7 or
higher on the scale signifying a major quake.)
The Coalinga quake puzzled many scientists. It
occurred on a previously unknown fault 40
kilometers (25 miles) east of the well-known San
Andreas Fault, and led to speculation that other
unknown faults may lie beneath California's rolling
hills. ■ Another strong quake in the United States
led scientists to wonder if quakes—even those
widely separated geographically—may sometimes
be "paired." On October 28 a strong quake hit
Challis, Idaho, and two days later a strong quake
killed more than 1,200 people in eastern Turkey.

*Lava scorches the Hawaiian landscape as Kilauea erupts.
It is considered the world's most active volcano.*

Ken Sakamoto/Black Star

Some researchers suggested that the Idaho quake might have upset stresses in the Earth's crust that triggered the Turkish disaster. ■ Yet another strong quake—6.7 on the Richter scale—struck the United States during the year, jolting the island of Hawaii in November, and a series of smaller quakes—up to 5.2 on the Richter scale—rattled parts of the northeastern United States and Canada in early October. ■ Other significant quakes during 1983 included a 5.7-magnitude quake that hit the Philippines 400 kilometers (250 miles) north of Manila, causing widespread damage and 13 deaths; a major quake in Popayán, Colombia, that killed more than 200,000 people; and a quake in Japan that led to tsunamis (giant seismic sea waves) that caused 32 deaths and great property damage.

Movements of Earth's crustal plates apparently trigger most earthquakes, but some California scientists believe there is a correlation between the behavior of the moon and earthquakes along California's San Andreas Fault system. Specifically, they think that the gravitational pull of the moon may set off the release of accumulated stress along the fault that triggers seismic tremors. Their study of lunar patterns suggests that a major earthquake could occur in southern California before 1990.

VOLCANOES AND VOLCANOLOGY

The world's most active volcano—Hawaii's Kilauea—erupted powerfully in 1983 but caused little damage. Starting in January and continuing intermittently throughout the year, Kilauea spewed out lava that is enlarging the island of Hawaii. It also provided a living laboratory for scientists who have analyzed the breathlike stretching and shrinking of the volcano's cone, the seismic activity and magma flows underlying it, and the electromagnetic and gas fluctuations associated with it. Meanwhile, Kilauea's neighbor, the huge Mauna Loa volcano, prepared to erupt. (It did so in early 1984.) Some scientists are considering the feasibility of constructing barriers to divert damaging lava flows from inhabited areas. The "hot spot" Hawaiian area is experiencing different effects from yet another volcano: Loihi—an undersea volcano—is building a new Hawaiian island. Located about 30 kilometers (17 miles) southeast of Hawaii, Loihi now has a summit about 970 meters (3,180 feet) underwater and is still growing.

Italy's Mount Etna was also active for several months during the year. In May, Italian volcanologists successfully used precision explosions and diversion barriers to control nearly 100 million cubic meters (130 million cubic yards) of lava flows from invading inhabited areas.

The relationship between volcanic eruptions and earthquake activity was illustrated in several ways. In November the eruption of Mount Oyama, which buried a village and killed three people in Miyakejima, Japan, was followed quickly by a 6.1-magnitude quake. In Antarctica the Mount Erebus volcano did not erupt but showed a significant increase in earthquake activity, suggesting movement of magma beneath the volcano. Mount St. Helens in Washington also remained seismically active, and some 20 other volcanic areas in the western United States bear watching, according to researchers who made a preliminary assessment of the area and its seismic activity.

EARTH'S AGE AND STRUCTURE

Scientists have discovered rocks more than 4 billion years old in western Australia. Radioactive dating indicates that these rocks are at least 300 million years older than any other rocks known by geologists. This find raises the hope that some of Earth's original crust may still exist. ■ Geologists are also optimistic about refining dating techniques: they have improved obsidian-dating methods and have developed a new dating method based on oxygen isotopes and periodic variations in Earth's orbit.

PALEONTOLOGY

Paleontologists and biologists took a new look at mass extinctions of prehistoric organisms. Fossils from the North Atlantic region reveal that a major extinction occurred some 650 million years ago when about 70 percent of all species of unicellular algae disappeared. Before the discovery of this Precambrian extinction, the oldest known mass extinction occurred about 450 million years ago.

Another mass extinction—that of the dinosaurs and many other species living about 65 million years ago—continued to generate controversy. Iridium is a metal more abundant in meteorites than on Earth. Its abrupt presence in rocks some 65 million years old has been used as evidence for the "asteroid impact" theory, which holds that the impact of an asteroid created conditions that led to the mass extinctions. Now new evidence suggests that global volcanic activity produced environmental changes resulting in the extinctions, and that iridium was not extraterrestrial but formed by the alteration of volcanic glass. Other geologists, however, presented evidence they say supports the asteroid impact theory—namely, the presence of osmium in the rocks of the period, which they believe to be of meteoritic origin.

The much more recent mass extinction of the giant mammals of the last Ice Age is also being debated. It has been suggested that the mammoths, mastodons, and saber-toothed cats died out because of spreading glacial climates. Opponents of this idea believe that these and 52 other species of large mammals were hunted to extinction by early humans.

WILLIAM H. MATTHEWS III

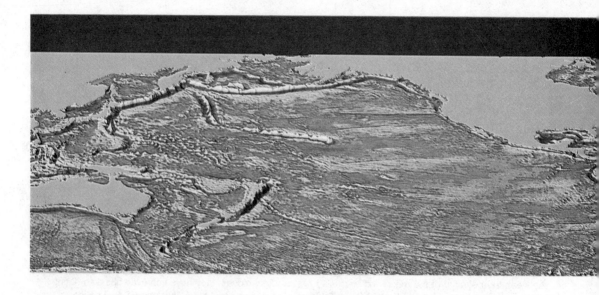

SEAFLOOR PANORAMA

by Marcia Bartusiak

"Iᵗ's as if he pulled a giant plug out of the sea bottom and all the water in the oceans of the world drained away."

That's the general reaction these days at Columbia University's Lamont-Doherty Geological Observatory when inquisitive visitors ask Lamont scientists what they think of their colleague William Haxby's latest endeavor.

The response doesn't end there. Interest in Haxby's work has quickly spread beyond the gates of the sprawling observatory complex that lies along the scenic banks of the Hudson River in Palisades, New York. Haxby has created a sensation among scientists everywhere.

What has Haxby done to engender all this excitement? After 18 months of painstaking computer processing, the young Lamont geophysicist has converted reams of satellite data into an exquisitely detailed panorama of the ocean floor—an elaborate color map that identifies underwater structures as small as 32 kilometers (20 miles) across. He did this by calculating how the seafloor topography gravitationally tugged at the water above, causing variations in the height of the sea surface that had been measured by satellite. The geophysics community views the feat with astonishment.

Worldwide Picture Now Complete

Although still in its infancy, such satellite-aided gravity mapping has become a powerful and multifaceted tool for oceanographers and marine geologists. Researchers Timothy Dixon and Michael Parke at the National Aeronautics and Space Administration's (NASA's) Jet Propulsion Laboratory in Pasadena, California, for example, have also generated similar maps from the same data. Their independent venture is enhancing the larger-scale features of the ocean floor.

Seafloor mapping, of course, is not a new endeavor. For decades, mariners have charted the submarine landscape by taking echo soundings of the depths below as their vessels crisscrossed the ocean surface. Even gravity maps of limited areas have been made from shipboard with precision instruments called gravimeters. But until the Lamont and Jet Propulsion Lab maps came spewing out of computers, the worldwide picture was far from complete. Large regions of the South Pacific and South Atlantic seafloors remained very sketchy. Many geologists have remarked, only half-jokingly, that the arid landscape of Mars was better known than the 70 percent of Earth's surface hidden beneath a watery veil.

"Let me put this into context," explains geophysicist John LaBrecque, Haxby's colleague at Lamont. "It would take several billion dollars and centuries of time for a research vessel to acquire the amount of knowledge contained in Bill's maps. They're virtually a television scan of the ocean bottom."

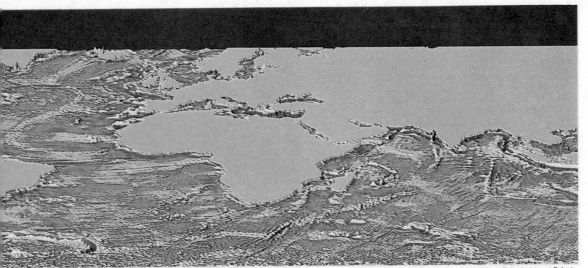

William F. Haxby/Lamont-Doherty

Crustal Collisions

Indeed, Haxby's maps of the world's seafloors reveal a terrain as diverse as any found on the seven continents. They also reinforce geology's most unifying concept: plate tectonics, an idea forged more than a decade ago. This revolutionary theory says that Earth's crust is divided into some 20 plates. Propelled by the turbulent motion of Earth's scorching interior, some of these giant slabs collide, thrusting mountains upward and digging trenches deeper. The kilometers-deep trenches that rim the western Pacific, for instance, mark where the huge Pacific plate is diving into the mantle with a vengeance. Other plates move apart, giving molten magma deep within Earth the chance to rise and form new seafloors. Such a process is going on right now at the prominent Mid-Atlantic Ridge, where the North American plate is inexorably separating from the Eurasian and African plates by about 2.5 centimeters (1 inch) each year.

Haxby's successful reproduction of these familiar seafloor landmarks gave him confidence in his technique, which he has dubbed "geotectonic imagery." But the real excitement grew when his relentless effort to sharpen the resolution of his maps began to reveal long-hidden features, especially in poorly charted sections of the ocean floor. New ridges, rifts, and underwater volcanoes (or seamounts) seemed to pop up everywhere. Needless to say, such discoveries are of more than academic interest. "The Navy is certainly interested in discovering new seamounts so their submarines won't run into them," quips Haxby.

Gold-toned land masses are distinguishable features in this unusual view of Earth. The computer-generated map shows ocean basins as if all the water were gone.

Fracture Zones Amazingly Clear

More intriguing were the long, dark streaks cutting across many of the seafloors, as if the muddy bottoms had been raked by a giant pitchfork. These are fracture zones, the scars left behind as the edges of the plates scrape past and away from one another. Costly ship surveys have tracked them before, but never so many or so clearly. "They're like railroad tracks," explains LaBrecque, "that now enable us to trace the past movements of the continents more accurately than ever before." Adds Haxby: "You can actually follow the motions back to Pangea," that single landmass that existed some 200 million years ago when the Americas, Africa, Antarctica, and Australia were fastened together.

"Take the fracture zone extending from the tip of Africa to the Falkland Islands off the coast of Argentina," suggests Haxby. "If you rotate the African continent along this fracture, you can see how Africa once tucked in quite nicely to South America."

Haxby credits the Seasat satellite for this astounding clarity. With the launching of the instrument-packed probe in the summer of 1978, oceanographers were obtaining their first worldwide monitor of ocean activity. Unfortunately, an electrical short circuit terminated the Seasat mission after only three months, but not before an onboard radar altimeter had measured

This color relief map of the ocean floor was created by computer processing of billions of measurements that were made by the Seasat satellite. Primary features of the seafloor include: the Mendocino Fracture Zone (1); the Hawaiian Emperor Seamount chain (2); the Eltanin Fracture Zone (3); the Mid-Atlantic Ridge (4); the Rio Grande Rise (5); the Falkland-Agulhas Fracture Zone (6); the Reykjavik Ridge (7); the Walvis Ridge (8); the Agulhas Basin Ridge (9); and the recently discovered India-Antarctica Fracture Zone (10).

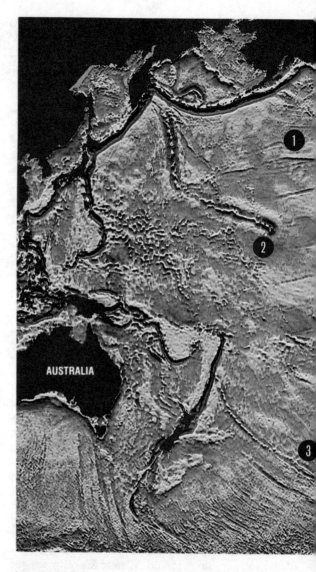

the distance between the spacecraft and the ocean surface to an accuracy of 5 to 10 centimeters (2 to 4 inches) over most of the globe.

"The altimeter sent out a thousand pulses each second and recorded the time it took for those radar pulses to bounce off the sea surface," says Haxby. "It did this continuously until the satellite failed." The resulting data—some 8 billion readings in all—enabled Haxby virtually to X-ray the ocean bottom. How? The answer begins with gravity's varying strength, determined by local variations in Earth's mass—both on the continents and beneath the seas.

Gravity "Signatures"

Scientists have long known that the sea would not be perfectly level even if the wind stopped blowing and the waves ceased rolling. Instead, the surface would subtly mimic the contours of the landscape below. This is because gravity tends to pile water above massive underwater structures such as mountains and ridges because its strength is greater near such large concentrations of mass. Conversely, the seas subside over objects of lesser mass, such as trenches and basins. "A 1,600-foot [490-meter] seamount, for example, elevates the sea surface by about 8 inches [20 centimeters]," notes Haxby. A kilometers-deep trench, on the other hand, causes the ocean to drop by dozens of meters. But don't look for these roller-coaster effects on your next voyage; such bulges and dips are spread out over many kilometers. Fortunately for Haxby, the Seasat satellite, from its lofty perch 800 kilometers (500 miles) out in space, was able to discern the broad variations quite easily.

The Seasat data arrived at Lamont on 24 computer tapes. "My original intention," recalls Haxby, "was to map only a small area, the poorly surveyed South Atlantic between Africa and South America." His method: converting the varied sea-surface heights recorded on the tapes into their corresponding gravity "signatures," the variations in the force of gravity due to the presence of underwater structures. Grav-

ity "highs" would pinpoint the mountains and ridges; gravity "lows," the trenches and fractures.

In fact, Haxby's processing revealed much more. His gravity maps also identified many features hidden beneath hundreds of meters of sediment, because the sediment has relatively little mass. This enables researchers to detect long-buried fracture zones not seen on conventional seafloor maps.

Ancient Ridge Revealed

Haxby displayed his first results in a standard but very crude form—as black-and-white contour maps comprehensible only to experts. Even then, the Lamont scientists sensed the power of the technique. LaBrecque remembers one night

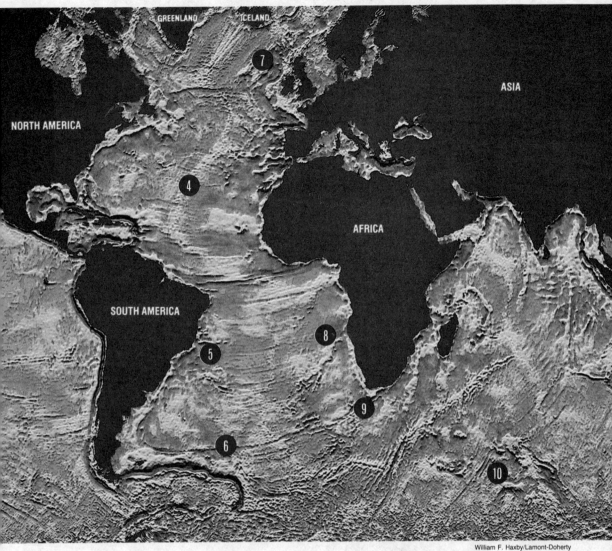

William F. Haxby/Lamont-Doherty

of computer processing in particular: "Bill was sitting at his terminal, cranking in finer and finer contouring. We asked him to crank it up some more. He replied, 'Well, there might be a lot of noise in there.' But to everyone's surprise, more resolution emerged. It was like focusing a microscope. Things came out of the ocean floor that we never had hoped to see."

The most exciting event for Haxby and LaBrecque during that all-nighter was finding a thin, sinuous line snaking southward in the Agulhas Basin, several hundred kilometers southwest of the tip of Africa. For the first time, geologists were viewing the ancient mid-ocean ridge that formed when South America, Africa, and Antarctica started to spread apart more than 100 million years ago, a time when dinosaurs

still roamed the Earth. The buildup of sediment had kept this fossilized seafloor-spreading center well concealed.

"The boundary between the plates later jumped somewhere farther west, so this ridge was left behind to subside," explains Haxby. But it continues to rest on the sea bottom as a vital piece of evidence that will help researchers trace the evolution of the oceans, a process crucial to the development of climatic patterns.

Color Graphics Add New Dimension

Such revelations in the South Atlantic prompted Haxby to expand the scope of his project. "We were perfectly happy with the contours," recalls LaBrecque, "but Bill was mumbling, 'Maybe I can make a picture out of this.' "

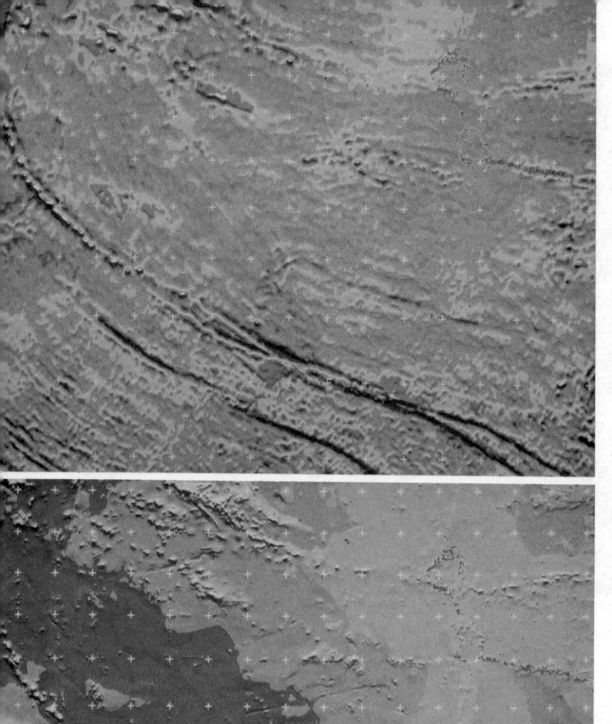

Indeed he could. Soon he was producing a single map of all the world's ocean floors, using a sophisticated color-graphics system. He chose to form his photolike image out of 8 million picture elements (pixels). Each of these pixels was assigned one of 15 colors to depict the gravity strength at that particular point, from dark blue (gravity low) to yellow or pink (gravity high). Haxby also used 256 degrees of shading to highlight the relief. "It acts as if there is a light source to the north," he points out.

Such touches are more than sheer theatrics. "Color and shading add a whole new dimension. They bring out features not seen in standard contour maps," says Haxby. Already, certain textures in the color image may provide a glimpse into Earth's interior. Haxby and Lamont geophysicist Jeff Weissel suspect that the faint lines running like a comb through the central Pacific seafloor may be the result of convection currents churning 48 to 145 kilometers (30 to 90 miles) beneath the ocean bottom in the upper mantle. In their model, each circulating loop consists of hot material rising and cooler material sinking back into the depths. "A single ship measuring this area would barely notice this feature," contends Haxby. Analyzing this relatively small thermal process might lead to better understanding of the monstrous convective cells suspected to be driving the plates themselves.

Accordion Squeeze and a Rubber Plate

Lamont has described geotectonic imagery as providing marine geologists with "a quantum leap in mapping power." It's an apt description. The inventory of structures that have been either discovered or better resolved with the observatory's maps increases daily:

• A newly revealed fracture zone in the southern Indian Ocean promises to shed light on India's break from Antarctica 200 million years ago. This 1,600-kilometer (1,000-mile)-long gash, located southwest of the Kerguelen Islands, was gouged out of the seafloor as the Indian continent inched northward. India's cataclysmic meeting with Asia 160 million years later forged the Himalaya Mountains.

• A strange series of east-west wrinkles in the ocean crust just south of India and Sri Lanka verifies that the Indian plate continues to push northward, resulting in earthquakes and surface deformations. "The entire crust is being squeezed in like an accordian," says Haxby. "This is something that is not well understood, since the plate was expected to be more plastic."

• The intense gravity low surrounding the distinctive Hawaiian Emperor Seamount chain in the mid-Pacific clearly indicates how a tectonic plate can bend like a rubber mat under the massive weight of a seamount. The crust beneath Hawaii (the southeast tip of the chain) deflects by as much as 10 kilometers (6 miles).

• The sharp bend in the lengthy Mendocino Fracture Zone that juts out of northern California confirms that the Pacific plate abruptly changed the direction of its motion millions of years ago. This may be one of the resounding effects of India slamming into Asia.

The Seasat satellite, launched in 1978, provided scientists with the first worldwide monitor of ocean activity.

NASA

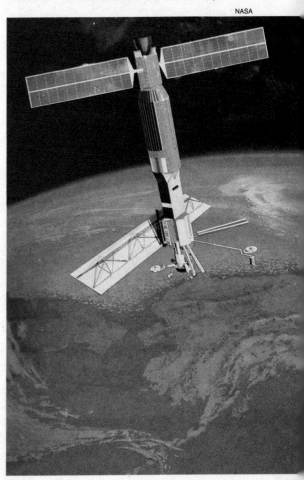

• The darkened area off the coast of Somalia in northwest Africa outlines what may be the oldest oceanic crust on Earth. This gravity low has formed as the aged plate sinks into the mantle.

• The gravity map confirms that the Reykjavik Ridge, southwest of Iceland, is indeed V-shaped—a matter of some controversy. This wedge seems to be forming as hot magma pulses up from Earth's interior.

Gravity Maps Cause Theory Refinements

Even a missing feature is very telling. The prominent East Pacific Rise—the long ridge system that separates the Pacific, Antarctic, and Nazca plates off the west coast of South America—barely shows up on the gravity map. "This is because the ridge is spreading relatively quickly, on the order of six inches [15 centimeters] a year—three times faster than the Mid-Atlantic Ridge," explains Haxby. "With this rapid spreading, the materials formed get distributed in such a way that the gravity signature gets smoothed out."

Scientists around the country are starting to use these gravity maps of the world's ocean floors to assess and refine certain geological theories. LaBrecque claims that the images have already made him change his mind about the evolution of two prominent features on the South Atlantic ocean floor.

Geologists have long believed that the Rio Grande Rise—a massive underwater plateau off the southeast coast of Brazil—and the Walvis Ridge near southwest Africa were built up long ago when both the South American and African plates moved over a "hot spot" beneath the ocean crust—a region near the Mid-Atlantic Ridge where molten magma could easily gush out of the mantle below. But the gravity map, contends LaBrecque, tells another story. It suggests that the Rio Grande Rise and Walvis Ridge were once joined, part of a fledgling mid-ocean ridge system that burgeoned as South America and Africa started to spread apart.

"The boundary between the plates jumped to the east, and these structures started to move apart some 100 million years ago," ventures LaBrecque. "That intense gravity low on Bill's seafloor map, south of the Rio Grande Rise, is the scar that was left behind." As the scenario goes, the Walvis Ridge continued to rift, "a rift that broke the African plate apart like a tear in a piece of paper," says LaBrecque. On the gravity map, this rifting, seen as a long rivulet of yellow, makes Angola look as if it had sprung a leak.

No one expects these new satellite charts of the ocean floor to replace on-site measurements. Most likely the satellite gravity maps will be used for reconnaissance, cheaply and efficiently identifying regions where detailed ship surveys of the ocean bottom are most needed. "The accuracy of a satellite radar altimeter will never match the accuracy of a shipboard gravimeter [an instrument that directly measures the force of gravity]," says Haxby. "Gravimeters can measure a millionth of the force that we normally feel pulling on us."

Geosat and Topex Data to Come

Haxby expects his maps to be updated and improved as new radar altimeters are launched into space by both NASA and the European Space Agency. He's the first to admit that his initial effort was far from perfect: "There were fairly large gaps in Seasat's orbital path; some of the gaps were as much as 120 miles [193 kilometers] wide." To compensate, Haxby had to identify dominant topographic trends and interpolate the data along those trends.

The Navy's Geosat satellite, launched in late 1984, will provide the first opportunity to fill in the missing pieces. Detailed altimeter information from the first phase of Geosat's mission will be classified (gravity variations can affect the flight of missiles), "but the second phase shows promise of improving the Seasat data," notes Haxby. And at the end of the decade, NASA plans to orbit a satellite called Topex (Topographic Experiment), which will be designed to sense sea-surface variations of less than 2.5 centimeters (1 inch). With such capabilities at hand, researchers should be able to produce gravity maps distinguishing seafloor features as small as 13 kilometers (8 miles).

For now, however, Haxby is anxious to put his computer "hacking" to rest so that he can concentrate on more interesting matters: joining his colleagues in interpreting the many intriguing features being revealed for the first time in these remarkable maps □

SELECTED READINGS

"The contours below," by Steve Olson. *Science 83*, July/August 1983.

"The globe in sharper focus" by Cheryl Simon. *Science News*, December 4, 1983.

"Geotectonic imagery: quantum leap in mapping power." *Lamont*, Fall 1982.

The Other Cold War

by Ellen Ruppel Shell

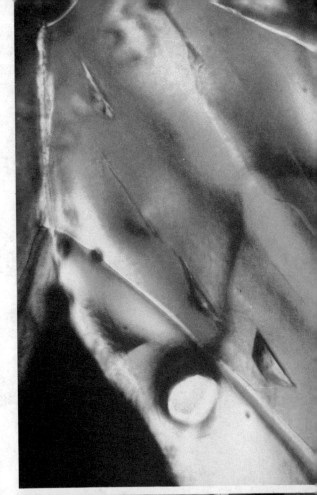

W hen the first crocuses break through the late-winter frost, most of us breathe a long sigh of relief. There are others, though, who at the vaguest hint of spring pack up their flannel shirts and quilted underwear and head out in search of more— more ice, more snow, more cold. They take field trips to Alaska, to remote regions in Norway, and to the hinterlands of northern Michigan. And returning home, they retreat from the summer sun into vast, refrigerated chambers that make winter of July.

Surprisingly, this small but influential group is a fairly ordinary bunch, though perhaps a bit thicker-blooded than average. Most are engineers and scientists, and all are committed to the purpose of taming and even making an ally of humankind's oldest nemesis—the cold.

Fascinating and Frustrating

"Many people talk about fighting the elements, but fighting doesn't get you very far," says cold engineer Albert F. Wuori. "We try to use the environment to our advantage." Wuori is chief of the experimental engineering division of the U.S. Army's Cold Region Research and Engineering Laboratory (CRREL) in Hanover, New Hampshire, the largest facility of its kind in the Western World. Although it is a warm Indian summer day, Wuori dresses for the worst in flannel shirt over a turtleneck, thick pants, and boots. Small, spare, with weatherworn features and a no-nonsense crew cut, he looks every inch the model engineer. And his nearly 30-year

Top: An ice sliver melting along its edges is seen in polarized light. The blue-green area is the thickest. Bottom: A CRREL scientist examines an ice-core sample.

Top: USACRREL; Bottom: Walter Bibikow

CRREL engineers built this precise model of the Ottauquechee River in Vermont to study ice buildup and its control.

career at CRREL, during which he has developed a slavish devotion to his work, does nothing to belie that impression.

"When water freezes, everything changes," he explains when asked why he chose this line of work. "Soil becomes as hard as concrete. You can walk on water. I find this fascinating."

The military finds it frustrating. It's an old army adage that both Hitler and Napoleon were defeated, at least in part, by the cold. The United States military establishment is not about to make the same mistake. Since two-thirds of the Soviet Union is considered "cold region"

(that is, covered with ice and snow a large part of the year), the Russians have had plenty of experience with frigid conditions. Their largest cold-research laboratory, in Leningrad, has 3,000 employees, 10 times the staff of CRREL.

"The Russians are ahead of us in both experience and technology," says CRREL commander Colonel Wayne Hanson. CRREL's purpose is to try to narrow the Soviet lead, in the interest of both national security and public health.

Anyone who's spent a 10-day weekend waiting out a blizzard will appreciate the valid-ity of this endeavor. Snow can turn a bustling urban center into a ghost town, a Sunday drive into a nightmare. The "unsinkable" *Titanic* was foiled not by foreign fire but by ice, as was the Boeing 737 that crashed in Washington, D.C., in 1982, costing 78 people their lives. CRREL specialists are often called upon to investigate the circumstances leading up to such dramatic events. But the bulk of their research is into the basic nature of the snow, ice, and frozen earth that are the ultimate perpetrators.

"We examine snow and ice on a molecular scale," explains Wuori, "at their bonding, chemistry, and physics. Our work is sometimes so basic that the bosses in Washington look askance. But we're convinced that we need a fundamental understanding of the stuff we're supposed to control, and that requires basic research."

Mimicking Mother Nature

Basic research means drilling out ice cores from glaciers, skimming snow samples from mountaintops, and burrowing deep beneath the seafloor for frozen sediments. These tidbits are brought back to the laboratory, where they are kept frozen in "cold rooms" chilled to as low as $-51°$ C $(-60°$ F). There they are examined by scientists who can read a column of ice the way a geologist reads rock strata.

But it is seasonal ice, not the polar variety, that causes cold scientists the most concern. Ice-clogged waterways are responsible for more than $100 million in damages and delays in the United States each year. And while they can't completely stop it from forming, CRREL engineers have found ways to control and even slow down ice formation.

The root of the problem, they've found, is the buildup of slushy "frazil" ice, which appears early in the winter and forms a natural barrier that thickens throughout the season, even in turbulent water. The solution is to stop the frazil buildup by placing wooden or metal booms across waterways. These booms act as a framework, allowing solid ice to form an insulating cover over the water, which inhibits the growth of frazil and thereby limits the ice's overall thickness.

To help them decide exactly where to place the booms, CRREL scientists build miniature versions of the waterway in question—complete with ice floes—in a 3-meter (10-foot)-deep, 9-by-37-meter (30-by-120-foot) test basin. The model room—its toy boats, miniature locks, and water pumps the embodiment of many a seven-year-old sailor's dreams—can be chilled to $-23°$ C $(-10°$ F), thereby closely mimicking the actual winter conditions under which ice forms.

Of course, Mother Nature is not always a cooperative modeling subject. A floating ice-control stucture that worked just fine in the lab was completely wiped out by an unusually large ice jam near Oil City, Pennsylvania, in 1982. "Models are certainly not perfect," Wuori admits. "There are problems of scale that we just can't account for."

Scientists blast a huge hole in the frozen surface of a river to prevent a major ice jam from occurring.

USACRREL

Of Tire Treads and Permafrost

Another elusive problem, one virtually impossible to model in the laboratory, is that of a tire spinning furiously on packed snow or ice. "There are many, many variables in a tire," intones CRREL engineer George Blaisdell, too many, he insists, to allow him to recommend any particular brand by name. But research he's conducted from the back of CRREL's specially equipped $200,000 Jeep Cherokee shows that when it comes to traction, all tires are definitely not created equal. For example, he's found that the best counter for soft, new snow is a slightly underinflated radial tire. He also discovered that so-called all-season tires do as well on snow as snow tires. On the other hand, no amount of tire tread will help on ice. "All that matters is the type of rubber compound used," Blaisdell explains. "Tires built for ice have lots of carbon." Tires high in carbon wear out quickly, but they may be the only way to get to Grandmother's house on a cold and icy Christmas morning.

Sometimes, though, it is easier—and cheaper—to cooperate with the dictates of nature than to attempt to preempt them. This is generally the case with another subject of CRREL scrutiny: permafrost. Permafrost, or perennial frozen earth, will melt and buckle if seriously disturbed, a process known as frost heaving. To understand the damage frost heaving can wreak, one need imagine no further than the lowly pothole, a frost-heaving consequence that's responsible for millions of dollars in damage each year. On a larger scale, frost heaves can be even more devastating, causing buildings to sink and roads to collapse.

Since 1965 CRREL scientists have studied permafrost and its preservation in a 100-meter (360-foot) tunnel dug 16 kilometers (10 miles) morth of Fairbanks, Alaska. It's an eerie place, filled with icy evidence dating back 40,000 years and permeated with the nauseating odor of decomposing Ice Age plants and animals. Metal measuring poles planted at one end of the tunnel are twisted and deformed by the creeping permafrost, bearing ominous testament to the power of shifting soil.

The tunnel was originally constructed to test the feasibility of underground installations in cold regions, places like Camp Century, a CRREL research post located beneath the

CRREL researchers devised a clever support system to keep the Trans-Alaskan Pipeline carrying hot oil from disturbing the frozen ground below it.

Greenland Ice Cap. Fundamental insights gleaned there have had far-reaching effects, one of the most sweeping of which was on the construction of the Trans-Alaskan Pipeline.

Roughly half of this giant metal tube's 1,290 kilometers (800 miles) crosses permafrost. The oil pulsing through it is hot—hot enough to thaw the earth beneath it, causing the pipes to break and an oil slick to form. CRREL was called in to head off that nightmare. Lead by civil engineer Fred Crory, the CRREL team held up construction until 80,000 vertical pipe supports, many equipped with a special thermopile arrangement to conduct heat from the ground to the air, were installed. So far, there has been no significant melting. CRREL is now consulting on a natural gas pipeline, another Alaska-based project with a similar purpose but the opposite problem of the oil line. Since the gas coursing through this line will be chilled below freezing, it could freeze the previously soft ground beneath it, causing the pipes to be jacked out of the earth. The solution, possibly better pipe insulation or deeper pipe burial, has yet to be decided upon.

The Iceberg Issue

While studies of things like tire treads, potholes, and melting permafrost remain its bread and butter, CRREL research does have far more glamorous moments. When ice flakes collected on the space shuttle *Columbia*'s fuel tank and damaged the ship's heat shields, for example, CRREL was there—recommending that a ground-based, hot-air fan be installed to prevent future icing. The laboratory's studies of snowflake configuration are aiding the design of weapons-guidance systems—which, no matter how sophisticated, are often confounded by even a gentle snow shower. And the region-specific, snow-load guidelines CRREL recently wrote should help prevent disastrous roof cave-ins like the one that put the coliseum at the Hartford Civic Center in Connecticut out of commission in 1978. And then there's iceberg transport.

"Icebergs are nonpolitical, they don't really belong to anybody," says Malcom Mellor, a CRREL glaciologist up to his knees in the iceberg issue. They are also very pure. This combination, he says, makes them very attractive to such freshwater-starved nations as Australia and, in particular, Saudi Arabia.

With no perennial rivers and an annual rainfall of less than 10 centimeters (4 inches),

the Saudis now get much of their water from desalinization plants. While effective, this is a very expensive process, both because its fixed costs are high and because sea salt quickly chews up the necessary machinery. Hence, the Saudis have become the biggest financial backer of research into various iceberg-hauling schemes.

About 4 billion cubic meters (7 million cubic miles) of freshwater—67 percent of the earth's supply—is frozen into antarctic ice, which is neatly packed in tabular slabs 180 to 240 meters (600 to 800 feet) thick. Theoretically, the bergs could be lassoed with cables and towed with powerful tugboats, and Prince Muhammad al-Faisal, who at that time was responsible for the Saudi water supply, banked heavily on the hope the theory would be realized.

Unfortunately, calculations made by Mellor and CRREL saltwater specialist Willy Weeks predict that an unprotected berg would dribble down to nothing long before it reached the Persian Gulf. Even the relatively short (two-month, 6,400-kilometer [4,000-mile]) hop from Antarctica to Western Australia would whittle the ice—and profits—down by half. So scientists are now busy figuring ways to cut their losses by insulating bergs in a saranlike substance that will minimize drag and exposure to warm currents.

"Icebergs are not stable, solid objects," says Weeks. "Anyone in their right mind stays away from them. Big pieces falling off without warning crush everything in their path. It's certainly possible that someday they could be used as a source for drinking water and swimming pools in Australia. But you're talking big money and big engineering."

So far, no one has even attempted to tow an intact berg, but they are, nonetheless, a very hot commodity. Chips of Greenland icebergs are sold for $5 a pound in Japan. The trapped air bubbles in a 100,000-year-old chunk of glacier make nifty crackling noises that have become the hit of the Far Eastern cocktail party circuit. As Wuori says, sometimes it's better to make use of the elements than to fight them □

SELECTED READINGS

"The cold truth about ice" by Richard Wolkomir. *National Wildlife*, December–January 1984.

The Physics of Glaciers by W. S. Paterson. Pergamon Press, Incorporated, 1981.

Dynamics of Snow and Ice Masses by Samuel C. Colbeck. Academic Press Incorporated, 1980.

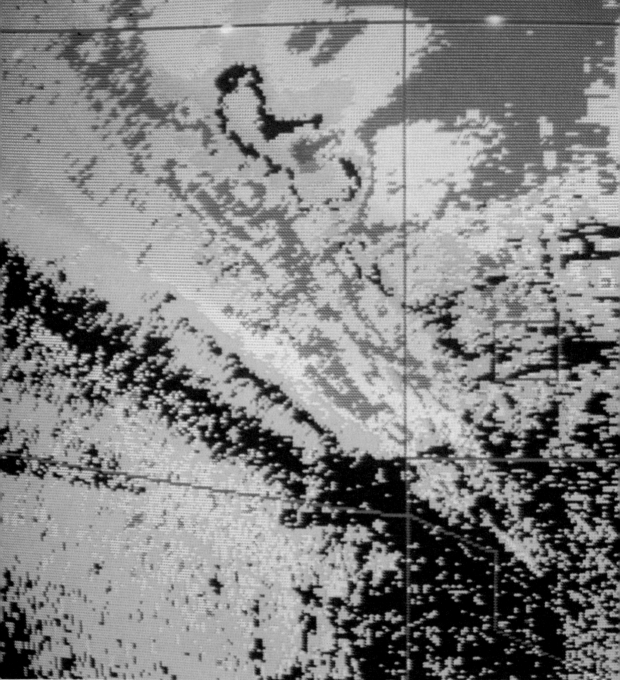

A New Eye on the WEATHER

by Don Witten

Steadily, the 9-meter (30-foot)-diameter radar dish rotated, sweeping the Oklahoma prairie with a narrow pencil beam of radio-frequency pulses. The pulses, measuring 10 centimeters (4 inches) in wavelength, penetrated a large thunderstorm some 80 kilometers (50 miles) to the southwest. Some of the energy was reflected back to the radar antenna by raindrops in the storm.

Inside the control room of the experimental Doppler weather radar at the National Oceanic and Atmospheric Administration's (NOAA's) National Severe Storms Laboratory (NSSL) in Norman, Oklahoma, the storm showed up as a color-coded mass on the radar's television screen.

Experimental Radar Well Received

Examining the monitor, NOAA researchers Don Burgess and Ken Wilk saw the closely spaced red and green areas near the edge of the storm. The bright green area identified wind-driven raindrops moving rapidly toward the radar. The adjacent red area revealed rain moving away from the radar.

The telltale sign was obvious to the two researchers: rain was spiraling in a tight vortex of wind that in minutes would become a tornado.

Speaking quickly into the telephone, Wilk briefed lead forecaster Joe Kendall at the Oklahoma City forecast office of the National Weather Service, just 32 kilometers (20 miles) away.

Kendall had already recognized the classic signs of a severe storm so common in the Oklahoma springtime. But the monitor of his 1957-vintage radar revealed no "hook echo," the rare sign of a forming tornado. This distinctive hook shape is produced sometimes by conventional radar signals reflected from swirling raindrops inside a storm.

Within minutes, however, Kendall issued a tornado warning. It was to be the first of 13 warnings issued that evening of May 11, 1982—all with an average lead time of 29 minutes. Each was based on the forecaster's skill and experience, plus the added information from the test Doppler weather radar at nearby Norman.

"Such advance tornado warnings, coupled with low false-alarm rates, will be possible on a routine basis once we make operational Doppler weather radars available to our forecasters," says Dr. Richard E. Hallgren, Director of the National Weather Service.

"Years of research and operational experience with experimental Doppler radars at NSSL and a few other locations have set the stage for the development of a new and sorely needed

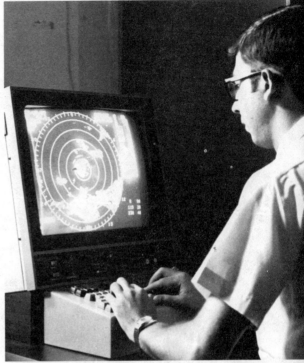

NOAA

Shades of color in a Doppler radar display help the operator determine the intensity of the observed storm.

national Doppler weather radar network," Hallgren adds.

The first operational units of a new Doppler weather radar are scheduled for installation in 1988. A national network of 160 new weather radars, many of them Doppler units, is planned for 1992.

The NEXRAD Program

This ambitious effort is called NEXRAD, the Next Generation Radar program. It is a joint undertaking by the U.S. departments of Commerce, Defense, and Transportation to replace the nation's weather radars. Most of these units are based on the vacuum-tube technology of the mid-1950's.

Organizations within these departments—the National Weather Service, the Air Weather Service, the Naval Oceanography Command, and the Federal Aviation Administration—provide the severe-weather warnings for the American public, military installations and areas of operation, and the civil airports and airways. All use weather radars that, for the most part, are becoming more unreliable and difficult to maintain with use and age.

Tony Durham (right), NEXRAD program manager, and his deputy, Sam Williamson, stand beneath a radar tower.

The Department of Commerce now has two U.S. high-technology firms under competitive contract to develop and verify the design of the new Doppler weather radar. One of these firms will be selected to produce about 10 Doppler units for early field evaluation and for full-scale production by 1988.

According to Tony Durham, the NEXRAD program manager, "joint use of the new radars by the three organizations will minimize the total number of units needed to provide coverage for this country. The 160 units planned under the NEXRAD program also will include

15 radars necessary for use at military installations overseas.''

There are now about 230 conventional weather radars deployed across the United States. Some 80 of these form a national network to provide coverage in most of the conterminous states for warnings of severe thunderstorms, tornadoes, significant rain- and snowfall, and hurricanes. Another 150 local warning radars supplement the national network in areas of high storm risk or where national network coverage is inadequate.

The Doppler Principle

Like conventional weather radar, NEXRAD's Doppler radar sends out repeated radio pulses and monitors their echoes from approaching storms as a way of measuring the storm's intensity and distance. By means of the Doppler principle, this new weather radar also will measure the shift in frequency of its signals reflected from raindrops. With this information, the direction of the storm's wind-drawn rain and, hence, the wind field can be detected to provide strong clues or signatures of developing tornadoes and other severe-weather phenomena.

Here's how the Doppler shift of rain-reflected radio signals occurs: Radio signals reflected from raindrops moving toward the radar are compressed due to the added effect of the forward movement of the rain. Conversely, radio signals reflected from raindrops carried by the wind away from the radar are stretched out. Compressed signals shift to a higher frequency; those stretched out shift to a lower frequency.

The most common example of the Doppler principle, named after the 19th-century physicist Christian Johann Doppler, is the pedestrian's observation of an automobile horn blast that rises in pitch as the vehicle approaches and drops as it passes.

Computers Enhance Capabilities

The new Doppler weather radars being developed under the NEXRAD program are total, computer-aided systems. They will process and display the reflected radar signals, including the Doppler data, into meaningful information. This will allow the forecaster to identify the type of storm, project where it is moving, detect early signs of tornadoes, and identify the expected amounts of rainfall.

Computer programs will be incorporated in the NEXRAD Doppler radar systems to estimate the amount of rain falling in severe storms and match that with the shape or capacity of local watersheds. This capability will allow meteorologists and hydrologists using the new Doppler radars to improve their forecasting of flash floods, which claim an average of 200 American lives annually.

Ten-Centimeter Signals Have Benefits

The NEXRAD Doppler weather radar will transmit signals at the 10-centimeter (4-inch) wavelength, the frequency judged best for use in a general-purpose weather-radar network. At this frequency, signals focused in a pencillike beam by a 7-meter (24-foot)-diameter dish antenna can penetrate nearby storms to spot those farther away, out to 160 kilometers (100 miles).

"Ten-centimeter radar signals are attenuated 16 times less than those at half the wavelength by atmospheric constituents such as water droplets, dust, and flocks of birds and insects," says Dr. Joe Friday, deputy director of the National Weather Service.

Although the limitations of shorter-wavelength signals have been known for some time, they were demonstrated tragically in 1976 by

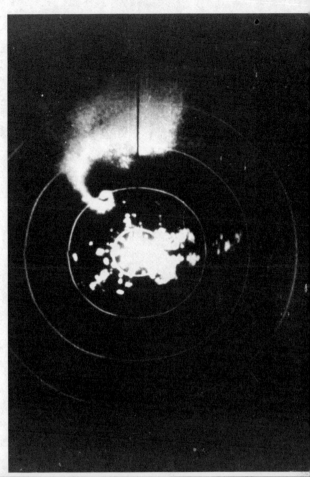

Right: The "hook" shape associated with tornado formation is seen on a conventional radar screen. Below: A meteorologist studies swirling wind patterns that were made with Doppler radar.

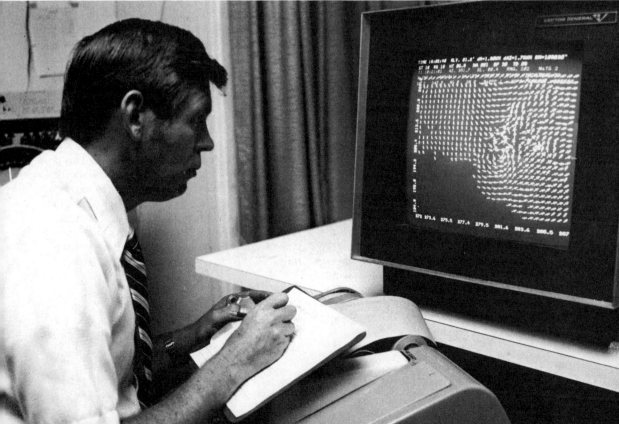

NOAA

Researchers monitor changing weather patterns inside the control room of the experimental Doppler radar at NSSL.

the crash of a commercial aircraft near Atlanta, Georgia. A short-wavelength radar in the cockpit indicated that one area in a line of thunderstorms ahead of the plane was thin. The pilots flew into that area to avoid the storm.

"It turned out that the radar had detected only the leading edge of an intense thunderstorm laden with rain. Although the plane made it through the storm, water stalled its engines. Over 70 people were killed in the crash," Friday recalls.

He added that the storm-penetrating capabilities of the 10-centimeter weather radar permit the use of only a quarter of the number of 5-centimeter units that would be required to cover the U.S.

NSSL began using a 10-centimeter-wavelength Doppler weather radar to study wind fields within severe storms as early as 1971. Since then, the existence of two useful severe-storm features with characteristic signature patterns have become apparent.

The first is that of a mesocyclone—a vertical column of rising rotating air that measures 10 to 19 kilometers (6 to 12 miles) across. This signature is seen first in the mid-part of the storm and descends to the cloud base as a tornado forms. Generally, it is in the region of the mesocyclone that a severe storm contains the largest hail, strongest winds, and tornadoes.

The second pattern, the tornado vortex signature, is produced by the tornado itself. This is the location of rapidly changing wind speeds contained in a region much smaller than the parent mesocyclone of which it is a part. Although detectable at relatively short range, this signature has the greatest utility for identifying a tornado's location within 0.8 kilometer (0.5 mile).

Radar Test Results Prove Positive

By 1977 the civilian and military users of weather radar pooled their talents to conduct the first operational test of a 10-centimeter wavelength Doppler weather radar at NSSL. Called the Joint Doppler Operational Project (JDOP), it had the added mission of outlining specifications of a new generation of weather radar for national network use. JDOP participants included NOAA's National Weather Service and Environmental Research Laboratories, the Air Force's Air Weather Service and its Geophysics Laboratory, and the Federal Aviation Administration.

Beginning with the 1977 severe-weather season, the JDOP conducted three years of intensive operational research with NSSL's experimental Doppler weather radar. During this effort, the JDOP meteorologists detected 70 percent of the tornadoes that occurred within range of their radar. The undetected twisters were the weaker and short-lived tornadoes that pose a lesser threat to public safety.

As part of their evaluation, the JDOP meteorologists prepared severe thunderstorm and tornado advisories based on the Doppler weather radar observations. These were compared to the public warnings issued for the same storms by the National Weather Service Forecast Office at Oklahoma City, which bases many of its warnings on public reports of tornado activity near or on the ground. A cadre of dedicated volunteer tornado spotters provide most of these reports.

"Lead time for the tornado advisories prepared with the Doppler information was 20 minutes on the average, compared with just over 2 minutes for the public warnings," says Dr. Edwin Kessler, NSSL director.

"In addition, the Critical Success Index for the test advisories was nearly twice as high as the public warnings. This index gives credit for a high probability of tornado detection and low false-alarm rates," adds Kessler.

During the JDOP tests, aircraft flew through the Oklahoma thunderstorms being monitored by the Doppler unit to make direct measurements for comparison. It was determined that Doppler weather radar can detect dangerous gust-fronts, wind shear, and in-storm turbulence. Consequently, NOAA researchers are confident that Doppler weather radars can also be used as a warning tool for aircraft.

Dr. Friday recalls one significant event during the JDOP that convinced the Department of Defense (DOD) that it should be involved in the NEXRAD program. As an Air Force officer then, he briefed Secretary of Defense Harold Brown on the following incident:

On May 2, 1979, a JDOP advisory was made available to Vance Air Force Base in Enid, Oklahoma. It provided two hours' advance notice that a severe storm with a mesocyclone was headed in the direction of the military installation. Based on this advisory, 52 T-38 training aircraft were moved inside hangers. As the storm passed, it spawned tornadoes on either side of the base and deposited large hail across the Air Force facility.

"Prevention of damage to the aircraft, valued at about $3 million each, saved nearly enough money to pay for DOD's participation in the NEXRAD program," Friday says.

Back to the Beginning

Radar—Radio Detecting and Ranging—dates back to the 1886 work of Heinrich Hertz, who first observed reflections of radio signals from solid objects.

The first practical radar was developed in the early 1930's by Sir Robert Watson-Watt of Great Britain and used to direct aircraft during World War II.

Although Doppler data were present in the first radars, they were considered a nuisance. One of the challenges during World War II, Dr. Friday remembers, was to eliminate this extraneous "weather noise" from the military radars.

By 1947, the U.S. Weather Bureau (now called the National Weather Service) recognized the value of radar for storm monitoring and began using converted military radars. The Weather Bureau's first radar designed specifically for weather monitoring was procured in 1957. Most of the weather radars in use today are of this vintage, although a limited number of today's weather radars were built in 1974, based on an improved design.

"It is ironic that the weather radar intended for use in the 21st century is based on the Doppler weather noise designed out of the very first radars developed," Friday says □

SELECTED READINGS

"The eyes of NEXRAD" by Robert Teitelman. *Forbes*, March 26, 1984.
"FAA [Federal Aviation Administration] planning to deploy weather radar. *Aviation Week and Space Technology*, January 2, 1984.
"First measurements of size and velocity of a violent tornado [use of Doppler radar by the National Severe Storms Laboratory] by L. Purrett-Carrol. *Weatherwise*, June 1982.

COLUMBIA:
A Glacier in Retreat

by Daniel Jack Chasan

The Alaskan sun is shining on the sheer face of Columbia Glacier, a 6-kilometer (4-mile)-wide river of ice that flows from the snow-packed Chugach mountain peaks behind it to the salt water of Prince William Sound. The seaward end of the glacier is a mass of crenelations (indentations), a forest of Gothic spires, the rock formations of the Southwest carved in ice, gleaming a pale turquoise blue.

Perhaps halfway up the face, 46 meters (150 feet) above the water, sunlight pierces a thin window in a vertical ridge, making the glacier glow a deeper, purer blue, like a piece of hand-blown glass held up to the light.

Suddenly, a tower of ice to the east gives way and topples into the water, then another and another, the great shards coming down through a rising plume of spray. The spray rises past the

Harald Sund

The Columbia Glacier, a massive river of ice in Alaska, is beginning to undergo a drastic retreat.

that fills almost the entire length of the fjord (an inlet bordered by steep cliffs) it formed. The fjords along the Alaskan coast were all carved by ice, but the glaciers that carved them have long since retreated; most of those that remain lie near the heads of the fjords.

Columbia, in contrast, completely fills its furrow in the coast. It is the last of its kind in Alaska, perhaps the last for hundreds of years to come. If it retreats inland, it will leave a new fjord about 40 kilometers long. Right now, it seems on the verge of starting to do just that. Columbia is "calving" so much ice during the summer and fall that it is receding during these times at a rate of more than 4 meters (13 feet) a day. And that seems to be accelerating. In 1981 it lost about 0.5 kilometer (0.3 mile) of ice. By 1984 it lost ice 8 times faster.

About 2 million years have passed since the beginning of the Pleistocene Epoch, during which glaciers covered much of the northern half of the continent numerous times, and roughly 10,000 years have passed since the last Ice Age ended in Alaska. No one is certain exactly when Alaska's surviving tidewater glaciers were formed, and no one knows how often each of them has advanced and retreated in the intervening millennia. All of the major ones have done at least once what Columbia Glacier appears ready to do now, the first of them more than 1,000 years ago and the last of them around the turn of the century. This will be the first drastic retreat to begin since then, and the first one ever that scientists can study in depth.

More than Academic Interest

Since most tidewater glaciers follow the same physical laws, Columbia Glacier will help glaciologists understand the others. It will provide insight into a process that has exposed the fjords of the world. Its collapse may be some kind of a model for the possible retreat of the West Antarctic ice sheet, which rests on the seafloor, its seaward edge supported by floating ice shelves.

One of the scientists studying the Columbia is Mark F. Meier, the senior glaciologist in the U.S. Geological Survey (UGS). Meier, who works out of the Survey office in Tacoma, Washington, believes that the glacier probably has receded past the point of no return. He also

window of deep blue and almost to the top of the ice. Then it subsides, and in its place, in an upwelling of darker spray, emerges a great, vertical, almost black-blue berg. If the glacier face is pale turquoise, this is dark lapis, and it heads steadily skyward like a breaching whale. It come up and up until its tip is level with the glowing window, then slowly sinks down again, lists to the horizontal, and drifts out into the brash (floating ice).

The Last of Its Kind in Alaska

There aren't any crows on this dramatic, mountainous coast, but as the raven or eagle or airplane flies, Columbia Glacier is less than 40 kilometers (25 miles) from the oil port of Valdez. Sixty-eight kilometers (42 miles) long, it is the only one of 52 Alaskan tidewater glaciers

Scientists from the U.S. Geological Survey study the glacier's terminus from a moraine on Heather Island.

believes that its retreat may have much more than academic interest. He has seen 100,000-ton icebergs created by the glacier, and has personally tagged and measured a couple of dozen that he has estimated at up to 50,000 tons. (Icebergs extending 30 meters [100 feet] and deeper usually cannot get over the moraine, the ridge of rock and gravel that the glacier had bulldozed before it down the fjord and out into the bay. The larger icebergs are trapped until they break up or melt down to a more mobile size.) Most of the icebergs created by the glacier drift out of Columbia Bay, turn west, and are soon out of sight and out of mind. Some, however, turn east and are soon in the oil-tanker lanes leading to and from Valdez.

No one knows exactly what would happen if a tanker hit an iceberg out there, and no one is eager to find out. Meier says that in the mid-1970's, before the Valdez terminal of the Trans-Alaska Pipeline was in operation, he gave presentations to warn about the increased possibility of ice hazards. The frequent response, he says, was that "they would say, 'Thank you,

Dr. Meier,' and dismiss me." Then, in the summer of 1981, the glacier started calving more rapidly, and enough ice appeared in the shipping lanes so that tankers were delayed more often. When Meier arrived in Valdez at the end of the summer, "they welcomed me with open arms." Actually, while the oil companies and the Coast Guard are well aware of the ice out there and give every sign of taking it seriously, no one seems terribly worried. The Coast Guard traffic-control center in Valdez does keep a log in which someone writes down every report of floating ice, and the Coast Guard relays those reports from one vessel to another.

Entries in the ice log are short and dry, but they suggest that the ice is not a figment of anyone's imagination. The fishing vessel *Ricky* reports ice "one story high and the size of a two-bedroom house." An Alaska state ferry and a tanker report chunks "ranging in size . . . to 15-foot [5-meter]-high and 50 foot [15-meter]-long ones." The ferry reports "southbound traffic lane blocked by ice." The tanker "is maneuvering to avoid ice."

Calving Rate Tied to Water Depth

Whether or not the floating ice ever becomes a serious navigational problem, the USGS was certainly right to predict it would be there. How did Meier and his colleagues know that Columbia Glacier would start dropping more ice into the water? Meier explains that in the early 1970's the tip of the glacier was resting on a moraine. It seemed likely, though, that in a period of low ice flow the glacier would retreat from the moraine. When that happened, the depth of the water against its tip could change from 21 to 213 meters (70 to 700 feet). It so happens that the rate of calving appears to be proportional to the depth of the water. As soon as the glacier was off the moraine, the rate of calving would increase dramatically, and the glacier would not be able to recover.

A Survey glaciologist named Austin Post intuitively grasped the relationship between calving and water depth by studying the behavior of numerous tidewater glaciers. Post has officially retired from the USGS but still works

for it part-time, and he could be found near the glacier in the summer on the research vessel *Growler,* a converted Navy launch with steel around its fiberglass bow for protection from floating ice.

There is Post in rubber boots, striding across the tidal mud of Heather Island, just across the icy water from the glacier face. He carries a camera bag slung over one shoulder and a battered wooden sextant (an instrument used to measure angular distances) case in one hand. The sextant is his own; it's a Tamaya, made in Japan, and he bought it as a young sailor in China at the end of World War II. The glacier looms over the ice-choked water to his right, gleaming blue in the morning light, but he doesn't give it a glance. He's in a hurry. He wants to reach an observation point on Heather in time to measure a couple of points on the glacier before they break off. The ice has been breaking off rapidly, so, feeling pressed, he climbs rapidly to an orange-wrapped pipe where he takes out his sextant and sets to work.

James Peter Stuart

When chunks of ice choke the waters of Columbia Bay and float into shipping lanes, oil tankers are often kept from traveling.

Heather Island

Heather Island is green with spruce and underbrush, a little hemlock, and a variety of low plants that grow from the muskeg and among the rocks. The end closest to the glacier is all damp sand and loose rock. On the beach, like so much driftwood, lie slowly melting chunks of glacial ice. More ice floats in the water offshore; when the glacier calves, huge pieces cascade down into a fountain of spray, the first waves spread quickly through the offshore ice, and the floating pieces click against each other like a forest of bamboo wind chimes.

The mound of rocks on the northern end of Heather Island is the glacier's moraine. When Post and other USGS staff began studying the Columbia Glacier a bit closer in 1973, the end of the glacier rested there. In late December of 1978 it broke away from the island. By now, it is a few kilometers away.

A Twist to Post's Conclusion

Standing in the pilot house of the *Growler,* looking out at the open water between Heather Island and a deep embayment in the glacier, Post says that the embayment is probably the deepest it has been since at least the 19th century. He would guess that that portion of the glacier did not rest on Heather Island in the 18th century, and a chart from Captain George Vancouver's voyages of exploration in the 1790's indeed shows the island as a separate entity with open water all around. Post reaches to the pilot house bookshelf, pulls out a photocopied edition of Vancouver's journals, and finds the chart. There is Heather Island, sitting out there by itself. There is no sign of the glacier—which has less to do with climatic conditions of the late 18th century than with Vancouver's perceptions. Post, who has collected photocopies or old typeset copies not only of Vancouver's maps but also those of Cook, Kotzebue, La Pérouse, and other explorers, thinks the captain did a splendid job of mapping but a wretched job of comprehending the nature of glaciers.

Post had used aerial photography to observe the upstream portions of glaciers, not just the termini (ends). He found that when glaciers "surge" forward, it isn't because of avalanching ice and snow. He also noticed some things about calving glaciers.

The tidewater glaciers that were discharging the most ice were also putting out the largest icebergs. Big icebergs need a lot of water to float in; the glaciers calving the most ice were also the ones that ended in the deepest water. Post deduced that there must be a connection between the depth of the water and the rate of calving. Columbia Glacier was a key. Here was a big tidewater glacier that was producing relatively small icebergs and relatively little ice. If Post's idea was correct, the glacier should end in shallow water. Yet soundings near its terminus suggested 185 meters (600 feet).

Soundings Provide Vindication

Post wasn't convinced. Those deep soundings hadn't been taken right at the glacier's face, where the moraine shoal, if one existed, would have to be. The first thing Post and other USGS personnel did in 1973–74 was to take soundings close to the glacier face. It turned out that the water was just as shallow as Post had thought—and that there was the potential for a drastic retreat.

The next step was to mobilize a field experiment and take various measurements, including the thickness of the ice upstream. To do that, the USGS took surveys with monopulse radar that bounced signals off the bedrock below the ice. The radar and subsequent computer analysis showed that the ice was very thick. (Sixteen kilometers [10 miles] above the glacier's terminus, the ice was 520 meters [1,700 feet] thick above sea level and 400 meters [1,300 feet] thick below.)

It seemed that Post's idea was correct. Presumably, all the tidewater glaciers that have been stable for any length of time have rested on moraines or at least in shallow water. No one knows why Columbia Glacier has apparently clung to its moraine so much longer than any of the others. As Meier puts it, "Someone had to be last."

Glacial Activity Not Totally Predictable

Post and USGS staff have been studying Columbia Glacier ever since. They have seen it back away from Heather Island, back away from all but about 300 meters (1,000 feet) on the crest of its underwater moraine. When its entire terminus is lying in deep water, it is believed, nothing will halt its rapid "collapse." Post and Meier both expect that to start happening soon. Lawrence R. Mayo and Dennis C. Trabant of the USGS's Fairbanks office, who have been

A jagged pillar of ice crashes into the water in a dramatic example of Columbia Glacier's frequent calving.

The glacier grinds and bulldozes rock and mounds of earth as it moves, sculpting landforms as it goes.

studying the upper portions of the glacier from helicopters, aren't entirely sure. Among other things, they say there is a huge reservoir of ice farther up, and even if the glacier backs off its moraine entirely, the surplus ice may flow down and keep it replenished. "We've had a lot of arguments with the Fairbanks people" over that, Post says. He concedes the ice reservoir is there but feels it probably won't come riding over the hill in time to save the glacier from collapse. But the only real question, it seems, is when, not whether.

Not that Post or anyone else claims to know exactly how the glacier works. No one can fully explain its normal variations in length or the waxing and waning of various sections of its face. Even the "law" that correlates calving with water depth isn't absolute. It works only if you look at the whole face of the glacier and average its behavior over the course of a whole year. It tells nothing about any single point on the glacier's face.

Nor has anyone been able to predict from day to day what Columbia Glacier will do next. Standing on the rocks of Heather Island, the rotor of his research helicopter thundering behind him, Trabant says, "You never know what it's going to do. You come back and wonder what it will have done since last time. That's part of the allure."

A Special Vantage Point

There is Post climbing out of a dory (a boat with flaring sides, a high bow, and a flat, narrow bottom) onto the lichen-encrusted rocks just west of the glacier, looping the dory's long rope around boulders set safely back from the water. The footing on the rocks is tricky, but he scrambles to a high point carrying a hammer, two strips of wood, and a piece of waterproof orange cloth. On the high point stands one of the USGS's surveying marks, a vertical pipe wrapped with orange cloth that has faded almost into invisibility. Post tacks his new orange cloth to the two strips of wood, then lashes them with copper wire to the upright pipe, creating a kind of square sail that will be visible from several kilometers away.

From these rocks, one can see the western spires of the glacier, including the part that still rests on the crest of the moraine. The glacier isn't calving now, but no one ever trusts the calm. Even when no ice is falling, people don't take boats very close to it.

James Peter Stuart

An awesome wall of ice towers over the research vessel Growler, *which is used primarily by U.S. glaciologists.*

When Post wants to take soundings at the glacier face, he uses a radio-controlled skiff. Today, as on most other days, such prudence makes sense: soon, to the east, great chunks of ice will cascade slowly down the face, then a whole tower will fall, sending up a fountain of spray that looks like the grand finale of a fireworks display on the Fourth of July.

This particular vantage point, Post says, "was one of G. K. Gilbert's photostations in 1899." He explains that Gilbert, a USGS geologist who came to Alaska with the turn-of-the-century Harriman expedition, was the first scientist to study Columbia Glacier. A whole series of people has looked at the glacier since

then. In the 1930's most of the work was done by William O. Field. Field "still comes to Alaska," Post says. "He hasn't come back to Columbia, though. That kind of hurts my feelings. It was a favorite of his. And it's the last."

Field does hope to return to Columbia Glacier. For those who care, it is special □

SELECTED READINGS

"Glacier on the move" by John Berendt. *GEO*, September 1983.

"Predicted timing of the disintegration of the lower reach of the Columbia Glacier, Alaska" by Mark F. Meier, L. A. Rasmussen et al. USGS Report 80-582, 1980.

ENERGY

ENERGY

"The U.S. isn't prepared to cope with an energy emergency such as was experienced in 1973," pronounced Senator James McClure (R-Idaho) in February 1983. The basis for his observation was a new study by the U.S. General Accounting Office, which found that in the ten years since the Arab oil embargo, the federal government still had not coordinated state and federal policies to deal with major energy-supply disruptions, nor had it addressed adequately how domestic stocks of oil would be distributed. Because of the country's continuing dependence on foreign-oil supplies, continuing violence in the Middle East, and the fact that reserves of U.S. oil were again falling, this report renewed concern within Congress and the energy community about whether another energy crisis was brewing. Meanwhile, the year was marked by mixed news about fossil-fuel reserves and some hopeful signs concerning renewable energy sources.

FOSSIL FUELS

At 15 million barrels per day, petroleum use in the United States remained close to 1982 levels, with imports again providing almost 28 percent of the total. However, while the rate of oil use remained stable over a two-year period, U.S. oil-reserve levels fell 5.3 percent. One reason for this drop was the low rate of discoveries of new oil reserves. There were 11.2 percent less discoveries reported than in 1982.

Discoveries of natural gas also fell during the year—16.1 percent from 1982. This was accompanied by a roughly 5 percent decline in U.S. consumption of natural gas compared with 1982.

In contrast, estimates of U.S. tar-sand resources (deposits of tarlike crude oil trapped in sand) rose in 1983 to 54 billion barrels, 10 times the previous estimate. The new estimates, prepared for the Interstate Oil Compact Commission, also identified 1,100 heavy-oil deposits. High viscosity makes both tar sands and heavy crude harder to extract than

The Salton Sea geothermal-electric project in California harnesses Earth's interior heat to produce electricity.

Union Oil Company

conventional light crude. However, it is expected that as reserves of light crude disappear, these viscous alternatives may become the predominant feedstock of liquid fuels for transportation.

Most fossil fuels—such as petroleum and coal—are burned to release their stored energy. But now there may be an alternative to the burning method—namely, fuel cells, in which hydrogen (from natural gas or naphtha) combines with oxygen from the air to generate electricity through a chemical reaction. An experimental 4.8-megawatt electric-generating station containing stacks of such cells began operation in New York City in December 1983.

NUCLEAR ENERGY

The year 1983 again saw the nuclear energy industry in very poor health. Since 1972, more than 100 power plants have been canceled— many of them after having been partially completed—and no new plants have been ordered since 1978. Probably most demoralizing to nuclear-energy advocates was the death knell sounded for the Clinch River Breeder Reactor project. After 12 long, agonizing years and an investment of about $1.5 billion, Congress refused to sink another penny into the controversial project. When the planning for the experimental reactor began, it was expected to cost $700 million and take no more than nine years to complete. But due to spiraling design and construction costs, most recent projections estimate that the plant would ultimately cost $3.2 billion and not be completed before 1990—at which time some experts feel it would already be obsolete, and furthermore, not needed by one of its expected prime power consumers, the Tennessee Valley Authority.

Recovery of the crippled Three Mile Island Unit-2 reactor also proceeded slowly during the year. Analysis of rubble removed from the reactor's badly damaged core revealed a radioactive mixture of crumbled fuel, shards of steel, and flakes of metal "skin" that once held the uranium-based fuel pellets in place. The owners of the plant, which has already been out of operation since 1979, now estimate that cleanup will take at least another five years to complete.

Further aggravating the nuclear industry's prospects was an April 20 ruling by the Supreme Court. At issue was whether California could ban nuclear-plant construction until the federal government approved a scheme for permanently disposing of high-level radioactive wastes. The Court ruled that because the California ban was ostensibly motivated by economics, it was indeed legal. By year end, seven more states had linked approval of future plants to resolution of the radioactive-waste issue.

SYNTHETIC FUELS

On June 30, three years to the day after the U.S. Synthetic Fuels Corporation (SFC) was created, the energy bank made its first legally binding award of financial assistance. The SFC was set up in 1980 to help with the development of a commercial industry to convert coal and oil shale into synthetic petroleum and natural gas. The first award went to the Cool Water coal-gasification plant in Daggett, California. As a loan guarantee, the $120 million award would not actually result in any government spending unless the price of the synthetic natural gas that the plant produces falls below an agreed-upon level. ■ A few months later, Union Oil Company completed a $260 million plant near Parachute, Colorado, to extract 10,000 barrels of oil daily from shale rock. Considering the product superior as a jet-fuel feedstock, the U.S. Department of Defense quickly signed up to buy the synthetic oil at $42.50 per barrel—$10.00 to $15.00 more per barrel than conventional oil. Later the SFC offered price supports, initially guaranteeing the company $60.00 per barrel for its oil.

RENEWABLE ENERGY RESOURCES

It has long been believed that an efficient process for splitting water into its component parts— hydrogen and oxygen—could make hydrogen, a promising automotive and jet fuel, inexpensive enough to foster its widespread use. A new process harnessing sunlight to split water was announced in 1983. Bell Laboratories in Murray Hill, New Jersey, used an improved liquid-junction solar cell that had an energy conversion efficiency of 16.2 percent, a significant improvement over earlier cells.

Harnessing Earth's interior heat to produce electricity also progressed during the year. January 19 marked the successful start-up in southern California of the 10-megawatt Salton Sea geothermal-electric project. Operated by Union Oil Company, the project sits atop a major geothermal reservoir. The new plant pumps hot, salty water up to the surface from wells 900 to 1,800 meters (3,000 to 6,000 feet) deep. At the surface, the water flows through vessels in which its temperature and pressure are lowered to encourage corrosive gases and minerals to settle out, forming a sludge that can be filtered out. ■ Meanwhile, construction began on a Heber, California, geothermal power plant, in which water once thought to be too cool to be useful in generating electricity will be tapped from wells up to 2,400 meters (8,000 feet) deep. Heat from the water—between 150° C and 200° C (300° F and 400° F)—will pass into pipes containing a solution of mainly isobutane, heating it to produce steam to run turbines and generate electric power.

JANET RALOFF

Peaceful Nuclear Explosives

by Malcom W. Browne

At precisely 9:00 A.M. on September 24, 1983, a jolt registering more than 5 on the Richter scale hit the vicinity of Astrakhan, a big Soviet port on the Caspian Sea at the mouth of the Volga River. (The Richter scale is used to express the magnitude of a seismic disturbance, with 2 indicating the smallest earthquake that can be felt. An 8.5 quake would be very severe.) In laboratories round the world, seismographic needles jiggled furiously, tracing jagged peaks on recording paper and alerting seismologists to an earth tremor strong enough to damage even reinforced concrete buildings.

Before scientists in Sweden, Britain, and the United States had time to evaluate their readings, five more Astrakhan shocks, at five-minute intervals, jarred their sensitive instruments. All six quakes apparently occurred no more than 80 kilometers (50 miles) east of Astrakhan, perhaps much closer.

A map shows Soviet peaceful nuclear explosion sites as dots, the Astrakhan blasts as a black asterisk, and nuclear weapons test zones as yellow rectangles.

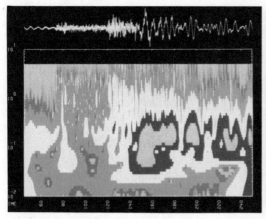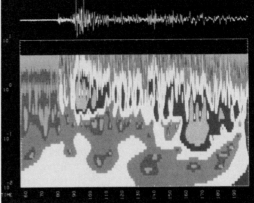

Color-coded seismograms, showing the energy transmitted by seismic waves from earthquakes, have distinct patterns that allow scientists to distinguish between a natural earthquake (at left) and a nuclear blast (at right).

Blasts for Peaceful Purposes

Had the quakes caused havoc in the ancient city? The Soviets remained silent about the tremors. But a few hours later, the Seismological Institute at Sweden's Uppsala University reported that nuclear explosions—not earthquakes—had caused the Astrakhan shocks. The city seemed safe. An underground nuclear blast can register an alarming Richter number without causing the damage that a natural earthquake of the same magnitude would. Astrakhan still stood.

But why would the Soviets set off nuclear blasts so close to a major city? Experts in the U.S. monitoring Soviet compliance with nuclear test treaties think they know why: the explosions were not nuclear weapons tests but blasts designed to create enormous underground storage chambers for natural gas. They helped confirm that the Soviet Union is finally realizing a long-standing dream once shared by the United States—the peaceful application of the atomic bomb.

That is the view of geologist Milo Nordyke, director of the Treaty Verification Research Program at Lawrence Livermore National Laboratory in California. Nordyke and his team at Livermore play a key role in monitoring Soviet nuclear explosions, by assembling evidence from global networks of seismological stations, atmospheric sensors, satellites, hints in the Soviet technical literature, and hearsay from Soviet emigrants.

The latest explosions, Nordyke said, demonstrate that the Soviet peaceful nuclear explosives (PNE) program is no longer merely experimental but has reached full-scale industrial development. "Nuclear explosives," he says,

"are now the standard tools of a mature and tested Soviet technology."

Since 1965 the Soviet Union has detonated more than 70 of these nonmilitary devices, 16 of them in 1983 alone, for a variety of purposes. Among them:

• To create underground cavities like those at Astrakhan to be used not only for storing fuels but also as permanent disposal sites for chemical wastes such as the brine produced by oil fields. Some nuclear-excavated cavities may also serve as dumps for radioactive waste.
• To generate seismic waves that help geologists prospect for minerals and fossil fuels, and learn more about the Earth's crust and mantle.
• To stimulate production in gas and oil fields by shattering rock and releasing trapped pockets of gas and oil.
• To extinguish stubborn fires in gas wells.
• To free oil from shale and sand deposits, and perhaps to tap heat from the Earth.

The Plowshare Program

Project Plowshare, the U.S. program to develop peaceful nuclear explosives, collapsed. Starved of congressional funds and beset by increasingly stringent antinuclear legislation in Colorado and other states, Project Plowshare was quietly scuttled by the Department of Energy in 1977 after it had produced only a dozen nuclear-test explosions in two decades. Since the last Plowshare blast in 1973, all U.S. nuclear tests have been for military purposes.

As recently as the early 1960's, the potential of peaceful nuclear explosives seemed unlimited. The United States, as well as the Soviet Union, was considering schemes for rearranging landscapes on a grand scale by rerouting rivers, creating artificial lakes, and piercing mountain ranges. In 1964, for instance, President Johnson signed a bill authorizing a study of the possibilities for using nuclear explosives in building a new sea-level waterway to replace the Panama Canal. But rising fears about the dangers of fallout, and the state laws banning nuclear explosions put an end to any plans for nuclear excavation by the United States.

Most of the opposition to the Plowshare program came from environmentalists, who contended that the testing was likely to contaminate the atmosphere, the oceans, subterranean water tables, and the entire food chain with dangerously radioactive substances. In fact, Plowshare explosions conducted at ground level did add to the atmospheric fallout problems of the 1960's, and several of the underground blasts vented radioactive gas.

Some people argued that any engineering project requiring large-scale blasting could be carried out with conventional rather than nuclear explosives. In any case, they maintained, some of the Plowshare explosions failed to achieve even their stated goals.

Supporters of a peaceful nuclear-testing program still feel that it was never given a fair chance in this country. They concede that radioactive debris was a problem in some Plowshare tests, but insist that better-engineered nuclear devices and better selection of explosion sites would have avoided the difficulties. Barred from carrying out enough tests, Plowshare scientists say they were unable to persuade critics of the usefulness of nuclear blasts in freeing oil and gas from the earth—technology the Soviet Union has apparently succeeded in developing.

Creating Storage Cavities for Gas

The Soviets have shown little hesitation in experimenting with atomic bombs as engineering tools. In 1965, for example, they set off a nuclear device near Semipalatinsk that created a crater large enough to span the river. The lip of the crater formed a dam that is now used as part of a flood-control project.

Soviet engineers once also seriously considered the use of nuclear explosives to divert water into the landlocked Caspian and Aral seas, both of which are drying up. Those plans seem to have been abandoned, however, probably because of the radioactive fallout produced by such explosions.

For other applications, Soviet experts clearly regard nuclear blasting as not only economical but also safer than conventional methods. Nordyke is particularly impressed by the rapid pace of the program at Astrakhan, which has been driven, in part, by political and economic necessity. "The Astrakhan natural gas field is not even in production yet," he says,

An artist's view of an underground cavity for storing gas and liquid hydrocarbons created by a Soviet nuclear blast.

George Kelvin/© 1983 Discover Magazine, Time Inc.

"but as part of their preparations the Soviets have stockpiled the bomb-excavated cavities they expect to need for storage and disposal."

From 1980 to 1983, the Soviets blasted out 13 underground storage cavities at Astrakhan. Each was created by a nuclear explosive with a yield of 15 kilotons—the equivalent of 15,000 tons of TNT (about the explosive power of the Hiroshima bomb). Livermore experts think the Astrakhan devices were detonated at depths of as much as 1,100 meters (3,600 feet) in mixed layers of salt and rock, and that each explosion created a cavity some 50,000 cubic meters (1.8 million cubic feet) in volume, with a diameter of roughly 45 meters (150 feet). "That's a lot of storage space," says Nordyke. "Furthermore, it's cheap, and since it's in salt, it's leakproof. It could even be used for storing the helium recovered as a by-product from natural gas production." (For lack of storage space, U.S. gas companies dump much of the irreplaceable helium that could be recovered from natural gas.)

Exploitation of the rich Astrakhan natural gas field, discovered in 1973, became an urgent necessity in 1978 when Iran cut off gas exports to the Soviet Union. To keep the towns and factories of the southern Caucasus supplied with fuel, Soviet authorities were forced to divert gas from the pipeline network that was recently built to supply West Germany, France, Austria, and Eastern Europe. That meant losing income from gas sales to Western Europe. What was needed was more gas, which could be supplied by tapping the newly discovered Astrakhan gas field. But that posed difficulties of its own.

According to Iris Borg, one of Livermore's experts in Soviet gas technology, the gas under the salt domes of the Astrakhan field is unusually "sour"—it is at least 25 percent corrosive, poisonous hydrogen sulfide and as much as 13 percent carbon dioxide, both of which must be removed before the gas can be used. Moreover, as it escapes the high pressures deep underground, the gas from Astrakhan's wells separates into two parts, both valuable: natural gas, and a liquid mixture of light hydrocarbons known as gas condensates. The liquid must be allowed to stand long enough for the gas remaining in it to bubble out.

Soviet engineers realized that the purification and standing processes required for raw Astrakhan gas would be a bottleneck in the delivery system. The processing and settling facilities would not be able to keep up with peak production from the wells, and vast, corrosion-proof storage spaces would be needed to handle the excess output. Engineers knew just how to create those spaces: by using nuclear devices to blast cavities in salt beds.

Controlling Radioactivity

While diversifying the uses of PNE, Soviet physicists have gone a long way toward reducing the radioactivity that results from nuclear explosions. Like their U.S. counterparts, they have mastered the art of building devices that produce very little tritium—a radioactive isotope of hydrogen that is extremely dangerous to human beings. This is especially important for nuclear explosives used in a gas or oil field, because tritium can react chemically with oil or gas to form radioactive hydrocarbon molecules, from which it cannot be readily separated. As a result, the gas or oil released by a tritium-rich blast would be dangerously radioactive.

The peril of releasing tritium effectively rules out thermonuclear (hydrogen) bombs as peaceful nuclear explosives, because they are partly fueled with tritium. Even fission (atomic) devices, which produce very little tritium when they explode, must be carefully designed to keep neutron radiation from their fireballs to a minimum. The problem is that strong bursts of neutrons can produce tritium and other radioactive substances in the surrounding soil.

Although nuclear explosions detonated above or on the ground produce radioactive fallout—solid particles of debris, consisting mostly of vaporized metal (from the bomb components) that has cooled and condensed—deep underground blasts do not. Underground, the radioactive particles condense on the walls of the new cavity, mix with molten salt and rock, and settle to the bottom as a puddle. The puddle eventually cools and hardens to a glassy solid, sealing off the radioactive material.

Thus, say the experts, a French housewife need not fear that Soviet gas, freed by a nuclear explosion or stored in a cavern blasted out by one, and then pumped to her kitchen from Central Asia, will make her bread radioactive.

Seismic Sounding

The Soviets' PNE deep-seismic-sounding project has enabled geologists to find mineral resources and geological formations that were previously unknown. The scientists monitor sound waves from a nuclear device set off below the surface, determining how the waves are bent, absorbed, reflected, or modified as they

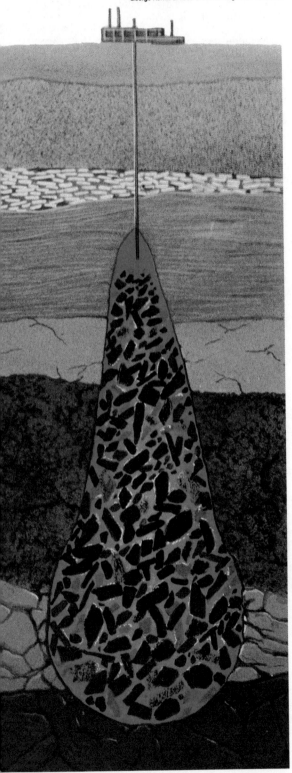

There is great potential in using underground nuclear explosions to release trapped oil from bedrock.

George Kelvin/© 1983 Discover Magazine, Time Inc.

pass through the ground. A computer analysis of the waves reveals the nature and structure of the material through which they have passed.

Although geologists have traditionally used dynamite for seismic sounding, nuclear explosives, according to Nordyke, have a two-fold advantage. First, sound waves from an atomic blast are much stronger, and can reach much farther and deeper into the ground. Second, a nuclear explosion takes place much more rapidly than a chemical explosion, producing patterns of sound waves that are more distinct and therefore more revealing. The Soviets have not reported using atomic blasting for geological surveys, but whenever they set off three or four nuclear explosions spaced at 800-kilometer (500-mile) intervals along a straight line, Western experts are certain that deep seismic sounding is going on.

Could nuclear sounding benefit the United States? "There's a great debate among geologists," says Nordyke. "Some say there are huge oil reserves at very great depths below the eastern edge of the Rockies. Others say there's no oil there. A sequence of nuclear explosions along a seismic line running from the Nevada testing flats to, say, North Dakota would probably settle the issue."

PNE's Other Potential

The Soviets' use of nuclear explosions to coax oil out of the ground is a technique that might also benefit the United States in some oil-starved future. Says Nordyke: "The kerogen hydrocarbons tied up in Colorado shale represent the petroleum equivalent of all the oil in Saudi Arabia. But strip mining and processing it on the surface would have very damaging environmental effects." A better and cheaper way, he says, might be to burn crushed shale deep underground and use retorts on the surface to condense its vapors into a liquid similar to crude oil. "We've tested that part of the technology," says Nordyke, "and it works. It could work even better if underground nuclear explosions were used to shatter the rock and free the oil."

Perhaps the most controversial of the proposed applications for peaceful nuclear explosions—one that may already be in use in the Soviet Union—is the creation of underground cavities for dumping dangerous wastes.

Highly radioactive waste, including spent reactor fuel, could be deposited in a rock cavity formed some 1,200 meters (4,000 feet) below

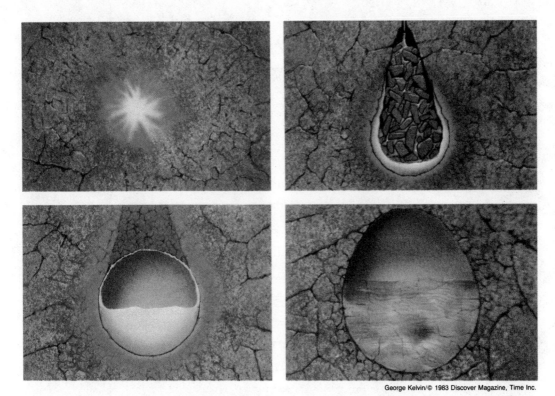

Nuclear blasts can create cavities for storing radioactive wastes, which melt, mix with molten rock, and solidify to glass.

the surface by a nuclear explosion. As Nordyke explains, the radioactivity of the waste would heat up the cavity, melting both the debris and surrounding rock into a glassy liquid. After about a hundred years, the resulting mixture would cool into impermeable glass, permanently sealing the radioactive waste from the environment.

U.S. Views and Options

Despite the apparent success and safety of Soviet PNE projects, the prospects for the revival of any kind of Plowshare program in the United States seem dim. In an atmosphere of frenzied—and sometimes irrational—opposition even to nuclear power plants, public acceptance of nuclear explosives for peaceful purposes is highly unlikely.

Even if the public mood should change, however, the United States could not benefit from the vast PNE experience of the Soviets. Moscow is always secretive about its nuclear explosions, and is under no legal obligation to disclose anything. Paradoxically, since 1976 the United States has had a lever that could be used to pry loose some useful Soviet secrets. In that year, both countries signed the Peaceful Nuclear

Explosions Treaty, which would have obliged them to disclose the details of every peaceful nuclear test—yield, location, depth, purpose, and results. But the Senate has failed to ratify the treaty.

In the years since, nuclear experts say, this country has missed out on information it might have obtained from the peaceful nuclear explosions in the Soviet Union. Even today, the Senate could ratify the treaty, at least theoretically obligating the Soviets to provide data on PNE.

Whatever course the United States may take, the Soviets seem to be unswervingly committed to exploiting the benefits of PNE. As long ago as 1949, a Soviet delegate boasted to the United Nations that "we are utilizing atomic energy for our economic needs in our own economic interest. We are razing mountains; we are irrigating deserts; we are cutting through the jungle and tundra; we are spreading life, happiness, prosperity, and welfare in places where human footsteps have not been heard for a thousand years."

In 1949 that claim was a gross exaggeration, but times have changed. The Soviet Union has transformed some of its boasts into practical nuclear technology □

LIMESTONE

COAL

FLUE GASES

STEAM

FLUIDIZED BED

WATER

STEAM

AIR NOZZLES

AIR

CALCIUM SULFATE AND ASH

PUMP

Rick Farrell

A New Way to BURN

by Walt Patterson

One afternoon more than a decade ago, a British engineer named Douglas Elliott tapped a bit of coal, ground to crumbs in a kitchen blender, into a tin can mounted on the wall. Below the can was an ordinary small open fireplace—ordinary, that is, at a casual glance.

For instead of burning coals, a red-hot bed of sand gently undulated like simmering lava. As the coal trickled out of the can onto the glowing bed below, each crumb exploded into flame with a pop and a blue-white flare. It burned only about 1 kilogram (2 pounds) of coal an hour, but the little fireplace glowed fiercely, delivering much more warmth than its size suggested.

An Innovative System

The fireplace, at Aston University in Birmingham, England, was no mere conversation piece. It was a simple, striking demonstration of a concept called fluidized bed combustion. Some say that FBC will transform the use of humanity's oldest technology—fire.

Like any combustion system, the end product of FBC is heat, which can generate hot water or steam or run turbines to produce electricity. Unlike other systems, however, FBC boilers can burn low-grade fuels, everything from high-sulfur coal to rice hulls. Suitably designed, an FBC system can use soaking wet coal mixed with rock. It can burn peat, heavy oil, and oil shale; wood and wood waste, including sawdust; urban and industrial trash; even, at one plant in Wales, sewage sludge that is half water. Yet emission of sulfur and nitrogen oxides is kept so low that neither tall stacks to disperse the flue gases nor fussy wet scrubbers to trap them in liquid are needed. If nothing else, FBC means that coal, our most abundant but naggingly pollution-prone energy source, could be exploited in almost any form, helping to stave off future energy crunches.

How It Works

The basic principle of FBC is simple: Blow air up through a box full of fine sand. When the air just lifts the sand grains, they churn and tumble like a boiling fluid. They have become a "fluidized" bed. Stirred with a spoon, it feels like thin oatmeal. A block of wood will float on it. Then the churning particles are heated red-hot,

Opposite: In a fluidized-bed boiler, a churning bed of sand is made red hot by a start-up gas before the burning fuel is fed in (coal here), which keeps the bed incandescent. The hot boiler turns water into steam.

to about 480° C (900° F). This can be done by injecting propane or natural gas into the airstream and igniting it in the bed. Once the bed is hot, the start-up gas can be shut off and fuel fed in. The heat released by the burning fuel will keep the bubbling bed of sand incandescent at about 815° C (1,500° F). The hot sand stores heat so well that even cold, wet fuel such as wet coal will not appreciably chill the bed. Heat is transferred through direct contact with the hot sand particles, so water tubes or air tubes immersed in the bed will collect heat perhaps five times faster than in an ordinary boiler. That means an FBC unit can be smaller than an ordinary boiler that puts out the same amount of steam, and cost less.

FBC controls worrisome pollutants—oxides of nitrogen and sulfur—simply and elegantly. An FBC boiler transfers heat so efficiently that it can operate several hundred degrees cooler than a conventional furnace or boiler. At this lower temperature little of the nitrogen in the air inside the FBC unit combines with oxygen to produce nitrogen oxide. If the fuel contains sulfur, crushed limestone or dolomite is fed in along with the fuel. Within the bed, the sulfur in the fuel reacts with the calcium in the red-hot stone to form solid calcium sulfate that is trapped in the bed and removed with the ash. An FBC unit can capture at least 90 percent of the sulfur in the fuel.

Acceptance at Long Last

For more than two decades, the official and commercial reaction to FBC amounted to studied indifference. Until the mid-1970's, the transatlantic community of FBC engineers was so tiny that most of them knew each other by their first names. Their tenacity finally has paid off. An FBC conference in Philadelphia, Pennsylvania, in October 1982 drew some 700 people from 17 countries, and major companies now promote FBC technology.

Only 10 years ago, FBC engineers were living hand-to-mouth. Douglas Elliott, probably the founding father of FBC, did much of his pioneering work in the 1960's at the laboratories of Britain's Central Electricity Generating Board—but the government body was flatly uninterested in pursuing his ideas. He resigned in 1968 to accept a professorship at Aston University.

Still a believer, in 1972 Elliott founded Fluidfire Developments with Michael Virr, a young British engineer. In a cramped workshop on a Birmingham back street, they worked to design and market FBC boilers and furnaces. When Elliott died of cancer in June 1976, FBC was still barely out of the lab. Virr hung on, scrambling for capital and orders. By 1982, Fluidfire—now Stone Fluidfire—had a full-scale factory, and a sister company, Johnston Boiler, had become the most active U.S. manufacturer of FBC equipment. One major customer is Campbell Soup, which uses FBC-generated steam to precook its products. Two small units have been installed at an Indiana public school to provide heat.

Another pioneering FBC firm, Foster Wheeler, has built a large boiler right under the noses of the U.S. government's air-pollution regulators, at Georgetown University in Washington, D.C. It heats campus buildings by burning cheap coal with a high sulfur content—2 to 3 percent compared to less than 1 percent in low-sulfur coal. After starting up in 1979, it turned out that the equipment used to clean up the combustion gases permitted the boiler to run at no more than 80 percent of capacity. Pushed beyond its limits, the filter system would break down. Restricting the boiler was a nuisance because it was 20 to 40 percent cheaper to operate than the oil- or gas-fired boilers the university had been using. But even before a larger filter system was installed in 1982, the stack gas easily met local air-quality standards. And with the new system in place, the university can run the boiler full blast.

Obstacles Restricted Development

With such signal virtues, why has FBC taken so long to catch on? The short answer is that until the Organization of Petroleum Exporting Countries (OPEC) forced the issue, scarcely anyone wanted to burn low-grade or troublesome fuel. Oil and natural gas were cheap, convenient, and plentiful; could be burned in simple, compact units; and left no residue. Why bother with coal, much less with trash?

Elliott and Virr were not the only determined enthusiasts of FBC. Raymond Hoy at the British Coal Utilization Research Laboratory had two large experimental FBC boilers running before 1970. But by 1972 cuts in government and industry funding had reduced his engineering staff from 80 to 6. Hoy persevered by taking on assignments from the U.S. government. When in 1967 Jim McLaren and his colleagues on the FBC team of Britain's National Coal Board discovered the sulfur-control effect, scarcely anyone else cared. The team, deprived of funds, dispersed. The first sizable FBC unit in the United States was built in Alexandria, Virginia, just outside Washington, D.C., by John Bishop, a mildly eccentric but brilliant engineer. His sudden death in 1972 came when FBC could ill afford to lose an advocate.

Foster Wheeler Corp.

Technicians monitor the FBC boiler at Georgetown University in Washington, D.C., which is used to heat the campus.

The mountains of culm (coal-mine refuse) behind these homes in Pennsylvania are being used to fuel an FBC unit.

OPEC's price hike later that year rekindled interest in coal, and heartened FBC supporters. But the recession that followed meant that FBC's two major potential markets—heavy industry and electric utilities—were in no mood to invest in new equipment. Then the first prototype FBC power station in the United States, at Rivesville, West Virginia, proved troublesome, reinforcing utility company doubts. Although it could generate 30 megawatts of electricity, enough for a city of about 20,000, it was dwarfed by a typical non-FBC generating unit of 700 megawatts. But the Tennessee Valley Authority (TVA), undeterred, went ahead in September 1979 with an experimental FBC boiler large enough for a 20-megawatt power station and fired it up in May 1982.

TVA Led the Way

"Back in the mid-1970's, TVA felt the need for new equipment that could use fuel that had been restricted from use because of environmental standards," says Roy Lumpkin, TVA's former manager of the pilot plant. "We were up against

the high transportation costs of low-sulfur coal, but we had cheap high-sulfur coal right on our doorstep. And the waste products from scrubbers were disconcerting; the waste from FBC— dry, granular stone and ash—appeared to lend itself to natural blending with the environment." TVA aims to have a 160-megawatt prototype FBC power plant operating by 1989.

Other electrical utilities began to take notice of FBC. "Utilities are interested in burning high-sulfur coal, in fuel flexibility, and in building coal-fired power stations in locations where you could not ordinarily burn coal," explains Shelton Ehrlich, manager of FBC programs at the Electric Power Research Institute in Palo Alto, California, and a former colleague of John Bishop. Northern States Power, a Minnesota utility, converted an old coal-fired boiler to an FBC unit that now burns wood waste from local suppliers. The burner started up in November 1981 and now runs 16 hours a day, generating 15 megawatts of power from fuels that include municipal solid waste, chopped-up railroad ties, old tires, and agricultural waste. The institute is

now gearing up for a tenfold jump, a demonstration power plant capable of generating 100 to 200 megawatts of electricity.

Alternative Designs

Most FBC boilers work at atmospheric pressure. But in 1967 Douglas Elliott and Raymond Hoy thought of putting an FBC unit inside an airtight shell and pumping in air under pressure. The greater volume of air would increase the amount of oxygen, greatly enhancing the heat output from an FBC unit of a given physical size. At his lab in Leatherhead, Hoy built in the 1960's what was for a decade the world's largest pressurized FBC unit.

Around 1975 the American Electrical Power Company and the Swedish turbine manufacturers Stal-Laval took a new look at the idea. Flue gas from a pressurized FBC unit, reasoned engineer Henrik Harboe of Stal-Laval, is not only hot but also at high enough pressure to be fed directly into a gas turbine to produce electricity. A coal-fired gas turbine, cheap to build and cheap to run, is an engineer's dream. The exhaust gases from the turbine can raise steam to generate more electricity or supply heat directly. American Electric, Stal-Laval, and the German boilermaker Deutsche Babcock have built a large test rig at Malmo, Sweden, to get enough information to design a prototype pressurized FBC power plant.

Engineers in the United States, Finland, and Germany are pursuing "circulating bed" designs. Here the air rushes through the bed at high speed, carrying the bed particles up and out of the combustion chamber into an adjacent hopper that returns them to the bed again. As with a pressurized design, the high rate of oxygen supply makes possible very high rates of heat release. The design also keeps fuel confined to the airstream more effectively than in an atmospheric boiler, so clogging of the filtering system is lessened. The design already is being used to heat water pumped down oil wells to stimulate production.

FBC engineers disagree, sometimes pungently, about the comparative virtues of different designs. And they concede bugs remain to be worked out. It is difficult, for instance, to remove dust and particles from the flue gas in a pressurized FBC system—junk that pits turbine blades, clogs the turbine, and corrodes metal parts because of alkali in the coal.

Technology Really Taking Off

At the October 1982 conference in Pennsylvania, the Russians revealed that they have mounted a major FBC campaign. Several industrial units are now operating, fueled by materials from anthracite to oil shale. At the previous conference, in 1980, the Chinese had revealed, without amplification, that the People's Republic had 2,000 FBC units in operation. This time the Chinese offered details. The FBC units are supplying electricity and heat to villages all over China. Four hundred of them burn lignite, a coal so soft it resembles peat; the other 1,600 burn what the Chinese call "stonelike coal," a gravelly material that is up to 70 percent ash.

The most advanced use of FBC is in Scandinavia. One unit operating at a Finnish paper mill burns peat and wood waste that is more than half water, generating 90,720 kilograms (200,000 pounds) of steam per hour to convert wood chips to pulp. Other units now running or due to start up soon will burn peat, wood waste, and even municipal refuse.

At the other end of the scale, perhaps the most unexpected impending development of FBC will be home heating. Stone Fluidfire's factory near Birmingham has built three prototype units, smaller than a family freezer and fully automated, to burn solid fuel and provide space and water heating. "We want to put 50 or 60 of them in houses in the U.S." as a demonstration, says Michael Virr of Johnston Boiler. The oil glut, he notes, has depressed prices, removing much of the incentive to develop coal-burning home furnaces and pushing back the market several years.

But FBC is in no danger of slipping back into near-oblivion. The crumbs of coal sifted into Douglas Elliott's tiny fireplace have become mountains poured into huge boilers. Prometheus would be proud □

SELECTED READINGS

"Burning coal mine waste" by Frank Harvey. *Popular Science*, December 1983.

"Germany's 'cool' coal fire—no pollution, no acid rain, no chimneys" by John Dornberg. *Popular Mechanics*, October 1982.

"Fluidized bed boilers" by Jason Makansi and Bob Schwieger. *Power*, August 1982.

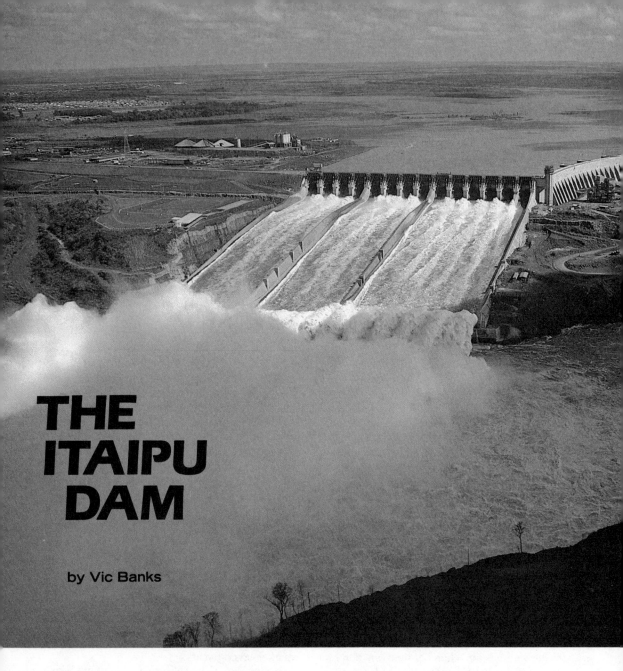

THE ITAIPU DAM

by Vic Banks

The tiny two-seat Hiller helicopter shudders against strong headwinds on our flight to see the world's largest hydroelectric dam—the mighty Itaipu. The dam is being constructed in South America where Brazil and Paraguay meet. Looking down through my open door, lush green forests and red, iron-rich soil abruptly give way to a view of countless tiny prefabricated houses. They form a veritable city in the jungle of western Brazil. The pilot points to the horizon and yells above the deafening roar of the rotor noise, "Look . . . the dam is at the end of a rainbow!"

A Dual Mission

Far in the distance I see a gray mountain of concrete that breaches Rio Paraná, the seventh-longest river in the world. On one side, an enormous floodland stretches beyond my sight. On the other side, a powerful spillway releases a torrent of water that thunders down a concrete chute and crashes into the riverbed, sending up a plume of mist nearly 300 meters (990 feet) high. It is a most awesome view that practically takes my breath away. The wind and sun have been just right—there is indeed a magnificent rainbow arching across the dam.

The mighty Itaipu Dam symbolizes Brazil's ambitious effort to achieve energy self-reliance.

There may not be a pot of gold at the end of this rainbow, but government officials here place great faith on Itaipu's dual mission: to power Brazil's rush to industrialization and to relieve her crushing dependence on Middle East oil. When all of the dam's turbines are generating electricity in the late 1980's, engineers estimate that Itaipu's total daily output will be equal to the energy released by more than 100,000 barrels of oil, or nearly 10 percent of the country's daily energy requirements. It's an enormous demand, and one that threatens to bust Brazil's budget.

New Role for a Nation

The country's power demand had its beginning during the late 1950's and early 1960's, when Brazil's President Juscelino Kubitschek articulated his vision for a new nation—one that would push Brazil past its status as a Third World, developing country and into the exclusive circle of modern, advanced nations. Increased growth and economic independence were his dual guidelines. He created a bold new capital city—Brasília—that would literally rise from the jungles and build upon the land's vast natural resources. It was a brilliantly modern city hewn from the country's interior high plains. A few critics, awed by Brasília's futuristic architecture, said it was too advanced for the Brazilian people who would have to leave the overcrowded Atlantic coast to work there. Nevertheless, Brasília emerged as a symbol—albeit controversial—announcing that Brazil had an important future.

In just the past 20 years, Brazil has experienced a remarkable rise to ascendancy. New superhighways crisscross the country, even penetrating the dense jungles of the Amazon river basin. The cities of São Paulo and Rio de Janeiro have a combined population of nearly 20 million, and they continue to grow steadily. Giant irrigation and agricultural projects have transformed vast wildlands into cultivated farms. Brazil's exports of iron ore and petrochemical products are among the largest in the world. Nothing in this country, it seems, is ever done on a small scale.

Moving Away from Oil Dependence

Brazil's hope to achieve rapid industrial independence was shattered in the mid-1970's when the Organization of Petroleum Exporting Countries (OPEC) placed a stranglehold on world oil supplies. Although Brazil has a wealth of many natural resources, oil is about the only resource that the country must seek outside its borders. As oil prices escalated with limited supplies, the nation's demand for oil grew ever stronger. It was obvious that something drastic had to be done.

A crash program for energy self-reliance was drawn up in 1976. Precious oil was sought along the country's continental shelf, while plentiful coal reserves were developed for industrial uses. A now-famous project to supplement automobile gasoline with alcohol—derived from sugar beets—proved successful as well as popular with the driving public. A major program to build 10 nuclear power plants by 1990 was instituted. But what really captured the attention of government planners was the abundance of waterpower available through the extensive river systems that flowed in the nation's interior. Brazil's hydroelectric potential, they thought, could conceivably make her South America's electricity czar.

Surmounting Start-up Problems

Itaipu will be the first of a series of giant water-run generators designed to meet growing power demands in the 1980's and beyond. Plans for this dam were completed in 1969, but the civil construction itself began in 1975, after oil prices had risen 400 percent. At that time the plant with its 14 turbines was expected to cost $2 billion to $3 billion. Then economic dislocation began. Skyrocketing oil costs drained the nation's financial resources and fueled inflation to a mind-reeling 120 percent a year. Rather than cut back on costly projects, the government decided that dramatic industrial growth would

create a better economic climate. Itaipu was expanded to house 18 turbines that would produce more electricity than Egypt's High Aswan Dam or the United States' Grand Coulee Dam: 12,600 megawatts. (This is about the same generating capacity that the state of California has for all electrical production.) The new price tag: $14 billion and rising.

Cost escalation throughout the planning and development phases for Itaipu has played havoc with scheduling exactly when the turbines will be operational. Best estimates point to 1986 for the first turbine to start producing electricity; the others will gradually join the power output in the late 1980's as each is completed.

In theory, power allocation will be a straightforward proposition: equal shares divided between Brazil and Paraguay. Brazil will use its portion to aid economic growth, especially in the giant industrial center of São Paulo, about 800 kilometers (500 miles) east of Itaipu. For tiny Paraguay, with just 3 million people as compared to Brazil's more than 120 million, the share of Itaipu's electricity will more than double current electrical capacity and cause a major surplus of power in the country.

A vast amount of concrete was required to construct the dam's immense foundation and power plant.

Thousands of workers were provided with free housing and facilities for education, recreation, and health care.

As of early 1984, Paraguayan officials admitted their economy would use only 30 percent of Itaipu's allotment. By special agreement with Brazil, Paraguay will sell excess power from the dam back to Brazil at significant commercial rates. This announcement, along with the high price for the dam, has caused serious public outcry regarding the practical use for this immense hydroelectric project.

A Town with All the Trappings

Everything about Itaipu is grandiose in comparison with typical engineering projects. The housing developments that I flew over provide shelter for more than 25,000 workers. During the early, heavy-construction phase of the project, more than 40,000 men came from Brazil and Paraguay—the tiny nation that shares the cost and labor-power for the dam on a fifty-fifty basis.

According to Reubens Naguera, information engineer at Itaipu, experienced skilled workers to operate the heavy equipment had to be imported into the region or trained on site. To attract laborers and technical personnel, the construction authority offered comparatively high wages and built a modern town complete with free housing and recreational, educational, medical, and dental facilities. Naguera points to this advanced, socialized community as the

ideal way to cope with the extraordinary numbers of people living in the hinterlands. Yet for some workers, this fantastic company town was simply beyond their realm of experience. They came from crowded urban *favelas* (slums) or rural *fazendas* (ranches) to make fast money. Many were not prepared to work for years under strange conditions, and some left. Those who stayed, well earned their rights to the site's amenities by risking their lives daily in the Herculean effort required to shift Rio Paraná from its ancient course.

An Unprecedented Effort

In preparing the site, nearly 23 million cubic meters (812 million cubic feet) of rock was blasted and dug from the stream bed. It was later used to build rock-fill dams and to manufacture concrete. On the east side of the river, a plant was constructed where the rock and other materials were processed to produce the more than 12 million cubic meters (425 million cubic feet) of concrete required to build the dam's foundation. It was a quantity, say officials, "sufficient to pave a major two-lane highway that would link Lisbon, Portugal, with Moscow in the Soviet Union." To properly set such a volume of concrete, engineers added crushed ice to the solution in order to maintain an ideal temperature of 2.2° C (36° F). The resulting

ice-making machine was the world's largest—capable of freezing water in the tropics at the rate of 50,000 kilograms (110,000 pounds) per hour.

Once the foundation was completed, the remaining concrete was used to build the mammoth power plant. This structure is without rival: 226 meters (743 feet) high, 34 meters (112 feet) wide, and an unbelievable 1,241 meters (4,072 feet) long. Inside its dimly lit recesses, my guide and I climbed a long flight of stairs to a skywalk. Peering through a web of steel cables, he pointed to a series of large snail-shaped tubes stretching the length of the room. These tubes made up 15 turbines; 3 others are housed in the diversion channel. When operational, water will fall through their twisting pipes to drive nine massive generators, which will each produce more than 1,000 megawatts of power.

Salvaging a Valuable Legacy

From the air, it was evident that the civil construction stage of the work had come to an end. Concrete and rock-fill dams had brought the powerful Paraná to a stop. Aside from Itaipu, a system of 20 smaller dams was built to control water flowing into Lake Itaipu—the reservoir that rapidly formed behind the main dam.

Before the floodwaters began rising, officials instituted a major survey and rescue operation for the archaeological and living heritage of the region. During this program, called *Mymba-Kuera* (local Guarani language for "fauna"), upward of 100,000 artifacts spanning 8,000 years of human habitation in the river area were reportedly collected. Most were materials of prehistoric hunters who camped along the river's edge.

A rescue operation was also set up for wildlife trapped by rising floodwaters on newly formed islands. Efforts were made to capture and translocate stranded animals to a special refuge located on the eastern bank of the main dam. Here they were to remain until the water reached its intended depth.

This temporary refuge is a tallgrass marsh some 2,630 hectares (6,500 acres) in size. At the entrance to the area, I encountered a series of pens and cages that contained a number of small animals to demonstrate the operation's success. I was told by engineers and animal keepers that anywhere from 3,000 to 14,000 deer, monkeys, tapir, peccaries, agoutis, snakes, wild cats, and caimans roamed the refuge. An animal keeper said to me that by returning these animals to island habitats, their fate will be far better than if they were to try to survive in the heavily hunted agricultural lands outside protected areas.

The lake has inundated vast sections of land, including hundreds of ranches and several towns, which forced a considerable number of the region's people to relocate. Environmental losses were more serious, and in some cases, irreplaceable. For example, the famed *Sete Quedas* (Seven Falls) cataracts—considered by many naturalists to be one of the most awesome sights in Brazil—was submerged.

A Boon to Tourism or a Bust?

Project officials say Lake Itaipu, and the power plant itself, will eventually draw huge crowds and become a major tourist and recreation resource. Visitors will be able to ride cable cars above the dam, view salvaged artifacts in a museum, and stay in deluxe hotels along the water's edge. Planners say Itaipu will be here for a long time. They predict the dam will be generating electricity for at least 600 years.

However, there are contrary voices to be heard. A zoologist I interviewed in Rio de Janeiro, who was familiar with Rio Paraná's turbid waters, claims siltation will be so rapid that in just 50 years Itaipu's capacity to generate power will be halved. In the meantime, he said incredulously, "we will have to be truly creative to use so much electricity." He did, however, take some solace in the fact that this great powerhouse could well have been substituted by seven nuclear power plants.

Whatever critics and advocates say about the dam, Itaipu is fast becoming a symbol for Brazil in the 1980's. It's a brave attempt to harness the raw power of South America's wilderness. And, if it succeeds, as most experts agree it will, there is always the Amazon□

SELECTED READINGS

"Dam the Amazon, full steam ahead" by Catherine Caufield. *Natural History*, July 1983.
"Dams: the high and the mighty" by Anne Millman and Allen Rokach. *Science Digest*, May 1982.
Ecological Aspects of Development in the Humid Tropics. National Academy Press, 1982.

Top: Archaeological treasures and wildlife were saved before this land was flooded. Bottom left: A caiman cruises through the water not far from the dam. Bottom right: Silhouetted at sunset, a heron stands in a marsh.

ICE PONDS

by Richard Woodley

The sign on the little pump house proclaimed: "World's First Ice Pond."
"My God!" Ted Taylor gasped in dismay as he stepped from the car in the back lot of the Kutter's Cheese factory near Buffalo, New York. It was a beautiful early-April day, strong sun, warm breeze. For him, dreary weather: it had stolen most of his ice. What a fortnight before had been a 6-meter (20-foot)-tall, artificial Matterhorn was now a molten lump sinking into an expanding pool of cold water. Meant to last the summer, the ice would barely survive the spring. "We missed it by two weeks because we didn't have the cover," Taylor said as he slogged dejectedly through the mud bordering the pond. "Well, at least we've proved the theory works."

Anybody can make ice. All you need is cold air and water. Enough chill in a winter, a snowmaking machine, or even a sizable water hose, and anybody can make several hundred tons of ice. Cover such a big pile with insulation, it will last the summer. As it slowly melts, all you need is a simple system of pipes and pumps; the ice water can be used to air-condition a building. Cheap, clean, safe energy.

From Atomic Bombs to Ice Piles

Ted Taylor has been making big piles of ice for several winters, experimenting. His theory is that you can make a business out of it, make it

Above: Ted Taylor first tested his energy-saving technique with this huge pile of ice at Princeton University.

pay. Making ice was easy, a matter of natural laws. Making it pay was proving dicier, a matter of mindfulness, materials, and manufacture.

Taylor is a theoretical physicist who has turned to applied physics and engineering. He is at his best when his mind roams free. His visions are based on natural laws, not engineering precedents or practicality. He dreams up better ways to make things work. He is not at his best making things work for himself. He is distracted by his thoughts. Waiting for water in a kettle to boil, he can tell you how to improve the efficiency of an ordinary kitchen refrigerator by 90 percent. Meanwhile, he has not turned on the burner under the teapot. Finally, he has to make instant coffee from the hot-water tap.

Taylor has always been fascinated by the shapes and forces and energy of things. Academics bored him. But he wanted to be a physicist. He joined the Navy when he was 17, attended the California Institute of Technology while in the Navy V-12 program, and graduated when he was 19. In 1946 he enrolled in the graduate school at the University of California, Berkeley, to study experimental physics, but he was more interested in theory. He preferred dreaming things up. "A lot of physics was a mystery to me," he said some years ago, "and still is." Twice he failed oral exams for his doctorate. His academic career seemed finished.

But one of his professors, who had worked on the first atomic bombs in the Manhattan Project at Los Alamos, got him a job there. While still in his 20's, he made his mark with new ideas for bomb design, leading to the smallest, largest, and most efficient atomic bombs then built. Taylor is now sorry he had anything to do with atomic bombs. Or with nuclear power in any form. "As far as I'm concerned," he says, "the world is a less safe place because of the work I did." In recent years he has lectured widely and testified before Congress against nuclear power and for nuclear disarmament. He served on the President's Commission on the Accident at Three Mile Island. Since then, he has turned his full attention to the design of renewable-energy systems derived from the power and rhythms of the sun. You can't wage war or terrorize a populace or contaminate the environment with a pile of ice.

Everything has heat in it. "Heat," as Taylor puts it, "is nothing but motion of molecules and atoms." You cool things by extracting heat. It was by thinking of heat that his life's work turned to ice.

Bill Ray

Physicist Ted Taylor once designed atomic bombs.

Birth of a Novel Idea

In 1975 Taylor was pondering the question of how to diminish pollution from use of pesticides and fertilizers in agriculture. He came up with the notion of massive greenhouses covering millions of hectares of farmland, in which you could grow protected wheat and corn and rice. You would have to heat these mammoth structures in winter and cool them in summer. Costs for using energy derived from fossil fuels would be ruinous. For heat, Taylor thought of using solar power, one way being to create giant hot-water ponds. "Giving attention to heating," he recalls, "we were always trying to avoid cold air. One afternoon, I said, 'Gee, is there a way of capturing the heat-absorption capacity of that cold air in winter and saving it until summer to solve our cooling problem?' As soon as I asked that question, the answer was obviously 'Yes.' You freeze water to make ice, and store the ice in icehouses."

You could harvest ice as it was done a hundred years ago, by chopping up the frozen surface of a lake. You could build up frozen surfaces by spraying on more and more water, like

Don Kirkpatrick leans on the snowmaking machine that he and Taylor (seated) used to make the ice mound behind them.

a skating rink. But temperatures in most places would produce insufficient thicknesses of solid glare ice. There was a more modern technology. On ski slopes, snowmaking machines were seasonally building up tons of synthetic snow. Taylor didn't want dry, fluffy snow because it takes too much volume. But if the flow of water fired through the machines at subfreezing temperatures were adjusted to make slush, excess water would quickly drain off and leave a much more compacted pile of ice.

Nothing came of the greenhouses. In 1976 Taylor came to Princeton University of New Jersey to teach, and his ideas for hot-water ponds and ice ponds (as he called his piles of ice) were expanded by students in his course on solar energy. Princeton's Center for Energy and Environmental Studies and its director, Robert Socolow, took an interest. Then, in 1979, the Prudential Insurance Company planned to build two three-story office buildings in Princeton, drawing upon the most advanced energy-saving design techniques. The building would be designed efficiently for heat, drawing directly upon solar energy. But the architects were attracted to the ice-pond idea for cooling one of the buildings. The company provided the Center

with a development grant of $150,000. Socolow brought in Taylor.

Experiment Shows Promise

The first experimental ice pond at Princeton in 1980 was simplicity itself. It was made in a hole 23 meters (75 feet) square at the top and 5 meters (15 feet) deep. Collaborating with Taylor was Don Kirkpatrick, a mechanical engineer and solar-engineering consultant who in private industry had worked on the mechanics of satellites. He seemed a good counterbalance to Taylor, to make sure that what Taylor dreamed up to do also got done.

They first tested making both solid glare ice and snow-machine slush, and decided on the machine. The pond was excavated near an old storage building, which they would try to cool with the ice melt. Pipes were laid and connected to make heat exchangers—like radiators—in the buildings. With the snowmaking machine planted on the ground like a machine gun, they made a low pile of a thousand tons of ice. They lost half of it to sun and warm breezes before they covered it with layers of straw and polyethylene sheets, weighed down by old tires. That summer, pumps sent water at 0° C (32° F) from the bottom of the ice pile into the building, where it absorbed heat, and returned it at 7° C (45° F) to be dumped back on the pile. The usually sweltering building was cool. The hottest days would drain away seven tons of ice. Taylor resigned from the university faculty to concentrate on ice.

To build the final Prudential pond in an area that had strict aesthetic architectural codes, they would need to make the pond, if not beautiful, at least acceptably less unattractive. Over a huge hole they would need a huge dome. As a step toward that, the next winter they constructed over a test hole a small white polyvinyl chloride tent, suspended over steel. For making snow, which requires the unobstructed passage of cold air to freeze the water droplets falling onto the surface of the pile, they could roll up the sides of the tent. To preserve the ice, they sealed the sides back down and covered the ice with blankets of polyethylene foam. By winter's end, they had made another pile of ice, their first small mountainlike shape under the tent. At long last, they were finally ready to start at Prudential.

Ice Flowers and Other Mysteries

Taylor was becoming ever more intrigued by the complexities of crystallization and compaction, the formation and behavior of ice. Using an ordinary garden hose propped up on a stepladder in the backyard of his home, he was making his own little piles of ice.

There are mysteries in even so common a procedure as making ice. Water droplets suspended in below-freezing air—say, −2° C (28° F)—do not necessarily freeze. The water in the spray may be supercooled but remain liquid until part of it freezes later. Some particle in the air is introduced into the spray (this process is called nucleation) to begin the process of crystallization. At first, Taylor says, "I had no appreciation of the role of nucleation and how to enhance it. The process is actually very poorly understood."

You could encourage crystallization by introducing silver iodide crystals (which, though they are used in cloud seeding, Taylor won't use because they are expensive), or even dust. He believes the best "seeds" to be other ice crystals.

In 1983 he built up a little pile 2.4 meters (8 feet) tall. He spent hours marveling at the wonderful pinnacles and icicles that were formed. One morning after a night of spraying he found delicate triangular cups of ice crystals strewn about—"ice flowers," he called them—the largest measuring 15 centimeters (6 inches). It never happened again. Nobody has been able to explain them. For two months he

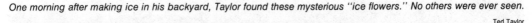

One morning after making ice in his backyard, Taylor found these mysterious "ice flowers." No others were ever seen.

Ted Taylor

kept the ice flowers in his freezer.

He experimented with different nozzles, different sprays. "Now," he asserts, "I can pretty much look at a spray and tell if it can make ice, or what type of ice it can make."

The main lesson from the garden spray was that in climates cold enough to freeze the larger droplets, you could use a hose to build a big pile of ice. That opened up vast new possibilities. Modest ice ponds could be cost-efficient for small businesses such as those in the northern dairy industry, with its extensive cooling needs for the processing and storing of milk and cheese.

The Prudential Pond

With Kirkpatrick, Taylor formed "TKI-Ice Energy Systems," a company to design and construct ice ponds for profit. They submitted a proposal to the New York State Energy Research and Development Authority (ER&DA) to survey the needs of industries throughout the state, for possible application of ice ponds. The key to everything was, of course, economics. "We will give you ice at a value which is competitive with cooling made by electric power—say six cents per kilowatt-hour," Taylor reckons. "That's the break-even point." He estimates that the energy-efficient Prudential building will save $15,000 a year by ice-pond cooling.

Construction of the Prudential ice pond began in the summer of 1981 and was scheduled for testing with ice in the winter of 1983. The pond was a one-fifth-hectare (half-acre) excavation 6 meters (20 feet) deep, with sloping sides and bottom lined with a thick polyvinyl-chloride sheet. The dome, spreading over an area the size of half a football field, and the most costly part of the construction, was a single span of aluminum arches covered with a light polyester-vinyl fabric. Held underneath the fabric by a net were fluffy pink rectangles of fiberglass insulation. Then another plastic sheet was sewn under that, making an insulation sandwich.

The entire dome was supported at the corners by four steel tripods on which were electric motors that could raise the dome 6 meters (20 feet) for the making of ice and lower it for preservation. Three huge blue snow machines—

This schematic shows how Taylor's system works. An ice pond made in the winter (left) melts in the summer (right), creating cold water that is used to cool a building. The circulated water is then returned to the pond to melt more ice.

Peter Tenzer/Wheeler Pictures

The dome for the Prudential ice pond is shown under construction. The pond will cool the building in the background.

with small high-power nozzles around the perimeter of a giant fan—would spew out 380 liters (100 gallons) of water a minute, mixing it with compressed air to produce more than 20 tons of ice per hour.

In this three-story-tall structure, a 12-meter (40-foot) mountain of ice would arise, 8,000 tons of it.

On to Chilling Cheese

While the work was going on, Taylor roamed north to the dairy country. Kutter's Cheese had expressed an interest. As cheese factories go, it has a modest output—3 million pounds annually. Cooling needs are large, both to chill heated cream in production and for storage of cheese. Winters around Buffalo, New York, are extremely cold and long. The location was a perfect place to make ice using just a hose and simple insulated cover.

The Kutter brothers, Tony and Dick, who run the plant, knew of Taylor's test ponds. "If Ted hadn't tried it before," Dick Kutter said, "I would have thought this was the worst harebrained scheme I ever heard of—to save winter ice for summer. I would have backed right off."

With ER&DA contributing half the initial costs, estimated to be about $30,000, the Kutters decided to put an ice pond in their back lot. The excavation would be about 15 meters square (50 feet square) and 2 meters (6 feet)

deep. In it, using a hose with a 98-liter (26-gallon)-per-minute spray and power from a 3-horsepower pump, they hoped to pile 650 tons of ice. The ultimate savings in cooling costs would be about $3,000 a year, Taylor estimated.

With both projects lined up, Taylor was buoyed and expansive. He envisioned ice ponds all over the place, large ones, small ones.

Dual Delays

You cannot, however, make a pile of ice whenever you feel like it. You need frigid weather. Construction on both projects lagged into the precious winter. In Princeton, time was lost because of the tricky construction of the dome, the likes of which had not been made before. It had to be anchored firmly enough so that a strong wind wouldn't take advantage of its aerodynamic qualities and lift if off and deposit it in the adjacent parking lot. Communications and chain of command in the bureaucracy among TKI, Prudential, architects, and contractors were not always clear or swift. At Kutter's, excavation and pipe laying got started late. December mud bogged down the laborers. The pump house sank in the mud.

But work proceeded anyway, Taylor explained, "because we would have lost another year of commercial demonstration if we didn't start up."

At Kutter's, they finally began making ice

in January 1983. They guyed an ordinary extension ladder out from a telephone pole, lashed the hoses to the ladder, and affixed two nozzles on their ends to help create the spray. The ice pile built quickly.

At the Prudential building, the three big machines also started making ice in January. But a week later the weather turned too warm. Ice was made in fits and starts after that. Under time pressures, they had started using the dome before it was entirely finished. Water got in between the plastic layers, soaked the insulating pads, and caused part of the inside layer to rip away and fall down on the ice. Ice production ended.

What greeted Taylor on a visit to Princeton in early April was not 8,000 tons of ice, but 2,500 tons—under a dome that could not be raised.

Although the experience gave Kirkpatrick serious misgivings about TKI's role as general contractor, Taylor was pleased with their first truly huge pile of ice, even though it wasn't all they wanted. The berg of 2,500 tons would allow Prudential to test the system. And probably it would last all summer anyway.

Later, he climbed up the ice pile, under the drooping innards of the dome, and pronounced: "A glacial era has come to Princeton." He added, chuckling, "That being defined as a pe-riod when snow from one season does not disappear before the snow of the next."

Meanwhile, ice making up at Kutter's had proceeded spectacularly, even in the shortened season, and there was a minor mountain of it. It was ready for the insulating cover. A local troop of Boy Scouts had stuffed 200 plastic bags with wadded newspaper to lay on the ice. A custom-made inflatable pad and sturdy plastic sheet to cover it had been ordered.

Undaunted Optimisim

Before making his climactic journey to Buffalo to supervise completion of the Kutter's job, Taylor paused for a reassessment. "I've just figured out the application," he said, leaning against the kitchen door waiting for water to boil over an unlit burner. "I've scaled it way down. Ice ponds are perfect for dairy farms, to cool milk. With a $5,000 investment, using a hose that pumps four to five gallons [15 to 19 liters] a minute—like in my backyard—a dairy farm with 300 cows could save $5,000 a year on cooling costs. There are little niches it will fit into. An ice pond would be ideal for a brewery."

But he was not yet finished thinking big. The computer industry, he noted, was already using large water-cooled systems. What an opportunity for ice ponds! And bigger. All of Long

An impressive winter ice pile awaits usage at Kutter's cheese factory in New York where cooling needs are large.

Anthony Kutter

Kutter's ice pond melted away much faster than expected, but its proponents were pleased to prove the concept worked.

Island was dependent on aquifers for fresh-water supplies. The supplies were becoming contaminated. If you make a big pile from sea-water, the salt and other impurities in unfrozen brine, through gravity, will naturally drain down and out. You're left with freshwater. He mentioned one Long Island town that had already expressed interest in this purification-desalinization process. "The pile of ice could be 70 feet [21 meters] high," Taylor sang, "and four acres [1.8 hectares] big. You wouldn't even need a cover."

While Taylor was speculating at Princeton, the sun was out. It was warm enough for a shirt-sleeve picnic. It was as warm in Buffalo, where the scene that met Taylor was of three-quarters of his ice pile gone.

"We had 400 tons of ice," Tony Kutter said, as Taylor stomped around the pond to assay the damage. "Ted didn't think it would melt that fast. But we've proved it will work. We used 32-degree [Fahrenheit] water from here all winter long in processing. We'll definitely do it again next year. You'll see these all over the country."

The dual setbacks at the two facilities did take a toll. Taylor and Kirkpatrick, citing some rather gentle disagreements in point of view and practicality, decided to head off in somewhat different directions, amicably dissolving their partnership in TKI. Kirkpatrick would remain as manager of the Princeton pond and consultant to Prudential, and would pursue such industrial applications for ice ponds. Taylor, detached from the Prudential pond, would form a new

company. Along with facilities like that at Kutter's Cheese, he would persist in the hunt for new uses of ice that intrigued him.

He has a dream beyond the current type of ice pile. "What would we really like to have? We'd like to have a big, flat pond full of solid glare ice all the way to the bottom. In places with very cold winters, you could see how economical this would be. But we don't know how to do it. The people who figure out how to do it will get rich. In a place like Minneapolis [Minnesota], where the temperature is often below zero, we'd like to supercool water and pour it into a big hole in the ground, keep it super-cooled and not frozen—meaning you don't produce that slightly magic set of phenomena, like that speck of dust or something, that forms the pattern on which the molecules can arrange themselves to form the magic word—or put in the magic thing—and WHOOF! It freezes solid, and you're all set.

"A hundred years from now people will know how to do this. I'll bet very heavily that people will be using ice ponds in which the ice is made in ways that we haven't even conceived. They will answer questions regarding the basic process of crystallization of ice that we haven't even had the sense to formulate."

A hundred years does not fall within the convenient scope of the operative optimist. For Taylor, blue sky is not an ethereal daydream but a real place where the water molecules await, and the motes of dust, and the rhythms of temperature; the ever-changing climatic place where all the action is □

THE
ENVIRONMENT

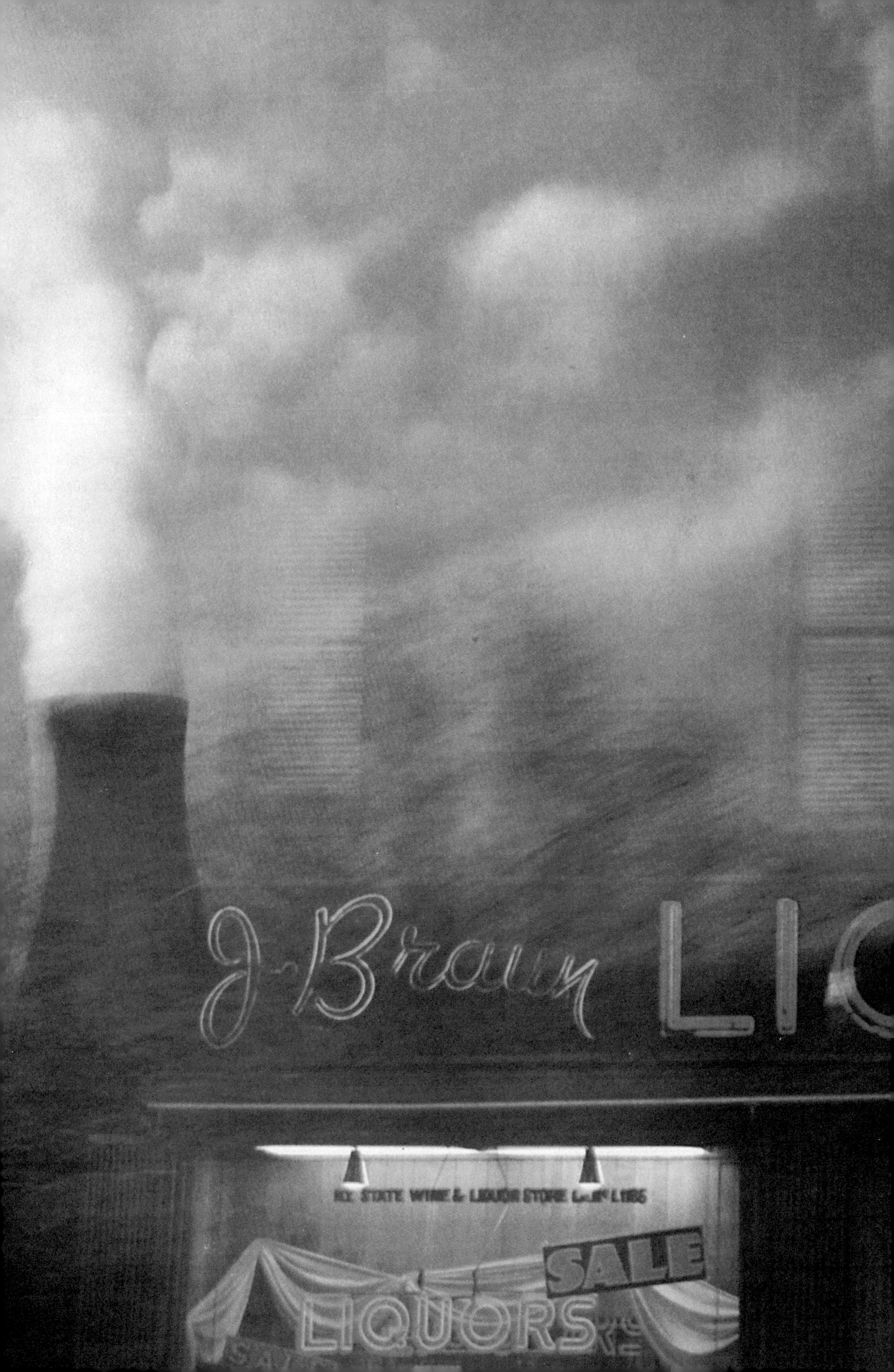
J. Braun LIQ
STATE WINE & LIQUOR STORE
SALE
LIQUORS

REVIEW
OF THE
YEAR

THE ENVIRONMENT

In the United States, environmental concerns about poisonous wastes, acid rain, and possible human-induced climate changes were coupled with disturbing conflicts over politics and personalities. On the international front, promising signs of cooperation among nations on safeguarding the environment shared attention with dispute over the sources of "yellow rain."

U.S. POLICIES AND PERSONALITIES

Policies and personalities on the environmental scene produced big news in Washington during 1983. The contentious programs and utterances of Secretary of the Interior James G. Watt, the Reagan administration's chief steward of natural resources, led to his replacement. An investigation of the Environmental Protection Agency (EPA) culminated in a major overhaul of its administration and in criminal proceedings.

The focus of congressional interest in the

EPA's activities was suspected political and corporate influences in administration of the $1.6 billion "Superfund" created by Congress in 1980 to speed the cleanup of hazardous-waste dumps. EPA Administrator Anne Gorsuch Burford resigned in March after being cited, then cleared, for contempt of Congress in withholding certain documents. Rita Lavelle, EPA director of hazardous-waste programs, was convicted on a federal charge of perjury in congressional testimony. Altogether, 24 high-ranking functionaries left the agency. To replace Mrs. Burford, President Reagan appointed William D. Ruckelshaus, who had headed the EPA from its inception in 1970 until 1973. He set out to restore agency morale, rebuild its staff, tighten enforcement, and bolster the budget.

Secretary Watt had angered the conservation community and many members of Congress with what critics considered exploitative policies on public lands, wilderness, oil and coal reserves, and national parks. But it was a public remark, construed as disparaging minorities, that finally tipped the scales against his tenure. He was succeeded by William P. Clark, a onetime California rancher and judge who had been a Reagan adviser on national security. Secretary Clark ended a Watt moratorium on national parkland acquisition and announced indefinite deferment of some portions of the

Opposite: An ominous sign blocks a roadway leading into the dioxin-contaminated town of Times Beach, Missouri.

offshore-oil-leasing program initiated by Mr. Watt. In addition, a Reagan administration program to raise money from the sale of "unneeded" federal lands came to a virtual halt—it was declared unpopular and a "political liability."

POISONOUS WASTES

The EPA raised its estimate of the amount of poisonous wastes produced annually in the United States fourfold—to 150 million metric tons. The agency also estimated the number of hazardous-waste dump sites throughout the country at 17,000, and enlarged to 546 sites its priority list for early cleanup under the Superfund. By the end of fiscal 1983, the EPA reported that nearly all of the $474.5 million Superfund money appropriated by Congress had been committed to cleanup projects. ∎ Meanwhile, the Occidental Chemical Company announced it had settled personal-injury lawsuits lodged on behalf of former residents of the poisonous-waste contaminated Love Canal area in Niagara Falls, New York. ∎ And federal authorities filed a record-size $1.9 billion suit against the Shell Oil Company over cleanup of wastes discharged at a chemical plant near Denver, Colorado.

The EPA launched an investigation covering hundreds of sites possibly contaminated by dioxin, which has been called "one of the most perplexing and potentially dangerous chemicals ever to pollute the environment." Dioxin is an unwanted by-product in the manufacture of the defoliant 2,4,5-T and of products containing it, including the herbicide Agent Orange, which was used in Vietnam from 1961 to 1972. It is considered a possible threat to human health in dilutions as small as one part per billion of any carrying medium. Debate about dioxin's toxicity to humans mounted among scientists because of inconclusive data: the chemical has appeared to be 5,000 times more toxic to some animal species than to others. ∎ Meanwhile, a pending class action lawsuit against seven companies that made Agent Orange contended that residual dioxin in the product caused genetic abnormalities among individuals—chiefly Vietnam veterans— exposed to it. ∎ The first major site of dioxin contamination was found in Times Beach, Missouri, where, in the early 1970's, dioxin-contaminated waste oil was sprayed on unpaved roads to control dust. The federal government spent $35 million to buy homes and businesses in the community. Medical tests on former residents were said to have shown "no meaningful ill-health effects."

ACID RAIN

Acid rain assumed ever-larger dimensions as a pressing environmental problem. Three U.S. scientific study panels reinforced previous findings that a major cause of acid rain in the United States is annual emissions of some 24 million metric tons of sulfur dioxide from coal-burning power plants in the Midwest—fumes later precipitated by rain into sulfuric acids harmful to waterways and plant and animal life. Bills were introduced in Congress to reduce allowable sulfur dioxide–emission levels by as much as 50 percent, but concrete remedial steps were stalled by regional bickering over who should bear the multibillion-dollar costs of fume-control equipment. Meanwhile, a United Nations report said that 5 million to 10 million square kilometers (2 million to 4 million square miles) of land in northwestern Europe and in North America had already been damaged by acid rain.

CLIMATE CHANGES

Possible changes in the Earth's climate from combustion processes stirred another debate. Combustion releases carbon dioxide into the atmosphere, where it acts like the glass or plastic roof of a greenhouse, trapping heat radiated from the Earth's surface and preventing it from passing out into upper atmospheric layers. This so-called greenhouse effect can cause a gradual rise in the Earth's average temperature. An EPA report issued in 1983 said that the buildup of carbon dioxide in the atmosphere was occurring at a rate that called for immediate steps to deal with the consequences. A general warming of as little as 5° C (9° F), it was said, could change global agricultural patterns and melt polar icecaps, which would raise ocean levels. A simultaneous report by the National Academy of Sciences said further study of the problem would be more useful than "near-term corrective actions."

ON THE INTERNATIONAL FRONT

A 16-nation European Treaty aimed at ending pollution of the Mediterranean Sea became effective on ratification by six countries.

And 17 American nations agreed on terms of a pact to combat pollution in the Caribbean; the pact will become effective when ratified by 9 countries.

"Yellow rain"—a possibly toxic powder appearing recurrently on vegetation in Afghanistan and Southeast Asia and locally blamed for many deaths and illnesses—stirred debate. Some scientists challenged the allegation that this represented Soviet chemical warfare, saying that the "rain" might be a natural phenomenon such as pollen.

GLADWIN HILL

Tropical Forests in Trouble

by Mary Batten

In jungle treetops, centipedes, earthworms, and even a snail-eating snake live in a bizarre plant that runs its own compost heap. Known as the "trashbasket plant," it anchors itself to branches of rain forest trees, catches falling debris, then turns it into humus. The aerial litter can that results is home for a mini-universe of creatures usually found only in soil. Until 1981, when botanist Robert Dressler of the Smithsonian Tropical Research Institute discovered it, nobody even knew that the unusual little plant existed.

In Central and South America, a plant species and a moth fight a battle for survival that leads to mass poisonings—and mass migrations. Caterpillars produced as part of the life cycle of the jewel-colored *Urania* moth gobble the jungle plants. In turn, the plants protect themselves by producing toxic chemicals. As successive generations of caterpillars eat the leaves, the plants become increasingly poisonous. More and more caterpillars die, until the surviving moths move as a group to lay their eggs on unattacked plants elsewhere. It was only within the past few years that Smithsonian biologist Neal Smith discovered this astounding interaction between plant and animal.

In the Amazon floodplain, fish that live in nutrient-poor rivers swim into forests during the two to four months each year when water is high. They eat seeds and fruit that drop from the trees, and, preposterous as it seems, they have developed flat-topped molars similar to those found in sheep, cows, and other animals that graze on land. In fact, their teeth are strong enough to crush hard nuts. This chewing ability, which explains how nutrients are transferred from forest to fish—and ultimately to the people who catch them—was unexplained until 1980, when Michael Goulding of Brazil's National Institute for Research on Amazonia first reported it.

A Wealth of Flora and Fauna

Such discoveries are routine in the tropics, where the extraordinary is commonplace and where so much is unknown that it is virtually impossible to do biological fieldwork without finding new species of animal or plant or uncovering layers of incredibly complex interactions. The grooved hairs on a sloth's back, the pulp of half-eaten figs—even the droppings of fish and mammals—contain intricate chapters of natural history. But for scientists engaged in a variety of tropical research projects, there is not always cause for celebration. The world's rain forests are disappearing so quickly that biologists may run out of time to ferret out their priceless secrets. Now, while there are jungles left, the rush is on to study them, to understand how they work, and how they might be saved. The stakes are staggering, the findings fascinating.

Tropical forests, from the monsoon jungles of Borneo to the cloud forests of Costa Rica, make up one of the basic life-support systems of the Earth, affecting not only worldwide rainfall but also the pattern of life's evolution. In a sense all modern living things are creatures of the tropics. Jungles are part of our biological heritage. For some 60 million years, tropical forests have been a hothouse for the planet's gene pool, incubators where the great diversity of land plants and animals, including early human ancestors, evolved. Not surprisingly, Charles Darwin and Alfred Russell Wallace independently developed their theories of evolution and natural selection while exploring the tropics. Modern studies of the natural history of tropical forests are helping scientists to learn more about how evolution works.

Unfortunately, many of the discoveries made by researchers in the tropics have been overshadowed by the depressing statistics of deforestation. Twenty million hectares (50 million acres) of the world's tropical forests—in an area about the size of Great Britain—are destroyed each year as forests are cleared for cattle pastures, timber, and human settlements. Burgeoning human populations in tropical countries need homes, food, and firewood. Tropical governments need the revenue from exports to industrialized nations. And the profits industrialized nations reap from tropical resources are many times greater than their investment in saving them. In the short run, there is more money

A steady battle for survival exists between the colorful Urania moth (left) and tropical plants. In response to heavy predation by immature moths, the plants produce such toxic leaves that adult moths are forced to leave the area to find edible plants on which to lay their eggs (below).

Above: James H. Carmichael/Bruce Coleman; Right: Neal G. Smith

in turning tropical forests into human settlements and grazing land than in preserving them.

But short-term gains from deforestation are often far outweighed by long-term consequences, and scientists concerned about lasting impacts are proving that the tropical forest is not the menacing, chaotic jungle depicted in movies and folklore. Rather, it is a tightly organized system of dazzling beauty and complexity. Discovery beckons at every turn—and so do potential, uncharted benefits for people.

Unlike the temperate zone, where coping with winter is a major problem faced by organisms, the crowded, competitive tropical forest poses a different challenge for living things— the need to get along with each other. As a result, some bizarre, seemingly impossible associations have evolved. For example, a number of birds specialize in following army ants as they forage along the forest floor. These antbirds, as they are called, depend for food on the insects flushed up by marauding ants. Antbirds have their followers, too—butterflies that feed on the birds' protein-rich droppings.

Such cooperative relationships are the mainstay of life in tropical forests. The payoff is that this ecosystem, or community of living organisms, has survived longer than any other on the Earth's landmass. Not surprisingly, therefore, much of the current research in the tropics is aimed at understanding the strategies of coexistence among various life forms. Consider a few of the recent finds:

Bugging a Sloth

The most curious thing about sloths is not their slow motion or the fact that they sleep most of the time, or even their habit of copulating while hanging upside down. These interesting quirks are hardly as unusual as the fact that sloths defecate only once a week, an event that is crucial to the life cycle of many creatures that make their homes on the sloth's back.

Sloths are bugged in every sense of the word. Nine species of moths, four species of scarab beetles, six species of ticks, and six or seven species of mites peacefully coexist in the hairy forest of the sloth's coat. Some of the mites live directly on the sloths, but at least two species simply hitchhike on beetles to get from one place to another. Mite survival depends on the maintenance of this peculiar association, says entomologist Jerry Krantz of Oregon State University, who is studying these tiny opportunists.

As many as 978 beetles have been found on a single sloth. They wait for the weekly toilet ritual, when the sloth slowly and laboriously clambers down from its lofty home in the canopy of tropical trees to the forest floor some 18 meters (60 feet) below, digs a shallow hole, and defecates. The virtually odorless droppings provide both food and choice egg-laying sites for many of the assembled company. An unknown number of algae also make themselves at home in the sloth's long, grooved hairs, adding a greenish tint to the creature's fur and camouflaging the animal as it sleeps in leafy treetops.

Nuts to the Bees

The Brazil nut, an important economic crop, produces a tightly closed flower that is difficult to open. Yet it must be opened and pollinated if it is to reproduce. That job falls to the carpenter bee, which lifts the flap of the flower to get at the pollen. Without the carpenter bees of the Amazon forest, the Brazil nut could not perpetuate itself.

Similarly, some 900 species of wild tropical figs depend for pollination on a group of tiny fig wasps. The wasps, in turn, rely on the figs for their own reproduction; they lay their eggs inside the fig's fruit. Nor is that the fig's only role. Figs are one of the most important food sources in tropical forests for a variety of animals worldwide, including orangutans, monkeys, and bats.

Nature's Medicine Chest

Tropical plants are getting the attention of biochemists seeking cures for disease and new weapons to combat destructive agricultural pests. Already, tropical plants yield L-dopa (a drug that treats Parkinson's disease), curare (a muscle relaxant), and certain steroids used in birth control.

In recent tests, one of the most potent pesticides known has been derived from the East African medicinal plant *Ajuga remota*. The plant contains hormones that mimic molting hormones within the bodies of insect larvae. When an insect eats *Ajuga remota*, the plant's chemical acts on cells inside the insect's brain to abort the molt, one of the developmental steps in the gradual transformation of a larva into an adult insect. Chemist Isao Kubo at the University of California at Berkeley discovered that feeding *Ajuga remota* to the larvae of two major cotton pests—the pink bollworm and the fall armyworm—prevents the creatures from shedding their hard external skin, or cuticle. Rather than being cast off, the old cuticle hardens over the insect's mouth and keeps it from eating. Thus, the deformed insects eventually starve to death.

Suicidal Trees

Big trees may seem indestructible, but *Tachigalia* trees kill themselves. For unknown reasons, these huge 48-meter (150-foot) canopy species found in Central America and northern South America have one reproductive event— one flowering in their lives—and then they slowly die. Approximately every four years, a group of mature *Tachigalia* flower, fruit, and die. "Somehow they count years. They have an internal clock," says plant ecologist Robin Foster of the Smithsonian Tropical Research Institute and Chicago's Field Museum, who discovered this mysterious behavior. Foster also discovered that *Tachigalia versicolor*, a species once listed as rare, is actually one of the more common trees in Panama.

John S. Dunning/Photo Researchers

Cooperative relationships between different living things abound in tropical forests. The antbird, for example, follows foraging army ants to feed on insects the ants uncover.

Michael Fogden

The extremely slow-moving sloth is host to a wide array of insects and some algae that live in the creature's coat.

These findings suggest that *Tachigalia* trees—which are fast-growing and which produce good fungus-resistant wood—could be cultivated on plantations and harvested for timber without destroying the natural forest. ''Basically it's a tree that naturally times its own harvest,'' says Foster.

Taming the Babassu

Scientists are also trying to identify, breed, and improve a variety of tropical plants with obvious economic potential. One such plant is the babassu oil palm—already the base of a cottage industry in Brazil and Bolivia.

Some 300,000 families in Brazil depend on the income they earn by harvesting this palm's nuts, which can be as large as a small coconut or the size of a plum. Without machines, they extract the kernels, which can be made into a cooking oil similar to coconut oil. In 1979 more than 250,000 tons of kernels netted more than $61 million, making babassu the largest oil-seed industry in the world that is dependent on a wild plant, according to Michael Balick of the New York Botanical Garden.

The babassu oil palm is also a source of animal food, charcoal, methyl alcohol, tar, ace-

tic acid, flour, thatch used to build tropical dwellings, and a milklike beverage. It can also be used in the production of plastics.

By Jeep, boat, and plane, Balick and his colleagues have led expeditions over thousands of kilometers of Amazon highways and Bolivian forests to collect specimens of the babassu oil palm. Along with Brazilian government scientists, Balick is trying to discover hybrids that could be cultivated. Seedlings are now growing experimentally on a Brazilian plantation, under the care of the government.

Incentives to Deter Destruction

Like much of the basic field research into the natural histories of animals and plants, Foster's investigation of *Tachigalia* could well lead to practical applications in the future. But the pressing problem is time. Scientists don't have the years to accumulate as much raw data as they would like. Consequently, they are trying to use what they already know. Various groups are beginning to translate research into short-term technologies for harvesting tropical resources without destroying any of the remaining forests.

Leading the effort is the Smithsonian Tropical Research Institute in Panama, one of the oldest, most distinguished tropical research centers in the world. It has recently launched a series of projects it calls "Alternatives to Destruction" aimed at sustaining human populations in the tropics without wiping out forests.

For example, the institute is trying to cultivate air-breathing fish. Many natural ponds in the jungle shrink during the tropical dry season, and the crowded conditions create shortages of oxygen in the remaining water. Air-breathing species can cope with this oxygen deficiency; in such fish, a mass of blood vessels on the creature's forehead works much like a human lung, pumping atmospheric oxygen directly into the bloodstream. Should the fish take to artificial ponds, they could provide a much needed native source of protein for people and animals.

Another institute project is a cooperative effort with the Brazilian National Electric Company to raise manatees—large freshwater mammals whose numbers are dwindling because of overhunting in the upper Amazon. The electric company would like to see more manatees in the reservoirs of its hydroelectric plants. There, aquatic vegetation flourishes and tends to wash down into turbines, causing problems in machinery. Since manatees graze on aquatic vegetation, both the animals and the power company stand to benefit if these creatures can be raised in the reservoirs. Biologists also have an interest in research on raising manatees; they'd like to be able to restock areas where these rare animals have disappeared.

The point of schemes like these is to make it more profitable to save the tropics than to destroy them, thereby creating an economic incentive to prevent their decline. But the success of such efforts depends on knowing more about the complex and often complementary lives of jungle creatures—so that scientists can figure out the best ways to help them flourish.

Protein from Pacas

The case of a large nocturnal rodent called the paca illustrates the possibilities. Pacas have the potential for becoming an important source of protein for tropical people. These rodents live in the dense forest, and their meat, which is rich and tastes like pork, sells for more than beef. But a meat industry can't depend on wild stocks.

Smithsonian biologist Nicholas Smythe has spent many hours studying wild pacas. He now knows enough of their natural history to

Michael J. Balick

Scientists are trying to propagate tropical plants with economic potential, such as the babassu oil palm above.

begin trying to domesticate them. The goal is to increase the number of paca offspring, and Smythe says the potential for a bigger brood is all in the animal's anatomy. Pacas typically have one or two young, but female pacas have four to six nipples. "All" Smythe has to do is discover how to increase the litter sizes to correspond with the number of available nipples. If he is successful, the protein yield from pacas could rival that from cattle in the lowland wet tropics. And people would have one more reason to preserve this creature's habitat—the rapidly disappearing jungles of the world □

SELECTED READINGS

"You can keep a good forest down" [effects of slash-and-burn agriculture] by Christopher Uhl. *National History*, April 1983.
"Nature's dwindling treasures: rain forests" by Peter T. White. *National Geographic*, January 1983.
"Teeming life of a rain forest" by Carol Hughes and David Hughes. *National Geographic*, January 1983.

THE ENVIRONMENT **195**

Plastics at Sea

by D. H. S. Wehle and
Felicia C. Coleman

Throughout the 1970's, a number of biologists studying the feeding habits of seabirds in different oceans of the world recounted the same story: the birds were eating plastic. Similar reports of plastic ingestion and of entanglement in plastic debris began to surface for other marine animals—fish off southern New England, turtles off Costa Rica and Japan, and whales in the North Atlantic. At the same time, plastic particles turned up in surface plankton samples from both the Atlantic and Pacific oceans; plastic debris was retrieved by benthic trawls in the Bering Sea and Britain's Bristol Channel; and plastic pellets washed ashore in New Zealand in such large numbers that some beaches were literally covered with "plastic sand." By the close of the decade, marine scientists around the world had become aware of a new problem of increasing ecological concern—plastics at sea.

Particles Widespread

Two forms of plastic exist in the marine environment: "manufactured" and "raw." Manufactured plastic material along beaches and adrift at sea is primarily refuse from transport, fishing, and recreational vessels. In 1975 the National Academy of Sciences estimated that commercial fishing fleets alone dumped more than 24 million kilograms (52 million pounds) of plastic packaging material into the sea and lost approximately 135 million kilograms (298 million pounds) of plastic fishing gear, including nets, lines, and buoys.

Raw plastic particles—spherules, nibs, cylinders, beads, pills, and pellets—are the materials from which products are manufac-

Plastic debris of all shapes and sizes is being dumped into the sea and is showing up on beaches everywhere.

tured. These particles, about the size of the head of a wooden match, enter the ocean via inland waterways and outfalls from plants that manufacture plastic. They are also commonly lost from ships, particularly in the loading and unloading of freighters. Occasionally, large quantities are deliberately dumped into the sea.

Plastics turn up everywhere. Along portions of the industrialized coast of Great Britain, concentrations of raw particles have reached densities of about 186 pieces per square meter (2,000 pieces per square foot) in benthic sediments. Near Auckland, New Zealand, 100,000 pieces of plastic were found for every linear meter (three linear feet) of beach. Particles have also washed ashore on beaches in Texas, Washington, Portugal, Colombia, Lebanon, and at such remote sites as the Aleutian and Galápagos islands.

Much of what we know about the distribution patterns and abundance of raw plastic in the world's oceans comes from plankton sampling of surface waters. Between 1972 and 1975, for example, the Marine Resources Monitoring, Assessment, and Prediction Program, a nationally coordinated program of the National Marine Fisheries Service, recorded plastic particles in plankton samples collected between Cape Cod and the Caribbean Sea. The majority of the particles were found to have entered the ocean from the coast of southern New England, and the highest concentrations were usually in coastal waters. Raw plastic, however, was ubiq-uitous in the open ocean and especially common in the Sargasso Sea. This suggests that winds and currents are instrumental in redistributing and concentrating particles in certain oceanographic regions.

A Proliferating Problem

Inevitably, many animals foraging in the marine environment will encounter and occasionally ingest these widely distributed plastic materials. One of the first records of plastic ingestion appeared in 1962 for an adult Leach's storm petrel collected off Newfoundland. Four years later, researchers in the Hawaiian Islands found that the stomach contents of young Laysan albatrosses contained plastic, apparently fed them by their parents.

For the most part, these early reports were treated as curious anecdotes included in studies of the feeding ecology of a few seabirds. During the 1970's and early 1980's, however, with the proliferation of such anecdotes, biologists began paying closer attention and were surprised to find how frequently plastic occurred in the stomach contents of certain procellariids (short-tailed shearwaters, sooty shearwaters, and northern fulmars) from the North Pacific and the North Atlantic, and alcids (parakeet auklets and horned puffins) from the North Pacific. Lower frequencies were reported for other Northern Hemisphere seabirds, including phalaropes, gulls, terns, and also other procellariids and alcids. The feeding habits of marine birds in southern oceans have not been studied as extensively, but plastic ingestion has been documented for several species of procellariids (petrels, shearwaters, and prions) in the South Atlantic, South Pacific, and subantarctic waters. As of 1983, approximately 15 percent of the world's 280 species of seabirds are known to have ingested plastic.

Seabirds choose a wide array of plastic objects while foraging: raw particles; fragments of processed products; detergent bottle caps; polyethylene bags; and toy soldiers, cars, and animals. Marine turtles, on the other hand, consistently select one item—plastic bags. Since 1968, plastic bags have been found in the stomachs of four of the seven species of marine turtles: leatherbacks from New York, New Jersey, French Guiana, South Africa, and the coast of France; hawksbills on the Caribbean coast of Costa Rica; greens in the South China Sea and in Japanese, Australian, and Central American coastal waters; and olive ridleys in the Pacific

This unfortunate gull may starve to death from one of a six-pack's plastic rings that is stuck around its head.

© Townsend Dickenson/Photo Researchers

These colorful plastic items were retrieved from the nests of a colony of albatrosses.

© Frans Lanting

coastal waters off Mexico. Evidence points to plastic ingestion in loggerhead turtles as well, based on liver samples containing high concentrations of a plasticizer (a chemical compound added to plastic to give it elasticity). Polystyrene spherules have been found in the digestive tracts of one species of chaetognath (transparent, wormlike animals) and eight species of fish in southern New England waters. They have also turned up in sea snails and in several species of bottom-dwelling fishes in the Severn Estuary of southwestern Great Britain.

Marine mammals are not exempt from participation in the plastic feast. Stomachs of a number of beached pygmy sperm whales and rough-toothed dolphins, a Cuvier's beaked whale, and a West Indian manatee contained plastic sheeting or bags. In addition, Minke whales have been sighted eating plastic debris thrown from commercial fishing vessels. Curiously, plastic has not been found in any of the thousands of ribbon, bearded, harbor, spotted, ringed, or northern fur seal stomachs examined from Alaska.

Plastic Mistaken For Food

The obvious question arising from these reports is, Why do marine animals eat plastic? In the most comprehensive study to date, Robert H. Day of the University of Alaska maintains that the ultimate reason for plastic ingestion by Alaskan seabirds lies in plastic's similarity—in color, size, and shape—to natural prey items. In parakeet auklets examined by Day, for example, 94 percent of all the ingested plastic particles were small, light brown, and bore a striking resemblance to the small crustaceans on which the birds typically feed.

Marine turtles also mistake plastic objects for potential food items. Transparent polyethylene bags apparently evoke the same feeding response in sea turtles as do jellyfish and other similar coelenterates, which are the major food item of leatherbacks and subsidiary prey of greens, hawksbills, loggerheads, and ridleys.

Seabirds, marine turtles, and marine mammals all eat plastic. So what? Perhaps ingesting plastic is inconsequential to their health. After all, cows are known to retain nails, metal staples, and strands of barbed wire in their stomachs for more than a year with no ill effects. For marine animals, however, the evidence is growing that in some cases, at least, ingested plastic causes intestinal blockage. George R. Hughes of the Natal Parks Board, South Africa, extracted a ball of plastic from the gut of an emaciated leatherback turtle. When unraveled, the plastic measured 3 meters (9 feet) wide and 4 meters (12 feet) long. There is little doubt that the plastic presented an obstruction to normal digestion. Similarly, a mass mortality of green turtles off Costa Rica has been attributed to the large number of plastic banana bags eaten by the turtles.

The twenty dead red phalaropes discovered on a beach in southern California, all with plastic in their digestive tracts, present a less clear case. Did the birds suffer an adverse physiological response after eating plastic, or were they already under stress because of a reduced food supply and eating the plastic in a last-ditch effort to prevent starvation? The same question applies to other instances of emaciated animals that have eaten plastic. At this time, we don't know the answer.

Effects Potentially Harmful

We do know that plastic is virtually indigestible and that individual pieces may persist and accumulate in the gut. Ingested plastic may reduce an animal's sensation of hunger and thus inhibit feeding activity. This, in turn, could result in low fat reserves and an inability to meet the increased energy demands of reproduction and migration. Plastic may also cause ulcerations in the stomach and intestinal linings, and it is suspected of causing damage to other anatomical structures. Finally, ingestion of plastic may contribute synthetic chemicals to body tissues. Some plasticizers, for example, may concentrate in fatty tissues, their toxic ingredients causing eggshell thinning, aberrant behavior, or tissue damage. When highly contaminated tissues are mobilized for energy, these toxins may be released in lethal doses.

Publication of data on plastic ingestion is in its infancy. As the problem gains notoriety, it will certainly be revealed to be even more widespread than is now recognized. There are already several known instances of secondary ingestion, in which plastic consumed by animals feeding at low trophic levels shows up in higher-level consumers. The remains of a broad-billed prion, together with the plastic pellets it had ingested, were found in the castings of a predatory South Polar skua in the South Atlantic; plastic pellets found in the Galápagos Islands were traced from transport vessels in Ecuadorian ports through a food chain involving fish, blue-footed boobies, and short-eared owls.

A more obvious effect of plastic pollution is the aesthetic one. Whether we venture deep into the woods, high atop a mountain, or out on the ocean to escape the trappings of civilization, our experience of the natural world is often marred by the discovery of human litter. Even more disturbing to the spirit is the sight of a young pelican dangling helplessly from its nest by a fishing line, a whale rising to the surface with its flukes enshrouded in netting, or a seal nursing wounds caused by a plastic band that has cut into its flesh. Unfortunately, such observations are becoming more and more common, another consequence of plastics at sea.

Mortality From Human-Generated Debris

During the past twenty years, fishing pressure has increased dramatically in all the world's oceans, and with it, the amount of fishing-related debris dumped into the sea. In addition, the kind of fishing equipment finding its way into the ocean has changed. Traditionally, fishing nets were made of hemp, cotton, or flax, which sank if not buoyed up. These materials disintegrated within a relatively short time and, because of the size of the fibers, were largely avoided by diving seabirds and marine mammals. With the advent of synthetic fibers after World War II, however, different kinds of nets came into use. These new nets were more buoyant and longer-lived than their predecessors, and some of them were nearly invisible under water.

The result of these changes in net materials has been a tragic increase in mortality of air-breathing animals. A few examples are sufficient to give an idea of the magnitude of the problem. During the heyday (1972–76) of the Danish salmon fishery in the North Atlantic, the incidental catch of thick-billed murres amounted to 750,000 birds annually; in 1980, 2,000 sea turtles off the southeastern coast of the United States drowned when incidentally caught in shrimp trawl nets. (Incidental catch refers to nontarget animals that are accidentally caught in an actively working net.) Another kind of net-related mortality is known as entanglement and refers to any animal caught in a net that has been lost or discarded at sea. Some government officials estimate that about 50,000 northern fur seals currently die in the North Pacific each year as a result of entanglement in fishing gear. Unlike working nets, which fish for specific periods of time, these free-floating nets, often broken into fragments, fish indefinitely. When washed ashore, they may also threaten land birds and mammals. In the Aleutian Islands, for example, a reindeer became entangled in a Japanese gill net.

Plastic strapping bands—used to secure crates, bundles of netting, and other cargo—are another common form of ship-generated debris. Discarded bands are often found girdling marine mammals, which are particularly susceptible to entanglement because of their proclivity for examining floating objects. The instances of seal entanglement in plastic bands has increased so remarkably in the past two decades that fur seal harvesters in Alaska and South Africa now monitor the number of ringed animals.

Seabirds that frequent recreational waters or coastal dumps are also subject to ringing by the plastic yokes used in packaging six-packs of beer and soda pop. Gulls with rings caught around their necks are sometimes strangled when the free end of the yoke snags on protruding objects. Similarly, pelicans, which plunge into the water to feed, run the risk of diving into yokes. If the rings become firmly wedged around their bills, the birds may starve.

This pelican died as a result of being hopelessly entangled in fishing line. Net-related mortality is a growing problem.

Above: A loggerhead turtle cruises through the sea with a fishing net wrapped all around its body. Right: If left unassisted, this California sea lion could strangle from the fish net caught around its neck.

Top: © C. C. Lockwood/Animals Animals; Right: Frans Lanting

Some Organisms Adapt

Not all encounters with plastic prove harmful to marine organisms. Some animals are incorporating the new material into their lives. Algae, hydrozoans, bryozoans, polychaetes (marine worms), and small crustaceans attach to plastic floating at sea; bacteria proliferate in both raw and processed plastic refuse. Plastic provides these organisms with long-lived substrates for attachment and transport; in some cases, hitching a ride on floating pieces of plastic may alter an organism's normal distribution. Several species of tube-dwelling polychaetes construct their tubes of raw plastic particles present in benthic sediments. Other invertebrates, such as sand hoppers and periwinkles, find temporary homes in aggregates of plastic particles they encounter on beaches. Marine birds all over the world incorporate plastic litter into their nests, but in this case, the use of plastic may be harmful because chicks can become entangled in the debris and die.

Instances of marine animals adapting to this new element in their environment do not alter the predominately negative effect of plastics at sea. The problem is global, and its solution will require international cooperation. Historically, the high seas have, in many respects, been considered an international no-man's-land. Recently, however, perception of the ocean as a finite resource has caused many nations to express concern for its well-being.

Controlling Disposal

In 1970 the U.S. Congress passed the National Environmental Policy Act, which, among other things, pledged to "encourage productive and enjoyable harmony between man and his environment." Subsequently, a number of laws on waste disposal were adopted, two of which affect pollution by plastics: the Federal Water Pollution Control Act (commonly known as the Clean Water Act) and the Marine Protection, Research, and Sanctuaries Act (Ocean Dumping Act). The Clean Water Act does not specifically address the problem of persistent plastics but does require all significant polluters of U.S. waterways to obtain a federal permit, under which limits are set on, among other things, discharges of solid matter. The Ocean Dumping Act prohibits the deliberate dumping of significant amounts of persistent plastic materials at sea. Having laws on the books, however, does not immediately solve the problem. Small-scale refuse disposal on the high seas is difficult to regulate; fishermen who claim to have unintentionally lost their nets at sea cannot be held responsible; and illegal large-scale dumping at sea is hard to detect. Granted, laws must be tightened, but enforcement is really the bigger problem.

On the international level, the problems of water pollution and litter in the oceans were highlighted at the United Nations Conference on the Human Environment held in Stockholm in 1972. The conference, with 110 nations represented, defined the need for international policy on marine pollution among coastal and maritime nations. Treaties to implement such a policy soon followed: the 1972 London Convention on the Prevention of Water Pollution by Dumping of Wastes and Other Matter (Ocean Dumping Convention), a part of which specifically prohibits marine dumping of persistent plastic material; and the 1973 London International Convention for the Prevention of Pollution from Ships (Marine Pollution Convention), which is broader in scope and regulates the control of oil pollution, packaged substances, sewage, and garbage. While neither of these treaties has been adopted by all nations, they represent a start toward global control of marine pollution.

United Effort Needed

In the meantime, the quantity of plastics in the world's oceans will undoubtedly continue to mount. Ironically, the very characteristics that make plastic appropriate for so many uses—its light weight, strength, and durability—lead to the majority of problems associated with its presence at sea. As organic material, plastic is theoretically subject to degradation by mechanical, oxidative, or microbial means. Owing to the strength of most plastics, however, mechanical degradation by wave action is generally restricted to the breaking of large pieces into smaller ones. Photooxidation and microbial action are limited by plastic's high molecular weight and its antioxidants, ultraviolet light stabilizers, and biocide additives, which effectively immunize it against degradation. The longevity of plastics in seawater is not known, but on the beach, particles may last from 5 to more than 50 years.

Given plastic's long life and projected annual increases in production, one thing is clear—the rate of plastic deposition in the marine environment will continue to be higher than the rate of disappearance. In a study of the accumulation of plastic on the beaches of Amchitka

Pelican-lover Ralph Heath displays jars of plastic material that he removed from the birds behind him.

Island in Alaska, Theodore R. Merrell, Jr., of the National Marine Fisheries Service, recorded that 250 kilograms (550 pounds) of plastic litter were added to less than 1.6 kilometers (1 mile) of beach in one year. He also found an increase of more than 250 percent in both the number and the weight of plastic items washed ashore over a two-year period.

Outside the realm of laws and treaties, solutions to the problem can come from both inside and outside the plastic industry. The technology to manufacture biodegradable plastics is available. In fact, one of the beauties of plastic is that its properties can be altered and its life expectancy prescribed. Alaska has already taken steps toward reducing plastic litter by requiring that plastic six-pack yokes be made of a self-destructing compound. Another, but perhaps less workable solution, given the logistics and expense involved and the degree of business and public cooperation required, lies in recyclable plastics. At the very least, all countries should require that the discharge of raw plastic particles from industrial plants be reduced by filtering outflow before it enters waterways. A recent decline in the uptake of plastic by marine organisms in southwestern England has been attributed, in part, to the efforts of one of the major contaminating plants to filter, collect, and reuse raw particles present in its effluent.

Consumers share with industry the responsibility to reduce the amount of plastic in the sea. Recreational boaters, beachgoers, and commercial fishermen all discard plastic refuse. Preferably, no trash plastic—bands, netting, or other debris—should ever be tossed overboard or left on a beach. If six-pack yokes or strapping bands must be discarded at sea, the rings should be cut first so that they pose less of a threat to marine animals.

The first step in combating plastic pollution is to alert both industry and the general public to the gravity of the problem and the need to do something about it soon. Education alone cannot solve the problem, but it is a beginning. Public awareness of the problem, combined with the resolve to correct it, can bring dramatic results □

Meeting Ground
for the Masses

by Peter Steinhart

I t is one of the few places in the United States where you can still see wildlife in astonishing numbers. There are sky-blackening swirls of ducks and geese, and water-thronging masses of grebes and shorebirds. At the peak of the migration season, more than 2 million waterfowl pass through the region. In late fall you may see 500,000 pintail ducks, 100,000 snow geese, and 80 percent of the world's Ross's geese. In winter the area boasts one of the largest populations of roosting bald eagles outside of Alaska. In spring 10,000 whistling swans fly into the basin. There are mule deer and antelope and muskrats. And the setting is framed with snowcapped peaks, sage-covered hills, and broad blue lakes.

Human Influence Improves Habitats

The place is the Klamath Basin, located on the Oregon-California border. Standing in this vast expanse of lake, marsh, and desert, you can easily lose your voice in the rush of wings and gabble of geese. You might think that this is exactly how the place looked centuries ago, before the first Europeans set foot on the continent. But you'd be wrong. In fact, many of these wetlands are largely the products of effective management efforts. Today a cluster of six separate habitats—known collectively as the Klamath Basin National Wildlife Refuges—is being managed as intensively as the farmlands that surround it. Like many other areas included in the federal refuge system, this basin was protected only after it had been altered by humans.

Robert Fields, trim and athletic at almost 50, superintends the 61,110 hectares (151,000 acres) that make up the refuges. There is something tight-lipped and apprehensive in Fields's manner. He speaks with the tone of a man who knows he must constantly keep bailing to hold the ship afloat. Fields stands on a recently built levee (an earthen bank to confine water) that encloses 240 hectares (600 acres) of marsh and open water, and points out a newly dredged channel through large bulrushes called tules. A few coots and scaups float on it. The levee is no engineering wonder, but it is symbolic of the work that goes on here.

Opposite: Thousands of snow and Canada geese fill the sky over the Klamath Basin during the fall migration.

Photos: © Tupper Ansel Blake

"It's really obvious that this is a man-created environment," says Fields. "Most of the people who say we should let nature take its course don't have an understanding of what would happen if it were turned over to nature: these wetlands would go through natural succession and would eventually return to upland areas." The marsh he is overlooking is itself fostered by the refuge managers. When the tules grow too dense to provide good cover for ducks and pheasants, the refuge workers will burn them off to make way for a younger marsh.

"Many of the refuges in the system were taken over from agricultural projects that didn't work or lands that were abused," notes Fields. "They weren't in a natural condition. To let them now revert to whatever state they would fall to if left alone would be to doom this magnificent waterfowl habitat."

Farmers and Waterfowl Share Resources

Fields points out the work activities going on around him. A World War II surplus dragline, still bearing a white star and olive drab paint, piles mud on a levee. New channels are dredged to provide more edges between marsh and water. New drains are put in. Old fields are flooded. The refuge's staff of 18 is constantly redesigning sections of marshland to improve habitat.

All around the open water, there are diked-off rectangles of green—fields of barley and alfalfa that have been reclaimed from ancient lake beds. Some are farmed by the U.S. Fish and Wildlife Service (FWS), which leaves the crops standing for autumn flocks to feed upon. Other fields are leased to local farmers, who harvest their entire crop. In the Tule Lake portion of the refuge, some 6,100 hectares (15,000 acres) are leased to farmers who harvest their entire crop but who agree to limit fall plowing and plant no more than 25 percent of the fields in row crops. Nearby, in the Lower Klamath Refuge area, 2,000 hectares (5,000 acres) of land are sharecropped; one-third of those crops are left in the ground for the waterfowl. "I'm sure there is no place in the refuge system where there is so much agriculture and so many birds using the same land," says Fields.

Sometimes a combination of farming and vast numbers of waterfowl leads to bitter conflict. In this case, however, birds and farmers get along because most of the waterfowl use the refuge in the fall and winter, when there is no farming. Most of the farming takes place in

Refuge manager Robert Fields keeps watch as a controlled fire spreads through a section of the reserve where marsh grasses have grown too dense to provide ideal habitat. New plants will soon sprout up and provide good cover for waterfowl and other wildlife.

summer, when the great flocks are in their northern breeding grounds.

Even livestock, a source of controversy in several western wildlife refuges, appears to be compatible with wildlife here. About 3,600 hectares (9,000 acres) of the Lower Klamath Refuge are grazed. Observes Fields: "We view it as a nonconflicting use." He explains that the cattle graze only in the fall and winter, when they cannot disturb nesting birds, and that the lands grazed are usually flooded again in the spring.

At First, Extensive Hunting and Drainage

Livestock and farmers are newcomers to the region. A century ago the basin was a vast network of lakes and meadows. Here the migration corridor used by birds in the Pacific Flyway narrows between the Cascade Range to the west and the arid Great Basin to the east, and in this bottleneck, waterfowl flying from Alaska and Canada in the fall concentrate. Early in the century, the waterfowl drew swarms of market hunters. In 1903 there were 30 separate camps of hunters on Lower Klamath Lake alone, and they sent 120 tons of ducks and geese to poultry markets in San Francisco, California. At the same time, plume seekers, drawn by the masses of grebes and terns, packed skins in bales and sent them by the thousands to New York milliners (makers of women's headwear). Despite those massive harvests, in 1905 an observer reported that the region "is the greatest feeding and breeding ground for waterfowl on the Pacific Coast."

By that time, however, human development had already begun to change the landscape. California and Oregon had ceded their lands in the basin to the federal government in the early 1900's, and the Bureau of Reclamation began to drain the lakes to create farmlands. The annual flooding of the Klamath River, which once filled the 360-square-kilometer (140-square-mile) Lower Klamath Lake, was eliminated by the construction of water-control gates in the channel, and the lake receded dramatically. By the 1930's, says Fields, it contained only about 400 hectares (1,000 acres) of open water.

Problems with the First Waterfowl Refuge

In 1908 President Theodore Roosevelt set aside 33,031 hectares (81,619 acres) on Lower Klamath Lake as a bird refuge—the nation's first waterfowl refuge. But the reserve was given no specific water rights, and the lands were left under the administration of the Bureau of Reclamation. Moreover, the executive order creating the refuge reserved only lands "unsuitable for agricultural purposes." The Bureau of Reclamation looked upon the refuge as an afterthought of development. And as the lake receded and more land suitable for agriculture surfaced, the size of the waterfowl reserve was reduced by successive laws. Today it comprises only 19,257 hectares (47,583 acres).

At the turn of the century, federal refuge biologists did not attempt to manage wildlife. Instead, officials concentrated their efforts almost exclusively upon controlling illegal shoot-

A male sage grouse puffs out air sacs on his breast and fans his tail in a spectacular display to attract a mate.

ing. Early wardens in the Klamath Basin had a hard time. The tradition of unregulated hunting was so strong that some people scoffed at the laws. There was no permanent federal officer on duty until 1928. Only after a warden named Alva Lewis arrested the mayor and three leading citizens of Merrill, Oregon, for poaching did area residents begin to respect the laws.

But the continuing loss of waterfowl habitat was beyond both the abilities of the managers and the scope of the law. The purpose of the reclamation project conflicted with the purpose of the refuge, and little was done to protect waterfowl habitat. "There were years when just nothing went on here," says Fields. "The refuge was just kind of on paper, and the Klamath project didn't pay much attention to it." In 1928, largely to make up for the reduction in

size of Lower Klamath Refuge, a second refuge of 4,170 hectares (10,300 acres) was established on federal lands at Tule Lake. In 1942 a pumping plant was built on the western edge of Tule Lake, and a 1,830-meter (6,000-foot) tunnel was excavated to allow excess irrigation water to be pumped into Lower Klamath Refuge. In 1965 Congress closed lands in the refuges to further homesteading. And over the years, more dikes, channels, pumping plants, and drains have been added to the refuge by the FWS.

Total Control of Water Levels

The new plumbing gives present wildlife managers remarkable flexibility to control water levels in the impoundments. They can cover the Lower Klamath Refuge with water or nearly drain it dry. "We can create the habitat we want. It's all subject to manipulation," observes Fields, whose first concern these days is the maintenance of proper water levels in the refuges. If Tule Lake, which is filled largely by runoff from agricultural fields, overflows during the nesting season—when farmers use the most water—the nests of ducks, geese, and other water birds may be flooded. Thus, during that time of year, excess water is pumped through the tunnel to the Lower Klamath Refuge, where it nurtures marshes. And in winter, when nesting habitat is not important but open water for flocking is needed, large parts of the refuge can be flooded. The Lower Klamath Refuge covers 7,200 hectares (17,700 acres) in nesting season, but 16,200 hectares (40,000 acres) in winter.

"It's a very healthy situation now," says Ralph Opp, a biologist with the Oregon Department of Fish and Wildlife. "It's very manipulated, but that's what it has to be, because we're using leftover water. The refuges play an important role in the Pacific Flyway. We're down to about 10 percent of the historic populations of waterfowl in the flyway, and we hate to lose another acre of wetlands."

Significant Population Declines

Despite the efforts of managers in the Klamath Basin, waterfowl populations on the Pacific Flyway continue to decline. Some of the most persuasive evidence comes from the Klamath refuges themselves, where the same people have performed the annual bird counts for over 20 years. Their tallies show that fewer ducks and geese arrive in the fall, and those that do arrive stay longer into the winter. White-fronted goose populations are down 50 percent from the early 1970's, and snow goose populations have declined 30 percent. Habitat loss is a possible cause, but nobody knows for sure. Healthy refuges such as these in the Klamath Basin should help provide new understanding of the decline. Says Fields: "As our knowledge expands, we'll probably find that the explanation is very complex."

While the managers at the Klamath refuges bemoan the decline, they concede that it does give them time to take better care of the present duck and goose populations, as well as to concentrate more on other species. Standing on a levee, Fields points to a flotilla of white pelicans on a nearby channel. The pelicans nest at Clear Lake National Wildlife Refuge, one of the six areas Fields administers. "This is one of three places in this part of the country where there are pelican colonies," he says. "They are very sensitive to human disturbance." He adds that several hundred bald eagles winter in the basin.

The Klamath Basin is an important feeding site for several hundred wintering bald eagles. Refuge managers are helping to preserve eagle roosts that are outside the reserve.

Besides waterfowl, refuge biologists attend to the habitat needs of other animals, including pronghorn antelope.

Expanded Role for Managers

"We're concerned about a wider array of species now," says Fields. "We're paying more attention to the habitat needs of shorebirds and raptors. We can't just be a duck and goose outfit." Nor can the managers confine their attentions to what goes on within refuge boundaries. Refuge officials have, for example, negotiated agreements with federal and state authorities and with private landowners to preserve eagle roosts outside the refuge. "The work here requires a lot of coordination," Fields observes. "We deal with three counties, several irrigation districts, and a myriad of state and federal agencies. We keep track of what's going on. You quickly realize that if you're going to be successful as a refuge manager, you can't be bound by the signs that mark your refuge boundaries."

Through five decades, the refuge managers at Klamath have seen their responsibilities expand from chasing poachers to modifying and managing habitat. They now face new challenges as a result of the shrinking federal budget. In 1979, before the Reagan administration took office, the FWS estimated that 54 percent of all national wildlife refuge facilities throughout the country were improperly maintained. Now there is a growing backlog of needed repairs of everything from pumps and ditches to levees and fences in the system's more than 400 refuges.

Budget Cuts Pose Growing Challenges

The Klamath refuges have been spared severe budget cuts, and to date, they have escaped any of the proposed moneymaking activities—such as oil and geothermal energy development, increased logging and grazing—that have been suggested for other federal wildlife preserves. But Fields and his staff have had to tighten their belts, cutting down on equipment maintenance and facilities and leaving positions vacant.

Since most of the costs at Klamath are in salaries, however, Fields is concerned about the possible effects of future funding cuts. "You let people go and you don't have the ability to run the ditches and clear the drains," he says. "This area requires a high level of maintenance. If a ditch fills with silt, the water stops running. You can't keep cutting back on things, because soon everything will stop working."

For Fields, though, there is plenty of incentive to face the mounting challenges ahead. "This is one of the premier wildlife areas in the whole system," he maintains. "It's really something to see in the fall—one huge, swirling mass of birds. I can't think of anyplace that's more exciting or more worth the hard work that's required to keep it going" □

AQUACULTURE by William J. Cromie

E veryone has heard of farmers who raise
corn and cows—but how about catfish
and shrimp? These and other aquatic
creatures are being cultivated like crops on
farms around the world.

Most of the catfish and rainbow trout eaten
in the United States are raised on farms. Ven-
turesome fish ranchers release herds of salmon
to graze for years on ocean pastures, and rely on
their homing instincts to guide them back at
"roundup" time. In Texas, shrimp are grown in
warm waters laced with treated sewage. Entrepreneurs in California are trying to commercial-
ize the work of researchers who raise lobsters
from eggs to market size in the laboratory.

Fish Farming to Aid World Hunger . . .

The raising of plants, fish, and other aquatic
edibles under farmlike conditions in either fresh
or salt water is called aquaculture. It is an
important industry in many parts of the world.
Asian farmers have raised fish in ponds—often
in combination with ducks, rice, and other
crops—for thousands of years. Today roughly
10 percent of all the fish consumed in the world
come from farms—most of them in Asia. Farm-
raised shrimp represent the second most important export from Ecuador. Half of the fish eaten
in Israel, and more than 25 percent of those consumed in China and India, are cultivated like
rice. The world's aquafarms produce as much as

9.5 billion kilograms (21 billion pounds) of food each year.

Aquaculture did not leap into world consciousness until the 1950's, after scientists and statesmen came up with the idea of feeding the planet's exploding population of starving and malnourished people from the oceans' storehouse. These thinkers viewed Earth's great bodies of water as vast tanks filled with fish and other high-protein food.

Unfortunately, things did not work out as planned. A careful estimate of the total amount of food that could be taken from the sea annually—without endangering future supplies—came to only 130 billion kilograms (286 billion pounds). This is hardly an inexhaustible food supply for the 4.5 billion people in the world, 3.3 billion of whom live in poor, underdeveloped nations. To make matters worse, with the present yearly catch of about 73 billion kilograms (160 billion pounds), stocks of popular seafoods already are seriously depleted.

Faced with limited natural food supplies and a rising demand for low-cost, low-cholesterol protein, experts came up with a new strategy for feeding the world—expanding aquaculture from a cottage industry in Asia to a mass production effort.

. . . But Not Quite as Expected

The hungry are still waiting. Lack of scientific knowledge, technology, and disease-control methods, coupled with environmental and political problems, have proved to be formidable enemies of the so-called "blue revolution." The hoped-for victory has not occurred; on the other hand, the revolution has not failed.

Much of the effort to date has gone into gaining the knowledge needed to successfully breed, feed, nurse, and nurture aquatic animals in captivity. Scientists had to discover new facts about genetics, nutrition, growth rates, and disease control. Although a great deal remains to be learned, aquafarmers armed with a new body of scientific and technical knowledge are turning aquaculture into a dynamic industry.

Cultivating Catfish

The largest, most successful such business in the U.S., catfish production soared from 8 million kilograms (17 million pounds) in 1975 to about 90 million kilograms (200 million pounds) in 1983. The Joint Subcommittee on Aquaculture, made up of representatives from federal agencies that support or regulate aqua-

Catfish production is the most successful aquacultural business in the U.S. A Mississippi fish farm is shown.

culture, put the value of the 1983 catfish crop as high as $130 million.

Most catfish are raised in diked ponds in Mississippi, Alabama, and Arkansas. Large growers maintain both nursery and production ponds, transferring 10-centimeter (4-inch)-long fingerlings (young fish) from the former ponds to the latter, where they reach a half-kilogram (1 pound) or more in weight. Smaller operators stock production ponds with fingerlings raised elsewhere. Pond yields have increased significantly in the past 10 years, with some farmers now obtaining as much as 1,100 kilograms per hectare (6,000 pounds per acre). According to Bille Hougart of the U.S. Department of Agriculture, the increase results from better nutrition, use of aerating devices to increase the oxygen content of pond waters, and continuous harvesting. Farmers once drained their ponds and collected the fish when they grew to market size. Now they keep fish of all sizes in one pond, and add one fingerling for each half-kilogram (1 pound) of catfish taken out. This allows three or four harvests a year, and the ponds need to be drained only once every three to four years.

Farmers in Idaho, California, and Nevada raise catfish in geothermal waters, which rise to the surface after contact with hot rocks deep in the earth. Farmer Leo Ray raises fish in Idaho's Snake River Valley with the help of geothermal springs that remain at 27° C to 29° C (80° F to 85° F) year round—optimum temperatures for the growth of catfish. Ray directs this water through a series of 3- by 7-meter (10- by 22-foot) raceways, each holding about 5,000 fish. A 7,600-liter (2,000 gallon)-a-minute flow maintains high oxygen levels and flushes wastes from the densely populated raceways. Ray and Hougart believe that catfish production can be tripled with the aid of better water-quality control and selective breeding of fish that will survive best and grow quickest in farm ponds and raceways.

Trout Farming

Idaho is better known for trout farms than for catfish cultivation. Ninety percent of the 1982 crop came from within 32 kilometers (20 miles) of Buhl, a small town in the southwestern part of the state. This area boasts the most important ingredient for successful trout rearing: endless volumes of clean water that gush from a vast underground aquifer at a uniform temperature year round. The 14° C (58° F) water allows trout to grow all year, even during winter months when the fish normally are dormant.

The largest trout farm in the world—the Clear Springs Trout Company—boasts its own hatchery, processing plant, and feed mill. After birth, juveniles grow to 8 centimeters (3 inches) in the indoor hatchery, then workers move them to outdoor raceways, each of which holds as many as 200,000 trout. A computer calculates the daily ration for each raceway, and the food is dispensed automatically. When the fish reach 25 to 40 centimeters (10 to 15 inches) in length, farmers harvest them and truck them to the processing plant. The entire process from birth to market takes only 9 to 12 months.

Salmon Ranching

Salmon ranching involves more time and natural hazards than other kinds of aquafarming. Ranchers raise juveniles in hatcheries until they reach the smolt stage—the time at which wild salmon are ready to leave freshwater, where they are born—and migrate to sea.

The smolts are taken to holding pens at the mouths of rivers. Each river's distinct odor, which comes from nearby plants and soils, becomes imprinted on the nervous systems of the fish when they are then released. This odor guides the salmon back to the pens after round-trip migrations to and from ocean feeding grounds. These trips may be thousands of kilometers in each direction, and the fish may stay in saltwater as long as five years before return-

The waterfalls at this Oregon trout farm help reoxygenate the water in these raceways where the bigger fish are kept.

Above: A typical Domsea ocean farm for salmon consists of a complex of floating net pens. Left: These salmon eggs will be raised in a hatchery until they are smolts.

ing home to spawn. How salmon find their way to their feeding grounds and return by scent to where they spend their early youth remains one of the greatest mysteries of animal behavior.

Aqua-Foods, a subsidiary of the giant Weyerhaeuser Corporation, has been trying since 1975 to raise salmon that will return to the company's pens. To make a profit, the company must get back 2 percent of all the smolts released. This has not yet happened. For example, Aqua-Foods recently released 4 million smolts and got back only 30,000 adults, or less than 1 percent.

The company will keep trying despite unprofitable returns and problems with environmentalists and commercial fishermen. The former worry that mass releases of cultivated fish will overtax food supplies and other resources of coastal estuaries. The fishermen claim that interbreeding between wild and domesticated salmon will produce hybrids that will be of lower quality and harder to catch. Conrad Mahnken of the National Marine Fisheries Service sees such opposition as "the primary obstacles to the success of salmon ranching in the U.S." It already has resulted in a ban on salmon ranches in the state of Washington.

Laws in Washington and other states, however, do not prohibit fencing in parts of the ocean for farming. Domsea Farms, a subsidiary of Campbell Soup Company, has done this in Puget Sound near Bremerton, Washington. Hatchery-reared smolts are placed in floating nylon pens and fed as often as 15 times a day for 6 to 12 months. When the salmon reach 3.5 to 5.5 kilograms (8 to 12 pounds) in weight, they are vacuumed into a barge and taken to a nearby processing plant.

At the Johnson Oyster Farm in California, a worker threads oysters on strings for hanging cultivation.

Oyster and Crawfish Husbandry

The demand for oysters and crawfish is less than that for salmon, but aquafarmers do a thriving business in these two crops. About one-third of the 36 million kilograms (80 million pounds) of oysters eaten in the U.S. are raised on farms. (The rest come from natural beds or are imported.) The farms—located on the East, West, and Gulf coasts—start with wild seed (very young oysters), seed from captive animals, or seed from commercial hatcheries. Growth to market size may take as long as five years. However, Long Island Oyster Farms in New York, one of the world's largest growers, succeeded in reducing the process by one year. The company raised its crops in the heated cooling-

water discharge of a power plant. Another advance has been selective breeding to produce hardier, faster-growing strains of oysters. Farms now produce more than 700 kilograms per hectare (4,000 pounds per acre), compared to 16 kilograms per hectare (90 pounds an acre) from natural beds.

While the consumption of oysters has remained fairly constant for the past 20 years, the demand for crawfish has increased dramatically. These small, savory shellfish, which resemble small lobsters, are raised in shallow farm ponds. The total area in the U.S. devoted to their cultivation has expanded from 2,400 to 46,500 hectares (6,000 to 115,000 acres) in the past two decades. Each year an estimated 25 million kilograms (55 million pounds) of crawfish, worth $27 million, are grown on farms, located primarily in Louisiana, Texas, and Mississippi. Many farmers grow them in flooded rice fields. Juveniles must be purchased as "seed," but once they are planted, the crawfish often do not require feeding. They eat vegetation and tiny animals that live in the ponds.

Maximizing Shrimp Crops

The extremely popular shrimp is a tasty crustacean and the most valuable U.S. fishery product. However, the fishery cannot keep pace with a rising demand because the amount of shrimp captured already approaches the maximum that can be taken from U.S. waters without depleting wild stocks. The difference must be made up by imports.

The imports include shrimp raised on farms in Ecuador, Panama, and Taiwan. Shrimp farming in Ecuador is heavily supported by U.S. corporations. It has been very successful because of cheap labor, a long growing season in seawater that stays above 27° C (80° F), and a plentiful supply of egg-carrying females and immature shrimp. The latter poses the highest hurdle to successful shrimp farming in the U.S. because federal laws prohibit capture of immature shrimp, and the supply of wild gravid (pregnant) females is unpredictable.

"The only way to obtain a reliable year-round supply of seed stock is to get adults to mate and spawn in captivity," declares Addison L. Lawrence of Texas A & M University. He and his colleagues have done so on a small scale for four different species. They used artificial insemination to crossbreed Gulf of Mexico white shrimp with a species that lives off the Pacific coast from Peru to Mexico. The resulting hybrid is better adapted for mating in captivity and living in farm ponds.

The Texans also developed a vaccine against a deadly bacteria that attacks farm-raised shrimp. And they discovered that removing one of the animal's eyestalks, which contains a hormone that inhibits sexual maturation, enhances growth and spawning. Applying these research results and those directed at developing a proper diet, Lawrence's group grew shrimp in ponds warmed by the cooling-water discharge from a Corpus Christi, Texas, power plant and artificially fertilized with treated sewage. In 1982 they produced shrimp crops of more than 400 kilograms per hectare (2,000 pounds per acre). Lawrence believes that such experimental efforts will make successful commercial shrimp farms possible in the U.S. by 1988.

Aquaculturists are attempting to achieve even higher yields by raising shrimp in raceways through which heated seawater circulates. In an experimental project conducted by the University of Arizona, researchers obtained the equivalent of 18,300 kilograms per hectare (100,000 pounds per acre), according to Wayne Collins of the university's Environmental Research Laboratory. The F. R. Prince Company and W. R. Grace and Company cosponsor a project in Hawaii to turn this experimental technology into a commercial reality.

Lobster Cannibalism Deters Commercialization

Like shrimp, lobsters are ideal candidates for aquaculture because U.S. appetites for them outstrip the annual catch of the country's fishery. Researchers at the Bodega Bay Marine Laboratory in California have successfully mated and raised lobsters year round in captivity. By manipulating water temperature and light, they coax the females to lay eggs at any time of year, rather than during one season. Raising the offspring in water temperatures higher than those in their natural habitats brings them to half-kilogram (1-pound) size in less than three years rather than the more than five years it takes to do so in the wild.

One formidable problem remains, however. The North American, or Maine, lobster prefers dining on lobster to the food pellets offered by farmers. Keeping each lobster in a separate compartment prevents cannibalism but raises costs to an uneconomical level. This difficulty has terminated a number of attempts to farm lobsters commercially. Yet entrepreneurs keep trying because of the potential for profit.

Problems But a Promising Future

In 1960 Congress passed the National Aquaculture Act ''to encourage the development of aquaculture in the United States.'' The Act called for writing a plan to outline the constraints and opportunities associated with this relatively new industry, and to recommend actions to solve the problems and achieve the potentials. In 1983 the Joint Subcommittee on Aquaculture submitted a National Aquaculture Development Plan to President Reagan and Congress. This document contains detailed descriptions of the status and potential of 12 aquacrops: largemouth and striped bass; catfish; crawfish; clams; mussels; oysters; freshwater prawns; salmon; shrimp; trout; and baitfish such as golden shiners, minnows, white suckers, and goldfish. It also lists more than 20 other species as being farmed or having a potential for aquaculture.

The plan acknowledges the significant progress made in basic research and applied technology in the past 15 years. It notes that the greatest obstacles to success are not scientific but social and political. Confronting every aquafarmer is a formidable wall of overlapping regulations and permits, governing everything from animal health inspections to dredge-and-fill permits. Besides this regulatory burden, aquaculturists face competition and opposition from commercial fishermen, residential and industrial developers, and environmentalists.

Despite these many problems, research advances continue and aquaculture keeps expanding. Factors such as declines in wild stocks, increases in the cost of fishing, and the rise in demand for fish and seafood create a mood of optimism and a willingness to invest. One market-research firm predicts that U.S. aquaculture will grow into a $1,600 million industry by 1989. On a worldwide basis, pessimists predict that annual production will triple by the year 2000 to about 27 billion kilograms (60 billion pounds). Optimists forecast a yield of 47 billion kilograms (104 billion pounds), or about half of the expected production of the entire world's fishery.

The bottom line appears to be that aquaculture is now on its way to fulfilling its promise as a mass producer of protein, a new source of jobs, and, in the U.S., a way to reduce the foreign trade deficit. However, even the most feverish proponent of aquaculture no longer believes that it will completely solve the world's hunger problem☐

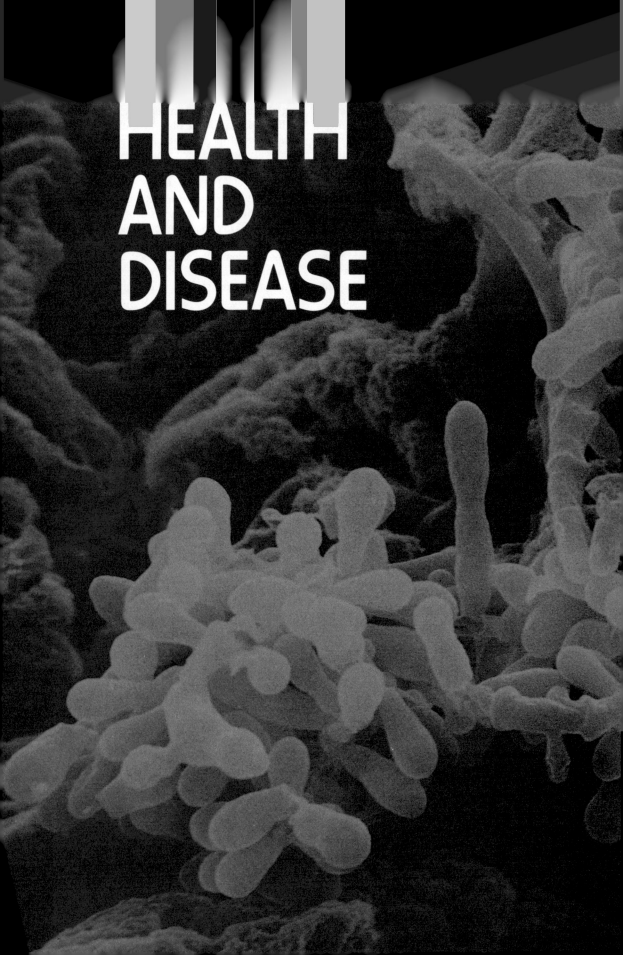

HEALTH
AND
DISEASE

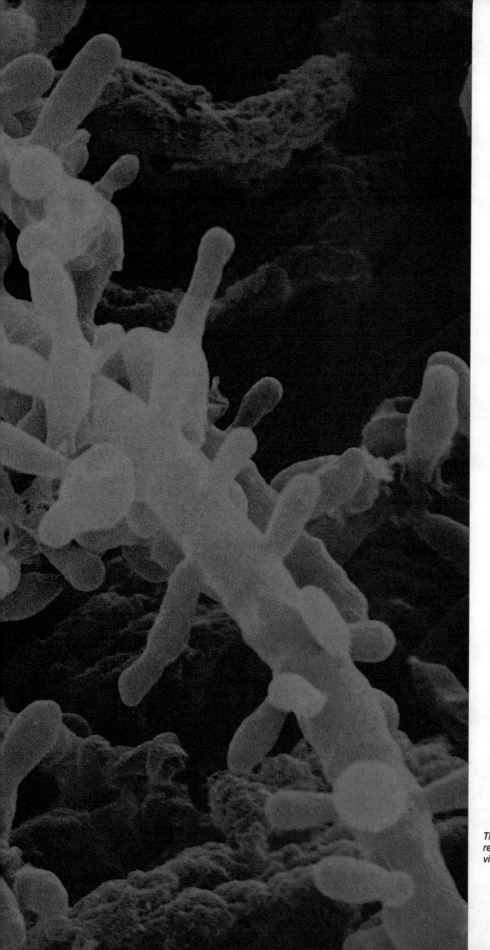

This photomicrograph represents a detailed view of a skin fungus.

REVIEW
OF THE
YEAR

HEALTH AND DISEASE

Medical news in 1983 included significant
advances in understanding AIDS; promising signs
in the treatment of heart and blood vessel
disorders; increased attention to the detection and
prevention of testicular cancer and a form of liver
cancer; the development of new techniques for
kidney stone removal; and the application of
molecular-biology tools to the possible early
diagnosis of Huntington's disease.

AIDS UPDATE

In April 1984 researchers announced that they
had detected the apparent cause of Acquired
Immune Deficiency Syndrome (AIDS)—a member
of the retrovirus family. French researchers first
reported detection of the virus in May 1983 and
called it LAV for lymphadenopathy-associated
virus. A year later American researchers at the
National Cancer Institute reported detecting a
virus they called HTLV-3 (believed to be the
same as LAV) from about 50 cases. Most
significantly, the American researchers developed
a way to grow large amounts of the virus, thus
opening up prospects of a diagnostic blood test
and the possibility of a vaccine.

AIDS emerged as a public health problem
throughout the world in 1983 with cases reported
on a spotty basis in 33 countries and all inhabited
continents. AIDS was first recognized in New
York and California in 1981, and since then it has
affected at least several thousand people, mostly
in the United States. The number of cases is still
small compared to the world population, and
AIDS is not believed to be sweeping the globe.
Nevertheless, most experts suspect that the true
incidence is much higher due to the undiagnosed
and unreported cases. And in 1983, preliminary
epidemiologic findings from studies in Africa

*Top: Dr. Robert Gallo holds a bottle containing a virus
called HTLV-3 that is believed to be the cause of Acquired
Immune Deficiency Syndrome (AIDS). Bottom: A nurse
checks a Venezuelan woman who is one of many in her
family afflicted with Huntington's disease.*

Top: Dennis Brack/Black Star; Bottom: Steve Uzzell III

suggested that at least in some areas, AIDS was not confined to the four high-risk groups described in the United States—male homosexuals, intravenous drug users, Haitians, and hemophiliacs. In Africa, AIDS seemed to be striking men and women in about equal percentages, perhaps because of the use of inadequately sterilized needles and syringes in everyday medical practice and because the disease may be spread heterosexually.

As is customary in medical situations when no lab test is available, doctors diagnose a case of AIDS on the basis of an arbitrary definition—in this instance, one established by the Centers for Disease Control (CDC) in Atlanta, Georgia.

So far it appears that the disease is invariably fatal, although the debilitating course may run several months to several years. It is characterized by repeated bouts of infections caused by organisms that usually do not affect people whose immune systems are intact. However, if AIDS obeys the rules of most other infectious diseases, doctors may confirm suspicions that many people have milder cases of AIDS and that some may recover without treatment. These would be the cases that do not meet the criteria set by the CDC and that were undetectable for lack of a diagnostic test.

AORTIC ANEURYSMS

Earlier diagnosis and surgical treatment of a ballooning of the body's main trunk artery—the aorta—could save the lives of at least 2,000 people in the United States and $50 million in patient-care costs each year, according to a statistical study by surgeons at the University of Rochester in New York. They analyzed data collected over a 27-year period about the ballooning condition—called an aneurysm—that occurs in the abdominal section of the aorta. Abdominal aortic aneurysms, which affect an estimated 3 percent of the U.S. population, usually result from the atherosclerotic process in which a buildup of fatty deposits weakens the walls of arteries. Death can come rapidly when an aneurysm bursts unexpectedly; when surgery is done under such conditions, the mortality rate is about 70 percent. However, when the same operation is done on an elective basis before the aneurysm bursts, the death rate is much lower—usually 5 percent or less.

STOPPING HEART ATTACKS

In 1983 about 1.5 million people in the United States suffered heart attacks—myocardial infarction is the medical term—and more than

550,000 died from them. Doctors in several countries are continuing studies of ways to stop heart attacks in progress. The exact cause of heart attacks is unknown, but most occur in individuals whose coronary arteries have been damaged by a buildup of fatty deposits on the arterial walls. (Coronary arteries surround the heart and carry vital oxygenated blood to it.) Many heart attacks are thought to result from the formation of a blood clot in a damaged coronary artery.

In 1982 enzymes such as streptokinase and urokinase were approved for use in dissolving clots during heart attacks. Doctors are now trying to determine the most effective way to inject the enzymes. The usual way is to squirt them directly at the clots in the coronary arteries. But could they be given just as effectively if injected into a vein in the arm? If so, the hope is that someday the enzymes could be administered wherever the individual suffers a heart attack—at home, at work, in a football stadium—and before arrival at a hospital. Preliminary results of ongoing studies have not been promising. They show that the enzymes are most effective when injected directly against a clot in a coronary artery. That is done by inserting a tube through a vein or artery in the arm or leg and guiding it through the circulatory system into the affected artery. This can be done only by specially trained cardiologists working in hospitals equipped with sophisticated X-ray machines.

However, hopes for a non-hospital-dependent way to dissolve clots were given a boost during the year by preliminary experiments carried out by American and Belgian researchers. They injected a different substance into the veins of a small number of patients with clots that impaired the circulation of their legs and hearts. The substance is a protein called TPA, for tissue plasminogen activator. TPA is a naturally occurring substance that stimulates the body's blood system to dissolve clots. Unlike streptokinase and urokinase, which affect many other parts of the body's complicated clotting and bleeding system, TPA seems to attack the clot without disrupting the rest of the intricate system. The crucial point is that TPA, like the enzymes, must be administered within the first few hours of a heart attack in order to prevent the death of heart-muscle cells.

TESTICULAR CANCER

Cancer experts began to focus attention on helping men test themselves for testicular cancer. For unknown reasons, the incidence of testicular cancer seems to be rising in the United States. About 5,400 men developed it during 1983, most between the ages of 20 and 40, and about 950 of them died of the disease. If testicular cancer is diagnosed in its early stages, it is one of the most easily cured cancers. But it can be one of the most fatal if it is diagnosed late, after it has spread elsewhere in the body.

Testicular cancer usually first appears as a hard, sometimes tender, nodule on one of the testicles. The testicles are smooth and free of lumps except for a soft, tubelike lump normally present along the back of the testicle; this is the epididymis, where sperm are stored. The best time to perform self-examination is during or after a warm bath or shower. The examiner should apply a small amount of pressure with the thumb and fingers of both hands to each testicle. Most abnormalities are found on the front or side of the testicle and are not attached to the covering scrotum.

A VACCINE AGAINST A LIVER CANCER?

Among the most common cancers in the world is one that arises in the liver called hepatocellular carcinoma. It cannot be treated effectively, usually is rapidly fatal, and kills at least 250,000 people each year, mostly in Asia and Africa. Epidemiologic studies in recent decades have shown a very strong link between hepatocellular carcinoma and past infection with the liver disease known as hepatitis B, which can be seriously disabling and often fatal. According to current medical belief, hepatocellular carcinoma develops from 20 to 40 years after hepatitis B infection and as a complication of the chronic liver disease that often follows the attack.

Now researchers are investigating the possibility of preventing hepatocellular carcinoma by immunization—due not to a vaccine against the cancer but to a vaccine developed in recent years against hepatitis B. Researchers in China and from the World Health Organization in Geneva, Switzerland, have begun a bold experiment in China using hepatitis B vaccine to prevent hepatocellular carcinoma. The experiment will take many years, if not decades, to complete. But the hope is that it will prove the widely held hypothesis that most cases of hepatocellular carcinoma can be prevented if an individual first acquires protection against hepatitis B infection.

HUNTINGTON'S DISEASE

Of the hundreds of genetic (hereditary) disorders, Huntington's disease is one of the most devastating. It is a fatal nervous system disorder that destroys the mind and motor function. There is no specific diagnostic test for the disease.

Initial symptoms usually do not appear until the fourth or fifth decade of life and are insidious—minor clumsiness or forgetfulness that progresses to dementia and major difficulty in moving. Each child born of a victim of Huntington's disease has a 50–50 chance of getting the disease. Because of the nature of the genetic pattern of the disorder and the late onset of symptoms, many victims unknowingly pass on the lethal gene to their children.

Now researchers have used molecular-biology tools to identify a marker for Huntington's disease. At present the marker test is used only in research and can be applied only to families where there have been at least ten persons afflicted with the disease. Within the family, researchers can identify the presence of the marker in affected individuals and confirm its absence in those beyond the age range for the onset of symptoms. Eventually, researchers hope to locate the Huntington's disease gene itself, to develop a diagnostic test with as close to perfect accuracy as possible, and perhaps find a way to correct the genetic defect. Meanwhile, researchers are exploring the ethical implications of a test that when available could detect the disease with nearly 100 percent accuracy—before an individual shows symptoms or decides about whether or not to have children. Such ethical considerations are bound to pose problems as markers for cystic fibrosis and other inherited conditions are identified.

NEW THERAPIES FOR KIDNEY STONES

Kidney stones, if they are not small enough to pass along on their own—painfully—with the urine, may have to be removed in a major operation. In this procedure, the surgeon cuts through layers of tissue to reach the kidney in the lower back and then removes the organ. Recovery often takes six or more weeks. Now European doctors have developed two techniques that can avoid the need for such major surgery.

One technique, being used in several hospitals in the United States, involves a simpler, less risky operation than the standard one—though the patient must still undergo general anesthesia. In the procedure, a tube is inserted through the skin and, under the guidance of fiber optics, burrowed from the skin to the kidney. The stone, often pulverized by ultrasound techniques, is then suctioned out through the tube.

The second technique is called extra-corporeal shock-wave lithotripsy: it uses shock waves to destroy stones. First, a needle is inserted into the back and a local anesthetic injected. Then the patient sits in a water-filled tank in such a position that the kidney stone is at the focal point of a shock-wave generator. (The

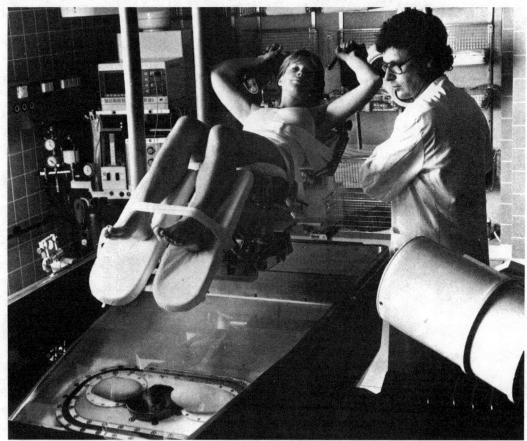

Using shock waves through water, doctors can destroy kidney stones without subjecting patients to major surgery.

focal point is determined by using several concave mirrors.) Then, while the patient is monitored by an X-ray device called a fluoroscope, from 500 to 1,500 shocks are given to destroy the stone. A few such devices are expected to be installed in hospitals in the United States in the near future.

PREOPERATIVE SHAVING

Mounting evidence on a simple but important problem has shown that the universal practice of shaving the skin around the site of an impending surgical incision may cause more harm than good. Studies with scanning-electron microscopes show that a razor shave by even skilled hands produces nicks—invisible to the naked eye—that can provide entry points for microorganisms, encouraging infection that may begin even before the incision is made. Instead, researchers have suggested the use of depilatory creams and sprays as a safer means of removing hair before surgery.

AUTOPSIES

Although autopsies are done after a person dies, at a time when it seems too late to do much good, these operations can save the lives of members of the deceased's immediate family, relatives, friends, and the population at large. Throughout medical history, autopsies have been one of the chief ways doctors have learned how to develop new therapies and to document the accuracy of their diagnostic skills. However, in recent years in the United States, autopsies have been performed less and less often. The decline of autopsies has now reached a point where many medical leaders are calling the situation deplorable. That judgment is based in part on recent studies that have shown that advances in expensive diagnostic technology have not reduced the value of the autopsy. In about 10 percent of the cases in one study, doctors were found to have misdiagnosed major conditions that could have been treated to prolong life.

LAWRENCE K. ALTMAN, M.D.

AEROBICS: HOW MUCH IS ENOUGH?

by Dianne Hales and Robert E. Hales, M.D.

"Aerobics," the system of regular workouts to enhance one's use of oxygen, is the word that has brought the world to its feet. But as it's become a buzz word of the better-body business, aerobics has turned into a fuzz word for fitness. Those three scientific-sounding syllables mean completely different things to many different people.

The Odd Couple

Consider, for example, that oddest of contemporary exercise couples: the sinewy marathon man on a cross-country trail, and the lady in bumble-bee-striped leotards, shaking her hips in her living room to the beat of muscle rock. Both are doing "aerobics," continuous exercise that works the heart, lungs, and limbs in concert. Strenuous bicycling, skiing, swimming, even fast walking qualify—in contrast to the stop-and-start, gasping agony of pumping iron.

Our leotard lady, trying for the lean and healthy look, believes that "aerobic dancing" can help her lose weight and boost her heart and lung capacity. But though her 10-minute dance sessions may feel good, they aren't hard, long, or frequent enough to do much for her body.

Her husband, logging 160 kilometers (100 miles) a week, is sleek as a cheetah, with low blood pressure and a healthy heart. But he's also running himself into the ground, pushing through rain, pain, and sprains—even though his knees and hamstrings beg for mercy.

Like Jack Sprat and his wife, these mythical partners are obviously out of balance. Yet if she's doing too little and he's doing too much, what about the estimated 90 million people in the United States who exercise between these extremes? Are their workouts inadequate, excessive, or just right?

Different people have different ideas when it comes to what aerobic fitness is really all about. Some people don't exercise long, hard, or often enough to do much good for their bodies, while others drive themselves to the limits of their endurance and beyond. Finding a happy medium to meet your fitness goal is the key.

Moderation the Key

The answer depends on who's exercising, how, and what they hope to achieve. The varied benefits of aerobic exercise come in at different levels of training. People who hope to lower their risk of heart disease, for example, may need to work out quite a bit harder than those who simply want to sweat away their tension. And for every benefit, there is an optimum level of exercise, which can be expressed as the number of calories burned per week.

In each case, moderation is the key. The trick is to do enough exercise to meet your goal, yet not so much that it becomes an obsession or cripples your joints.

A decade ago, exercise enthusiasts were pushing inactive people to get moving. Now some of these same experts have begun to tell people when to stop. Some of the strongest warnings have come from Dr. Kenneth Cooper, the former Air Force physician who virtually invented aerobic exercise in the 1960's, and now runs a fitness clinic and research center in Dallas, Texas. "If some exercise was good, I thought more had to be better," Cooper says today. "I was wrong. There is a point of diminishing returns." Muscle and joint injuries in runners pushing past 40 kilometers (25 miles) a week have "overwhelmed" the staff at his Aerobics Center, he says.

Psychological studies of aerobic athletes support the case for moderation. Just a few minutes of exercising the body can make the mind more calm and alert. At California State University in Long Beach, Dr. Robert Thayer has found that people who take a 10-minute walk on a treadmill report feeling more energy for half an hour afterward. A short walk outdoors—around the block, say—gives a boost for up to two hours. (A candy bar gives you only a quick boost, then an energy drop.)

Up to a point, greater efforts bring greater rewards. Stanford University studies indicate that running 13 to 16 kilometers (8 to 10 miles) a week helps transform people who are compulsive Type A's (a behavior pattern characterized as being achievement-oriented and competitive)

© Warren Morgan/Focus on Sports

When a person exercises hard enough to achieve "the training effect," dramatic physiological changes occur.

The researchers, who are long-distance runners themselves, stress there's nothing inherently crazy in logging a lot of kilometers. But if your main concern is simple fitness, that kind of extreme effort is beyond the call of duty.

The Training Effect

For most people, exercise begins to pay off when they work hard enough for the "training effect." The body becomes trained with exercise for at least 20 to 30 minutes every other day. Each session must be intense enough to push the pulse to 70 to 85 percent of its maximum rate (the maximum is 220 minus your age). You should start feeling the difference the training effect makes at about 1,000 calories of exercise a week.

These changes are profound. Heart muscles thicken and become stronger, while the pulse rate slows down. Red blood cells increase in number, while blood pressure may drop slightly. The lungs take in more oxygen, increasing their capacity (a powerful predictor of longevity). Body temperature rises by several degrees, helping kill bacteria. Aerobic training also makes the body's cells more sensitive to insulin, a boon to diabetics.

Since many of these changes begin with around three half-hour sessions a week, that level of activity has become the standard prescription for aerobic exercise. But depending on what you want to achieve, you may need to push your body into a higher gear—or you may be able to get by with an easier schedule.

Even light exercise specifically benefits women, for example, by preventing the slow loss of calcium that can end in osteoporosis—the brittle bones that more than 11 million women in the United States suffer after menopause. Some evidence suggests that younger women may be able to prevent osteoporosis through moderate stretching and calisthenics, even if they're doing only half the amount for the training effect.

Although more exercise can further improve bone strength, very strenuous workouts may actually increase the risk of premature bone deterioration by interfering with hormonal balance. Some women who run much more than 32 kilometers (20 miles) a week may develop amenorrhea (stop menstruating). Scientists think this occurs because body fat drops below critical levels, leading to a drop in estrogen production; and low estrogen levels are linked to bone degeneration.

into more easygoing Type B's (typically tranquil, lacking underlying hostility and tension-producing characteristics). Three half-hour aerobic sessions weekly will help fight depression and anxiety, according to researchers at Duke, Purdue, and the University of San Francisco.

But certain athletic overachievers become compulsive to the point of obsession. Psychiatrists at the University of Arizona draw an unhealthy picture of some middle-aged long-distance runners. Like anorexic (self-induced starving) teenage girls, these men pursue "a grim asceticism," marked by self-denial and an obsession with being extremely thin.

Benefits from Substantial Aerobics

What are the best reasons for pushing above the training effect? Two major ones are trying to lose weight and to ward off heart disease. In each case, the standard three-times-a-week prescription just begins to provide the potential benefits of aerobics.

Other kinds of exercise are not nearly as effective as aerobics for long-term weight loss. An exercise that uses muscles in short bursts, like weight lifting, mostly burns glycogen, a form of starch that provides quick energy. But as you do continuous aerobics, your body switches from burning glycogen to burning fat (and some protein). And losing fat is the key to losing weight.

Aerobic exercise seems to raise your metabolic rate, so you're burning more calories not only while you're on the track or in the pool, but for hours afterward. And it also helps bring your appetite into line with your body's physical needs.

How much do you need for these benefits? Five minutes of toe touching with Richard Simmons, the frizzy-haired ex-fatty who's made flab the stuff of daytime television, does no good at all. The American College of Sports Medicine prescribes more substantial aerobics three times a week to trim the torso.

But four or five weekly sessions may be a better standard. Researchers at the Western Psychiatric Institute in Pittsburgh, Pennsylvania, recently analyzed a decade's worth of studies. They found that people who exercised four or five times a week lost almost a quarter of a kilogram (half a pound) a week—three times more than those who had three weekly bouts of exercise.

Keeping the heart in top shape may require the same fairly hefty dose of exercise. Last year researchers at the universities of Washington and North Carolina probed the health habits of 163 Seattle, Washington, men ages 25 to 75, with no history of heart disease, who were victims of sudden cardiac death. When these men were compared with still-living community members, it turned out that "vigorous" exercisers—those who spent more than 20 minutes a day walking, running, chopping wood, swimming, playing singles tennis or walking up stairs—had a 55 to 65 percent lower risk of sudden death. Mild to moderate levels of exercise, burning anything under 2,000 calories a week, did not lower the danger.

Unlike aerobic activities in which muscle cells use oxygen to produce energy at a slow, steady rate, weight lifting is an anaerobic exercise in which muscle cells use no oxygen but provide usable energy at a much faster rate.

© G. Rancinan

These findings echo the earlier observation of Dr. Ralph Paffenbarger, Jr., of Stanford University. In a widely quoted study of 16,936 Harvard University alumni, he found that the more active men (those burning at least 2,000 calories a week in exercise) had one-third less heart attack risk than those who were less active. More-moderate activity offered less protection.

Aerobics may reduce the risk of heart disease by raising levels of high-density lipoprotein (HDL), the "good guy" substance that may help keep fats from building up in the arteries. In general, the higher the HDL level, the lower the likelihood of coronary heart disease seems to be.

What isn't clear is how much exercise is enough—and whether you have to enter the danger zone for injuries—to boost HDL significantly. A few studies show improvements in HDL with running only 16 to 18 kilometers (10 to 11 miles) a week, though others set about 32 kilometers (20 miles) as the weekly minimum. One New Zealand study found that running 56 kilometers (35 miles) a week for at least four years was necessary to increase HDL.

Not Easy to Maintain

Even all-out aerobic efforts, however, can't guarantee protection from heart disease. In 1982 physicians at the Naval Regional Medical Center in California, who found a 99 percent coronary artery blockage in a highly trained 48-year-old marathoner, warned that the "tempting illusion of invincibility in the runner" can be dangerous. It's even deadly when it "replaces a measure of common sense with self-delusion."

There's another problem with all-out exercise: It's hard to maintain. And for true health benefits, the commitment to activity has to last a lifetime. In a Swedish study of inactive women who had once been top-level swimmers, fitness levels declined drastically over the 15 years since they had quit swimming. Other studies show that inactive ex-jocks have more heart disease than college classmates who started as bookworms but became exercise converts.

Conventional aerobic programs, like running and jogging, may be hard for people to maintain because they do little for the muscles that most of us use the most. Some exercise physiologists point out that only four groups of people need to be able to walk or run considerable distances: postal workers, protective services personnel, police, and fugitives. "For almost everyone else, life is not running around a track," says Dr. Barry Franklin, an exercise physiologist at Sinai Hospital in Detroit, Michigan.

Unknowns in Fitness Process

Franklin thinks it's time to look up—to the arms and shoulders—for the other half of the total fitness story. Even runners with tremendous thigh power, he's found, are no stronger above the waist than inactive individuals. Raking leaves, lifting a typewriter, or shoveling snow can take their breath away and send their blood pressure and heart rate soaring, possibly to hazardous highs.

Along with a growing number of researchers, Franklin believes that the skeletal muscles of both the upper and lower body may be critical zones in aerobics. "Until now we've assumed that aerobics acts primarily on the heart," he explains. "But tremendous changes occur in the muscles being exercised. It could be that because of these changes, the skeletal muscles make fewer demands on the heart—and that's why it pumps more easily and efficiently."

Many scientists believe that upper-body conditioning is just as critical in aerobics as lower-body fitness.

© Jane DiMenna

Many people maintain an exercise program throughout their lives not because of potential health benefits or to try to live longer but for the simple fact that it makes them feel good.

After many years and much sweat, scientists are still uncertain about the ways in which aerobics changes us cell by cell. But the father of aerobics is unfazed. "I look at declining heart disease deaths, at increased lung capacity and HDL, at lower blood pressures," says Cooper. "I don't need to know how they happen, as long as they do."

To some extent, it is possible to run from the risk of disease. But it's not possible to run—or swim, walk, or jog—toward everlasting life. Ultimately, most people keep exercising because it makes them feel good, not because they're searching for immortality. At the end of his talks, Cooper asks his audiences whether they'd continue to exercise even if research showed that aerobics could not add to longevity. More than 90 percent say yes.

And Dr. Jere Mitchell, an exercise physiologist at Southwestern Medical School, notes that very active cross-country skiers in Scandinavia live only two years longer than their sedentary countrymen. "I figure that I'll spend two years of my life just dressing for a run, driving to the track, warming up, and showering afterward," says Mitchell. "That's not enough of a benefit to motivate most people. The reward has to lie in the act itself" □

SELECTED READINGS

Aerobics Program For Total Well-Being: Exercise, Diet, and Emotional Balance by Kenneth Cooper. M. Evans and Company, 1982.

Aerobic Dancing by Jackie Sorensen and Bill Bruns. Rawson, Wade Publishers, Incorporated, 1979.

Mᵃᵉⁿ Martinet, 172, r. Rivoli et 41, r. Vivienne   Lith Destouches 28 r Paradis Pᵗ

Paris grippé

THE FLU

by Stephen S. Hall

In February 1980, in a small isolation room located at the Institute of Experimental Pathology just outside Reykjavik, Iceland, a highly significant—and thoroughly unplanned—exchange of viral information took place. At the time, Robert G. Webster, a visiting influenza researcher from the United States, was staring down into the mouth of a rheumy-eyed, flu-infected harbor seal while an Icelandic laboratory technician struggled to hold the rambunctious animal from behind.

The restive seal bucked, whirled around, and then sneezed in the face of the lab technician. "I tried to play it cool, but I was very concerned," recalls Webster, who was aware that this was far more serious than a breach of interspecies etiquette.

From Bird to Seal to Scientist

The seal had been experimentally infected with a flu virus known as H7N7. This particular virus had been ravaging seal populations along the New England coast, causing the most devastating lung damage Webster had ever seen in flu-infected mammals. H7N7 also is associated with an exceedingly lethal form of avian influenza known as fowl plague, which, in lab parlance, "knocks over" chickens in 48 hours by destroying the central nervous system. So Webster's concern was obvious: Could a seal pass this potentially deadly virus on to a human? The answer, both fascinating and frightening, appeared to be yes.

"Have a look at this." Webster steps across his office at St. Jude Children's Research Hospital in Memphis, Tennessee, in search of a slide. Fifty-odd years old, a native of New Zealand, Webster is regarded as one of the world's leading flu researchers. His office is comfortably cluttered with the signs of souvenirs of a flu chaser: a print of Daumier's *Paris Grippé* over the desk; a postcard of a harbor seal looking cute

and harmless; thick binders full of lab results; and a map of the world, which almost teases with its infinite hiding places for an invisible but very well traveled pathogen.

Webster finds the slide and places it on a light table. It shows the work of the seal virus on the lab technician's eye, which is streaked red and bleary with conjunctivitis only 48 hours after that ill-timed sneeze. High concentrations of the seal virus were found swimming around in the infected eye; fortunately, that was as far as the damage went.

And now the kicker: genetic tests at Webster's lab determined that the same virus, capable of killing seals in New England and causing eye infections in humans, originated in birds. It was the first time bird-to-mammal flu transmission had ever been documented. "That virus had all eight of its genes come from various avian sources and was still able to cause serious disease in mammals," says Webster, who believes it requires only a short leap of imagination to theorize that avian viruses could create similar pathogenic havoc in people. "If that

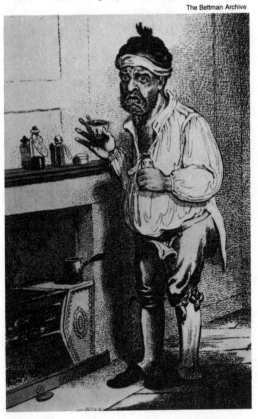

Taking Physick *by Gillray*

The Bettman Archive

virus had been as virulent in humans as it was in seals. . . .'' He shudders and shakes his head. ''My God!''

Pandemics Still Possible

More than fifty years after the first flu virus was isolated from a pig, influenza continues to cause shudders in the medical community, and with good reason. It is a family of shifty viruses that abide by no rules. By changing molecular makeup frequently, the viruses confound the body's immune system and have a habit of making today's vaccines obsolete within a few years. By swapping genes with fellow viruses on rare occasions, a flu virus can create new and deadly hybrids that, like a spark in a forestful of human tinder, send disease sweeping through susceptible populations. The same virus that nestles comfortably in the gut of wild ducks, causing nary a sniffle, can turn the respiratory tracts of other species to curd. To scientific researchers who stalk it, influenza is the viral equivalent of the leopard that changes its spots.

In an age when the flu is viewed by the public as a three-day nuisance rather than as a killer, the memories of past pandemics (unusually severe and widespread epidemics) are always humbling reminders. The shadow of the 1918–19 pandemic—with its 20 million dead, its estimated 2 billion global sufferers—falls on the United States, Australia, the United Kingdom, the Soviet Union, and every other place where flu research is going on today.

No one knows why it killed as efficiently as it did, but the 1918 virus was unique in its deadliness. Healthy, able-bodied young people fell as swiftly as influenza's customary victims—infants, the elderly, and the infirm. The mystery of the virus' virulence is so compelling that some years ago a party of scientists exhumed 1918-vintage Eskimo corpses from the frozen tundra of Alaska, hoping to isolate the virus that caused such overwhelming devastation. The effort failed.

Could such a pandemic happen again? ''It could happen tomorrow,'' argues John R. La Montagne, who administers the $6 million influenza program for the National Institute of Allergy and Infectious Diseases in Bethesda, Maryland. La Montagne's job is to worry about such dire possibilities, of course, but he is hardly alone. ''It's not being a Cassandra to worry about another pandemic,'' says Edwin D. Kilbourne of Mount Sinai School of Medicine

in New York. ''It's being realistic to keep that in the back of one's mind all the time.''

Flu researchers are understandably gun-shy about making predictions, particularly after the swine-flu episode of 1976, when an outbreak of influenza similar to the 1918 strain was reported at a U.S. Army base in Fort Dix, New Jersey. A nationwide inoculation campaign was launched in anticipation of a pandemic that never materialized, and the program was canceled amid considerable controversy when hundreds of people receiving the vaccine developed the rare Guillain-Barré syndrome, resulting in illness, paralysis, and, in some cases, death.

The swine-flu virus defied almost every scientific assumption about it—including the widespread feeling that a major new flu strain was ''due.'' Even now, though, there is talk of a disquieting calm on the flu front. The seal incident suggests the way a new pandemic might start, but it gives no clues as to timing. No new influenza virus afflicting humans has made an appearance since the Hong Kong strain arose in 1968, and variants of that strain appear to be ''changing their spots'' far less readily.

''It seems to be losing its oomph,'' observes Webster. ''During an outbreak in Australia last winter, the circulating strain was very similar to the one in 1979. Very similar. So maybe Hong Kong flu has reached the end of its tether.'' Then, realizing his folly, Webster leans forward and tries to wave away that prediction with a cautionary gesture. ''I had better not say that,'' he adds hastily. ''You really have to leave the door open with flu. I've found that out.''

The Mace Metaphor

In the hallway that runs through the department of virology at St. Jude, several large freezers line the wall, quietly maintaining temperatures of about $-70°$ C (-94 °F). Inside the freezers, encased in metal sleeves, are hundreds of small white cardboard boxes. Each one contains small sealed ampules full of viruses.

There are nearly 7,000 different kinds of flu virus on ice at St. Jude. Most of those viruses would bring hardly a tickle to a host's throat; others could cause serious disease and death—in turkeys, chickens, horses, and pigs as well as in humans.

All the death and all the misery stem from a virus so small that 2.5 million of them in a line would take up 2.5 centimeters (1 inch). Flu viruses fall into three types: A, B, and C. Type

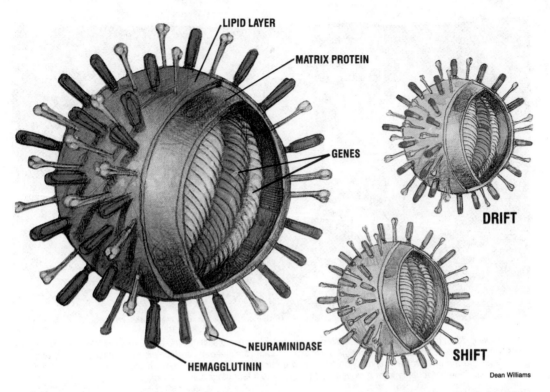

LIPID LAYER

MATRIX PROTEIN

GENES

DRIFT

NEURAMINIDASE

HEMAGGLUTININ

SHIFT

Dean Williams

The flu virus, left, consists of eight coiled genes in a lined sac. It is recognized by the human immune system by its surface proteins, hemagglutinin and neuraminidase. Flu tricks the body defenses by changing these proteins from one season to the next. At top right, a point mutation (red spot) in the hemagglutinin gene results in a slightly altered "spike" on the virus' outer covering, and antigenic drift occurs. At bottom, antigenic shift results when a virus takes on a totally new gene (blue) to create a completely different hemagglutinin "spike."

A, the most variable, causes pandemics as well as regular seasonal outbreaks; type B causes smaller outbreaks and is just now receiving greater attention; type C rarely causes serious health problems.

In appearance, a flu virus somewhat resembles the head of a medieval mace—a ball of iron studded with spikes. These spikes are two surface proteins called hemagglutinin (HA) and neuraminidase (NA). Inside the virus is a thick tangle of genes. In many other viruses, a number of different genes fit onto one strand of nucleic acid; but each flu gene is a separate segment of ribonucleic acid (RNA)—eight threads in all.

The mace metaphor provides a crude but vivid picture of the influenza virus at work. Hemagglutinin is the substance that in effect bashes into a cell during infection and allows the virus access to the cell interior, where it can replicate. Neuraminidase permits all the viral offspring to break free of the host cell once replication is complete.

So specific are the structure and function of these two surface molecules that all A-type viruses can be classified into 13 distinct HA subtypes. The Hong Kong flu of 1968, still circulating in human populations, is classified H3N2. The only other type-A strain now circulating among people, the erroneously named "Russian flu" (it actually reemerged in northeast China in 1977 after a 27-year hiatus), is H1N1.

Within the past few years, the hemagglutinin molecule has been scientifically undressed to the point where Webster calls it "one of the best-known biological molecules we have." In 1981 John J. Skehel of the National Institute for Medical Research in London, England, reported success in crystallizing hemagglutinin. That enabled Don C. Wiley and Ian A. Wilson of Harvard University to produce three-dimensional X-ray diffraction pictures of hemagglutinin crystals of the Hong Kong strain. The pictures showed that hemagglutinin is a trimer, a molecule with three legs, planted on the surface of

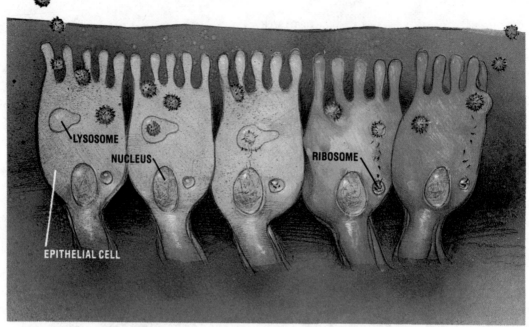

After traveling down to the trachea, a flu virus makes its way into an epithelial cell (far left). The virus is surrounded by the cell's lysosome, which releases acidic chemicals that "unsheath" the invader's surface proteins. When this happens, viral genes spill into the cell proper (center) and travel to the nucleus where they command the ribosome to produce viral components. The components form new viruses that go off to infect other cells (far right).

the virus a bit like a tripod, with clublike projections on top. A drawing of Wiley's dramatic illustration hangs on the walls of a lot of flu labs, displayed with a prominence normally reserved for family snapshots. In 1983 researchers in Australia gave a detailed picture of the neuraminidase molecule as well.

Antigenic Drift and Shift

The two surface proteins seem to be the key to influenza's uncanny ability to change from year to year. On each leg of the hemagglutinin molecule, for example, there are four "hot spots." These are short segments of amino acids that display an unusual degree of instability and change.

The hot spots function as antigens, the sites on a foreign molecule that the body's immune system identifies, "fingerprints" for future reference, and uses as a molecular template to make antibodies. Antibodies are one of the body's major lines of defense against outside invasions. They seek out antigens, interlock with them in a kind of fatal biochemical handshake that never lets go, and thus neutralize their activity.

But the influenza virus always stays one step ahead. In effect, it changes the whorls of its fingerprint just enough that the immune system

doesn't recognize it immediately, and thus fails to neutralize it completely. When this change is minor, it is called antigenic drift, resulting in slowly evolving variants that cause seasonal outbreaks of flu each winter. Epidemics from antigenic drift might affect anywhere from 5 to 20 percent of a local population. When the change is major, it is called antigenic shift, a variation so dramatic that major pandemics usually ensue, rapidly spreading disease among billions around the globe. Three such shifts have occurred this century: the great pandemic of 1918, the Asian flu of 1957, and, most recently, the Hong Kong flu of 1968.

Antigenic shift is a perennial worry to scientists because it obliges no timetable, although it was mistakenly believed during the swine-flu incident that shift was occurring every 11 years. Antigenic drift, however, is easier to follow. "We know exactly how antigenic drift occurs," says Webster. "After the Hong Kong strain appeared in 1968, if you got a single amino acid change in only one of these four sites, you got a new variant." The changes are prompted by point mutations—changes in a single chemical unit out of more than 1,800—in the gene that codes for the 550 or so amino acids in the hemagglutinin molecule. Such mutations, in neuraminidase as well as in hemagglutinin, occur in one out of every 100,000 viruses in the course of one generation in the lab.

"An accumulation of these changes makes it sufficiently different almost every year," says Webster, "and then it can cause an epidemic."

The Hazards of Hybrids

But an epidemic due to antigenic drift is small potatoes compared to the global mayhem associated with antigenic shift. The explanation for the swift and mysterious transformation resides in the unique nature of the virus' genes.

Because all of the flu's genes are separate units of RNA, they can be scrambled and rearranged easily. That is what creates antigenic shift. When two different flu viruses happen to infect the same cell, there are suddenly 16 different genes floating around that can be recombined in 256 different ways.

The effect is a little like shuffling the cards from two different decks together. Mix-ups and incompatibilities occur. But mix-ups also can give the virus a competitive advantage—the genetic equivalent of five aces, as it were—by creating a new hybrid.

Hence, a recombined virus may emerge from that overcrowded cell with a new gene that creates a completely different hemagglutinin molecule. When that happens, entire populations are rendered immunological virgins. No one, young or old, has developed antibodies to the new surface protein, and the virus can be highly infectious.

William J. Bean Jr. and Clayton Naeve, members of the St. Jude team, have been using a variety of sophisticated new lab techniques to gain an intimate glimpse at influenza genes. Working with a technique known as RNA-RNA hybridization, Bean has determined that the virus that decimated the New England seals, for example, carried genes that originated in birds. These finds support his belief, shared by many others, that "influenza is essentially an avian disease which by fortuitous accident went into mammals." Birds, along with pigs and even humans, can be the site of the kind of reassortment that leads to antigenic shift. For sheer numbers of viruses, however, the most obliging hosts for such genetic reassembly seem to be wild ducks.

Ducks Have Major Role

"Among avian viruses, ducks hold a special niche," says Virginia S. Hinshaw, a St. Jude researcher. Cradling a coffee cup labeled "Le Canard," she warms to one of her favorite topics: the ecology of influenza. "Ducks seem to be a very natural reservoir. They have every subtype, no disease, and a lovely transmission device—feces."

Hinshaw concerns herself with where flu viruses turn up, how they interact with hosts, how they spread. And the trail keeps leading back to ducks and other wild waterfowl. Every hemagglutinin and neuraminidase variant is found in ducks. The viruses grow happily in the cells lining the intestinal tract, causing no illness at all, passing easily from duck to duck in pond water. And because ducks migrate great distances, the virus has traveled the globe.

Wheel of Misfortune

The St. Jude team has taken its search for influenza viruses to some of the more interesting corners of the world, not to mention some of the more interesting corners of animal anatomy. No fleshy niche or cranny is safe from their probing swabs—the hindquarters of ducks; the blowholes of whales; the throats of turkeys, horses, and pigs; the noses of ferrets; and the brains of seals.

There is a drawing in Hinshaw's office that sums up her theories about influenza and the

A Cure for a Cold *(uncredited)*

The Bettman Archive

way it spreads. A duck forms the hub of what might be considered a wheel of misfortune. Around it are figures of a human, a pig, a horse, a turkey, and now a seal. There is strong circumstantial evidence that ducks are in a perfect position to spread influenza to all these other creatures.

Often the problem comes home to roost with humans: For instance, genetic studies of the H3N2 virus, which appeared in 1968, indicate that while seven of its genes came from the previous Asian strain H2N2, the eighth gene—the one that codes for all-important hemagglutinin—probably came from a duck. H3N2 is the Hong Kong flu, which plagued millions of people in the U.S. in the winter of 1968.

Webster believes that the Hong Kong virus arose when a strain of duck flu shuffled genes with a strain of human flu, resulting quite by chance in a powerful new combination. "A single cell, somewhere, has been simultaneously infected by a duck-strain virus and a human virus," says Webster, describing one of the more likely scenarios. "It probably occurred in a human, because the virus retained most of its genome [one haploid set of chromosomes with the genes they contain] from the human strain. In China, where this presumably happened, it would have required that someone was infected with Asian flu and was working with a duck that was also infected."

China, with its huge human and duck populations, makes the odds slightly more favorable for what is admittedly a very unlikely event. In such a genetic exchange, for example, the human can't have too many antibodies. The odds of finding such a person are better in a large population such as China's. "It's just a numbers game," says Webster. "You don't get mutant strains with small numbers."

Live-Virus Vaccines

Brian Murphy also talks about "the numbers game," but he is not referring to Chinese recombinants. He is talking about vaccines and how they work. Murphy, along with Robert M. Chanock, has spent more than a decade developing and testing live-virus vaccines at the National Institutes of Health labs in Bethesda, Maryland. Unlike the conventional flu vaccines, which contain killed viruses delivered by injection, these carry weakened but still living viruses—and they are administered in the form of nose drops to mimic flu's natural route into the upper respiratory tract.

In many ways, the experimental flu vaccines are similar to the live-virus vaccines used for measles, mumps, and polio. During a flu infection, either the body's defenses can overwhelm the invading virus or, as more frequently occurs, the rapidly replicating viruses spread and destroy cells before the immune system can really kick into gear. The principle of live-virus vaccines is simple: it attempts to tilt the odds of this numbers game in favor of the body.

"What we're doing is exactly what the influenza virus does in nature," says Murphy, "but we're doing it to our own advantage. The goal is to get hemagglutinin and neuraminidase into the body, where the antibody immune response can develop without the rapid replication of a virus that can cause heavy infection. Essentially what you're trying to do is put a governor on the replication of that virus."

This is accomplished by turning down the biological thermostat and in effect training the viruses to grow at lower temperatures. Called cold-adapted mutants, these viruses, developed by H. F. Maassab of the University of Michigan, have been adapted to grow at a chilly (for viruses) 24° C (75° F). In adapting to growth at this temperature, the viruses experience mutations that also have the effect of slowing down the rate of replication. The result: at any temperature, the cold-adapted mutants are real plodders.

Cold-Adapted Vaccines Most Effective

Then, using a sophisticated array of chemical screens, Maassab manipulates genetic reassortments in the lab, pairing his own mutants with live disease-causing viruses that are in general circulation. He not only shuffles the genetic decks of these two different viruses together just as in nature, but he picks out exactly which genetic hand he wants to deal to the offspring. The progeny viruses, in fact, receive the immunity-producing HA and NA genes from the disease-causing virus and all the other genes from the mutant that has been trained to plod along. This "seed virus" goes into the vaccines that Murphy tests. After inoculation, this live-virus vaccine produces a low-grade infection without symptoms. In other words, little or no disease, only immunity at a level expected to be far stronger than the response triggered by traditional, killed-virus vaccines.

But does it work? In a study prepared with collaborators Maassab, Mary Lou Clements of the University of Maryland, and Robert Betts of

Show Me Your Tongue *by Daumier*

the University of Rochester, Murphy says that tests showed the cold-adapted, nose-drop vaccine significantly outperformed conventional vaccines in college-age volunteers. In the tests, volunteers not only received the vaccines, but were "challenged" a month later with a virulent form of H3N2 known as A/Wash/80, the same virus used to prepare both vaccines. The recipients of cold-adapted vaccine showed greater resistance than those who got the conventional one. Murphy's carefully worded reaction: "Very encouraging."

Making Progress

Influenza teaches people to hedge their bets, so Murphy is enthusiastically pursuing an alternative strategy with Robert Webster and Virginia Hinshaw of St. Jude. The Memphis group has contributed a number of avian viruses that replicate slowly in mammals, according to tests with primates. Murphy is using the slow-growing viruses in recombinations with current human strains that cause disease. By controlled reassortment, the viral offspring inherit the surface proteins of the human strain and the sluggish replication rate of, for example, a duck strain. The first testing of this virus began in late 1982, and the seed virus replicated just as sluggishly in humans as in test animals.

But all the live-virus vaccines may be merely prelude. Researchers in Australia, England, Israel, and the United States have reported the first tentative steps toward a synthetic "universal vaccine," one that would confer long-term immunity against all type-A flu viruses, the ones responsible for most of the disease.

The theory is simple. Researchers propose to isolate a portion of the hemagglutinin molecule, for example, that is common to all type-A flu but is not one of the hot-spot areas and therefore is not prone to mutation. The body would then be provoked to manufacture antibodies that react against such a stable segment but are unaffected by changes elsewhere on the molecule. While theoretically feasible, a universal vaccine is now viewed as a long-term prospect.

Other goals are only slightly more modest. "What our dreams are for the next 10 years—and it would probably take at least 10 years—would be to really understand influenza," says Webster. "What genes are responsible for what properties, what combination of genes causes epidemics, why it can kill a chicken in two days. By understanding those things totally, we'll have the ability to produce stable and reliable vaccines." Webster fiddles with his "Workaholics Anonymous" mug and allows a broad smile to spread across his face. "And put ourselves out of business" □

Eric Preau/Sygma

New Hope for the Disabled

by Laurence Cherry and Rona Cherry

According to the most recent data compiled by the National Center for Health Statistics, the number of Americans with severe physical disabilities soared by more than 49 percent between 1970 and 1981. Some of the disabled are born with such handicaps as cerebral palsy, spina bifida, and Down's syndrome. But five out of six of the handicapped develop impairments much later in life, including survivors of spinal-cord injuries, burn victims, and many of the elderly whose proportion in the population has swelled from 3 percent in 1900 to 11 percent in 1980.

New Field of Physiatry Helps Handicapped

One of the paradoxical consequences of the life-saving medical advances of the past few decades is that most Americans—some studies predict almost three-quarters—will eventually suffer some physical limitation. "Every family has someone affected by a disability," says Dr. Howard A. Rusk, founder and former director of the New York University Medical Center's Institute of Rehabilitation Medicine.

Unable in varying degrees to walk, talk, see, or move about, the disabled have become one of this country's most hidden minorities.

Often ostracized by society and, until recently, largely ignored by medical science, they have been exiled to public institutions, private nursing homes, or the care of relatives. One quadriplegic, hospitalized in California, would rather die than continue her life of pain and total dependence on others.

This grim picture is beginning to brighten, however, thanks to one of the youngest and least-known medical specialties—rehabilitation medicine. Also called physiatry, or physical medicine, this field has already helped hundreds of thousands of the estimated 32.3 million Americans who are disabled to some degree. And new research, plus a dazzling array of new devices, are also helping to make life easier for a still greater number of people.

As the number of handicapped Americans grows, the field of rehabilitation medicine has escalated from being a marginal specialty of last resort to an often vitally important stage of treatment. "Other medical specialties may help add years to our lives, but rehabilitation medicine will be the one that helps us to truly live them," says Edward Eckenhoff, president of the National Rehabilitation Hospital in Washington.

Soldiers Prompt Rehabilitation Efforts

Until a few decades ago, many of the severely handicapped quickly died. Those who survived serious accidents or disease often succumbed to such secondary infections as pneumonia. The introduction of life-saving antibiotics just before World War II changed that. Within a short time, the death rate for spinal-cord injury, for example, plummeted from more than 80 percent to less than 10 percent.

Yet few doctors had any inkling of how to treat patients with long-term handicaps. Medicine has traditionally sought rapid cures: stemming infection, setting broken bones, removing life-threatening malignant growths. The disabled became grim reminders of medicine's limitations. "Like everyone else, we felt uneasy around the disabled," says Dr. Henry B. Betts, medical director of the Rehabilitation Institute of Chicago (in Illinois) and chairman of the department of rehabilitation medicine at Northwestern University Medical School.

That attitude changed somewhat during World War II, when wounded soldiers returned from the front as heroes. "There was a feeling that something had to be done to help those brave young men," remembers Dr. Rusk, who headed the Army–Air Force convalescent-care program during the war.

It was only a short logical step from rehabilitating soldiers to treating civilians, and in 1947 a professional certifying body—the American Board of Physical Medicine and Rehabilitation—was established. At first, many doctors disdained the field, believing that medicine should confine itself to curing acute ailments. Even now, physiatrists rank low in the medical hierarchy, and the need for more of them is great. Yet approximately 75 rehabilitation hospitals have been established around the country, and many more medical centers have created extensive rehabilitation units.

Ideas Abound for New Aids

Technical aids for the disabled have been few until recently. Today the number of available devices multiplies steadily. With increasing sophistication, the latest aids have assisted the blind, the deaf, and those paralyzed or suffering from speech impairments. Even the wheelchair has been updated. A new, more maneuverable, three-wheeled model constructed of lightweight titanium (a strong metallic element) has been developed by the Theradyne Corporation.

Such research has become a vital part of rehabilitation medicine. To encourage it, Congress established the National Institute of Handicapped Research in Washington in 1978. The Institute's budget goes to supporting research at academic or medical centers around the U.S.

New ideas come from many sources—rehabilitation engineers, small medical companies, and increasingly, amateur inventors. In 1981, for example, to mark the United Nations International Year of the Disabled, Johns Hopkins University organized a national search for computer-based designs to help the handicapped. "Some people scoffed," says Paul L. Hazan of the Johns Hopkins University Applied Physics Laboratory, who directed the program. "They were very dubious about our hope that a little inventor in Nebraska or Montana could turn up something that hadn't occurred to a Ph.D. at Stanford, M.I.T. [Massachusetts Institute of Technology], or here at Hopkins. We proved them wrong." At the judging ceremony, more than 100 finalists demonstrated scores of relatively inexpensive devices. Some of them are already in use.

Innovative Communication Systems

One of the search's prizewinners, Howard F. Batie of Herndon, Virginia, devised an inexpensive communication system for a 50-year-old cerebral palsy victim who could not speak, use a pen, or even use a mouthstick (a tool that when grasped in the mouth can be used to press buttons and switches). Mr. Batie rigged an ordinary home computer to a specially designed keyboard with five-centimeter (2-inch)-wide buttons that could be depressed by the palms of the hands. The device gave the woman her first opportunity to communicate other than by grunting or blinking her eyes. After only a few minutes of instruction, she eagerly sent the inventor her message: thanks, thanks, thanks.

A more elaborate computerized system, developed by a team at the Rehabilitation Institute of Pittsburgh (in Pennsylvania), allows young paralyzed patients to speak. The children stare at an electronic grid, and an infrared video camera follows their gaze to the words they want, responding with synthesized speech.

A pocket-sized telecommunicator created by Harry Levitt, professor of speech and hearing sciences at the City University of New York, allows the deaf to communicate to the outside world by punching out messages that move across a display panel like a stock-exchange ticker tape. The device also transmits and receives messages by telephone.

Technology has also been harnessed to give the blind access to a huge body of acquired information. In 1980 the Department of Justice in Washington opened its Andrew Woods Sensory Assistance Center for blind lawyers. A "talking terminal," invented by a law professor, hooks up to the department's computerized law library and reads the material aloud. A nearby machine can transform it into Braille.

Below: Lee Foster/Bruce Coleman; Bottom: Ken Seibert/The Rehabilitation Institute

Left: A special device developed by Canon allows speechless people to communicate by printing out their messages like a ticker tape. Below: This eye tracker communication system follows the gaze of a nonvocal boy as he stares at the words he wants to use, and it responds with synthesized speech.

Major Advances in Environmental Control

In the area of environmental control, the advances have been just as remarkable. One experimental robotic arm, developed at Johns Hopkins' Applied Physics Laboratory, attaches to a worktable and enables quadriplegics to maneuver and feed themselves. It also facilitates the use of typewriters. Robotic arms are in experimental use in several states.

A team of rehabilitation engineers at the Palo Alto Veterans Administration Medical Center and Stanford University in California is attempting to make a similar robotic arm even more sophisticated. The arm, mounted on a mobile device and equipped with a video camera, can be programmed to move through a user's home. For example, watching a television monitor, a user can vocally command the machine to retrieve a book from a shelf or fill a glass with water. This device may be commercially available within a few years.

Personal robots now on the market, mass-produced by R.B. Robot Corporation in Colorado, can, for example, turn on switches as well as fetch food from the refrigerator. Specific programs for the handicapped are now being developed, and costs could run upward of $2,500.

''Sip-and-puff'' environmental systems already permit quadriplegics and the bedridden to control their surroundings. By sucking on a long command tube attached to a wheelchair, they can operate a switch to as many as 16 appliances.

Computer-aided Motion

Many researchers are using computers in attempts to restore muscle movement to paralyzed limbs. A study team from Yugoslavia, supported by the National Institute of Handicapped Research in the United States, reports considerable success in using computer-guided jolts of electricity to restore movement to paralyzed leg muscles. But in the United States, most attention has been focused on the work of Dr. Jerrold S. Petrofsky, director of biomedical engineering laboratories at Wright State University in Dayton, Ohio. Dr. Petrofsky and his team have spent more than 12 years attempting to prove that computers can mimic cerebral impulses that fail to reach the limbs because of spinal-cord injury.

In a much-publicized demonstration in 1982, Nan Davis, a 22-year-old senior at Wright State University and a paraplegic, took a few jerky steps down a 3-meter (10-foot) runway, assisted by Dr. Petrofsky's microcomput-

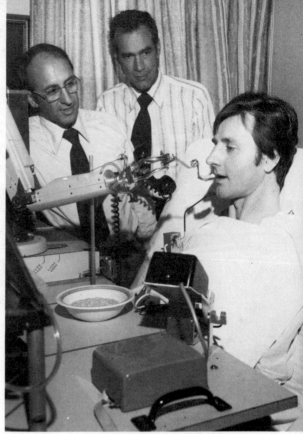

The Johns Hopkins University Applied Physics Laboratory

Researchers at Johns Hopkins' Applied Physics Laboratory look on as a client operates a robotic arm to eat.

ers. Dr. Petrofsky believes he can refine his computer program to allow a person to stand and sit with some grace. His optimism is catching; his subjects are confident they will eventually walk again.

Others are more guardedly optimistic. ''There's a big difference between actually walking in the world and taking a few lumbering steps in a carefully controlled setting,'' says James Reswick, associate director for science and technology at the National Institute of Handicapped Research in Washington. ''Petrofsky's work has caught the public's attention, and that's excellent, because it will encourage other, more cautious researchers. But any disabled people who plan to be walking soon are going to be very disappointed.''

Meanwhile, at Case Western University in Cleveland, Ohio, Dr. Hunter Peckham and colleagues have inserted tiny electrodes into nine key muscles of the hands. By controlling electrical signals from a computerized transmitter, small jolts of electricity make the muscles move in unison and computer programming gives a

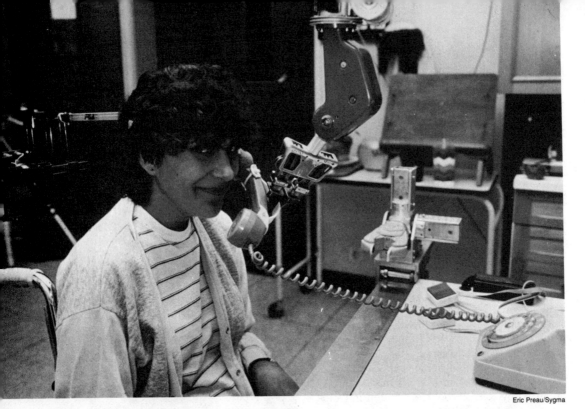

This telemanipulator called "Spartacus" allows quadriplegics to perform many ordinary daily tasks on their own.

rough semblance of fine movement. Some patients have become proficient enough to comb their hair, use a telephone, feed themselves, and in one case, even change a bladder catheter. Dr. Peckham, who regards the device as experimental, still believes it can eventually become far more sophisticated.

Equipment Costly; and Scarce

While many of the new devices to help the disabled are as yet experimental, the commercially available equipment is extremely expensive—a sip-and-puff wheelchair costs about $10,000, a specially equipped van approximately twice as much.

"The sources to help the handicapped pay for this equipment are now increasingly limited or have dried up altogether," says Simone Nathan, spokesman for the Rehabilitation Institute of Chicago, in Illinois. "Help varies enormously from area to area." Although some states offer the handicapped aid to purchase such devices, these funds are scarce, and the waiting lists are long. Private insurance plans are another source if a handicapped person can demonstrate that the equipment is necessary in order to earn a living. But insurance companies differ widely in reimbursement. Some of the disabled turn to local charitable organizations for assistance or appeal to hospital-administered private donor funds.

A Personal Experience

Most people are unaware of rehabilitation medicine and the assistance available—until tragedy strikes. For Gregory T. Boehm, in his early 30's, this abrupt revelation came several years ago. While driving in Connecticut, his car skidded and smashed into a guardrail. In only a second, the accident severed his spine, paralyzing him from the neck down. "I couldn't believe it," Mr. Boehm recalls. "I'd always been very active. I enjoyed a fast-paced social and active sexual life. I couldn't imagine any other way of living. I just looked at the doctor and cried."

Beginning rehabilitation quickly is essential in cases of spinal-cord injury. Two months after the crash, Mr. Boehm transferred to the Institute of Rehabilitation Medicine in New York. In the months to come, he would experience the kind of basic learning young children must undergo—how to handle objects, move from place to place, and care for his bodily needs as much as possible.

The process of rehabilitation takes place in stages: Many victims must learn to cope with

Gregory Boehm uses a "Sip and Puff" tube to activate a switch that controls the movement of his wheelchair.

pain. Reduced mobility can cause poor blood circulation and lead to excruciating skin ulcers. Bowels and bladder may not function properly, and breathing may become labored. Catheters or a mechanical respirator may be required. Infections are common and must be quickly controlled if they are not to become fatal.

Patients like Gregory Boehm must discover how to maneuver their wheelchairs, adjusted to their degree of disability. They must learn to substitute for their paralyzed fingers in order to perform basic tasks. Within a few weeks, for example, Mr. Boehm learned to use a long mouthstick to dial telephone numbers and operate an electric typewriter.

Treatment Requires Team Effort

All of these physical factors—along with the emotional ones—become the concern of the rehabilitation team, the key element in rehabilitation medicine. A physiatrist, who coordinates the team, usually orders tests to determine what capacities remain and which, if any, can be improved; a physical therapist helps increase strength with carefully designed exercises; a rehabilitation engineer may design a special aid; and a rehabilitation nurse monitors the patient's condition and helps educate the family on patient care. In addition, a respiratory therapist can teach quadriplegics to breathe more efficiently, and a speech therapist can help stroke victims speak more coherently.

While this essential treatment goes on, occupational therapists and social workers try to prepare the patient for his discharge. Psychologists and chaplains help the patient deal with feelings of anger and despair. Each expert reports back to the physiatrist, who maps out long-term treatment. In Mr. Boehm's case, for example, eight rehabilitation experts saw him separately. "All the observations are put together in regular brainstorming sessions," says Dr. Willibald Nagler, physiatrist-in-chief at the New York Hospital–Cornell Medical Center.

With the help of the rehabilitation team, most patients begin to adapt to their new lives. Today Mr. Boehm rents a two-bedroom apartment in Clinton, Massachusetts. Mr. Boehm, who graduated from the Culinary Institute of America in Poughkeepsie, New York, and now works part time as a counselor for other handicapped people, hopes to open his own food-service firm this year.

"Of course, I can't claim that my life is a happy one," he says reflectively. "I've got a lot to contend with every day. I've lost friends who can't bear to see me the way I am, but there are others who are very proud of the way I've coped. Rehabilitation helped me understand that you can make your disability seem like hell, or view it simply as an inconvenience."

Self-help Programs Gain Favor

But cures are still in the future. While continuing their research, practitioners of rehabilitation medicine must teach the disabled to accept their

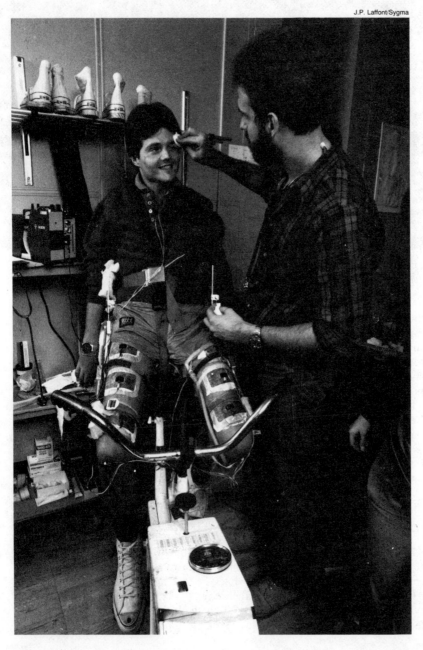

J.P. Laffont/Sygma

Dr. Jerrold Petrofsky uses computer-guided jolts of electricity in an effort to restore muscle movement to the paralyzed limbs of this young man.

state. But for some, all counsel to "accept" their state sounds cruelly hollow.

"Wanting to give up is a common danger," says Dr. Joseph Goodgold, director of medical rehabilitation services at the Institute of Rehabilitation Medicine. "How many of us wouldn't do the same . . . ? But to make the disabled want to stay alive, you have to give them a reason to live."

For some, one of the best sources of hope is the Independent Living movement, whose astonishingly rapid growth has been a cheering note in the long and doleful history of the disabled. A mix of 1960's campus radicalism and old-fashioned self-help, it began in 1972, when a group of disabled students set up the Center for Independent Living Incorporated on a quiet side street in Berkeley, California. "We were the first organization in the world to be run for and by the disabled," says Judith E. Heumann, former deputy director of the center. "We wanted to challenge all of the barriers that kept us out of circulation." To encourage the handicapped to become more mobile, the group offered on-call wheelchair repair, revamped vehicles so they could drive, and an attendant-referral service.

Today about 150 Independent Living programs, financed by the U.S. Department of Education's Rehabilitation Services Administration and some state and private agencies, help the handicapped cope with a host of problems—from finding suitable housing to hiring attendants to learning how to successfully navigate an able-bodied world. Most of their members' problems are intertwined: lack of finances and mobility. "We're trying to get the disabled back into the work force, but often they can't even get to work because they can't jump the step on their neighborhood bus," says Marilyn E. Saviola, director of the Center for the Independence of the Disabled in Manhattan, New York. "If they can't get to work, they can't have a job."

Serious Problems Remain

Some of the disabled have begun to criticize rehabilitation medicine only a little less vociferously than they do the government's lack of support. "Some feel we are not responsive enough to their needs and don't hesitate to let us know it," says Dr. Betts of the Rehabilitation Institute of Chicago. Many criticize rehabilitation medicine for striving hard to help patients while hospitalized but abandoning them once they leave. "Too often, rehabilitation medicine doesn't

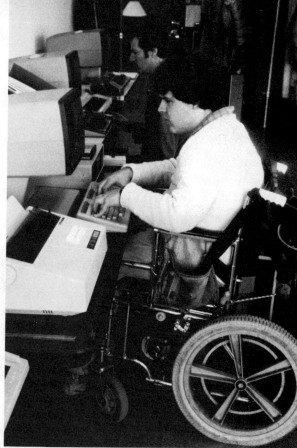

Jerry Stoll

At the Center for Independent Living in San Francisco, California, the handicapped learn valuable skills.

bridge the gap from what goes on inside the hospital to what you're going to confront when you leave," says a young quadriplegic who prefers to remain anonymous. Even though certain hospitals have outpatient programs, many of the disabled are released without being given enough counsel about housing, attendants, equipment, and finances.

More trying than the struggle against governmental or institutional neglect is that against the deep-rooted and pervasive antipathy toward the handicapped. "This is the ultimate barrier for the disabled, and the most difficult one to clear away," says Dr. Douglas A. Fenderson, director of the National Institute of Handicapped Research.

No matter how impressive the new technology, or how elaborate the rehabilitation system, until this last hurdle is overcome, the disabled will continue to suffer. Too many of them face dismal lives of hardship and pain. For this reason, most physiatrists temper their pride in their field's achievements with a sober awareness of how much remains to be done □

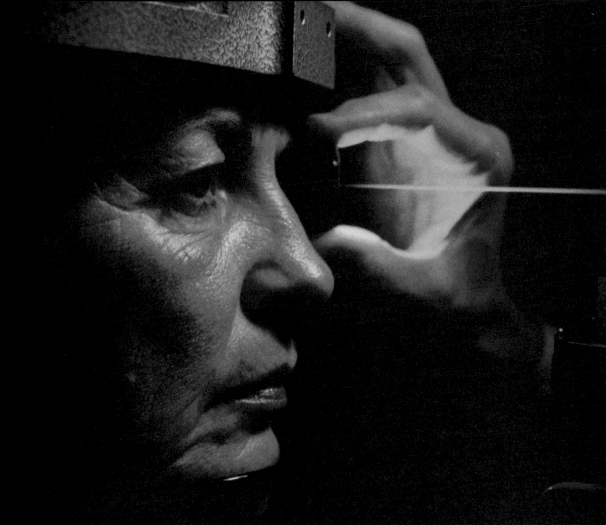

Charles O'Rear

LASER SURGERY

by Benedict Leerburger

For centuries the surgeon's basic tool has been the scalpel. In recent years, however, the development of the laser has opened new avenues in surgical techniques. It is not only replacing the familiar scalpel in many surgical procedures, but is also allowing doctors to perform specialized operations never before possible.

An Amazing Tool

Specifically, laser surgery enables a surgeon to cut sensitive tissue without doing harm to surrounding tissues. (This is most important in some types of brain surgery when damage to adjacent tissue can occur easily, especially if a scalpel is used.) Heat from a laser's cutting beam of light can be used to cauterize (sear) an operative area, thus reducing a patient's bleeding and lessening the possibility of infection. The precise laser beam also allows doctors to operate in highly localized parts of the body, such as the ears, eyes, throat, and vagina, without the additional cutting usually required to expose the surgical area.

By focusing a beam of light to a point smaller than the period at the end of this sentence, a surgeon can vaporize living cells in the light's focal spot. Surgeons employing the laser in microsurgery can now penetrate tissue as tiny as a vein.

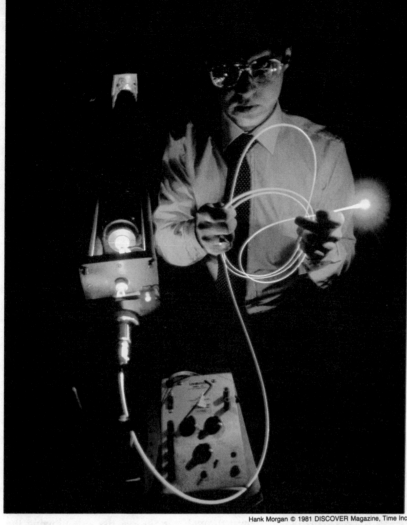

Hank Morgan © 1981 DISCOVER Magazine, Time Inc.

Opposite: A laser beam vaporizes abnormal blood vessels in a woman's diseased eye. Left: This argon-laser apparatus at Sinai Hospital in Detroit, Michigan, is used to perform a variety of specialized operations.

Surgical lasers are stationary, bulky machines. A special arm attached to the device looks like the arm of a dentist's drill. It transmits the laser beam through a series of mirrors to a hand-held tool used by the surgeon. A foot switch turns the laser on and off.

Cutting with Light

Light, like sound, travels in waves. The wavelength of light depends on the illuminated substance. Scientists discovered that certain substances can be electronically stimulated, or "lased," to emit a highly concentrated beam of light, usually of a single color or wavelength. Most surgical lasers (the word is an acronym for Light Amplification by Stimulated Emission of Radiation) use as the lasing substance inexpensive gases—carbon dioxide, argon, or a mixture of helium and neon—or use the crystal neodymium: yttrium-aluminum garnet, known simply as Nd:YAG, or YAG.

The laser selected by the surgeon is often dependent on the type of surgery. Argon-laser energy, for example, is easily absorbed by hemoglobin and therefore is used to coagulate blood from small, ruptured blood vessels. It is also used to "weld" tissues, such as affixing a detached retina to the back of an eyeball.

YAG-laser energy is not selectively absorbed by tissues. It is used to penetrate blood clots and reduce the size of tumors. Since the light projected by both argon and YAG lasers can be transmitted through very small quartz fibers, these lasers can be used on parts of the body not easily accessible.

The carbon-dioxide laser is the most widely used in surgery. However, since it cannot be used in a fiber-optic system, its laser beam is only applied by directing the spot of light on the specific point to be cut. Its prime advantage is its low power requirements and its ability to cut tissue and vaporize moisture in tissues, thus enabling a surgeon to make a very small incision or destroy a tumor without damaging healthy tissues nearby. Carbon-dioxide lasers are used for surgical procedures in the ears, mouth, nose, and throat.

Early Uses Expanded

The earliest use of the laser beam by a physician was reported in the early 1960's by Dr. Leon Goldman, a Cincinnati, Ohio, dermatologist. He reportedly used a laser beam to reduce the size of a birthmark. Prior to placing a patient "under the beam," however, Dr. Goldman performed experiments on his own skin to détermine the possible harmful effects. Goldman found that he could remedy many skin disorders with the laser. He also showed how the laser could be applied to other medical disciplines. Goldman, now president of the American Society of Laser Medicine and Surgery, says "For operations on the larynx, inoperable birthmarks, and certain brain tumors, lasers are *required*."

Laser surgery is now being used to remove birthmarks, warts, skin cancers, and even unwanted tattoos.

J.P. Laffont/Sygma

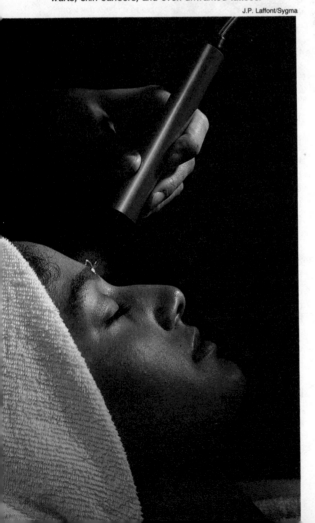

As a result of continuing technology, laser surgery is used for a wide range of disorders—from brain tumors and lung lesions to infertility. Low-powered laser beams are even used to stimulate the pressure points of traditional acupuncture. In 1983 the laser unit at the University of Utah Health Sciences Center treated an average of 102 patients a month—almost 80 percent on an outpatient basis. That same year, Detroit's Sinai Hospital treated some 4,000 patients with lasers. Says Dr. Hugh Beckman, chairman of Sinai's ophthalmology department: "With a laser I can do in five minutes under a topical anesthetic what one formerly had to do with an in-patient under surgery."

Aiding Eyes

One of the most widespread applications of laser surgery is in the treatment of ophthalmological, or eye, disorders. For example, a laser beam can vaporize in seconds the milky-white membranes known as secondary cataracts that often develop behind implanted lenses. This relatively simple procedure avoids the risk of costly and potentially dangerous major eye surgery.

The laser's effect is most dramatic in glaucoma surgery. Glaucoma, a disease caused by the buildup of fluid pressure in the eye, blinds approximately 5,400 Americans every year and affects about two percent of all people over the age of 40. With an argon laser, a surgeon can burn through certain eye tissues, allowing the trapped fluid to drain out. Unlike traditional major eye surgery, there is no bleeding and no risk of infection. The procedure takes less than 15 minutes and can be performed in a doctor's office. Says Dr. Charles D. Belcher, an ophthalmologist at Harvard Medical School, "The laser is a miracle that has changed the life of people with glaucoma."

Another important optical use of the laser is in the treatment of senile macular degeneration (SMD). The disease causes approximately 16,000 Americans to lose their sight each year. It is the major cause of blindness among the elderly. SMD is a disease of the retina in which blood from tiny, abnormal blood vessels in the back of the eye leaks into the portion of the retina's center called the macula, destroying vision cells.

In treating SMD, a hair-thin beam from an argon laser is focused on a tiny portion of the retina. Heat from the laser ray vaporizes the abnormal blood vessels and prevents further

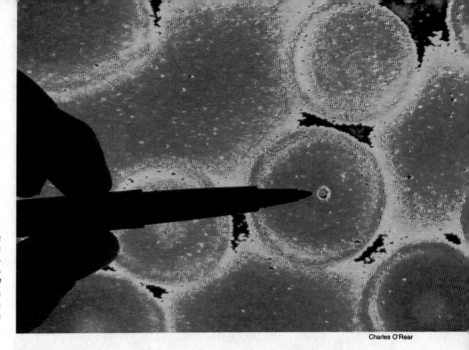

This tiny borehole in a red blood cell was made by a laser in experimental genetic surgery at the University of California at Irvine. The hole is about 160 times smaller in width than a human hair.

Charles O'Rear

leakage of fluid. The National Eye Institute has stated that 90 percent of the blindness caused by SMD can be prevented with laser treatment. The technique has also proved effective in treating other eye disorders, including myopia and diabetic retinopathy.

Laser Microsurgery

Since a beam of light can reach places a surgeon's fingers can't, lasers are becoming more widely used in the delicate field of microsurgery. Although laser microsurgery was only used to a limited extent in 1984, its future is extremely promising. Dr. Rodney Perkins, a California surgeon, reported that he restored the hearing of about 20 patients by using an argon laser to vaporize defective stapes bones in the inner ears of his patients. The stapes bone transmits sound vibrations to the inner ear. If this delicate bone hardens, it can become locked in place, severely impairing hearing. The standard procedure is to remove the locked bone with tiny surgical picks and chisels. This procedure, however, can cause permanent hearing loss and prolonged dizziness.

In the field of gynecology, surgeons restore fertility in some women by using the laser to reconstruct blocked or tied fallopian tubes (the tubes that carry eggs from the ovary to the uterus). The surgeon uses a special microscope and controls that work like the joystick on a computer to direct the laser's rays. Researchers are testing a similar technique to restore fertility to men who have had vasectomies.

Ridding Birthmarks and Tattoos

Although still in its infancy, the use of laser treatment to remove birthmarks, warts, and skin cancers was performed by about 30 doctors in the United States in 1983. In treating the deep red birthmark called a "port wine stain," an argon laser is used to close, or coagulate, many of the blood vessels. These often disfiguring birthmarks are caused by an abnormally dense network of blood vessels close to the skin's surface. Reducing the number of active blood vessels often causes the port wine blemish to fade. Researchers are also experimenting with the laser to remove certain types of scar tissue on the skin.

The same basic technique used to treat birthmarks is being used to remove tattoos. The laser breaks down, or bleaches, the dyes in the tattoo, causing the unwanted design to fade or disappear. Unfortunately, the use of lasers in the treatment of skin disorders is not without its problems. Undesired skin coloration or scars may result, particularly in fair-skinned people.

Unblocking Clogged Blood Vessels

One of the most promising forthcoming applications of laser surgery may be in the treatment of heart disease. Atherosclerosis, a process in which a buildup of fatty deposits weakens the walls of arteries, appears to lend itself to laser treatment.

Atherosclerosis is responsible for slowing the flow of blood to and from the heart. When fat globs called plaques accumulate in the chest,

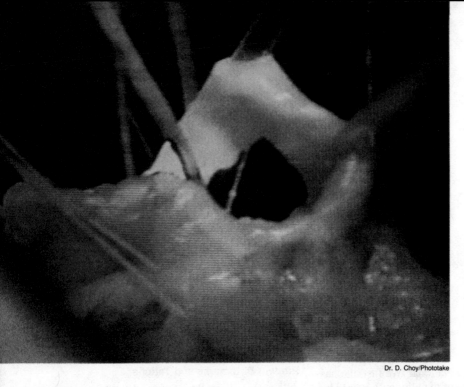

An argon laser is used to open passageways through clogged arteries by burning away fat deposits called plaques that restrict the flow of blood. Lasers may someday replace expensive surgery in the treatment of some kinds of heart disease.

they may cause a heart attack; in the head and neck, they may result in a stroke. The most widespread treatment for severely narrowed arteries is bypass surgery. In 1984 more than 100,000 Americans experienced bypass surgery. Of these, about 2,000 failed to survive the operation, and in about 10,000 cases, patients had their grafted blood vessels close up within a year.

A new technique called laser angioplasty that clears blocked arteries was successfully administered in both France and the United States in 1983. At Stanford University, cardiologists used a laser beam to clear plaque from the thighs of two patients whose femoral arteries were almost totally blocked. The first patient "is doing fine," according to Dr. Robert Ginsburg at Stanford. "The procedure didn't work as well for the second patient. The vessel closed again several weeks later and necessitated a bypass." Although doctors continue to experiment with laser angioplasty, the government has not as yet cleared the procedure for general use.

One important new laser-angioplasty technique has been developed by Dr. Garrett Lee, a cardiologist at the University of California, Davis. He perfected an experimental method by which a tiny tube called a catheter is threaded through a clogged blood vessel. Within the flexible tube is a bundle of optical fibers. Light is projected through one set of fibers, allowing doctors to view the catheter's journey within the blood vessel through an eyepiece or by projection onto a television screen.

When plaque is detected, the surgeon can direct a YAG-laser beam through the optical fibers to focus right on the fatty deposit and literally burn it away. A suction tube in the fiber bundle vacuums away any residual particles once the lasing is complete.

Although Dr. Lee's sophisticated laser is still experimental, he expects to use it on humans in 1985. Many believe laser surgery will become an important, and perhaps safer, alternative to coronary bypass surgery. Says Dr. George S. Abela of the University of Florida, "It is only a matter of time, not technology."

Zapping Cancer

The use of lasers to treat certain lethal cancers is also undergoing extensive experimentation. One of the remarkable features of the surgical laser is its ability to affect one tissue without affecting adjacent tissues. This ability is being put to use to destroy cancer cells. A patient is first injected with a substance called hematoporphyrin, or HPD. The chemical collects in cancerous tissues and ignores healthy cells. When the HPD-treated tissue is exposed to certain kinds of laser light, a form of oxygen is released that destroys the diseased cells. The treatment was tried successfully in 1983 on a patient with choroidal malignant melanoma, a common form of eye cancer. The technique has also been tested on terminally ill patients. Doctors say that the technique shows promise for early treatment of lung and bladder cancers.

Problems Hamper Widespread Use

Despite the progress laser surgery has made in the past few years, there are still obstacles to its widespread use and acceptance. There are, and probably always will be, safety problems with lasers. Although lasers can be used with a great deal of precision, it is nevertheless possible to injure adjacent tissue if the laser's power is too strong or if the laser is not used properly. Other potential hazards to patients and operators alike are injury to the eyes from laser beams reflected off shiny surfaces and risks of fire or explosion of combustible or chemical materials because of the heat produced by some lasers.

The U.S. government has established elaborate safety regulations on all medical instruments used in the United States. These measures, coupled with the decline of federal research funds, had previously caused many doctors and hospitals to be extremely cautious before acquiring laser equipment. Although these restrictions help protect both technicians and patients from unexpected side effects, they also limit the development of effective new treatments.

In Europe and parts of Asia, laser technology is making rapid progress. Commenting on the slow development of medical laser technology in the U.S., Dr. John Dixon, director of the laser program at the University of Utah's Medical School, states: "We're going to get left behind if we don't get our act together."

Costly Equipment

One of the nonmedical problems associated with the use and acceptance of lasers in the operating room is their cost. A complete medical system may cost anywhere from $18,000 to more than $150,000. "Most surgeons are not used to spending a lot of money for tools," says Dr. Joseph H. Bellina, director of the Laser Research Foundation in New Orleans, Louisiana. "Most of them simply use a saw, a knife, and something to stop the bleeding. And $2 or $3 for a knife is considered excessive."

Advocates of the greater use of medical lasers argue that $100,000 is really not a lot for a hospital to spend on equipment that actually treats disease. They point out that hospitals are quite willing to spend over $1 million for medical scanners that are only used for diagnosis.

Charles O'Rear

At the University of Southern California Medical Center, a surgeon checks a laser before a lung-cancer operation.

Another potential stumbling block to the greater acceptance of laser medicine is the lack of a consistent reimbursement policy by medical insurers for approved laser applications. This "slowdown" by insurance companies in accepting the laser as a standard surgical device is ironic considering that costs for laser-surgery procedures fall well below those of conventional surgeries □

SELECTED READINGS

"Laser heart surgery teams chemistry and medicine" by Rebecca L. Rawls. *Chemistry and Engineering News*, March 12, 1984.

"The laser: a splendid light" by Allen A. Boraiko. *National Geographic*, March 1984.

" 'Poof' goes the plaque with experimental laser angioplasty." *Journal of the American Medical Association*, December 16, 1983.

"How laser surgery is moving medicine light years ahead." *Business Week*, October 17, 1983.

"The exploding field of laser-beam surgery" by Abigail Trafford. *U.S. News & World Report*, July 25, 1983.

FUTURE

REVIEW
OF THE
YEAR

PAST, PRESENT, AND FUTURE

FOOD AND POPULATION

In 1983 the production of cereals—the major staple—decreased worldwide by about 5 percent, to 1.6 billion metric tons. This was due mainly to a sharp reduction in the U.S. corn crop that resulted from adverse weather and from idling nearly one-fourth of the corn acreage. Food trade, however, did not decline, and as a result the world's carryover food stocks fell to less than 44 days of consumption—significantly less than the 62-day standard that the U.N. Food and Agricultural Organization (FAO) considers the safe minimum. The U.S. share in the world food trade fell below 50 percent for the first time in more than a decade.

Population growth continued at an annual rate of 2.6 percent, adding about 115 million people to the world total, the vast majority in food-deficit regions. The number of seriously malnourished people continued to hover around 500 million, and upward of 16 million children died unnecessarily from an injurious synergism of malnutrition and easily preventable or treatable diseases.

Sub-Saharan Africa continued to be the most obvious region of food insecurity, with the least likelihood of achieving self-reliance in food in this century. An international effort is under way to meet the short-term African food crisis, but logistical problems, lack of rural infrastructure, and deficiencies in management constitute major obstacles to the relief and development effort. ■ There are, in addition, millions of hungry people in southern Asia, but there is no food deficit in that region. In Latin America the Andean countries (for example, Bolivia and parts of Colombia) are particularly hard hit. There are pockets of poverty-related hunger in the industrialized world, particularly in the United States, where there was a sharp increase in the

Top: Scientists examine the coral-encrusted anchor from the Monitor—*an ironclad ship of the Civil War—which was recently recovered. Left: A skeleton bedecked with rings was one of many uncovered at Herculaneum.*

Top: NOAA; Left: Jonathan Blair/Black Star

number of people needing food assistance. All of these statistics underscore the fact that food insecurity is more a problem of structures and policies than of weather conditions and climate.

As 1983 ended, the United States—the major exporter of grain (that enters the international market)—had more than 30 million metric tons of it in farmer-owned reserves, 4 million in the government's emergency grain reserve, and nearly 10 million tons of government-owned dairy products that it could not dispose of without jeopardizing the market. Thus, the international community still wrestled with the dilemma of how to bring food security to a world in which farmers in the richest producing country face surpluses while millions of people suffer hunger because of economic, social, and political conditions that prevent them from growing or buying the food they need.

MARTIN M. MCLAUGHLIN

ANTHROPOLOGY AND ARCHAEOLOGY

The most widely discussed event in anthropology in 1983 was publication of Derek Freeman's *Margaret Mead and Samoa: The Making and Unmaking of an Anthropological Myth.* Margaret Mead was probably the best-known anthropologist of this century. Because of her fame among both anthropologists and the general public, and her role in the development of anthropology in the United States, Freeman's challenge to her work was given much attention in both the popular press and scholarly debates. Freeman claims that in her classic study, *Coming of Age in Samoa,* published in 1928, Mead incorrectly portrayed Samoan society as peaceful and noncompetitive, and Samoan adolescence as tranquil and conflict-free relative to the adolescent years in the lives of Americans. Besides challenging her early work that established her reputation as an anthropologist, Freeman's book raises broader questions about how anthropologists conduct their fieldwork and how they present their findings. The lively and far-ranging debates generated by the book show that these problems have been on the minds of many anthropologists. A summary of the scholarly reaction to the book so far might state that although Mead's research was weak in some respects (which is not surprising, since she was a pioneer in the early days of anthropology), on the whole it was accurate and thorough. Freeman's arguments are based upon his own fieldwork carried out between 1940 and 1970 in Western Samoa, whereas Mead's field studies were done in 1925 in American Samoa. Cultural patterns may have been different in the two regions, and

Samoan society probably changed in the decades between Mead's and Freeman's fieldwork because of the rapid pace of Westernization. The debate over Freeman's book is far from over, but from the early reactions of most anthropologists, it appears that Mead's reputation as a pioneer in anthropology will stand strong.

Research on the origins of humans and their relationship to apes continued during the year. The classification and dating of the famous "Lucy" skeleton and of *Australopithecus afarensis* and other hominids continued to be controversial. ■ At the same time there was renewed interest by some zoologists in what some think may be the best living model of the last common ancestor of humans and apes—the pygmy chimpanzee (*Pan paniscus*), now rare in equatorial Africa.

New archaeological discoveries have revealed the ancient predecessors of two modern urban centers. During subway construction under Mexico City, workers came upon the remains of the Great Temple of the Aztecs. This temple stood at the heart of the Aztec empire in the ancient capital city of Tenochtitlán, and archaeologists anticipate that the area around it will yield many artifacts from the early metropolis. ■ Chinese archaeologists have discovered the original settlement of Peking. City walls and ancient graves found 69 kilometers (43 miles) southwest of the modern capital are those of Ji, capital of a fiefdom of the Western Zhou Dynasty around 1000 B.C. Excavations in the southwest region of modern Peking have revealed structures suggesting that the city was transferred to its present location around the third century B.C.

An exciting event in underwater archaeology was the recovery of the anchor of the Civil War ironclad ship the *Monitor* from the sea bottom off Cape Hatteras, North Carolina. The battle between the Union *Monitor* and the Confederate *Merrimack*, fought on Chesapeake Bay on March 9, 1862, was the first between ironclad ships.

What is believed to be the oldest stone fort in America is under investigation near Castine, Maine. Fort Pentagoet was built by the French in 1635 of stone shipped from France. It was small, housing just 25 men and 16 cannons, but it played an important role in early French-English rivalry on the Maine coast.

Analyses of skeletons from the Roman city of Herculaneum, destroyed with Pompeii in the eruption of Mount Vesuvius in A.D. 79, have revealed that some of the people may have suffered from chronic lead poisoning. It has often been hypothesized that lead poisoning was one of the factors that contributed to the decline of the Roman Empire, and this new evidence may provide support for that idea.

PETER S. WELLS

BONSAI

by David Halberstam

Space is of the essence. That is perhaps the foremost fact about Japan. As in no other industrialized nation in the world, people live on top of one another. In the rural areas, the tiniest clump of arable land must be exploited. Foreigners have suggested that Japanese farming is closer to gardening than to farming. "In Montana," says United States ambassador Mike Mansfield, the former Montana senator, "we have 793,000 people. In Japan there are 118 million people, and in Montana we have 1,314 more square miles."

Bonsai Embodies Japanese Spirit

Inevitably, the national sense and definition of beauty is about being small, and about the delicacy of lines and the simplicity of proportions. In a restaurant the portions are small; the aesthetic seems to inhere as much in the way the food is served as in the taste. Japanese houses are small, and even quite prosperous Japanese, who live in more westernized homes with far larger rooms, usually manage to keep at least one small room in a traditional Japanese style. In it they place the minimal number of objects lest they detract from an essential, understated symmetry. The haiku, a treasured form of poetry, is all of 17 syllables long. In contrast to the United States, with its high expectations and its fierce desire to consume and accumulate, Japan is a nation that has been built around lower expectations and, in many ways, on restraint and delicacy.

No wonder, then, that the Japanese treasure the bonsai, or dwarf tree. It is an almost perfect reflection of their culture. Bonsai are small, at the most 0.6 to 0.9 meter (2 to 3 feet) tall. They reflect the beauty of nature, which is so much prized, particularly as it becomes diminished by the smog of intense industrialization. In a society that is extremely reverent of tradition, the bonsai are part of a cultural and historical link with the past. A tree can easily be 100 or 200 years old, and some precious ones may be as much as 500 years old, although they are likely to be in the Imperial Palace. Thus, a bonsai reminds a harried Japanese today of the continuity of the past, present, and future.

An Old and Delicate Art

Bonsai is the art of miniaturizing trees. Other, more modern Japanese now excel at miniaturizing electronic equipment or cars, but the bonsai masters work with nature. The trees are potted and carefully pruned, the branches are wired,

Above and opposite: Eberhard Grames

Above: A realistic forest of dwarf hinoki cypress trees. Opposite: A miniature red maple over 100 years old.

and every couple of years the roots are systematically trimmed in order to keep the trees small. It is a delicate art, for it requires a hand both artistic and supple so that the tree stays small and at the same time lives. The master may on occasion impose too much of himself on the tree, thereby creating an artificial shape. There was a period in which this was fashionable, but the results were oddly without a sense of harmony. It is now called octopus style, and it has since given way to a more natural look.

As early as the 14th century, bonsai started showing up in Japanese scrolls. They are also part of national myth. In a favorite No drama (a classic Japanese dance-drama), a poor but honest samurai (warrior) receives a visitor on a bitterly cold and snowy evening. In order to comfort and warm his visitor, the samurai puts his favorite bonsai tree on the fire. Fortunately for him, the visitor turns out to be a powerful shogun (a military governor) traveling incognito and seeking just such selfless deeds.

There is, I think, something moving about the entire act of growing bonsai. Most of the trees existed long before the incumbent generation first touched them. Many others will become truly beautiful only after the present generation has passed from this earth. Yet they are attended with constant care and subtle passion. The bonsai is the symbol of a patient society that places a high value on continuity.

Masters Aim to Enhance Nature's Will

Everywhere I look, there is the bonsai. The prime minister has a bonsai smack in the middle

The first blush of autumn color graces the leaves of the Japanese maples in this bonsai planting.

of his office. A group of Toyota auto workers are judged to be the most efficient workers at their plant, and are given a handsome prize. They in turn decide to spend their prize money for bonsai, a gift to future generations and a reflection of their confidence that Toyota will be in the auto business for a very long time. An extremely successful Japanese businessman, involved in the most modern of communications companies, returns to his surprisingly small apartment every night, and with an intensity just short of frenzy, busies himself with his bonsai tree. He explains that it has been in his family for 50 years, that it once belonged to a neighboring family before it passed to his. From his pride, it is clear that he not only believes in the beauty of the bonsai but also feels that his family got a good deal on it. A 100-year-old bonsai may be worth as much as $10,000 today.

Here bonsai are regarded as nothing less than art, and the bonsai masters, the men who grow them professionally, are considered to be artists. Saburo Kato, one of the foremost contemporary masters, adds that bonsai masters are different from other artists in two critical respects. The first, he says, is that the bonsai artist

works with a living organism. These plants are alive. Thus, the master, unlike other artists, is directly connected to nature and is more aware of how much people depend on nature, for he sees it firsthand with his trees. His hands are always on the living. The second difference is that the work is never complete. A painter finishes a painting. A writer completes a play. But a bonsai master will work with a tree and shape it and help it, and in the end he will be gone and someone else will come to take over his work. So master or not, famous or not, his role has a built-in modesty to it.

The bonsai master, says Kato, must also understand the individualism of each tree, and he must abide by that unspoken will of nature. His art must have immense amounts of self-restraint built into it. Otherwise, he may force the tree into his vision of it against its own will. An attitude like this will destroy the tree. He is, says Kato, trying to walk a very delicate line. He loves nature and wants to improve on it. Yet there is always the danger that this desire will prove to be immodest. The key is to sense the will and the direction of the tree. A good master will be able to do that, which is one reason that working with trees is so humbling. Another is that they will always outlive you.

Kato's Perspective

Kato knows. He is the son and grandson of bonsai masters, and three of his sons hope to be masters as well one day. He lives in Omiya, a suburb of Tokyo, in an area where there are now 13 major bonsai gardens and perhaps as many as 100,000 seedlings just beginning to grow.

The bonsai business is better now than it has ever been before, Kato says. There is a bonsai boom, and he is not completely happy about that. He is pleased that more and more people (and no longer simply rich older gentlemen) are taking up the art. But he is unhappy with the reason: many Japanese are doing it because they

have become increasingly cut off from nature in their daily lives. Japan is an urban society now, he says. It is far more materialistic, everyone is always in a rush, and there is much less concern with nature. He does not think that this is a good thing, and he has seen it happen before his eyes. Once the Japanese were a people who were at ease with nature. Now they live in cities, work in offices, commute long hours, and never have time to think. They never see anything green. As a last resort, he says, many of them are turning to bonsai, particularly young people. ''They now try to escape to nature by bringing it right into their homes,'' he says.

Kato is almost 70. As a sign that he is an artist, he wears a beret and a sweater. Surely he is the only successful man in Tokyo on this day

Many people in Japan are becoming interested in bonsai as a way of keeping in touch with nature while living in an urban society.

Eberhard Grames

Noted bonsai master Saburo Kato applies his expertise and artistic eye in shaping each tree to enhance its beauty.

who is not in a dark-blue business suit. He has lived through the war. He and his brothers all survived, but his father, who tried to keep the family's trees alive, was pilloried by neighbors for caring more about trees than about the war effort. He tried in desperation to water his trees at night, but neighbors would spy on him and report him. The family lost almost everything it had during the war, including most of the wonderful old trees. It is one thing, Kato says sadly, to lose possessions. That is not important. It is another thing to lose something that was entrusted to you.

In Omiya, Kato is surrounded by his trees. Each has its own special beauty and perfect proportions. Each seems to be in perfect harmony with its setting. Visiting a bonsai village is not like visiting some wonderful greenhouse, where everything is vibrantly alive. There is a different feeling here, a feeling of magic. This is not so much about greenery as about scenes and about imagination. What is astonishing is how evocative of a moment or a mood each tree seems to be. It is a stop in time, a moment in which a scene from the past flashes through the mind of the beholder, transporting him instantly to another time. Kato, son of a nation where poetry is much prized, says that every tree is a poem, but for me, lover of movies and plays, they are like theatrical or cinematic scenes. We talk about

this, and we decide to play a game. He will show me a tree, and he will tell me what he feels; then I will tell him what I feel. We shall see, he says, about East and West.

Each Tree Tells a Story

The first is a pine tree, wonderfully wrought and, in its main branch, bent almost parallel to the ground. Its trunk is still handsomely scarred from an avalanche that occurred more than 100 years ago. The tree is now several centuries old. It spent the first 300 years in the wild, then it was transplanted and brought indoors. That would be illegal today.

Kato says that this tree tells the story of the hardship of life. The tree has been near a cliff, and that is why it is so disfigured. He looks at it and thinks of a terrible storm on a dark night.

I tell him I see it as a scene from a Hollywood movie set in the U.S. Southwest. If it is a soppy, sentimental movie, the boy will kiss the girl under the tree in the last scene. In a better, more honest movie, a dark deed will happen under it. At first Kato is slightly puzzled by my description; clearly he is not a fan of Hollywood. Then Nobuku Hashimoto, my interpreter, translates the last part of my thought, and Kato comes alive. "Yes," he says, "the Japanese have always known about that, about dark deeds and long nights."

Lots of time and care
went into the skillful
pruning of these trees
in a bonsai nursery.

The next is a clump of maples. These trees are the most beautiful of the bonsai I have seen in Japan. Their beauty is stark. They are only 70 years old. I am absolutely captivated by them; it is as if they are the only things in this room that are real, and everything green is false. Kato tells me he thinks of a wide field and a lot of space. A man sitting near these trees is comfortable because he does not feel crowded. He has been crowded too long now, and he finally feels released from being around other people.

I tell him that I am filled with absolute melancholia. For me, the maples mean that the summer and early fall in Nantucket, Massachusetts, where I do my writing each year, are over and I must return to the city. I must be more serious, wear suits and ties, and do a lot of traveling on jet planes. The sweet part of my year is over; I must become a grown-up again. Of all the answers I have given, this one pleases Kato the most; it is as if his tree is truly appreciated. "You must understand," the bonsai master says, "that in Japan we believe very much in the beauty of melancholia. We are very comfortable with it."

We decide to do one more tree, a relatively small one that is about 70 years old and has lovely limbs reaching outward. It is a winter jasmine with yellow flowers that are just beginning to bloom. "We are in a wide field," Kato explains, "and there are children here, and all you have to do to be young is listen to the children—you become like them." He asks me what I think.

"Easy," I say. "The first day of spring. No more overcoat. Life is going to get better."

Again he is pleased with the answer. "The Chinese name for this tree," he says, "is the flower that welcomes spring." He pauses, for the talk has gotten a little somber. "It is a very good reminder that everything can always be reborn" □

SELECTED READINGS

The Essentials of Bonsai by Shufunotomo Editors. Timber Press, 1983.

Suiseki: Japanese Viewing Stones and Their Use with Bonsai by Vincent T. Covello and Yuji Yoshimura. Charles E. Tuttle Company, Incorporated, 1983.

Bonsai Techniques by John Y. Naka. Dennis-Landman Publishers, 1982.

Uncovering a Lost Spanish Mission

by Benedict Leerburger

Archaeologists and anthropologists are constantly searching for new discoveries that may link early civilizations and cultures with our own. It is unusual, however, when a major discovery occurs in the United States.

In 1979 a group of archaeologists from the American Museum of Natural History in New York set out on an expedition to a small, remote island off the coast of Georgia. They went in search of the remains of a Spanish mission that was established on the island less than 75 years after Columbus sailed to the New World. After diligently digging up many samples of soil, the scientists were successful. By 1984 they had uncovered 140 prehistoric and historic sites.

The Occupation of St. Catherines

Located 6 kilometers (4 miles) off the coast of Georgia is a virtually uninhabited island—approximately 35,000 hectares (14,000 acres) in size—about as big as Manhattan, New York. The island, named St. Catherines, is near the midpoint of a chain of islands along the coast of South Carolina, Georgia, and northern Florida, called the Sea Islands. It is covered with a dense, almost impenetrable, forest of hardwoods, briars, and palmetto thickets.

For years St. Catherines was known only to the Guale Indians, who had inhabited this region of the Atlantic coast for centuries. However, shortly after Columbus discovered the New World, the peace and tranquility of the area was permanently disrupted. In 1521 a Spaniard named Lucas Vásquez de Ayllon sailed from the Spanish territory Hispaniola (an island in the West Indies) to the Americas in search of Indian slaves. One of the many captives taken by Ayllon told him of the wealth of Indian settlements along the coast. Tempted by tales of these Indian communities, which meant a steady supply of profitable slaves, Ayllon returned to the Americas in 1526 to establish a Spanish colony. The settlement was probably located near either Santa Elena Sound or Port Royal Sound off the coast of South Carolina. Soon French settlements followed the Spanish, and communities sprang up along the coast. With the growth of European coastal settlements came the influx of Christian missionaries.

By the early 1600's, missionaries had finally urged the Guale Indian population to inhabit the Sea Islands. Before the end of the century, St. Catherines was colonized by a small group of Guale Indians and several Spanish missionaries. A Christian mission was constructed, and the island's inhabitants tried to establish a permanent community, which was named Santa Catalina de Guale.

Unfortunately, the islanders' efforts were short-lived. British forces, exerting their claim to coastal lands, attacked Spanish and French outposts in the region. Yuchi, Creek, and Cherokee Indians soon allied themselves with the British troops. In 1680 the sleepy island of St. Catherines felt the brunt of the combined British and Indian forces.

Forced into Abandonment

Though few in number, the islanders fought bravely. Captain Francisco de Fuentes commanded a pitifully small band of 5 Spanish soldiers and 15 untrained Guale Indian musketeers. It was his responsibility to defend the Santa Catalina de Guale mission from the hundreds of Yuchi Indian attackers that were led by dozens of well-trained, well-armed Britishers who tried to storm and conquer the island. Although the captain and his ragtag militia managed to stave off their attackers, they felt they had little tactical choice but to abandon the mission and head south to the Spanish-controlled town of St. Augustine, Florida.

The abandonment of Santa Catalina de Guale marked the beginning of the end of Spanish power and influence in the southeastern U.S., and the beginning of British domination in the area. Three years after the Spanish retreat from the mission, a small British force visited the island. The commander of the force, Captain Dunlop, kept a journal in which he noted that ". . . where the great Setlement [sic] was we see the ruins of several houses which the Spanish had deserted." Captain Dunlop's journal was the last known account of the Santa Catalina mission. The "great Setlement" disappeared under the forces of nature—hurricane-tossed soil, dense-growing saw palmetto and other vegetation, and the ever-wearing process of erosion over the years.

By examining soil samples from all over the island, archaeologists were able to zero in on the mission's location.

Scientists Intrigued by Island

Archaeologists and anthropologists have been fascinated by the Sea Islands and have been aware of the lost mission for many years. In addition to Captain Dunlop's journal, other historic documents—many still housed in Spanish archives—have described an island and its buried mission. The extensive shoreline along the islands has been studied, mapped, and searched for buried treasure that is believed to have been hidden there during the period when pirates sailed the region.

Scientists were first attracted to St. Catherines in 1896, when the American archaeologist C. B. Moore conducted excavations of prehistoric burial mounds on the island. From 1967 to 1970, the University of Georgia followed Moore's work and also examined several ancient burial mounds. Then, in 1974, the American Museum of Natural History initiated major fieldwork on St. Catherines under the direction

of Dr. David Hurst Thomas, now Curator of Anthropology at the museum. The museum entered into an agreement with the John Noble Foundation to assist with funding and to "encourage and facilitate scientific research" on the island. Over the past decade, more than 100 scientists and advanced students have been conducting research on various aspects of the cultural and natural history of the island.

Tedious Tasks Reap Rewards

Thomas and his crew studied many 16th- and 17th-century documents to confirm their belief that St. Catherines was the site of the buried mission. Their problem: Where do you begin to excavate for a buried mission, about the size of a football field, on an entire island? The island's topography complicated their problem by providing numerous physical and natural barriers. In addition to the thick vegetation covering the island, scientists surveying the land encountered huge fallen trees, palmetto and briar patches, large swamps, pesky insects such as ticks and chiggers, and poisonous rattlesnakes and cottonmouth snakes.

According to Thomas, "It would have been impossible to explore the entire island, so we began our search by random sampling. This method, which involved following the compass heading and walking through, over, or under any obstacle encountered without deviating from the transect, forced us to look in the most unlikely, inaccessible places."

The Thomas team used a 20 percent random sample, trekking over and sampling 1 out of every 5 square feet [0.1 out of every 0.5 square meter] on the island. At each sample site, a group of archaeologists dug test pits and hand-screened the soil in search of any clue to the past. It was a long and tedious task that took several years to accomplish. Said Dr. Thomas: "From the air the island looked like Swiss cheese." Despite the underbrush, insects, and snakes, the team's tedious efforts paid off. They discovered at least 140 prehistoric archaeological sites.

Spanish Artifacts at Last

A detailed analysis by Thomas and his co-workers established that St. Catherines contains between 650 and 700 archaeological sites. The sites represent 12 distinct periods of human development on the island covering some 4,000 years. However, of all the sites examined, only one contained any artifacts representing 16th- or 17th-century Spanish culture. This was the first tangible evidence of the mission's general whereabouts.

By concentrating their search in the region bearing the Spanish finds and by using statistical-sampling patterns, the researchers honed in on a target area roughly the size of 30 football fields. They used a gasoline-powered post-hole digger to excavate a series of holes. In less than 60 seconds, the noisy earth borer could dig a 1-meter (3-foot)-deep hole and provide the scientists with a soil sample for detailed analysis.

Dirt from the majority of the holes contained many shards of Indian pottery but no clues to indicate the location of the Spanish mission or the early settlement of Santa Catalina. However, the scientists knew they had a major breakthrough when they unearthed small pieces of Spanish pottery and rusty iron fragments from sample holes concentrated in a 100-meter-square (330-foot-square) area. They then concentrated their efforts to uncover Santa Catalina in this area so rich with Spanish artifacts. Dr. Thomas knew what to look for: "Since Santa Catalina had served for a century as the northernmost outpost of Spanish hegemony [leadership] on the eastern seaboard, we were looking for a church, buildings to house the soldiers and priests, and dozens of dwellings, for hundreds of Indians."

Clues from Baked Mud Lead Way to Site

During the 16th and 17th centuries, buildings were constructed of wattle and daub (also called stud and mud). Timbers were placed vertically along walls and laced with twigs or tree branches. These "wattle walls" were then plastered, or "daubed," with a mixture of marsh mud, sand, grass, and Spanish moss.

Although wattle-and-daub buildings crumble and disintegrate over a period of years, leaving ruins behind, wattle and daub that burns leaves significant traces for future archaeologists to uncover. The mud is fired and hardens

Dr. David Thomas, seen here excavating a daub wall, directs the research being conducted on the island.

Dennis O'Brien/AMNH

The Proton-Precession Magnetometer

Once an archaeologist knows or suspects that a metallic metal is located beneath the ground, a proton-precession magnetometer can be used to search for magnetic anomalies beneath the surface. The device measures the strength of magnetism between Earth's magnetic core—the north pole—and a section of the ground being "read" by a built-in sensor on the machine. After taking several hundred readings across a defined grid, a computer plotter generates a magnetic contour map that highlights both the shape and the intensity of subsurface anomalies.

The receiving portion of the instrument, about the size of a transistor radio, is usually strapped to the chest so the operator can monitor the readout while walking over the site being surveyed. The receiver is attached by a long cord to a sensor—an instrument about the size of a coffee can, filled with alcohol or another hydrocarbon-charged fluid—that is attached to an aluminum pole. Protons in the sensor's fluid act like tiny spinning magnets; they can be temporarily aligned, or polarized, by applying a uniform magnetic field from a wire coil within the sensor. When the current is withdrawn, the spin of the protons causes them to *precess* in the direction of Earth's magnetic field, much like a spinning top rotates about a gravity field. (Precession is the relatively slow gyration of a spin-

Left: Deborah Mayer O'Brien directs the proton-precession magnetometer along a tape measure while Dr. Evan Garrison records the device's readings. Below: Fired daub walls are detectable by the sensor.

Deborah Mayer O'Brien/AMNH

3 cm.

in much the same way that pottery bakes in a kiln. Fired daub is nearly as indestructible as potsherds—the mainstay of archaeological chronology.

In recent years, scientists discovered that when daub plaster is subjected to intense heat, the microscopic iron particles that are contained in the marsh mud line up so that all the particles point toward magnetic north the way compass needles do. "It was this magnetically anomalous [unusual] orientation of fired daub walls that led us to Santa Catalina de Guale," said Thomas. He and his crew worked with a specialized remote-sensing team from Texas A & M University to find the fired walls and other subsurface magnetic anomalies.

Many simple and ornate religious artifacts were uncovered. These finds are associated with high-status burials.

ning body's rotation axis about another line intersecting it in a motion that describes a cone.) From this precession it is possible to measure Earth's magnetic intensity at the exact spot of the reading. Thus, a magnetic "X ray" may reveal exactly what lies beneath the surface of the ground.

The proton-precession magnetometer has allowed important advances in archaeological exploration in the 1980's. However, magnetic anomalies discovered beneath the surface still must be "ground truthed"—that is, physically examined by laborious excavation—to separate real finds from natural sources of magnetic "noise," such as underlying rocks.

Significant Discoveries

A major subsurface magnetic anomaly led to the unearthing of a large iron ring. As the archaeologists continued to dig, a second ring was discovered beneath the first, and a third beneath the second. At 2.7 meters (9 feet) below the surface, they hit the water table. Further excavation revealed a well-preserved oak casing from a well. Thomas and his crew had literally hit pay dirt—they discovered the Santa Catalina mission well.

For an archaeologist, the discovery of a well is an extremely important find, as it is a first-rate artifact trap. In addition to a tin plate and the bones of a deer that had fallen down the shaft, the Santa Catalina well offered more: Indian and Spanish potsherds typical of the 16th and 17th centuries. A second magnetic anomaly led to the discovery of a fired daub wall. Beneath the wall were hundreds of Spanish artifacts including cups and plates, cooking utensils, and charred bones. According to Thomas, "Clearly this was a mission structure, possibly the *convento*, the modest dwelling of 17th-century Franciscan friars."

Another anomaly exposed a daub wall more than 12 meters (40 feet) long. Since only a few artifacts were unearthed at the site—none of which were like those found near the other wall—and because of its large size, the archaeologists concluded that the wall was part of the *iglesia*, or church, of the mission.

According to Spanish custom at the time, churches also served as cemeteries, with people buried beneath the floor. Excavation beneath the wall revealed the remains of hundreds of Guale Indians, all buried with arms crossed on the chest and feet facing the church's altar.

It was also the custom for Spanish Catholics to reserve the altar area for burying the most esteemed members of the community. Digging near the altar, the team unearthed the remains of an individual wearing a massive necklace made of thousands of blue-and-white glass beads suspending two copper bells. Another skeleton found closer to the altar was wearing a pendant of 24-karat gold bearing the inscription: *sin pecado*, meaning "without sin." The pendant was a *venera*, a highly prized Spanish religious artifact. Other uncovered treasures included a gold-plated, silver medallion featuring the Blessed Mother. It was next to a cross draped with a shroud. An infant had been buried in the shroud. Dr. Thomas noted that "the interaction of silver and copper set up an unusual chemical microenvironment that preserved swatches of the infant's 300-year-old shroud cloth." Near the entrance of the church, archaeologists found the pitiful remains of the poor—crude crosses, a stone arrowhead, and a few glass beads.

All the religious items discovered represented an extremely rare find, as they contain invaluable evidence about medieval Catholic beliefs. Anthropologists have found many examples of conflict between pre-Christian attitudes and Catholic teaching at the mission.

The human bones unearthed from the Santa Catalina cemetery may prove more valuable to the scientists than the various artifacts. Physical anthropologist Dr. Clark Spencer Larsen, from Northern Illinois University, is part of the museum's team at the site. Dr. Larsen hopes that the bones will reveal information about the sexes and ages of the Santa Catalina mission dwellers as well as data on the nature of their diet, evidence of disease, and perhaps even their causes of death. During the summer of 1983, for example, Larsen excavated bones that indicated that most of the mission's inhabitants had rarely exceeded 1.7 meters (5.5 feet) in height, a fact that may indicate a dietary deficiency that affected growth.

Archaeological and anthropological work at Santa Catalina will continue well into the 1980's. According to Dr. Thomas, "We have quite literally only scratched the surface of this impressive site. Thorough examination of Santa Catalina will take a few more years of digging by teams of trained specialists. It will take that long again perhaps to complete our analysis of our finds. Once we have finished our work, all Guale Indian skeleton material will be respectfully reburied in accordance with current Catholic custom" □

Hundreds of Guale Indian skeletons were found beneath the church, which served as a cemetery. The most respected members of the community were buried near the altar; the remains of the poor were near the church's entrance.

Clark Spencer Larsen & Lara Regan/AMNH

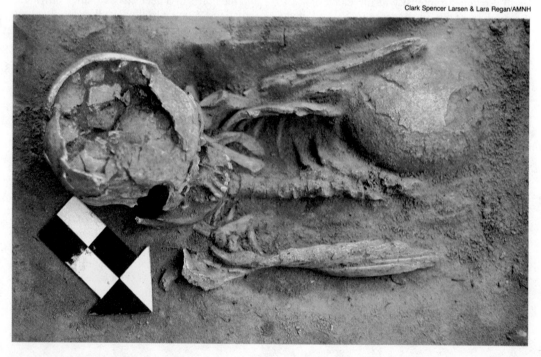

John Olson

The Perpetrator at Piltdown

by John Hathaway Winslow
and Alfred Meyer

On December 18, 1912, Charles Dawson and Arthur Smith Woodward announced to the Geological Society of London and to the world the discovery of the remains of an early human fossil, *Eoanthropus dawsoni*. It was found in a shallow, unimpressive gravel pit near the village of Piltdown in the County of Sussex in Great Britain. It became popularly known as the Sussex Man, Dawn Man, and then Piltdown Man, the name that stuck.

Dawson (far right), Woodward (center), and an unknown worker excavate the quarry where the Piltdown Man's remains were uncovered. The site was well within walking distance of Doyle's home.

British Museum of Natural History

Famous Fossil Really a Fake

In an age when the Empire was still expansive and when the great antiquity of human evolution was still a fresh idea, the find seemed to confirm British primacy, or at least British longevity. For Piltdown Man was widely regarded as the earliest known human fossil, older than anything the French or Germans or anyone else had yet dredged up.

Some 40 years later, J. S. Weiner and his colleagues published an equally startling discovery. Piltdown Man, once the pride of British science, was an out-and-out fake, fabricated by a party or parties unknown. Its mandible wasn't human at all, coming instead from a juvenile female orangutan. The molars and canine tooth associated with it had been artificially filed, and the condyle, or hinge, was apparently purposefully broken to prevent the discovery that it did not properly articulate with the skull. The skull fragments recovered from the site were, on the other hand, definitely human, although of unusual thickness and construction.

The bones, teeth, and antlers of a variety of extinct mammals were also found at Piltdown—elephants, mastodons, rhinoceroses, hippopotamuses, and beavers. Blatantly enough, in retrospect, they ranged in age from the early to late Pleistocene, a stretch of approximately half a million years. Many were typically British fossils, but a few now appear to derive from the Mediterranean area. There were also a number of primitive tools and crudely flaked flint stones

known as eoliths, which were then being found in great numbers at various sites in southern England. All in all, the Piltdown find was an extraordinarily mixed bag.

Piltdown Discoverer Blamed

Once the hoax was uncovered, Charles Dawson, the discoverer and one of the two principal excavators of the site, became the prime suspect. Woodward, his co-excavator, was the British Museum's leading paleontologist and therefore considered above reproach, as was Father Pierre Teilhard de Chardin, the French Jesuit who had assisted in excavation on several occasions and who went on to become a geologist, a paleontologist, and a philosopher. Each was properly credentialed. Dawson, on the other hand, was a solicitor by profession. Although knowledgeable and enthusiastic in geology and paleontology, he was nevertheless regarded as an amateur, a status that made him all the more vulnerable to innuendo when the hoax was finally announced, some 37 years after Dawson's death.

His reputation suffered further with the publication in 1955 of Weiner's book, *The Piltdown Forgery*. Weiner had amassed a great deal of information that led him to point the finger directly at Dawson.

Since Weiner's book, several other individuals have been proposed as the hoaxer, either as a co-conspirator with Dawson or independent of him. These include Sir Grafton Elliot Smith, a neuroanatomist; W. J. Sollas, a geologist; and Teilhard, who has been favored by Louis Leakey and, more recently, by noted anthropologist Stephen Jay Gould. There is little doubt, however, that in the eyes of many experts, Charles Dawson is still the front-runner.

Master of Criminology Suspected

But there was another interested figure who haunted the Piltdown site during excavation, a doctor who knew human anatomy and chemistry, someone interested in geology and archaeology, and an avid collector of fossils. He was a man who loved hoaxes, adventure, and danger; a writer gifted at manipulating complex plots; and perhaps most important of all, one who bore a grudge against the British science establishment. He was none other than the creator of Sherlock Holmes, Sir Arthur Conan Doyle.

That Doyle has not been implicated in the hoax before now not only is a testament to the skill with which he appears to have perpetrated it, but it also explains why the case against him is circumstantial, intricate, even convoluted. For to be on Doyle's trail is, in a sense, to be on the trail of the world's greatest fictional detective himself, Sherlock Holmes. And Holmes, as his admirers know, was not only a master of deduction, he was also a forensic genius as expert in chemistry as he was in pharmacology, as familiar with human pathology as he was with anthropology. His exploits would become required reading for the police forces of several nations, and his creator, Doyle, is still regarded as a pioneer of modern criminology.

Earlier Hoax First Link to Doyle

What led one of us, Winslow, to first embark on that labyrinthine trail was an earlier hoax that contained similar elements. It had been perpetrated by the eccentric English naturalist Charles Waterton many years before Piltdown. In 1825, in *Wanderings in South America*, Waterton claimed to have come across and killed an ape-man. Because of the beast's burdensome weight, Waterton explained tongue in cheek, he had severed its body and carried only the head and shoulders out of the rain forest and back to England. These he then preserved using his own unique methods of taxidermy. Anyone doubting the veracity of his tale was welcome to gaze upon this ape-man, which he called Nondescript.

Waterton actually had taken the head and shoulders of a red howler monkey and shaped its facial features to give it a humanoid appearance. His creation stirred some mirth and some indignation.

Could there be a connection between Waterton's ruse and the hoax at Piltdown? Because of the great gap in time between the two, there is clearly no direct link. Waterton was long dead and gone by 1912. But an indirect connection provided the first notion that Doyle was more than casually exposed to the Waterton tradition and might be involved in Piltdown as well. It turned out that he attended the same Jesuit preparatory school from which Waterton had graduated many years before.

That Doyle's name surfaced in this way seemed, at first, merely coincidental. But in reviewing the limited roster of those who enjoyed access to the Piltdown site during excavation, and who knew Dawson and Woodward beforehand, his name came up again and again. Not only did he live in Crowborough at the time, only several kilometers from the site, he also appears to have visited it openly in 1912. Dawson seemed gratified by his attention. In a note to Woodward, he stated, "Conan Doyle has written and seems excited about the skull. He has kindly offered to drive me in his motor anywhere."

This taxidermic joke called Nondescript that Charles Waterton created may have inspired the Piltdown hoax.

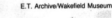

E.T. Archive/Wakefield Museum

Harry Price Library

Sir Arthur Conan Doyle, creator of Sherlock Holmes, was a doctor, a fossil collector, and a practical joker.

The Case Grows Stronger

A medical doctor who no longer practiced, Doyle was at this stage reaping the rewards of being a successful author. Moreover, his interest in paleontology had recently been stimulated by his discovery of several fossilized dinosaur footprints and bones close to his house in Crowborough. Dawson and Woodward met with Doyle to examine these discoveries.

Doyle was also a prodigious walker who thought nothing of setting forth on long jaunts, "geologizing" as he went along. There can be little doubt that he often visited the relatively unguarded site either by walking up the driveway that passed next to it or by peering over the hedge to observe the progress of excavation. Since most of the remains were found on or near the surface, it required no great feat on the part of the hoaxer to insert them into recently exposed cuts or toss them onto the spoil heaps where their discovery could be assured. All one had to do was keep an eye on the excavators. The real trick was to concoct a convincing creature, stain it to match the color of the Piltdown gravel, and surround it with acceptably appropriate fossil remains and implements. Unquestionably, Doyle possessed the anatomical competence, knew sufficient paleontology and chemistry, and had more than ample opportunity to do so. The case against Doyle, however, becomes more convincing in light of associations that can be established between him and the actual remains found at Piltdown.

The Jawbone Connection

Perhaps the single most expertly contrived part of the plot is the selection and modification of an orangutan's jawbone to make it resemble a primitive human mandible, one that appeared to "fit" the skull. Not only were the condyle and chin areas broken off, but the remaining molars were filed in such a way as to simulate patterns of human wear. During his first days as a practicing doctor in Southsea, a suburb of Portsmouth, Doyle moved into a house whose previous occupant had been a dentist. Heaped about in great numbers were the casts of human jaws. It may be partly for this reason and partly because many of his first patients had jaw ailments—having been referred by a dentist neighbor—that Doyle developed an abiding interest in human jaws.

When it was dated many years later, the Piltdown jaw proved to be young, but not that young—500 or 600 years old. Where could the hoaxer have obtained the jaw of an orangutan who lived somewhere in the East Indies and died approximately in the 12th or 13th century?

Among several possible sources, East India travelers and collectors whom Doyle may have known, Cecil Wray is the most likely. A former neighbor of Doyle's, Wray in 1906 had just returned from the Malay Peninsula, where he had worked both as a magistrate and a collector. He was also a fellow of the Royal Anthropological Society. Wray's brother, moreover, was head of the Malay museums and specialized in excavating caves, an ideal environment for preserving bone remains. One of his museums had recently purchased a large collection of animal specimens from Borneo. Orangutans live only in Borneo and Sumatra.

Doyle Linkable to Piltdown Remains

Several of the fossil mammal remains that were salted into the Piltdown gravel pit have since been identified as coming from the Mediterranean. The likely sites include Malta and a fossil cache in the Ichkeul area of Tunisia, which was not known to paleontologists until 1946. Whoever was the hoaxer had to have access to such exotic materials. In 1907, some two years before any fossils were discovered at Piltdown, Doyle visited archaeologist Joseph Whitaker, one of the few scientists who had frequently been to the Ichkeul region.

A few months after that meeting, Doyle and his bride—it was his second marriage—honeymooned for two months in the eastern Mediterranean. In all probability, they went ashore at Malta, a British port, in late November or early December on their return voyage. Coincidentally, the *Daily Malta Chronicle* announced on November 16 the discovery of the fossilized remains of a hippopotamus by workmen excavating a limestone fissure on the island. One of the planted items at Piltdown was a hippopotamus tooth whose form and chemical content indicate it came from a limestone chamber in one of the Mediterranean islands, Malta being regarded as the most likely.

Beyond these examples, it is possible to link Doyle with virtually all the physical fragments that, taken together, comprise the find at Piltdown—from jaw and skull to fossil animal remains and flints. Unsurprisingly, it is also possible to link the hoax with what Doyle did best: fiction.

A Further Tie with Fiction

In his exposé of the Piltdown hoax, Weiner mentions Doyle only incidently, speculating that perhaps Doyle had gained some inspiration for his novel *The Lost World* from witnessing the excavation. To be sure, an examination of that book reveals many suggestive points of resemblance. But it is the timing of its conception and publication that is truly relevant.

Consider a few touch points between Doyle's fictional adventure story and the Piltdown hoax:
• The statement by one of his characters that "if you are clever and you know your business you can fake a bone as easily as you can a photograph."
• The observation by another character that the practical joke "would be one of the most elementary developments of man."

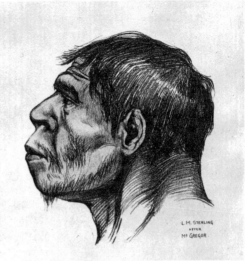

J. Kirshner/AMNH

An artist's rendition of Piltdown Man, whose remains were once regarded as the earliest-known human fossil.

• The occurrence of a time-mix in the living fossil animals of the story and those at Piltdown.
• In the story, a tribe of shaggy, red-haired, nest-building ape-men is discovered in the upper Amazon not too far from where Waterton's red-haired Nondescript ape-man was supposed to have lived. The description of these creatures makes them appear to be most closely allied with the orangutans of Borneo and Sumatra. There are also several references to early man and the "missing link" in the story.

Weiner did not consider the possibility that instead of Doyle's book being inspired by the Piltdown excavation, the Piltdown hoax was inspired by, or developed hand in hand with, the plot of *The Lost World*. In April 1912, Doyle's story came out as a serial in *The Strand Magazine*, and in December, Dawson and Woodward made the first announcement of their momentous discovery. The timing is crucial, for the seeds of *The Lost World* appear to have been planted in Doyle's mind long before Piltdown was a site of recognized significance in anyone's mind.

A Motive Comes to Light

Doyle, it must be said, genuinely believed in the scientific significance of the evidence pointing to the existence of early man. What, then, could prompt him to perpetrate a hoax and, as it proved, to make fools of the individuals who discovered, excavated, and interpreted the Piltdown remains?

The central characters in the Piltdown hoax are depicted in this 1915 painting, including Woodward (top right), Dawson (top second from right), and Doyle's arch adversary, Edwin Ray Lankester (bottom right).

Besides being fascinated with science, Doyle was also a believer in Spiritualism. He so declared himself as early as 1887 in a letter to the journal *Light*, and he spent much of his time, energy, and money furthering the Spiritualist cause. He came, for example, to believe in such things as the existence of fairies and other diminutive folk, as well as in the idea that an imminent apocalypse would be brought about by a wrathful "Central Intelligence," resulting in the death of most of the world's population.

These spiritualist and unorthodox views are central to the question of motive. Here we may look to the formidable figure of Edwin Ray Lankester.

Lankester was a dedicated Darwinian evolutionist. Besides holding Spiritualists up to ridicule, he encouraged scientists to wage war on them. This is precisely what he did with the American medium, "Doctor" Henry Slade, the rage of the Spiritualists of Britain in the mid-

1870's. Lankester arranged to attend a seance with Slade, the purpose of which was to communicate with a spirit. During the seance, he obtained evidence that Slade intended to cheat and defraud him. The medium left England as expeditiously as he could. In a single stroke, Lankester became the Spiritualists' greatest foe.

Doyle first made reference to the Lankester-Slade affair in a short story, "The Captain of Pole-star," published in 1883. The message was that even if mediums such as Slade had been guilty of fraud, one does not condemn Spiritualism for this reason alone. But Lankester had used just this kind of reasoning. Piltdown would provide a chance to reverse the tables by applying the same kind of logic: If science swallowed a scientific fraud like Piltdown Man, then all of science, especially the destructive and arrogant evolutionists, whom Doyle called the Materialists, could be condemned.

Lankester a Perfect Target

In a number of subsequent writings, Doyle made frequent reference to his adversaries, the Materialists, and he singled out Lankester as one of their most flagrant representatives. Not long before the hoax was set in motion, Lankester chose to remind the Spiritualists of his unabated contempt for them. Cerebral disease, he stated in his book *The Kingdom of Man*, may account for their beliefs.

Ironically enough, it was Lankester who provided the hoaxer with the recipe for what would prove a believable and long-lasting paleontological meal. Between 1906 and 1909, Lankester had made known his views on the kind of discoveries the prehistorians of the future might be expected to make. Man, he declared, emerged quite early, "perhaps in Lower Miocene times." Furthermore, he believed the cranial capacity of early man, contrary to the prevailing view, most likely would be remarkably large. He arrived at this prediction by extrapolating the results of a comparison between paleolithic skulls and those of contemporary "savage races."

Lankester also predicted that other less crude man-modified rocks would shortly be found in pre-Pleistocene deposits. In 1911 he argued in a paper presented before the Royal Society that such rocks had indeed been discovered. He dated them from the Pliocene or possibly earlier.

Unwittingly, Sir Ray had set himself up. He provided a list of objects to be discovered or verified as being man-made, and the hoaxer obliged him on every count. Piltdown Man's skull turned out to have a large brain capacity considering its apparent antiquity as determined from the age of the fossil mammals with which it was associated. Some of the fossils were identified by Lankester as being of Pliocene and possibly even Miocene age. Piltdown Man, as he saw it, represented a large step in the direction of his hypothetical Lower Miocene ape-man. It was a case of self-fulfilling prophecy, with the hoaxer providing the wherewithal.

So the case against Doyle is made. Besides the necessary skill, contacts, knowledge, and opportunity to qualify as the hoaxer, Doyle also had sufficient motive and an inviting target—Lankester.

Did Doyle Try to Reveal Hoax?

That the blame for the Piltdown hoax fell largely on the shoulders of Dawson can now be regarded as an injustice. It appears that Dawson was the scapegoat for a deeply embarrassed segment of the scientific community.

To have fooled science for so long may thus be regarded as something of a Holmesian triumph, though perhaps a dubious one considering how much time was wasted, how much confusion was created in our understanding of human evolution, and how many reputations were tainted as a result. Dawson, Woodward, Lankester, Teilhard de Chardin—they were all duped. But why didn't Doyle ever reveal—anonymously, at least—so exquisite a hoax?

He may well have tried to do so, for Doyle was a sportsman as well as a jokester. As he was an expert cricketer who had played on some of the country's top amateur teams, what could be better than to place a cricket bat "in the hands of" Piltdown Man? In 1914 a portion of a fossil elephant femur was discovered at the site. When it was formally described at a Geological Society meeting, a scientist rose to state that "he could not imagine any use for an implement that looked like part of a cricket bat." He further believed in the possibility "of the bone having been found and whittled in recent times." But most of the scientists either ignored it or preferred to believe that the object was a genuine paleolithic tool, though no one could assign it a plausible function.

Piltdown II

The following year, 1915, another fossil deposit was discovered by Dawson just a few kilometers from the original site. Called Piltdown II, this site also may represent an attempt by Doyle to strain the credulity of scientists to the point where they would question the authenticity of both deposits. Among other oddities, Piltdown II contained a skull fragment and a molar tooth that appeared closely related to the jaw found at Piltdown I. (In fact, it is now known to have come from that jaw.) For many believers, the tooth confirmed that Piltdown Man was not an aberration but a legitimate early fossil—and that apparently there was more than one. Even some who had doubted that the human skull and apelike jaw at the first site belonged to a single creature were convinced when they saw the same combination—an apelike tooth and human skull—at the second site. The effect of the tooth, like that of the cricket bat, was probably the reverse of what Doyle had intended.

Such gullibility must have exasperated Doyle, or made him howl with laughter□

DAEDALUS:
Design for a Starship

by Paul Patton

Over the past quarter of a century, space technology has opened up the solar system to direct exploration. Sophisticated robots from Earth have been hurled deep into interplanetary space, probing five of the Sun's nine major planets. These voyages have taught us more about our solar system than observers bound to the Earth could have ever hoped to learn; they have prepared the way for eventual manned interplanetary expeditions. Yet the solar system is only an infinitesimal speck in the vast swarm of several hundred billion stars that is our galaxy. Peering across the void of inter-

stellar space with powerful telescopes, we have learned that the stars are other suns. Many may possess their own retinues of planets—entire worlds about which we know absolutely nothing. Can we ever hope to send robot emissaries to explore these remote worlds?

Team Takes on Major Study

In the opinion of a group of British scientists and engineers, the answer is yes. The scientists, working under the auspices of the British Interplanetary Society, took part in a five-year design study for such a starship, called Project

Daedalus. They concluded that a robot starship to explore nearby stellar systems might be built by the middle of the 21st century.

Named for a brilliant craftsman who, in ancient Greek mythology, fashioned wings of wax and feathers to escape from the labyrinth of King Minos, the project is the largest single study of interstellar exploration ever undertaken. The core of the Daedalus study team was a group of 13 scientists and engineers under the leadership of Alan Bond of the United Kingdom Atomic Energy Authority; about 50 others also contributed. The goal of the study was to develop a design for a spacecraft capable of carrying out the simplest possible interstellar mission, an unmanned flyby. The project members specified that, whenever possible, the design should use present-day technology; reasonable extrapolations of near-future technological capabilities might be made when necessary.

The Daedalus study was not intended to produce the blueprints for a starship—since no real starship will ever consist of a hodgepodge of 20th- and 21st-century technology. Instead, it sought to show that interstellar flight can be discussed in sensible terms within the framework of established science and technology. Nevertheless, the Daedalus study provides what may be our best glimpse of the spacecraft that may someday probe the mysteries of the stars.

Destination: Barnard's Star

The target chosen by the Daedalus team is the Sun's second-nearest neighbor in space, a red dwarf known as Barnard's star. (A red dwarf is a dying star.) Barnard's star was chosen because certain controversial photographic evidence suggests that it has planetary companions. It lies 5.91 light-years from the Sun. (A light-year equals approximately 9,500,000,000,000 kilometers or 5,878,000,000,000 miles.)

Reaching Barnard's star will be a formidable undertaking. If our entire solar system (bounded by the orbit of Pluto) were reduced to the size of a 25-cent piece, the distance to Barnard's star would still stretch longer than a football field. Present-day interplanetary spacecraft such as Voyager or Pioneer would take 100,000 years to cross such a distance. The Daedalus team decided that an interstellar mission could be considered feasible only if it were completed within 50 years or less. To reach Barnard's star in that time, a spacecraft would have to travel at a speed of 36,000 kilometers (22,400 miles) per second, or 12 percent of the velocity of light. (A craft moving at this velocity would cover the distance between the Earth and Moon in about 10 seconds.) The chemical-powered rockets that propel present-day spacecraft are utterly incapable of reaching such speeds, as are the rockets propelled by solar power and nuclear fission that are proposed for manned interplanetary travel. Only one form of energy is likely to yield the necessary velocities in the near future—nuclear fusion.

Fusion Power

Nuclear fusion is the energy source of the stars. In a fusion reaction, two lightweight atomic nuclei join, forming a single, more massive nucleus and releasing enormous quantities of energy. The powers of nuclear fusion were first unleashed by humans in 1951 in the detonation of the first hydrogen bomb. For more than 30 years, physicists have struggled to find a way to release fusion energy in a more controlled manner. In the early 1970's, studies began on a promising new method called inertial confinement fusion.

In inertial confinement fusion, a tiny fuel pellet of frozen hydrogen is bombarded from several directions by powerful laser or electron beams. The beams compress and heat the pellet to conditions comparable to those found at the core of the Sun, which triggers a small fusion explosion. A rapid succession of such mini-explosions could provide a continuous source of power. The members of the Daedalus team believe that, given the projected technology of the 21st century, a nuclear pulse rocket based on fusion can provide the velocities necessary for practical interstellar flight.

Fuel from Jupiter's Atmosphere

Fueling a fusion-powered starship would pose serious problems. Only one specific type of fusion reaction, the deuterium/helium-3 reaction, appears feasible for Daedalus. Deuterium, an isotope of hydrogen, is reasonably abundant on Earth and can be extracted from ordinary water. Helium-3, however, is rarer and more valuable than gold. If Earth were the only available source for it, the Daedalus starship—which requires 30,000 tons of helium-3—could never be economically feasible. By the mid-21st century, however, manned interplanetary flight may be routine. Might we find greater supplies of helium-3 elsewhere in the solar system?

Earth and the other terrestrial planets are small, rocky, metallic globes, whose feeble gravity cannot retain much helium. The gas-giant planets of the outer solar system, however, are made primarily of hydrogen and helium. The solar system's largest planet, Jupiter, is a swirling ball of gases 318 times more massive than the Earth. An estimated 17 percent of Jupiter's atmosphere is helium, and as much as one ten-thousandth of this may be helium-3. To obtain the vast quantities of helium-3 needed for their starship, the Daedalus team has boldly proposed mining Jupiter's atmosphere.

Unquestionably, this is an ambitious project. Jupiter has no solid surface and a gravitational pull 2.7 times that of Earth. The Daedalus study envisions fleets of automated "aerostat factories" plunged from space into Jupiter's atmosphere and suspended beneath giant "hot hydrogen" balloons. About 130 such factory-ships, operating over a period of 20 years, would be needed to extract the necessary amounts of helium-3. The precious fuel would be retrieved by manned nuclear-powered shuttlecraft that would visit the factories periodically. Perhaps the most spectacular—and difficult—mission to which any 21st-century astronaut will ever be assigned will be the fiery glide from space down into the churning clouds of Jupiter.

Starship Design

The Daedalus starship itself would be a two-stage, 190-meter (625-foot)-tall interstellar rocket. Constructed in space and destined never to enter a planetary atmosphere, the vehicle would lack the streamlined appearance of a Saturn V or the space shuttle. Instead, the ungainly assemblage of fuel tanks, equipment, and engines would most resemble a gigantic bunch of grapes.

An enormous hemispherical chamber 100 meters (330 feet) in diameter lies at the base of the ship. Wrapped in magnetic-field coils and encircled by a ring of 75 powerful electron "guns" and an induction coil, this is the first-stage engine.

Above the engine chamber would be a cylindrical 12-sided open frame structure—the first-stage service bay. The service bay would contain several of the engine's subsystems and be the structural backbone of the first stage.

The base of the Daedalus starship's second stage sits nestled within the upper portion of the first-stage service bay. Its engine would be a smaller version of the first-stage engine, and its chamber would double as the antenna for the spacecraft's radio-communications system. The second-stage service bay would contain engine subsystems, the ship's communications system, and a small reserve tank of fuel pellets for final maneuvers as the craft approaches the Barnard's star system.

Above the service bay is the ship's payload bay, a 450-ton cylinder containing scientific experiments and the master computer. The computer would be located at the central core of the payload bay, with scientific experiments and their support equipment on four decks surrounding it. The uppermost deck, the probe deck, would contain 18 detachable planetary and stellar probes. The second deck, the astronomy deck, would contain two 5-meter (16-foot) reflector telescopes, Schmidt cameras, and other optics, as well as two large radio telescopes. These instruments are to be used as the starship approaches Barnard's star to locate and track its planets. The third deck, the probe communications deck, would be crammed with the equipment and antennas necessary for communication with the detached probes. The lowest deck would contain probes designed to study the interstellar medium, and two mobile robot repairmen known as wardens. The wardens are controlled by the ship's computer and can make repairs and deploy experiments using multiple robot arms.

From Callisto to Craft's Target

The Daedalus team envisions the starship as being assembled at a space manufacturing facility somewhere in the inner solar system. It might take 20 years to design, manufacture, and test—the time needed to mine its fuel from Jupiter. After the ship is assembled, it would be taken to Jupiter for fueling. The ship could be fueled while orbiting Jupiter's moon Callisto. The outermost of the four Galilean moons and the only one that lies beyond the most intense portion of Jupiter's deadly radiation belts, Callisto is a likely base for conducting Jovian mining operations.

Before its final departure for Barnard's star, the ship would be sent on one or more test

Opposite: Daedalus accelerates away from the fueling base on Jupiter's moon Callisto and begins its journey to remote worlds beyond our solar system.

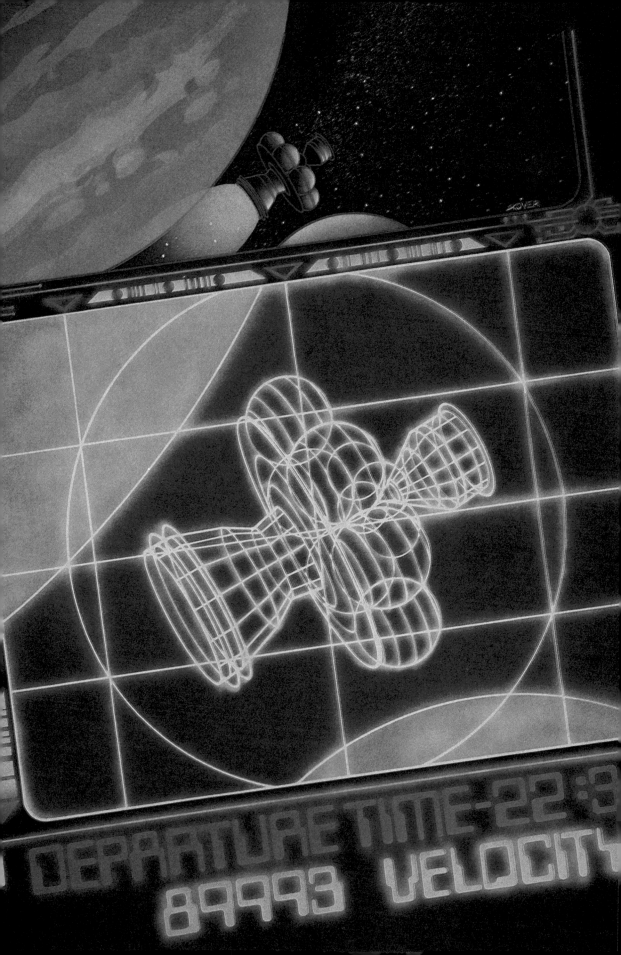

DEPARTURE TIME=22:3
89993 VELOCITY

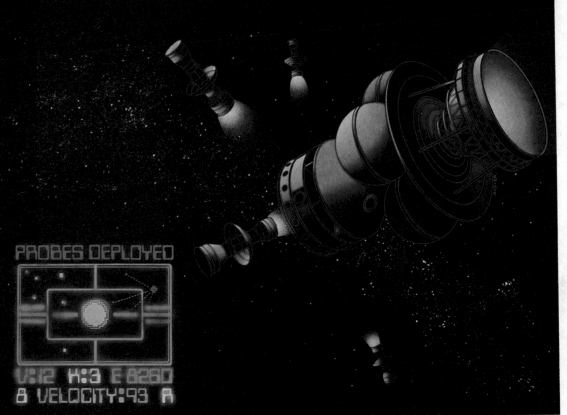

PROBES DEPLOYED

V:12 H:3 E 8260
8 VELOCITY:93 A

Daedalus deploys four of its data-gathering probes to prepare for encounter with Barnard's star and nearby planets.

cruises beyond the limits of the solar system. When the great day finally arrives, high above Callisto, the ship's mighty first-stage engine would ignite. A jet of superhot plasma would stream from the ship's reaction chamber at 10,000 kilometers (6,200 miles) per second, driving the 54,000-ton craft forward. The engine would continue to fire, with only brief pauses for jettisoning empty fuel tanks, for more than two years. At first-stage shutdown, the ship would be traveling at 7.1 percent of the velocity of light. Explosive bolts would fire, and the now-empty first stage would be cast adrift. The second-stage engine would then ignite, pushing the remaining spacecraft for 1.8 years more. Three years and 290 days into the mission, the second-stage engine would fall silent. The ship, now down to 908 tons and consisting of the second-stage engine and service and payload bays, would be traveling at its interstellar cruise velocity of 12 percent of the velocity of light. The 47-year coast to Barnard's star lies ahead.

At second-stage shutdown, the spacecraft would be 0.21 light-years from the Sun. A radio signal transmitted by the ship, traveling at the velocity of light, would take 2½ months to reach Earth—and Earth's reply would not be received by the ship until more than 5 months after the message was sent. By the time the ship reached the vicinity of Barnard's star, 5.91 light-years from Earth, the lag between query and response would grow to nearly 12 years. Starships can never be controlled effectively from Earth. Instead, such craft will need to rely entirely on the intelligence of their computers.

Ship's Computer Will Command

Advances in computer technology are just as important to the feasibility of the Daedalus starship as are advances in rocket propulsion. Present-day spacecraft computers are capable of little more than storing and executing lists of instructions received from their programmers on Earth. Some computer scientists working on artificial intelligence hope to develop computers capable of behaving in a far more flexible and intelligent manner. They believe it will be possible to endow future computers with a human-like ability to reason, learn, recognize patterns, make decisions, and solve problems. By the mid-21st century, Daedalus' designers believe that we will have developed computers intelligent enough to replace mission control and take command of a starship.

The challenges facing such a computer would be formidable. During the long interstellar journey, it would have to operate, maintain, and even (using the robot wardens) repair the starship, coping with any problems that arise. As the ship approaches Barnard's star, the computer will use data collected by its telescopes and instruments to build up a detailed image of the Barnardian planetary system.

Mission Has Enormous Potential

Long before the Daedalus ship approached Barnard's star, it could begin to reward the enormous technological efforts of its makers with important scientific data. The Daedalus team has identified a number of significant scientific studies that the starship might perform during its five-decade interstellar coast. Through most of the long voyage, Daedalus' large optical and radio telescopes will be available for a variety of uses. They would be uniquely suited for any astronomical observations requiring long, uninterrupted observation time or a radio-quiet environment. Precise measurements of stellar positions, made with Daedalus' 5-meter optical telescope, when compared with similar observations made near Earth, would permit great refinements in our knowledge of the distances to the stars. With such parallax measurements, astronomers could determine the size of the galaxy to an accuracy of much better than 1 percent. Daedalus' robot wardens, equipped with their own propulsion systems, could unfurl giant lightweight radio telescope antennas to float freely at distances of many kilometers from the starship. The antennas would be used for wide-spectrum astronomical observations at long wavelengths, far from the radio interference of the solar system.

As Daedalus passes the halfway point in its mission, 25 years after launch and 3 light-years out from Barnard's star, it would begin regular observations of the star. Telescopes and sensors would start the vital search for planets and potentially hazardous concentrations of interplanetary dust.

Interferometry May Detect Planets

Daedalus' optical and radio telescopes would be arranged in opposing pairs across its diameter for interferometry, a technique in which pairs of telescopes can resolve objects too closely spaced to be resolved by either telescope alone. It is likely that Daedalus' first detection of any Barnardian planets will be made by infrared interferometry. With present-day technology, gas-giant planets could be detected a few hundred days before encounter and terrestrial planets about fifty days before encounter. The Daedalus team anticipates that advances in instrumentation will improve these figures at least tenfold by the mid-21st century. Early detection and tracking of the Barnardian planets are vital to the starship's preparations for encounter.

At encounter the Daedalus starship would perform the most important part of its scientific mission, closeup study of Barnard's star and its presumed planetary system. Many questions about other stars and their planets can be answered only by a starship. The Project Daedalus study has attempted to identify the instruments and exploration strategy necessary to answer them.

Ship's Probes to Encounter Planets

Traveling at 12 percent of the speed of light, Daedalus' stay within the Barnardian system would be measured in days. To ensure full coverage during this brief flyby, the starship would dispatch the 18 probes stored on the uppermost deck of its payload bay. The probes would be sent to encounter various planets in the system as well as Barnard's star itself.

During and after encounter, data from the probes would flow back to the Daedalus command ship to be stored in its computer. These data could be beamed toward earth by a transmitter operating continuously for a day and a half. Anxious scientists on Earth would have to wait nearly six years for the radio waves carrying the data to cross the intervening void. It would be received by a large array of radio telescopes on or near our home planet. With its arrival, the direct exploration of the stars will have begun.

The Project Daedalus study has shown us that once controlled nuclear fusion, computer intelligence, and jovian mining become feasible, interstellar exploration becomes possible. The structure, communications system, and even much of the payload for an interstellar flyby spacecraft can be designed on the basis of only present-day technology. A robot starship would open nearby star systems to direct probing by instruments from Earth and vastly expand our knowledge of the universe in which we live. It is a tremendously exciting endeavor □

PHYSICAL SCIENCES

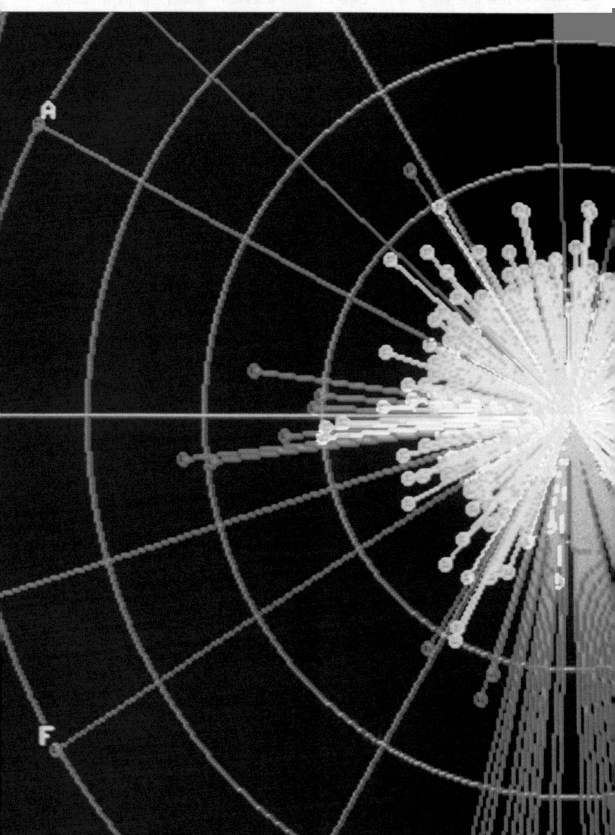

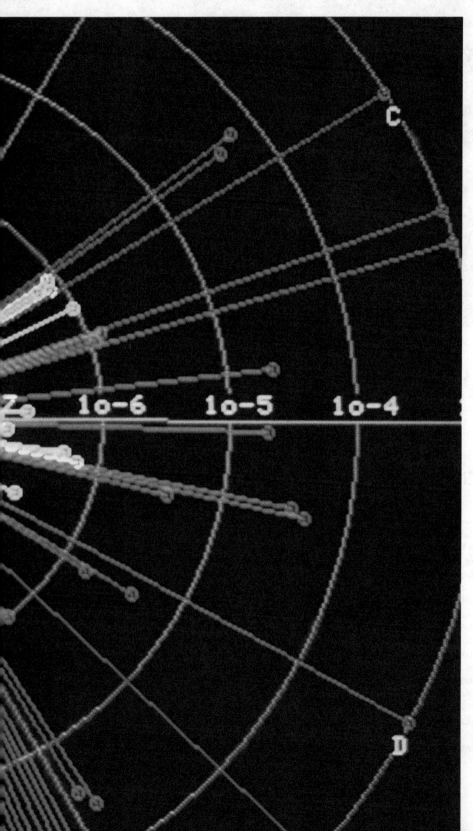

This burst of color appears on one screen of a complex computer system that monitors superconducting magnets at the Fermi National Accelerator Laboratory in Illinois.

REVIEW
OF THE
YEAR

PHYSICAL SCIENCES

PHYSICS

Scientists have long been trying to unite nature's four basic forces—gravity, electromagnetism, the strong force that holds the nuclei of atoms together, and the weak force that mediates both beta decay in atoms and some reactions that occur within the Sun—into one so-called Grand Unification Theory (GUT). According to this theory, the four basic forces are all manifestations of a single force that existed at the beginning of the universe. What may be an important step in this search for unification was made during 1983 with proof that electromagnetism and the weak force are two manifestations of a single force, called "electroweak." An international team of physicists at CERN, the European Laboratory for Particle Physics in Switzerland, provided the proof by discovering all three subatomic particles (called intermediate vector bosons)—the W^+, the

W^-, and the Z° particles—that had been theorized to carry the so-called weak force.

Protons are the positively charged components of the nucleus of an atom, which until recently were believed to have infinite lifetimes. Then, in the early 1980's, came evidence that protons decay. The existence of proton decay could mean that the universe will ultimately end with all matter turning into energy. New findings now suggest that earlier estimates of the lifetime of a proton must be changed—to greater than 5 times 10^{31} years. To put that another way, in a batch of 5 times 10^{31} protons, one decay a year can be expected. The new estimate came after researchers from the University of California at Irvine, the University of Michigan at Ann Arbor, and the Brookhaven National Laboratory at Brookhaven, Long Island, using 8,000 tons of water as a proton source, failed to detect a single emission of the Cerenkov light that is theoretically emitted with proton decay. Researchers doing similar work using a proton source of metal plates reported that they had detected only a single possible proton-decay event.

In the international particle physics research competition, the United States is gaining a leading position—thanks in part to the newly built accelerator, called a tevatron, at the Fermi

National Accelerator Laboratory in Illinois. The accelerator has been pushed to 512 billion electron volts, propelling protons to nearly the speed of light. The tevatron may also reduce by half the cost of accelerating protons: it uses superconducting magnets that greatly reduce electric-power demands.

A team of U.S., Chinese, and Finnish researchers working at the Lawrence Berkeley Laboratory of the University of California have detected a previously unknown form of radioactivity—known as two-proton decay—in the nuclei of the unstable isotopes aluminum 22 and phosphorus 26. The decay may involve a pair of protons flying off as a single particle. The discovery is the fifth known form of radioactivity, the others being emission of alpha and beta particles, spontaneous splitting of uranium atoms, and single-proton decay.

Scientists continued their efforts to make nuclear fusion a practical energy source. A major step in this effort is achieving break-even—the point at which the energy created by the reactor equals the energy used to sustain the fusion reaction. Now, in what physicists called a milestone experiment, the doughnut-shaped tokamak fusion reactor at the Massachusetts Institute of Technology achieved two of the three criteria for break-even—namely, the necessary plasma density and confinement within the reactor (the third criterion is temperatures in excess of 200,000,000° C [360,000,000° F]).

CHEMISTRY

Chemists at the University of Minnesota made the first reliable prediction of a chemical reaction ever accomplished when they determined how fast one atom of hydrogen would break its bond to another hydrogen atom and bond instead to an atom of muonium, a lighter variant of hydrogen. The reaction, which involves just three electrons, was so mathematically complicated that the researchers used a mathematical shortcut to describe the motions of nuclei. The technique will be useful in helping to predict such reactions as freon interaction with atmospheric ozone, combustion, and biochemical reactions that occur inside cells.

A new way for synthesizing large and useful organic compounds may be developing. Scientists at the University of Alberta and the University of California at Berkeley succeeded in activating the single-carbon compound methane. Activation of saturated hydrocarbons occurs when certain metal complexes are added to the compound. Until now, methane had been exceptionally resistant to activation. If another carbon compound called a carbonyl group can be added to the methane part of the activated complex, scientists may be able to use the activated complex for building useful compounds.

Scientists are reassessing their ideas about hydrogen. A chemical reaction initiated by scientists at Los Alamos National Laboratory led to the "capture" for the first time of a hydrogen molecule (H_2) by binding it, intact, to a metal complex. Hydrogen molecules normally tend to break apart when they interact with metals. The ability to study the bound intact molecule is expected to aid research in the processing of hydrocarbon-based materials such as coal, petroleum products, synthetic fuels, and polymers. ■ Meanwhile, chemists are rethinking the phenomenon of hydrogen bonding. In the bifluoride ion, two fluoride ions are bonded to a single hydrogen atom, with the bond between each fluoride atom and the hydrogen atom being the same strength and about ten times greater than theoretical calculations suggest that the strength of hydrogen bonding should be. Chemists now believe that the hydrogen atom in the bifluoride ion behaves contrary to the rules of chemistry; they think it shares its single electron simultaneously with both fluorine atoms. If this phenomenon is more widespread than has been observed to date, fluorine may be able to interact with hydrogen atoms in water or other chemicals within the body. Researchers at Oregon University and in Great Britain are studying the possibility that this type of reaction might affect biochemical processes in cells.

Other discoveries made in chemistry during the year may lead to technological developments. Certain elements heavier than plutonium are being considered as possible power sources for remote power stations, as radiation sources, and as diagnostic tools in medicine. A preliminary step toward these goals was made in 1983 when scientists at the Los Alamos National Laboratory measured for the first time the physical and chemical properties of one of these elements— Einsteinium (element 99). ■ New and better ways of making superconducting organic crystals for computers, microcircuits, and superconducting magnets may result from an Argonne National Laboratory discovery. Chemists there found an important clue to how certain crystals that contain no metal atoms—but are called "organic metals"—are able to conduct electricity with virtually no resistance when cooled to nearly absolute zero (−273.15° C or −459.67° F). ■ A more efficient way of "growing rubber" may result from the finding that when a natural plant substance called DCPTA is applied to the desert shrub guayule, it significantly increases the rubber content of the plant.

MARC KUSINITZ

Connoisseurs
of Chaos

by Judith Hooper

A bout 40 years ago, the late poet Wallace Stevens wrote a poem, titled "Connoisseur of Chaos," that unwittingly prophesied the birth of a whole new science: *A great disorder is an order. Now, A/ And B are not like statuary, posed/ For a vista in the Louvre. They are things chalked/ On the sidewalk so that the pensive man may see. . . .*

Stevens's lines run through my head as I look out the restaurant window. "Only once or twice in a millennium is there a true scientific revolution, a real paradigm shift—Newtonian mechanics and the invention of calculus in the 17th century was the last one," my host, mathematician Ralph Abraham, notes between generous helpings of spicy vegetables. "The current scientific revolution will synthesize the whole intellectual discourse of the species."

Chaotic Dynamics

This is cosmic talk. And here in downtown Santa Cruz, where casualty cases of the 1960's wander the streets like possessed souls, it's hard to grasp the sense of his words. But up the hill, at the University of California, Abraham is one of a small band of scientific revolutionaries who aim to reinterpret the universe through the new science of *chaotic dynamics*. You've probably never heard of Ralph Abraham or any of the other connoisseurs of chaos, but you will. Though they work far from the public eye, in realms abstract and otherworldly, some have already begun to speak of their esoteric discipline as the science of the 21st century.

Unlike quarks or gluons, the business of chaos can be seen with the naked eye and understood even by nonscientists. Why can't your chatty television weatherman predict next weekend's weather? Why can't you outwit a roulette wheel? Why does a normal, regularly beating heart suddenly go on the fritz, killing its owner? How does orderly creation rise out of the chaos of the Big Bang? How does a unique human being develop out of a cluster of common cells?

The answers involve a bold new understanding of these creatures of chaos—and what we call chance. According to traditional Newtonian mechanics, if you know certain things about a system—all the forces acting on it, its position, and the velocity of its particles—you can describe, in theory, all its future states. Here is how the turn-of-the-century mathematician Jules Henri Poincaré once described the Newtonian view: "If we knew exactly the laws of nature and the situation of the universe at the initial moment, we could predict exactly the situation of that same universe at a succeeding moment." But Poincaré's own work and that of his heirs was to topple that secure assumption. Anticipating the weird world that Abraham and his colleagues now routinely confront, the French mathematician said, "These things are so bizarre that I cannot bear to contemplate them!"

A simple example: Suppose you're sitting beside a waterfall watching a cascade of white water flow regularly over jagged rocks, when suddenly a jet of cold water splashes you in the face. The rocks haven't moved, nothing has disrupted the water, and presumably no evil spirits inhabit the waterfall. So why does the water suddenly "decide" to splash you? Physicists studying fluid turbulence wondered about this kind of thing for several hundred years, and only recently have they arrived at what they consider to be satisfying answers.

A Conclave of Contemplators

It boils down to this: Randomness, or chaos, is not merely a matter of complexity. Many physical systems, including some very simple ones, have pockets of randomness built into them. And that's why the most supreme scientist, wielding impossibly perfect tools, can never accurately predict the weather three days hence or mark the final destination of a ball in a spinning roulette wheel. We're condemned to live with chance, and even now some scientists cannot bear to contemplate such things.

Others, like theoretical physicist Rob Shaw of the University of California at Santa Cruz (UCSC), can't bear not to. While still a graduate student in late 1977, he heard about something called chaotic dynamics, a field that few physicists outside of the tiny chaos fraternity knew existed. An analog computer, a remnant of a defunct engineering department, materialized at just the right time for Shaw to try out the new equations. He dropped his nearly complete thesis on superconductivity to contemplate chaos and the crazy, seductive shapes its equations generated on his computer screen. The rage soon spread to his UCSC friends: Doyne Farmer, who set aside astrophysics; Norman Packard, who left statistical mechanics; and James Crutchfield, who changed his undergraduate studies. From these men, the Dynamical Systems Collective—colloquially dubbed the "chaos cabal"—was born. Abraham and the

Opposite: Friedrich Hechelmann

PHYSICAL SCIENCES   **285**

other chaos elders attended their informal conclaves to trade mathematical know-how.

In a lab whose nether corner suggests a jet plane's cockpit, Shaw shows me a mundane but elegant model of chaos: an ordinary dripping faucet.

Other people tinker with car engines; Shaw builds chaotic appliances. In this case, he has constructed a water tank with a spigot dripping droplets through a ruby-colored laser beam on their way to a bucket. The intervals between drops are transmitted as pulses to the analog computer. That mathematically transforms them into spiraling video patterns.

The faucet, says Shaw, is a microcosm of chaos. By Newtonian law, the spacing of the drops ought to be regular and predictable, but it's not. "The fascinating thing about a standard faucet," he explains, "is that even though the [water] flow is constant, the spigot doesn't move, and nothing perturbs the system, the pattern never repeats itself. It's got a random element in it."

Strange Attractor the Cause

Where does that random element come from? The same place as the waterfall's sudden, random splashes. Not from some imperceptible jiggle, as scientists had long supposed, but from the inner dynamics of the system itself. Behind the chaotic flow of turbulent fluids or the shifting cloud formations that shape weather, to give

but two examples, lies an abstract something that physicists now call a strange attractor.

O.K., so what is an attractor and what makes one strange? Suppose you put water in a pan and shake it up; after a time it will stop swirling and come to rest. That state of rest—the equilibrium state—can be described mathematically as a fixed point, which is the simplest kind of attractor. (All of the mathematics describing a system's motion are inexorably drawn toward the attractor like iron filings toward a magnet.)

Now imagine the periodic movement of a metronome or a pendulum swinging from left to right and back again. Geometrically speaking, this motion is said to remain within a fixed cycle forever. That is the second kind of attractor, the limit cycle. There are many different kinds of limit cycles, but they all share one characteristic: regular predictable motion.

But the third variety, the strange attractor, is a breed apart. It is irregular, unpredictable, quirky. In a word, strange. For example, when a heated or moving fluid moves from a smooth, or laminar, flow to wild turbulence, it switches to a strange attractor. While strange attractors' geometric shapes are multitude, they have certain distinct characteristics in common.

The Baker's Transformation

One is an abstract structure that the Santa Cruz cabal likens to filo dough, the intricately folded

Stephen Pope, Department of Physics, UCSC

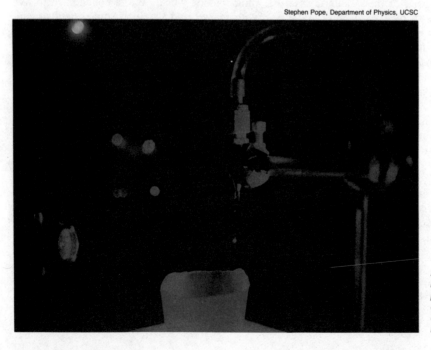

At the University of California at Santa Cruz, physicists examine the chaotic nature of a faucet with a ruby-colored laser beam.

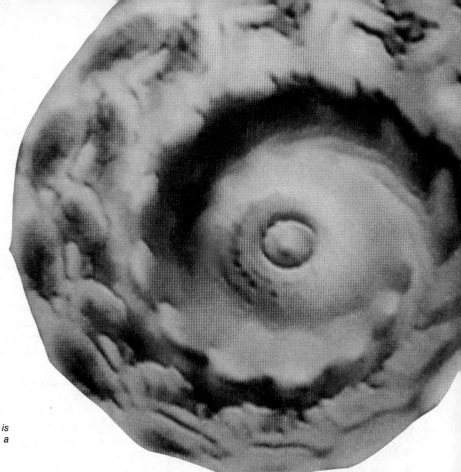

This logarithmic spiral is the result of chaos in a video-feedback system.

Jim Crutchfield

pastry of baklava. "Suppose a baker puts little dots of ink in his dough," Shaw tells me. "He stretches or rolls out the dough, then folds it in half, stretching and folding it over itself again and again. Pretty soon the dots of ink could spread anywhere." Shaw and his cotheorists call this the baker's transformation. This is more than a metaphor. The mathematical models behave like the stretching and folding dough; the spreading ink dots represent the fuzziness of any initial measurements.

Similarly, mathematical points—representing such successive measurements as velocity, temperature, amplitude, or whatever—may start close together with a strange attractor and then drift ever farther apart, like the ink dots in the filo dough. So you can neither predict what something will become by looking at its initial situation, nor reconstruct its origins from its present position. With enough stretching and folding, randomness will reign. In the lingo of chaos, this phenomenon is described as "rapid divergence of nearby trajectories."

But why this maddening lack of precision? Well, that brings us to another chaotic-dynamic catchphrase: "sensitive dependence on initial conditions," the very thing Poincaré had pinpointed as the hidden mechanism of chance.

Why can't we foresee the entire series of physical processes that lead into the future? Because, as Poincaré observed, "a very small cause that escapes our notice determines a considerable effect that we cannot fail to see, and then we say that effect is due to chance." When you spin a roulette wheel, for example, a twitch of your finger controls the final outcome. Because of this sensitive dependence on initial conditions, tiny errors compound exponentially into colossal ones. Or take weather. Barely perceptible convection-flow events at a given moment can dictate rain, shine, sleet, or hail days later; so it's not always the weatherman's fault when the partial cloudiness he predicts becomes a downpour by next Wednesday.

Whimsical as strange attractors are, however, there's a method to their madness. Wander and loop they may, but they will never fly out of their mathematical envelope of "phase space." Within the phase space, though, thickets of chaos grow. Every time you fold the cha-

Jane DiMenna

A plume of cigarette smoke is chaotic as it travels straight up into the air and then suddenly goes erratic.

Otto Roessler, the father of the elegantly simple "Roessler attractor"; the University of Maryland's Jim Yorke, whose equations helped demystify chaos; and Massachusetts Institute of Technology meteorologist Edward Lorenz, the father of the Lorenz attractor—the first strange attractor ever discovered.

Fluid Turbulence

Back in 1963, Lorenz had been working on ways to perfect weather forecasting, when his equations turned up a startling truth: Long-range weather prediction was impossible. Examined mathematically, convection currents moved in a bizarre new way. This was the first strange attractor, but it went unnamed until 1971. Then mathematical physicist David Ruelle of the Institute of Advanced Scientific Studies near Paris, France, and mathematician Floris Takens of the State University of Groningen in Holland, set their sights on deciphering a perennial enigma: fluid turbulence.

"Before Ruelle and Takens, most scientists thought fluid turbulence was just too complicated to predict," explains Farmer, who left the Santa Cruz brotherhood last year to spread the seeds of chaos to Los Alamos National Laboratories in New Mexico. "The old idea was that when you 'force' a fluid beyond equilibrium by heating it or turning up the flow, it goes from a rest state to periodic motion—like the regular rolling patterns you see in heated frenchfry grease. Then, the thinking went, you get more complicated periodic motion: doubly periodic (with two independent frequencies), triply periodic, quadruply periodic, and so on. But Ruelle and Takens discovered that after doubly periodic motion, something totally different occurs." That totally different something was the work of the strange attractor—which Ruelle and Takens finally named. Four years later, fluid experiments by physicists Jerry Gollub of Haverford College in Pennsylvania and Harry Swinney of the University of Texas at Austin proved what many mathematicians had predicted. The strange attractor was indeed a fact of nature.

Visual Mathematics

You don't have to be a mathematician to fall under the spell of strange attractors. Computers breathe life into them, setting equations to dance forever as shifting shapes in the thousands of future states that could never be calculated by hand. Aiming a creaky school projector at a wall, Shaw shows me the movies the

otic filo dough back on itself, you get a minuscule gap between one layer and the next. Soon there are multitudes of gaps, and numerical fuzziness overtakes order.

Given all this, it's not hard to see why it took the Dynamical Systems Collective months to persuade their faculty advisers that strange attractors weren't *Alice in Wonderland* objects. Luckily, by 1977 or 1978, the work of the older grand masters of chaos began to infiltrate such established fields as solid-state physics. And as fate or sensitive dependence on initial conditions would have it, the young UCSC graduate students soon found themselves in the chaos mainstream. When a scientist's missed plane left an agenda gap at an international conference in 1978, Shaw was invited to talk. And Doyne Farmer once found himself addressing such chaos experts as German theoretical chemist

The Belousov-Zhabotinsky Reaction shows chemicals oscillating in a chaotic and bizarre manner.

Fritz Goro

Dynamical Systems Collective made of its computer-generated images. "This is our local compulsion," says Shaw, as circles swell and elongate like stretched rubber on the screen, galaxies spiral into space, and delicate filigrees grow increasingly ornate.

In another dreamlike film, shot in grainy, early-art-film black and white, subtitles announce: THE BELOUSOV-ZHABOTINSKY REACTION. The movie shows chemicals pouring into a great black vat, where they are mixed by magic forces and then sent gushing out the other side. "When chemicals react, they can do one of three things," Shaw says. "They can get mixed and just sit there at equilibrium—that's a fixed-point attractor. Or their concentrations can oscillate periodically in a limit cycle. Or the oscillations can get bizarre and chaotic, as is the case with these chemicals."

If you know where to look, chaotic-dynamics people tell us, the same patterns are all around. "There are only a few movies, a few dances," Abraham observes. "And everything around us is the working of one of these."

Some Order in Randomness

Why do the same geometrics turn up in chemical reactions, waterfalls, and the climate? There is a deep-seated order to this strangeness, according to the creative work of Cornell University physicist Mitchell Feigenbaum. Even when dynamical systems make transitions, or bifurcations, from smoothness to chaos, certain rules and numerical values always hold. That chaos has some order to it is one of the key breakthroughs of the chaos connoisseurs.

It isn't that no one ever dared look at the face of chaos before; it's just that the old guard saw it differently. "The pre–World War II crowd used equations for total randomness as a model for chaos in nature," Farmer says. "In that case randomness is simply postulated. The models work very well to describe certain things, like the statistical fluctuations in a gas in which the molecules are pretty evenly distributed.

"However," he continues, "there are many situations in nature where orderly things happen in the midst of great chaos. For those physical systems, *deterministic chaos*—with its strange-attractor structure—is the best model."

Farmer's Ph.D. dissertation is aptly titled "Order in Chaos"; its frontispiece is emblazoned with Wallace Stevens's "*Connoisseur of Chaos.*" The seeming paradox lures the young physicist. Chief among his insights is: "Some systems have a 'clock' inside them that goes on keeping perfect time in the midst of very chaotic stuff."

Maybe, he speculates, evolution did not proceed wholly at random; maybe the genetic code employs some orderly clocks in the midst of wild Darwinian randomness.

"According to our current model of creation," he adds, "out of a formless, miasmic blob, the Big Bang, you eventually get very orderly structures, like biological organisms. Or think about conception: There's enormous randomness involved in sperm reaching the egg. But once you get conception, you make a human being, a highly ordered structure. Right from the beginning your life is profoundly influenced by chance events caused by sensitive dependence on initial conditions."

Qualitative Dynamics

Think form, not content; quality, not quantity. At least that is how Alan Garfinkel of the Crump Institute for Medical Engineering at the University of California at Los Angeles (UCLA) sees it. "Poincaré is the father of this field," he says. "More than half a century ago, he saw that you can't get exact numerical values for many phenomena, and that even if you could, they wouldn't tell you what you wanted to know. If you're studying the motion of the Earth around the Sun, for instance, it's more important to know its path or topological shape than the exact distances it travels. Is the orbit a closed ellipse or a very, very long curved line that doesn't close—in which case the Earth might eventually spiral off into space? So Poincaré invented topology, the science of forms of motion.

"Much of the current work in chaotic dynamics exists in a never-never land of unproven mathematics," Garfinkel adds, drawing explanatory diagrams from time to time in front of him. "For example, there's still no mathematical proof for the first strange attractor discovered, the Lorenz attractor."

Like many of the chaos connoisseurs, Garfinkel is young, extraordinarily articulate, and interdisciplinary: a mathematically inclined phi-

losopher of science who is a professor of kinesiology at UCLA. "Topology, or qualitative dynamics," he points out, "is the perfect mathematics for biology. All humans have the same form, yet we differ in details. Why?

"I think sensitive dependence on initial conditions in the embryo is what makes us individuals," he says. "As two cells develop into zillions of cells, there is a distinct sequence of qualitative changes. You have epochs of smooth change and then—bang—qualitative change and differentiation."

This view of nature—called catastrophe theory—runs counter to the smoothly changing, continuous world described by calculus. Nature does not make jumps, the 17th-century inventor of calculus, Gottfried von Leibniz, proclaimed. "But nature does make jumps," Garfinkel tells me. At a certain critical value, metals snap, fluids make a sharp, qualitative leap into turbulence, chemical concentrations and animal populations turn abruptly chaotic. In biology, too, deterministic chaos prevails. What are heart fibrillations, two Canadian scientists propose, but a bifurcation from normal periodic oscillation to

The Lorenz attractor, the first strange attractor ever discovered, is one kind of random element in some systems.

Rob Shaw

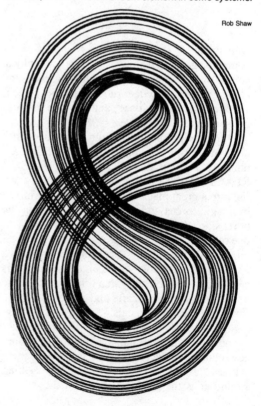

The slime mold's self-organizing, two-phased life is a perfect example of "emergent order," a patterned state that arises out of chaos.

a fatal, chaotic one? Likewise, Garfinkel and others view schizophrenia as chaos in the brain's chemical-feedback systems.

Slime Mold Shows Emergent Order

And while biological chaos sometimes makes for pathology—Parkinson's disease, schizophrenia—it can also be useful. Garfinkel is enamored of something called slime mold, to which he has devoted a long, fascinating paper. The green slime that coats the surface of stagnant ponds is one of nature's most splendid examples of "emergent order," an idea that very much intrigues the chaos crowd.

"When I first saw it, I said, 'Alan, stop what you're doing; this is the most beautiful thing in the world,' " he recalls.

"The [slime mold] creature has two life phases. In the first, it's a single-celled amoeba that crawls around, leading its own little life. But when deprived of food—bacteria—it undergoes a radical transformation. By pulsing a messenger chemical called cyclic AMP, it signals to the other amoebas, and they all cluster into colonies of 10,000 cells. Then the cells undergo differentiation to become one animal.

"The front part becomes a head; the back, a stalk. Then the body becomes spores covered with hard cases. The spores break away, their cases crack open, and out come individual amoebas, completing the life cycle."

An equation describes this process. The patterns the slime mold forms are its solution. The equation apparently has a self-organizing property built into it. It's the same one underlying the Belousov-Zhabotinsky Reaction, and some suspect it's at the heart of embryonic cell differentiation—turning a single zygote (a fertilized egg) into an animal composed of variegated organs and tissues.

What's more, societies may self-organize in the manner of slime mold. It so happens that fireflies flash and crickets chirp in concert. Human females living together fall into the same ovulatory rhythm. Social insects cooperate like a single organism, and individual, chaotic human beings form societies, nations, kingdoms, and economic structures.

Slime mold, Garfinkel proposes, is a model for the emergence of social systems. Why is it people agree to observe laws and conventions, such as driving on the right side of the road? How does one explain altruism, the sacrifice an individual makes to a group?

"The total state of the system will move to a certain attractor—say, cooperation—even if the individual doesn't consciously intend that," says Garfinkel.

The dance of randomness shaping itself into patterns is called emergent order. It fascinated Nobel physicist Ilya Prigogine, who sees it in termite colonies and the human body; just as it fascinated Doyne Farmer, who sees it in the cosmos; Alan Garfinkel, who sees it in the slime mold and society; and others in the chaos vanguard, including, perhaps, Wallace Stevens, who saw it everywhere:

The squirming facts exceed the squamous mind,/ If one may say so. And yet relation appears,/ A small relation expanding like the shade/ Of a cloud on sand, a shape on the side of a hill . . . □

At 10 a.m. this spot was at the pendulum.

THE CORIOLIS EFFECT

by Alan Linn

A pig's tail, to some people, is a microcosm in which they can discern rigid laws governing the universe. These people, who include more than a few hog farmers, say that if you compare the tails of porkers from all over the world, you will see that tails north of the equator curve to the right (as seen from the snout end), while those in the Southern Hemisphere curve to the left. They will tell you that water running down a bathtub drain will form counterclockwise whirlpools in the Northern Hemisphere; clockwise in the Southern.

At 10 a.m. this spot was at the pendulum.

© Kim Nielsen

The Foucault pendulum at the National Museum of American History demonstrates the Coriolis effect. The line described by the pendulum seems to be turning, but actually the Earth rotates underneath it.

A Natural Phenomenon

Such beliefs turn out to have little or no basis in fact. Yet the "force" implicitly invoked affects everything from ocean currents and weather patterns to the trajectory of a ballistic missile. Physicists can show that an object moving freely over the surface of the Earth tends to curve to the right in the Northern Hemisphere and to the left in the Southern. If pigs could fly, they would do the same. No actual force is involved; what happens is better understood as an effect—an effect of the Earth's rotation.

The phenomenon has been recognized by physicists and philosophers for centuries. It was little understood, however, until an inquisitive assistant professor of analysis and mechanics at the École Polytechnique in Paris quantified the natural law involved in 1835. The man's name was Gaspard Gustave de Coriolis, and his main preoccupation in life was the dynamics of moving objects. He was the first to use the words "work" and "kinetic energy" in their modern scientific sense.

Sir Isaac Newton, the "discoverer" of gravity, it is said, was inspired by falling apples. In the case of Coriolis, it may have been billiards that provided the inspiration. He realized that any object moving over the surface of a rotating body, be the body a billiard table or a planet, would appear to be deflected from its appointed path.

In 1835 Professor Coriolis submitted to the Polytechnique's journal a paper titled "On the Equations of Relative Motion of Systems of Bodies." In it he demonstrated that on a rotating surface an imaginary force appears to act at right angles to the direction of any moving body. As the Earth we live on is a rotating body, the implications are profound.

For example, if you fired a rifle at a target 120 meters (400 feet) away, the bullet would curve to the right by about 0.25 centimeter (0.10 inch). This assumes you are firing in the middle latitudes of the Northern Hemisphere on a windless day. You cannot blame your poor aim on such a small deflection, but it can give artillerymen fits.

In World War I, German gunners manning the "Big Bertha" shelled Paris, France, which was 122 kilometers (76 miles) away. At that range and at the velocity at which the shells traveled, the Coriolis effect deflected them about 1.6 kilometers (1 mile) to the right. However, the crew of 35 manning the gun corrected for the effect. Of their first seven shots, only one of the 120-kilogram (264-pound) shells fell outside of the walled city. The greater the distance, the larger the effect. A missile (without its own guidance system) aimed at Eureka, California, from the North Pole would fall into the ocean about 160 kilometers (100 miles) to the west of its destination.

Many people have the notion that pigs' tails curl one way in the Northern Hemisphere and another way in the Southern. Although this belief has no basis in fact, the phenomenon described affects weather and ocean currents.

An Effect, Not a Force

On Earth, the curve of such moving objects is unmistakable. But if you watched the projectiles from a space station, they would appear to travel in an absolutely straight line. The explanation of this paradox strikes at the heart of the Coriolis phenomenon: the curve results from a rotating frame of reference. It is an effect, not a force.

To understand the Coriolis effect, you should reproduce it. You can see the Coriolis effect vividly on a merry-go-round. Throw a ball at a companion riding with you. He need not worry about being hit. The ball will curve mysteriously from its target. If the merry-go-round is turning counterclockwise, as is the Northern Hemisphere, the ball will curve to the right. Stand on the ground beside the merry-go-round and watch two riders play catch. You will see that the ball travels in a straight line—and you will wonder why the pitcher cannot connect with the catcher.

To a rider the ball appears to curve, because during its flight the merry-go-round rotates beneath it. Similarly, during your rifle bullet's half-second flight, the target moved 0.25 centimeter (0.10 inch) counterclockwise. In 1918 the Big Bertha shells were in transit for three minutes, time enough for Paris to move about 1.6 kilometers (1 mile).

Just like the inside and outside horses on a merry-go-round, different places on the Earth's surface move at different speeds as the whole thing rotates. On a phonograph turntable rotating around its post at 33⅓ revolutions per minute, it is easy to see that the outside of the turntable is moving faster than the inside, even though every part of the turntable is making 33⅓ rpm. The outside travels a greater distance, so it moves faster.

Now imagine that the turntable is rubber and we can pull the center portion up the post until the once-flat turntable becomes a perfect half sphere, now rotating counterclockwise as a model of the Northern Hemisphere of the Earth. Exactly the same conditions prevail. The edge of the turntable is still farthest from the post (now called the axis of rotation) and travels fastest. Moving up toward the post, we find points on the turntable moving slower and slower as their distance from the axis decreases, until finally, at the post itself, the motion is slowest of all.

Quito Moves Faster Than New York

On the real Earth, Quito, Ecuador, is moving east at roughly 1,600 kilometers (1,000 miles) per hour. It has to. Take the circumference of the Earth at the city's latitude and divide by 24 (for the number of hours in a day) to calculate the speed of the city's motion. New York City moves east at only 1,200 kilometers (750 miles) per hour. It is closer to the Earth's axis and travels a shorter distance in 24 hours. The farther north you move, the closer you are to the Earth's axis and, therefore, the slower you move.

Anywhere that you may be in the Northern Hemisphere, everything south of you is moving faster than you are and everything north is moving slower. This is just as true of the room you are sitting in as it is of continents. The south wall of your room is infinitesimally farther from the Earth's axis and, therefore, moves infinitesimally faster than the north wall.

Yet neither your room nor the continent breaks apart. They are constantly turning, or "twisting," in exact compensation for the differences in velocities. Imagine two runners side by side, somehow rigidly harnessed to one another, and imagine that the runner on the right is just slightly faster. They will continually turn to the left in order to remain side by side.

In the Northern Hemisphere, the earth beneath us twists to the left all the time, very slowly near the equator but faster and faster with increasing latitude. Anything moving across that twisting surface, whether bullet or ocean current, will appear to stray to the right. But it is only appearance: the bullet flies in a straight line while the target moves to the left.

A Growing Suspicion About the Earth

Our prehistoric forebears may have suspected that the Earth provided what today would be called a moving frame of reference with respect to the stars. Anyone watching the sky on a clear night will see that the stars curve in their flight across the heavens. About 270 B.C., Aristarchus, a Greek astronomer on the Aegean Island of Samos, became probably the first to suggest that the Earth moves in an orbit around the sun. The Polish astronomer Copernicus refined the idea nearly 2,000 years later, but scientists could not devise any experimental proof.

Hope sprang anew in the late 1600's when Newton proposed his law of inertia—the tendency of moving objects to keep moving straight, unless acted on by an outside force. Experimenters realized that they could demonstrate the rotation of the Earth by dropping a round ball down a deep mine shaft. The Earth would turn while the ball was falling, and the ball would hit ahead of the spot directly beneath it at the moment it was dropped.

Enter Jean Bernard Léon Foucault, a Parisian medical student who became more interested in physics than in physiques. Foucault was a devout tinkerer. As a child he had been adept at devising mechanical toys. As a young man he realized that a pendulum would tend to swing to the same point in space, true to Newton's law, while the Earth turned beneath it. If he could demonstrate that, he would prove that the Earth rotated.

The young Foucault went to his cellar and built a pendulum using a 2-meter (6.5-foot)-long wire. He carefully set it swinging by burning through a restraining bit of string. Voilà! The line of the pendulum seemed to turn slightly clockwise with every swing; in fact, the Earth was turning underneath it. Foucault later built larger pendulums for the public, including one at the Panthéon in Paris in 1851. It was a big hit. Foucault put a pin on the bottom of the weight, so it drew lines in wet sand spread beneath it, graphically illustrating the turning. He demonstrated the Earth's rotation by showing the twisting that is the essence of the Coriolis effect.

Major Consequences of Coriolis Twisting

Pigs' tails notwithstanding, the most significant result of the Coriolis effect is on our weather. As masses of air move from high- to low-pressure areas, they take curved paths because of the Coriolis "force." This initial twist makes nearly all weather systems form in circular cells—storms, hurricanes, typhoons, cyclones (low-pressure systems), anticyclones (high-pressure systems), and the aptly named twisters, or tornadoes. Expanding anticyclones form clockwise spirals in the Northern Hemisphere; contracting systems—cyclones, hurricanes, tornadoes, and waterspouts—are almost always counterclockwise.

"If there were no Coriolis effect, there wouldn't be any storms," says Dr. William L. Donn, a senior research scientist at Columbia University's Lamont-Doherty Geological Observatory in Palisades, New York, and former head of the meteorology section of the U.S. Merchant Marine Academy. "Anytime the sun's heating of the Earth caused a low-pressure area, air would move in and that would be the end of it."

Near the equator, where the Coriolis effect is weakest, is the area of calms and light winds known as the doldrums. Thunderstorms do occur, but large differences in atmospheric pressure are quickly smoothed out. Hurricanes generally do not form closer to the equator than about five degrees of latitude.

The Coriolis effect also influences the ponderous movements of ocean currents. In spite of what your high school science teacher may have told you, water does not always seek its lowest level; oceans and other large bodies of water are not level. They are "piled up" by the winds and Coriolis phenomena.

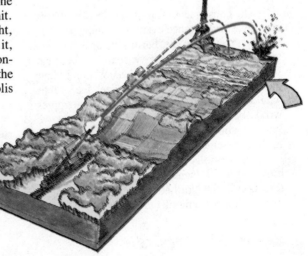

When German gunners manning "Big Bertha" shelled Paris, France, during World War I, they adjusted their aim to compensate for the deflection caused by the Coriolis effect.

When a soldier on the equator (bottom) is carried forward by Earth's rotation to the position once held by the soldier before him, he will not have turned at all. A soldier in the far north will have turned to the left due to Coriolis twisting.

The consequences of these Coriolis effects are of vast importance even to landlubbers. The ocean currents are the Earth's natural thermostat. The currents can carry vast amounts of heat from the equator to the poles. Without their moderating effect, continent-size ice cubes would accumulate at the poles. The tropics would become parched wastelands. Life as we know it would be possible only in narrow bands in the temperate zones.

The Coriolis effect also causes icebergs to move 30 to 50 degrees to the right of the wind direction in the Northern Hemisphere. Because of their immense momentum, the icebergs may

continue to move—and curve—after the wind stops, making them even more of a threat to shipping. Even ocean currents generated by winds blowing along the surface move in a direction to the right of the wind; the velocity of such a current drops with depth but continues to turn to the right.

According to many geologists, Coriolis twisting produces lopsided valleys. River water is piled up against the right banks (looking in the direction of flow) in the Northern Hemisphere. In a 1.6-kilometer (1-mile)-wide rapid river, the difference in water level would amount only to about 2.54 centimeters (1 inch), but even in smaller, sluggish streams, a slight difference would cause more erosion to one bank than to the other. In fact, many rivers in the Northern Hemisphere do have steep right banks and gently sloping left ones, particularly the Missouri River and a number of rivers in Alaska, southern Long Island, and Siberia. The effect is disputed by some geologists. In many areas local features, such as rocky outcrops in the riverbed, overshadow any Coriolis effect.

Insomnia, No—But Bathtubs, Maybe

Scientists have their theories, and you and I have ours. Besides pigs' tails and whirlpools, some people say Coriolis twisting affects impotence and insomnia. They maintain that this can be counteracted by orienting your bed east-west with the Earth's rotation. (Others, just to be obstinate, insist that north-south does the trick.)

The whole question of whether water draining out of a tub whirls in different directions in different hemispheres has been taken up at least once in the scientific literature. In December 1962, Ascher H. Shapiro at the Massachusetts Institute of Technology described his theory and experiment in the British journal *Nature*. He found that normally the Coriolis force is much smaller than other forces present in a tub of water, including residual motion from the original filling, air currents in the room, or even differences in temperature around the vessel. The Coriolis force is about 30 million times weaker than the force of gravity at middle latitudes, so if the tub is not perfectly symmetrical, the Coriolis force is overwhelmed.

Shapiro went on to build a special tub, however, and found that when all these other forces are removed, the water always does form a counterclockwise whirlpool in Cambridge, Massachusetts.

During the years I spent as a farm editor in the midwestern United States, hog owners from Kansas to Pennsylvania were willing to bet big money that pigs' tails generally twirl to the right in the Northern Hemisphere. They swore that the favored direction south of the equator was to the left. I could not check tails for myself because most breeders dock (amputate) them when the piglets are about a week old to eliminate tail biting in the ranks. But Dr. Hobart W. Jones, professor of animal science at Purdue University, has been asked about tail curls so often that he has taken to noticing the curl of the

Surface winds generate ocean currents (represented by fish) that curve more and more to the right in the Northern Hemisphere as they weaken with increasing depth.

countless newborn pigs of eight major breeds that he has seen during a long career: "I can assure you that in this section of the country, the tails curl randomly in both directions. I see no reason whatsoever why the same would not be true south of the equator."

The possible effect on humans is a much thornier question. Little if any research has been done. Professor James E. McDonald, a physicist at Iowa State University who wrote on the subject in the May 1952 *Scientific American*, believed that people constantly and unconsciously correct for the Coriolis effect when walking. McDonald calculated that a person walking 6 kilometers (4 miles) per hour on a nearly frictionless surface, such as ice, but who could somehow still keep from falling flat, would drift about 75 meters (250 feet) to the right for every 1.6 kilometers (1 mile) walked in the Northern Hemisphere. Normally, however, this does not happen.

A Force That Shrinks to Insignificance

Even penguins do very little walking on ice. Usually, one foot or the other is solidly on snow or the ground and twisting with it; no correction is needed. When you consider that the force required to correct for the Coriolis effect on a 9,100-kilogram (20,000-pound) jet plane moving at 1,000 kilometers (600 miles) per hour is only about 25 kilograms (55 pounds), the effort needed for a human walking at 6 kilometers per hour shrinks to insignificance.

Similarly, the theoretical Coriolis drift on your car, if you are going the legal speed limit, amounts to nearly 3 meters per kilometer (15 feet per mile), but the frictional bond between tires and road easily negates the effect. Locomotives in motion exert about 140 kilograms (300 pounds) more weight on the right rail than the left (in the Northern Hemisphere).

The next time you are on a jet, you may wonder how the pilot avoids getting lost in the face of the Coriolis twisting. "Nowadays, we don't even think about the Coriolis effect," says Captain David Leney, a Concorde pilot for British Airways. "All corrections are made by our computerized inertial-guidance system. Even in the days when we navigated by sextant, only a slight correction was needed. Pilots are never aware of any need to constantly bank left to correct for a Coriolis drift to the right."

The Coriolis effect is sometimes invoked to explain the widely held belief that lost people tend to wander in circles. But the belief is true only some of the time, and even then the circling arises from other causes. "Rarely, lost people do wander in circles, particularly at night," says Chief Master Sergeant Bob Moran, a survival training manager at the U.S. Air Force Survival School in Washington State. Moran has seen thousands of lost men wandering north and south of the equator during nearly three decades as a survival-training instructor and tracker of lost persons. "When people circle, it's usually to the right, no matter what hemisphere they're in," Moran continues, "I think it's because people are taught from adolescence to drive to the right and pass objects on the right. Interestingly, lefties tend to circle to the right less than right-handed people."

The Coriolis effect also may influence geologic features on some planets. Dr. Anthony Dobrovolskis at the Jet Propulsion Laboratory in California has suggested that the ray pattern of debris blasted from meteorite and volcanic craters on Mercury and other planets may reflect the bodies' Coriolis effects at the time of the explosions. "These patterns may be basic to understanding the entire geophysical evolution of a body," Dr. Dobrovolskis says. If enough distinct ray patterns are found, they may reveal long chapters in the history of our solar system, in much the same way that tree rings can tell the history of a climate.

One day Professor Coriolis' imaginary force may help unravel some of our solar system's remaining riddles □

In a Victorian reenactment of Coriolis' inspiration, a rotating table causes the cue ball to miss its target.

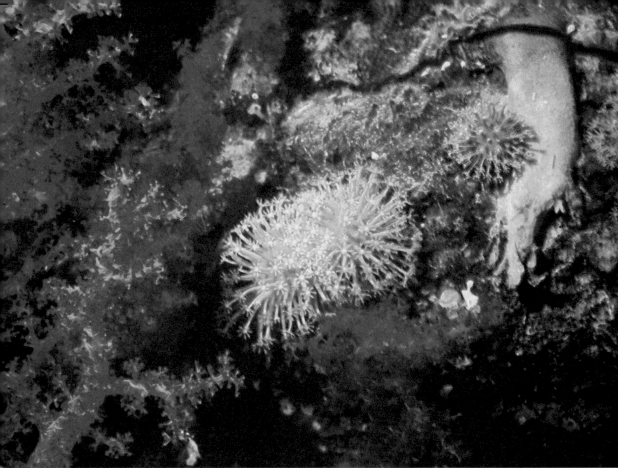

Jane Shaw/Bruce Coleman

Chemical Treasures from the Sea

by John Langone

Osamu Shimomura nimbly leaps atop a workbench in his cramped quarters at the Marine Biological Laboratory in Woods Hole, Massachusetts. He takes a large, dusty jar from a high shelf, and after checking its stained and faded label—GENUS: CYPRIDINA. YEAR: 1925—he pulls out the stopper. "It should work," he mutters. He scoops a spoonful of a dry, brownish substance from the jar into a mortar and grinds it to powder.

Now Shimomura is ready. He flicks off the lights and adds water to the powder. Instantly, the mixture begins to glow with a blue light that dimly illuminates the scientist's face. "Japanese naval officers used to read maps with this stuff during wartime blackouts at sea," he says.

"And jungle soldiers used to put a little of it on their backs, then moisten it with saliva, so men behind them could keep in file at night."

Learning from Marine Glowers

The phenomenon that Shimomura has just demonstrated is called bioluminescence—the production of light by many animals and plants—including the familiar firefly. What is remarkable is that the glow from the tiny dried crustaceans in the laboratory jar can still be rekindled to shine as bright as it did in 1925,

Above: These red and white soft corals from the Red Sea are two of the many marine species that yield valuable chemicals which are made into useful drugs.

Dr. Osamu Shimomura demonstrates the bioluminescent glow of dried crustaceans that were caught in 1925.

ular, the pharmaceutical possibilities of seaweed, sponges, corals, and algae remain largely unexplored.

Anti-cancer Substance in Sea Mat

Shimomura and other marine biologists have already begun scouring that vast frontier. Aided by sensitive new instruments that have improved their ability to detect minute concentrations of interesting chemicals, they have extracted from various forms of marine life an impressive and fast-growing list of powerful substances that may soon find uses as cancer drugs, antibiotics, and cardiovascular, neuromuscular, and anti-inflammatory medicines.

In December 1982, for instance, chemists from Cornell University Cancer Research Institute reported the discovery of a remarkable anti-cancer substance in *Bugula neritina*, a species of common marine animal known as ''sea mat'' or ''false coral.'' These creatures, which abound in the eastern Pacific and attach themselves to the hulls of ships, contain a compound that bears a chemical resemblance to some antibiotics. Working in collaboration with the National Cancer Institute, the scientists found that when the compound, called bryostatin-1, was administered to mice in extremely low doses, it was highly effective against leukemia. As little as one ten-millionth of a gram doubled the lifespan of some of the diseased mice (the drug does not appear to work against other forms of cancer, however). While the results have raised hopes that the substance might be effective against human leukemia, bryostatin-1 is not easy to come by.

Fortunately for marine biologists, who must spend much of their time at sea—or in it—many of the possible useful organisms they study live in warm waters. There, untold numbers of species thrive, and competition is fierce. Life is especially hard for algae and soft-bodied invertebrates—like gorgonians (soft corals) and opisthobranchs (mollusks with no external shells)—and they have developed elaborate chemical defenses to keep from being eaten. The chemicals that make these soft creatures unpalatable to predators are also what makes them interesting to scientists. Says William Fenical, an organic chemist at the Scripps Institution of Oceanography in La Jolla, California: ''There are 200 or more species of gorgonians in the Caribbean. It's likely that they contain toxic substances because it's obvious they are being left alone.''

when the Cypridina were fished from the sea. But to Shimomura, a marine biologist, the cypridinoid glow is much more than a laboratory curiosity. For years he has been studying these creatures and other marine organisms that glow—certain squid, worms, jellyfish, snails, clams, and bacteria. His purpose is not only to learn more about the unique chemical reactions involved in bioluminescence, but also to help detect human diseases.

Shimomura belongs to a new breed of marine explorers, who roam the coasts and seas searching not for oil, minerals, or sunken gold but for biological treasure—exotic chemicals that can be extracted from sea life and used in medicine. Plants and animals have long been sources of useful drugs. Aspirin, for example, was first extracted from the bark of willow trees, penicillin from bread mold, insulin from animal pancreases, morphine from opium poppies, curare from a vine, and certain tranquilizers from the Indian snakeroot plant. But while scientists have cataloged and classified most of the plants and animals on land, and have extracted many beneficial chemicals from them, they know far less about marine life. In partic-

Promising Discoveries

Using this kind of reasoning to help narrow their search among the myriad species of the sea, Fenical's team of scientists, from Scripps and the University of California, has outpaced all other groups of marine biologists in probing sea life for new drugs. They have purified and tested hundreds of chemicals, many of which show great promise. Among these are 32 that inhibit cell division, which makes them candidates for the treatment of tumors and fungal and viral infections, as well as for use in male contraceptives. Fifteen other marine chemicals studied by Fenical's group reduce inflammation, and may someday be effective in treating arthritis and rheumatism, and in preventing the rejection of transplanted organs. Other finds: a cardiovascular drug that could reduce blood pressure or regulate heartbeat; a neuromuscular drug for treatment of muscle weakness; and a neurotoxin named Lophotoxin (LTX). The neurotoxin is being used by five laboratories to study the transmission of chemical signals between nerves and muscles. The results could shed light on disorders like Parkinson's disease, which is associated with faulty transmission of these signals.

LTX, extracted from the Mexican gorgonian, is a deadly poison. It mimics the action of some snake venoms by blocking impulses between nerves and muscles, and can kill. But at much lower doses, LTX may one day become the basis of a drug to treat strabismus—crossed eyes caused when muscles on either side of an eye exert unequal pulls. Fenical suggests that a tiny quantity of LTX might be injected into the stronger muscle, to block some of the neural impulses that activate it. This would allow the eye to shift to a normal position.

Another highly active drug in Fenical's catch is stypoldione, a bright-red crystalline compound found in a species of Caribbean seaweed that seems impervious to predators. Because most seaweed in these waters are eaten by animals, Fenical reasoned that this one must contain some powerful repellent. He was right: when he casually put some of it in an aquarium, almost immediately the lone fish in the tank began frantically bumping into the glass walls, and finally leaped out. After analyzing the seaweed, Fenical found that the stypoldione it contained not only upset the fish but also inhibited cell division. Because of that effect, it is being investigated as a possible cancer drug.

Marine biologists are exploring the seas in search of medicinal chemicals that can be extracted from ocean organisms.

Risks and Great Rewards

Because marine biochemists often handle toxic substances, and sometimes rely on marine folklore as leads to promising chemicals, their work involves a certain amount of risk. After tasting a purified (therefore concentrated) extract of a type of seaweed used as food in Hawaii, Fenical became dizzy, and a colleague found him reeling in a hallway. Says John Faulkner, an organic chemist at Scripps who does not eat seafood because he is allergic to it: "It's a standing joke that I avoid seafood because I work with the toxic chemicals in it."

Paul Boyle, a microbiologist at the New England Aquarium's Edgerton Research Laboratory in Boston, Massachusetts, and Ralph Mitchell, a microbiologist at Harvard University, are intrigued by another form of sea life. They are studying a wood-eating marine crustacean with a digestive tract that, unlike the exterior of its body, is totally free of bacteria. Boyle thinks there may be some substance in the creature's gut that acts as a natural antibiotic. "We may be looking at something that's very bor-

ing," he says, "or it may be a new source of antimicrobial agents, something we might put in an ointment or in animal feed."

One of Shimomura's luminescent chemicals—a protein from a Pacific jellyfish called *Aequorea*—has proved valuable in medical diagnoses. Injected into the muscle cells, the protein (aequorin) glows when the muscle con-

Below: P. Boyle; Bottom: OSF/Animals Animals

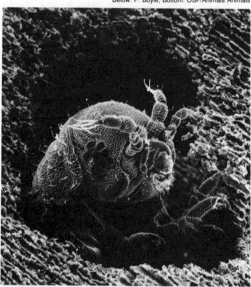

Right: An unusual marine crustacean, whose gut may contain antimicrobial agents, burrows through wood in this micrographic close-up. Below: A protein from this Pacific jellyfish is used in medical diagnoses.

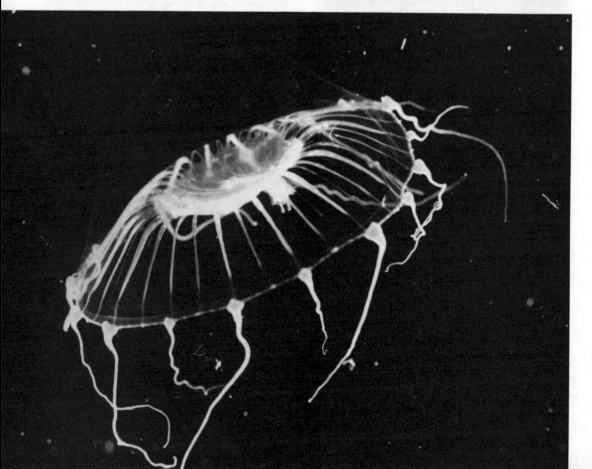

tracts, but not when it relaxes. Why? Aequo-rin's bioluminescence occurs only in the pres-ence of calcium, which regulates such activities as muscle contraction and cell division. This property has made aequorin useful for measur-ing tiny changes in the amounts of calcium in body fluids or cells. Such changes may provide early warning of several different kinds of dis-orders: the destruction of cells, the onset of can-cer, a slowdown in the growth of bones, abnor-mal functioning of the parathyroid glands (which control calcium metabolism), and distur-bances in the rhythm of the heart. Furthermore, a little aequorin can go a long way. Says Shi-momura: ''With only a milligram of aequorin, one can do thousands of tests.'' (One milligram equals 0.00004 ounce.)

Several other valuable marine extracts are being used or studied for possible use as chem-ical detectors of disease; still others serve as lab-oratory tools in basic biomedical research.

Horseshoe Crab Yields Useful Enzyme

One researcher who has benefited financially from marine biochemistry is Stanley Watson, a senior scientist at the Woods Hole Oceano-graphic Institution. In 1975 Watson started a business based on a frequent visitor to bays and beaches from Maine to the Yucatán: the horse-shoe crab. Watson's company, Associates of Cape Cod, extracts an enzyme called limulus amebocyte lysate (LAL) from the crab's creamy, bluish blood (the color is due to an oxygen-carrying system based on copper rather than iron). Solutions of this enzyme have a unique property: they clot in the presence of a class of poisons called endotoxins, which are produced by bacteria. This makes LAL so reliable a detec-tor of these poisons that other companies li-censed by the Food and Drug Administration also produce it, although Watson still controls 80 percent of the market.

Watson's customers use LAL in a test for endotoxin contamination of pharmaceuticals and such clinical devices as syringes, catheters, needles, and kidney dialysis machines. It has supplanted the more expensive and time-consuming rabbit test, in which suspect solu-tions were injected into the animals, which would develop fevers if the solutions contained endotoxins. Researchers are now looking into the possibility that LAL may also prove effec-tive in testing for endotoxins in blood poison-ing; gonorrhea; spinal meningitis; colitis; perito-nitis; and infections of the eye, ear, and urinary

Dan McCoy/Rainbow

An enzyme in the horseshoe crab's bluish blood is a reli-able detector of bacterial poisons called endotoxins.

tract. One promising use may be in detecting bacterial infections in newborns and premature babies who show no obvious signs of them.

The horseshoe crab has remained virtually unchanged for some 200 million years, and the LAL entrepreneurs are making sure that they will not endanger its existence. Watson's com-pany, which gets the crabs from local fisher-men, drains each animal of 30 percent of its blood and then returns the crab to the ocean (where it soon regenerates its lost blood). Each crab supplies enough LAL for as many as 150 tests, and Watson's company can bleed 500 creatures a day.

The recent successes of marine scientists in extracting drugs from sea life may soon bring more players into the game. Among those now eying marine creatures with new interest are several large pharmaceutical companies beset by the ever-rising costs of developing drugs in the laboratory. As Ralph Waldo Emerson wrote: ''I wiped away the weeds and foam, I fetched my sea-born treasures home'' □

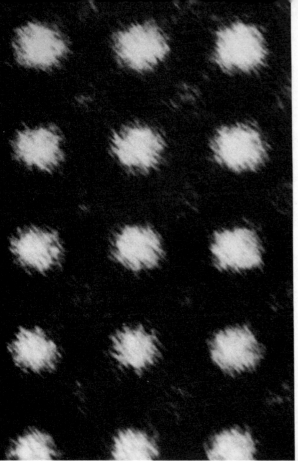

The researchers who designed it and built it are calling it the "moon-shot" of electron microscopy: an instrument powerful enough to see individual atoms in solid material. Atoms have diameters of less than 2 ten-billionths of a meter, or 2 angstroms, and the new Atomic Resolution Microscope at the University of California's Lawrence Berkeley Laboratory is the first instrument capable of resolutions that fine. It can operate in the realm between 1 and 2 angstroms.

Finest Resolution Ever Achieved

Scientists have seen single atoms before, and have even made motion pictures of them, but only under special experimental circumstances. These views did not reveal atoms in the natural conditions that determine a material's physical properties. Now microscopists can explore metals, semiconductors, ceramics, and even organic materials atom by atom at resolutions below what has come to be called the 2-angstrom barrier.

Breaking the 2-angstrom barrier, like getting to the moon, did not come easily. The pioneer of atomic microscopy, German physicist

Photos: Lawrence Berkeley Laboratory, University of California

by William J. Cromie

Above: Zirconium atoms appear as large, bright dots and oxygen atoms as smaller, more diffuse spots as seen by the powerful Atomic Resolution Microscope at right.

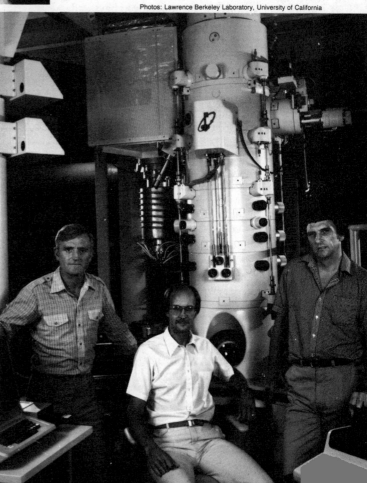

The Atomic Resolution Microscope is housed at the National Center for Electron Microscopy in the silo at right.

Erwin W. Müller, laid the groundwork with his invention in 1936 of the field emission technique and his later use of it in field-ion microscopy. The main part of a field-ion instrument consists of a thin needle of the material, usually a metal, to be studied. The needle and a fluorescent screen sit in a glass tube filled with helium at low pressure. An electric field gives the sample a positive and the screen a negative charge. Helium atoms lose electrons, becoming positive ions, and are repelled by the slender sample. The ions stream to the screen, fanning out to create an enlarged image of bright dots that mimic the arrangement of atoms on the sample's surface. Müller used such a microscope to take the first photographs of the pattern of atoms on a tungsten surface in 1951.

Researchers still employ this technique to study defects in metals, the effect of heat, cold, and corrosive agents on surfaces, and the reactions of various atoms with metal surfaces. With the atom-probe field ion microscope, invented in 1967, they can determine the chemical identity of a single atom.

Filming Atoms in Action

Albert V. Crewe, a University of Chicago physicist, in 1970 combined field emission with scanning and transmission electron microscopy to photograph individual atoms. Scanning instruments, in which the electron beam scans the specimen in a pattern of close parallel lines, depict topographic features with brilliance and clarity. Transmission microscopes, in which the beam illuminates the entire specimen at once, provide the highest possible resolution.

In Crewe's instrument, which combines the advantages of both, a negative electric field repels electrons from a tungsten point about 1,000 angstroms in diameter. Magnetic lenses focus the electrons to an intense beam that is both scanned across and transmitted through a thin carbon film with uranium and thorium salts on its surface. The organic molecules of the film hold the atoms of uranium and thorium specific distances from each other. Electrons colliding with and scattered by the atoms produce the images as bright spots on a dark background. Crewe achieved a millionfold magnification in these experiments.

During that work, he noticed that thermal motion caused the atoms, as he puts it, "to walk around." And in 1976 Crewe and Michael Isaacson, now at Cornell University, put together 100 successive still photos of this motion to make the first film of atoms in action. It shows uranium atoms reacting with the 50-angstrom-thick film and with each other. Such sequences make it possible to study the formation of molecules by observing directly how atoms hook together. A wandering atom joining a nearby agglomeration, Crewe says, is "the beginning of the growth of crystals. We would not have seen this in still photos; only when you run them as a movie do you see what happens."

Crewe now employs this capacity to trace the movements of single atoms in such processes as crystal growth, organic-molecule formation, catalysis, and surface diffusion. He uses a microscope with a resolving power of 2.4 angstroms to study heavy atoms ranging in size from thorium down to silver. Because the instrument's resolving power falls short of the critical 2-angstrom level, Crewe cannot see atoms as they exist naturally, tightly packed together in metals or other crystalline materials. He still must separate them and spread them over the thin carbon film.

Bridging the Gap

The resolving power of electron microscopes ordinarily depends on the wavelength of the electrons, which in turn depends on the voltage used to accelerate them from their source to a target. Electrons with energies of a few KeV (KeV = 1,000 electron volts) possess wavelengths comparable to, and should be able to resolve, the diameters of atoms. To achieve this resolution in a transmission microscope, energetic electrons should pass through a specimen, be focused by a magnetic lens onto a photographic plate or fluorescent screen, and reveal the specimen's structure. But lenses are the knot. In practice, they do not perform as efficiently as optical lenses do, and they create a gap between what should and what can be seen.

The gap can be narrowed by increasing the accelerating voltage, which decreases the wavelength of the electrons. And engineers around the world have responded by building machines with voltages as high as 3 million electron volts (MeV).

None of these microscopes breaks the two-angstrom barrier; combinations of lens aberration and difficulty in keeping voltages constant get in the way. The aberration stems from the fact that electrons passing through the center of the lens are not focused at the same point as those passing nearer the edges. Voltage fluctua-tions of a few volts in a million produce electrons of slightly differing frequencies, which results in an image that appears to jump in and out of focus.

A case in point is a recently completed 1.5-MeV instrument located near the smaller, more precise 1-MeV Atomic Resolution Microscope at the Lawrence Berkeley Laboratory. Still the largest high-voltage electron microscope in the United States, the former has been in operation since May 1982. Aberrations and voltage instabilities limit its resolving power to 3 angstroms.

Although they do not resolve individual atoms, the 1.5-MeV and other high-voltage instruments possess distinct advantages. The beam of the 1.5-MeV Berkeley microscope, for example, is energetic enough to penetrate thick specimens, and the specimen chamber is large enough to accommodate complete experiments. In other instruments, materials to be studied must be thinned to 1 micrometer or less to permit beam transmission. (A micrometer, or micron, equals one-millionth of a meter.) If a specimen is too thin, most of its atoms may lie close to the surface, raising the question of whether the sample exhibits the same properties as the bulk material. "The 1.5-MeV microscope allows us to image entire structures, such as a semiconductor device, with more certainty that what we see is not dominated by surface effects," says Gareth Thomas, director of Berkeley's visitor-available National Center for Electron Microscopy.

Breaking the Barrier

Microscopists can choose one of three paths around the obstacles to higher resolving power: They can correct lens aberration, increase volt-

These movie frames show atoms of uranium (which appear as white dots) in motion.

Mitsuo Ohtsuki/Enrico Fermi Institute, University of Chicago

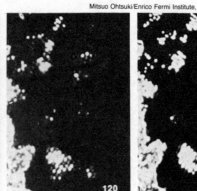

3-atom-thickness
2-
1-

10Å T=0sec

60

120

180

ages, or learn to stabilize high voltages. None is easy.

Increasing voltages means constructing large lenses to handle more-energetic electrons and producing large beam tubes to hold the lenses. (The 1.5-MeV microscope at Berkeley occupies an 18-meter [60-foot]-tall silo and weighs more than 20 tons.)

For the Atomic Resolution Microscope, Berkeley scientists chose a compromise between high voltage and high stability. This enabled them to get a resolution of better than 2 angstroms with a 1-MeV instrument in which the accelerating voltage remains steady to within 0.1 volt. This raised the cost of the machine to $3 million, however, compared to $1.7 million for the 1.5-MeV microscope. "The added expense is well worth it, because we broke an important technological barrier," says Ron Gronsky, manager of the Department of Energy–supported project.

In its initial tests, the Atomic Resolution Microscope achieved resolutions of 1.7 angstroms. (The tests were conducted on what Gronsky called "noisy ground" in Japan, where the instrument was built to the Berkeley scientists' specifications by Japan Electron Optical Laboratories, Limited.) Thomas and Gronsky were overjoyed at the results; they are confident of improving them with the instrument more carefully mounted and cushioned at its permanent home in California. "With computer enhancement of the images, we might even break the 1-angstrom barrier," Gronsky says.

Computer enhancement involves digitizing the various shades of gray in an image and reconstructing it to enhance its contrasts and reduce the fuzziness produced by lens aberration and other causes. "Computers also will be used to work out subtle variations in image intensity, by which we hope to identify individual atoms," says Gronsky. Atoms such as uranium and iron should appear brighter than carbon or oxygen atoms, which have less mass and fewer electrons to reflect the instrument's electron beam.

The more common method of identifying atoms involves measuring the energy loss of electrons that are scattered by the atoms. These "inelastically" scattered electrons can be magnetically isolated from those that lose no energy; they can then be recorded separately. This requires detectors that have not yet been installed on the Atomic Resolution Microscope, how-

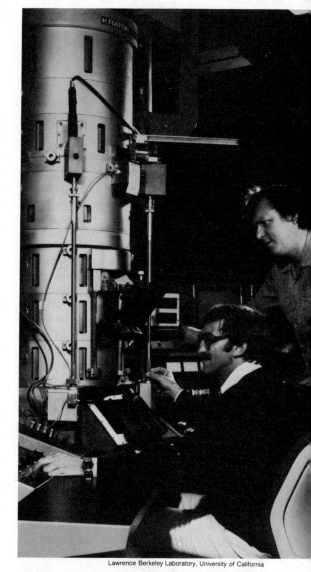

The high-voltage electron microscope (HVEM) accelerates electrons to energies of 1.5 million electron volts.

ever. "We have a setup for installing them and may do so later," Gronsky explains, "but for the present our top priority is to see atoms. Identification gets us into a whole new set of problems."

Pushing the Limits

Electron microscopists have also developed techniques that enable them to push the barriers to higher resolution without high voltages. (Only microscopes with energies greater than 500,000 electron volts are considered high-voltage.) Researchers at the Facility for High

Resolution Electron Microscopy at Arizona State University are especially adept at this. "Our 200-KeV transmission microscope resolves points 2½ angstroms apart," says Ray W. Carpenter, director of the Arizona center. "However, by looking at perfect crystals aligned parallel to the electron beam, we obtain images of atomic planes 2 angstroms or less apart."

Near-perfect crystals contain regularly spaced planes one atom thick, Carpenter explains, and these can be seen as parallel lines when a crystal is viewed edge-on. These lattice-plane images, as they are called, reveal distortions in plane periodicity that represent defects in a material. The planes are easiest to photograph in rocks, where they may lie 50 angstroms or more apart; they are hardest to photograph in metals, where atoms are packed close together. Semiconductors and ceramics fall in between.

"Images taken with microscopes having an instrument resolution of 3 or 4 angstroms often contain fine detail 1 angstrom or less in size," notes John Cowley, a physicist at Arizona State University. "We can use computer reconstruction techniques to extract information from this detail."

Researchers doing such work at Arizona State and elsewhere also employ computers to make model images of a material's structure to compare with observed images. When a model contains features beyond instrument resolution, a good match between observed and calculated images provides information about details that cannot be seen directly.

Another technique, called compositional analysis, takes advantage of the fact that inelastically scattered electrons undergo energy losses characteristic of the atoms they strike, or ionize. As ionized atoms decay back to a neutral state, they emit X rays whose wavelengths can be used for identification. Electron microscopes fitted with extra lenses, energy filters, and detectors are capable of X-ray analysis of the composition of clusters containing fewer than 20 atoms.

At Arizona State, compositional analyses also are done by a 100-KeV scanning transmission microscope equipped with a field-emission source, a tungsten wire tip 50 angstroms across. Combined with a lens that concentrates the electron beam by a factor of 10, this source produces an intense electron spotlight only 5 angstroms in diameter. Scanning such a beam across a specimen and recording the energy losses of inelastically scattered electrons enables researchers to determine elemental composition of a 10- to 20-atom cluster in a material. Ondrej L. Krivanek, associate director of the facility, believes that the sensitivity of this technique has been improved "to the point where identification of single atoms in a solid does not seem a long way off."

The bright 5-angstrom spotlight can be moved a fraction of an angstrom at a time across a crystal to yield diffraction patterns containing detail half an angstrom in size. And when researchers move a beam in half-angstrom steps across a space between atomic planes, they see distortions that indicate properties of a material.

Information from these overlapping images requires sophisticated interpretation. The images consist of patterns of spots, the intensity and spacing of which contain information about the structure of the material. But waves diffracted or bent around features reinforce and interfere with each other, blacking out atoms or producing spots where no atoms exist. Says Cowley: "X-ray diffraction has been a standard way to study crystal structure since 1912. The theory of interpretation has been worked out and computerized to the point where positions of atoms can be determined to within hundredths of an angstrom. We can probably do the same with electron diffraction, when the theory and computer programs became fully developed."

A Holographic Way

Field-emission guns make workable an idea originated in 1949 by Dennis Gabor, the physicist who invented holography. (Holography is the process of making holograms. A hologram is a record on photographic film of the interference pattern between laser light reflected by an object and a direct beam from a laser. Once made, the film presents no recognizable image until it is illuminated with a laser beam. Then the object is re-created with startling realism in three dimensions.) Gabor suggested that intense beams of electrons crossing each other in front of a specimen would generate shadow images with a magnification of 10 million. "We did not possess the technical ability at the time, but today we have the required lenses and intense electron sources," Cowley notes. "Using a crossover spot to illuminate a specimen does not produce a direct image; rather, you get a hologram that must be interpreted." He plans to do this by digitizing the shadow images and using special computer programs to reconstruct them

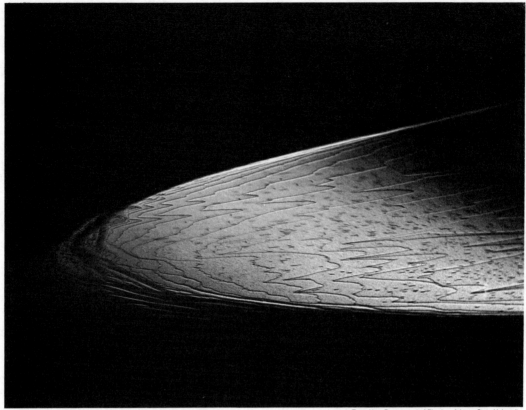

An electron beam reflected from a gold crystal reveals overlapping atom-high edges of the material's atomic planes.

as black-and-white pictures. "The distribution of intensities in a hologram taken with a 4-angstrom-resolution microscope contains spatial information on a 1-angstrom scale," Cowley says, "and the intensities are related to the identity of the atoms. We know that information about single atoms is there and that computer reconstruction works. It's just a matter of time, and not too much of it, until we start obtaining results with this technique."

Imaging and analysis can be improved by fitting field-emission guns to higher-voltage microscopes than those used at Arizona. Japanese and French researchers plan to do this with 1-MeV machines. However, no immediate plans exist to install a gun on the Atomic Resolution Microscope, which uses a special heated filament to produce electrons. The heating currents in these filaments may fluctuate in intensity, and electrons may boil off them at different regions. Either occurrence can blur the images and reduce resolution. "We are working to minimize this problem," notes Gronsky. This has

priority over installing a field-emission source, which involves other emission unsteadiness problems, Gronsky explains.

Attacking Lens Aberration

Crewe is planning a low-powered microscope, fitted with a field-emission source, that he believes will have greater resolving power than the Berkeley atom-viewer. He intends to sidestep the problems and cost of high voltage by solving the lens-aberration problem.

Focusing electrons requires a doughnut-shaped arrangement of magnets with the beam passing through the "hole" of the doughnut. Such symmetrical lenses introduce spherical aberration, the failure of electrons from the edges and center of the hole to meet at the same focal point. Theoretically, it is possible to correct the aberration by introducing an asymmetry in the lens.

Physicist Otto Scherzer, working at the University of Darmstadt in Germany, devoted many years in the first half of this century to

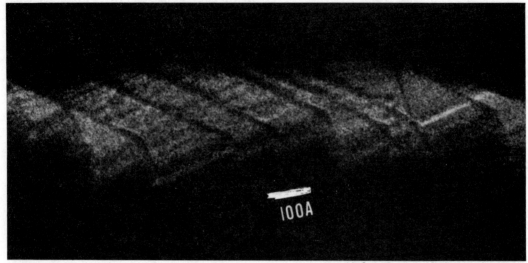

The steps in a magnesium-oxide sample are 100 angstroms wide and are crossed by steps one atom high.

finding an arrangement of magnets that would accomplish this. His solution, worked out in 1946, requires a lens containing 40 pole pieces machined and positioned to a tolerance of a thousand angstroms. "We devoted 10 years to building and testing such a corrector," Crewe says, "but success eluded us." He gave up the project until, in the late 1970's, "I stumbled on another way to make the correction," he recalls.

Scherzer's solution had called for a series of quadrupole magnets, magnets having two north and two south poles. While teaching a graduate course in electron optics, Crewe worked out the theory of fields produced by sextupoles—three south and three north poles—arranged in a circle. "No one had done this before," he recalls. "I found out that sextupoles can, in theory, correct the spherical aberration. Such a corrector would consist of 12 rather than 40 pole pieces, and the required tolerances are not forbidding."

Utilizing this corrector theory, Crewe has designed a scanning transmission microscope with an expected resolution of half an angstrom. It will need an electron-accelerating voltage of only 200-KeV, and it will be small enough to fit inside an ordinary laboratory. He received a $990,000 grant from the National Science Foundation to construct the major components of this instrument, and International Business Machines has offered him $1 million worth of computer equipment. If things go as planned, the microscope will be completed in 1986.

The Frontier

These electron-microscopes are putting their designers on the frontier of a new type of exploration—the direct study of the atomic structure of solid materials. "With this ability, we can take a fresh look at all the problems and questions of materials science," Gronsky declares. "For instance, how do carbon and iron atoms combine to form steel? What happens on an atomic level in alloys exposed to heat, cold, and strain? What . . . arrangement of atoms determines the electrical properties of semiconductors?"

For the study of biological materials, it is hard to say whether high- or low-voltage machines will be better. Electron microscopy is notoriously destructive to organic materials. At high energies, sample damage can be minimized by limiting the time a specimen must be exposed to irradiation. Electrons with energies of a million volts or more spend less time inside a material, reducing the opportunity for collisions that break chemical bonds.

At the low-energy end, scanning transmission microscopes with field-emission guns provide an intense beam that can be quickly scanned across organic material, keeping exposure times short. Crewe, for example, is using a 35-KeV scanning transmission instrument to study the structure of oxygen atoms in hemoglobin molecules. With the planned half-angstrom machine and suitable staining techniques, he believes, it may even be possible to read directly the sequence of atoms in a DNA (deoxyribonucleic acid) molecule □

THE 1983 NOBEL PRIZES
Physics and Chemistry

by Barbara Tchabovsky

Americans were awarded the 1983 Nobel Prizes in Physics and Chemistry, completing a United States sweep of the 1983 Nobel science prizes. The physics award went to two astrophysicists: Subrahmanyan Chandrasekhar of the University of Chicago and William A. Fowler of the California Institute of Technology for their work on how stars age and collapse. The chemistry prize was awarded to Henry Taube of Stanford University for his pioneering work in inorganic chemistry, specifically on the mechanism of electron-transfer reactions in metal complexes.

Chandrasekhar's Contribution

Black holes, white dwarfs, neutron stars, supernovas, and scenarios of how stars of different sizes form, mature, function, explode, and collapse are now fundamental concepts in stellar astrophysics. They were not, however, several decades ago, before Chandrasekhar and Fowler, working independently, began their studies of stellar evolution and composition.

In 1930, when he was only 19 years old, Chandrasekhar began to wonder what happened to large stars as they burned out. At that time it was thought that a star's gaseous envelope held up as long as heat was generated in the core, but that once the internal fuel was used up, the star collapsed like a punctured balloon. Relatively small stars—those not much larger than the Sun—were said to collapse into objects called white dwarfs. White dwarfs were believed to be so dense that, as British astronomer Sir Arthur Eddington put it, one ton "would be a little nugget that you could put into a matchbox."

But what about more-massive stars? What became of them when their internal fuel was exhausted? On a long voyage to England from India, Chandrasekhar used his mathematical skills to calculate the possibilities. He found that if the mass of a star is more than 1.4 times that of the Sun, the dense matter resulting from its collapse could not withstand the pressure to simply become a white dwarf. Such a star, he

said, "cannot pass into the white-dwarf stage, and one is left speculating on other possibilities." He believed it would keep on shrinking. Sir Arthur rejected this idea, pointing out that if such a massive star kept on shrinking, it would become so dense that no light could radiate from it. "I think that there should be a law of nature to prevent the star from behaving in this absurd way."

Chandrasekhar's ideas prevailed, however. The Chandrasekhar limit—1.44 times the mass of the Sun—has become a standard in modern astrophysics. Stars below that limit collapse quietly to become white dwarfs; stars above that limit undergo supernova explosions. After exploding, those stars two or three times the mass of the Sun collapse to become what are now called neutron stars; those stars even larger become black holes, from which no light can escape.

Chandrasekhar's theoretical work led to the concepts of the black hole and neutron star, and opened new avenues for astronomical observation. As Massachusetts Institute of Technology physics professor Philip Morrison said, Chandrasekhar's "mathematical insight and its elegance has been responsible for most of what we know about stars."

Fowler Studies Element Synthesis

Supernova explosions that herald the death of certain stars mark the birth of some chemical elements. In the 1950's William Fowler investigated how heavy elements are synthesized in stars. It had been proposed at that time that only hydrogen and helium were produced in the Big-Bang birth of the universe. Elements heavier than these were believed to be synthesized through bombardment reactions in the cores of stars, which gradually caused the buildup of elements whose nuclei contained more and more particles. There were many problems with this theory, however, and William Fowler and his colleagues finally came up with the right answer. They unraveled the complex succession of steps whereby stars—during their lifetimes and in their death throes—produce all the chemical elements.

Under the heat and pressure conditions of a star's core, the nuclei of light elements collide and merge with one another to form new nuclei having more and more particles—thus synthesizing progressively heavier elements. However, once conditions necessary for the synthesis of iron are present in the star, this energy-generating, collision process is suppressed. Then the star is ready to collapse.

When a large star collapses, the supernova explosion that results creates enough ferment to forward element synthesis right to the development of very heavy elements. In some cases very massive, unstable elements form and later decay into stable elements. Elaboration of these steps eventually explained the formation of all the basic substances from which the universe is made.

Much of Fowler's theoretical analysis and laboratory work on the origin of elements was done with three British colleagues: Dr. E. Margaret Burbidge; her husband, Dr. Geoffrey R. Burbidge; and Sir Fred Hoyle. Their joint paper, outlining the major processes involved in the stellar manufacture of chemical elements, was published in 1957 and has since become a classic of astrophysics, known popularly as the BBFH paper.

Subrahmanyan Chandrasekhar was born October 19, 1910, in Lahore, India (now part of Pakistan). He is the nephew of Sir Chandrasekhar Venkata Raman—the 1930 Nobel laureate in physics for his work on light scattering. After early education in India, Chandrasekhar traveled to England to study at Cambridge University, where he earned his Ph.D. in physics. In 1936 he moved to the United States, and became a U.S. citizen in 1953. He spent his long research and teaching career at the University of Chicago and the Yerkes Observatory in Williams Bay, Wisconsin.

In addition to his studies of stellar collapse, Chandrasekhar investigated nuclear reactions in stars and the chemical composition of stars. He authored *An Introduction to the Study of Stellar Structure* (1939) and *Principles of Stellar Dynamics* (1943). From 1952 to 1971 he served as the editor of the prestigious periodical called the *Astrophysical Journal*.

William Alfred Fowler was born in Pittsburgh, Pennsylvania, on August 9, 1911. He

Astrophysicist William Fowler holds a shirt presented to him on the day of his Nobel prize announcement.

California Institute of Technology

Photos: Wide World

Astrophysicist Subrahmanyan Chandrasekhar

Inorganic chemist Henry Taube

received his bachelor of science degree from Ohio State University and his doctorate in physics from the California Institute of Technology, where he then spent his entire career. Fowler has been the recipient of numerous awards, including the National Medal of Science in 1974 and the Eddington Medal of the Royal Astronomical Society of London in 1978.

The Prize in Chemistry

While preparing lectures for an advanced chemistry class, Henry Taube noticed that there were many opportunities for research in inorganic chemistry. He went on to become a pioneer in the field, and was called by the Nobel Committee "one of the most creative contemporary workers in inorganic chemistry."

Working with metallic solutions, Taube studied exactly how chemical reactions take place—specifically, how an electron gets from one place to another in a reaction. It was then thought that ionic reactions simply involved a transfer of electrons, but there were puzzles: why, for example, does the speed at which reactions among similar metals and ions occur vary so much?

Taube found that electrically charged atoms (ions) react through the migration of atoms and groups of atoms, or, in other words,

that a chemical bridge of atoms was necessary for electron transfer to take place. As one former student put it, "He bridged the gap between descriptive inorganic chemistry, which is all those reactions one has to memorize in high school and college, with the basic principles of thermodynamics and kinetics."

After hearing of his award, Taube said that the prize was in recognition of his field, not just himself. However, as the Nobel Committee pointed out, it is a field Taube has dominated, making at least 18 major discoveries. Or, as Stanford chemistry professor James Collman put it, "Taube built an edifice, which *is* the field."

Henry Taube was born in Neudorf, Saskatchewan, Canada, on November 30, 1915. He received his bachelor's and master's degrees from the University of Saskatchewan, and a Ph.D. in chemistry from the University of California at Berkeley in 1940. He then served as a professor at Cornell University and the University of Chicago. In 1961 he joined the faculty of Stanford University as professor of chemistry. Taube has been the recipient of many awards, including the National Medal of Science in 1977 and the American Chemical Society's Award for Distinguished Service in the Advancement of Inorganic Chemistry □

TECHNOLOGY

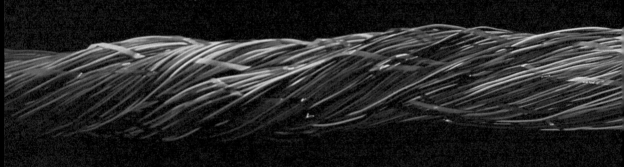

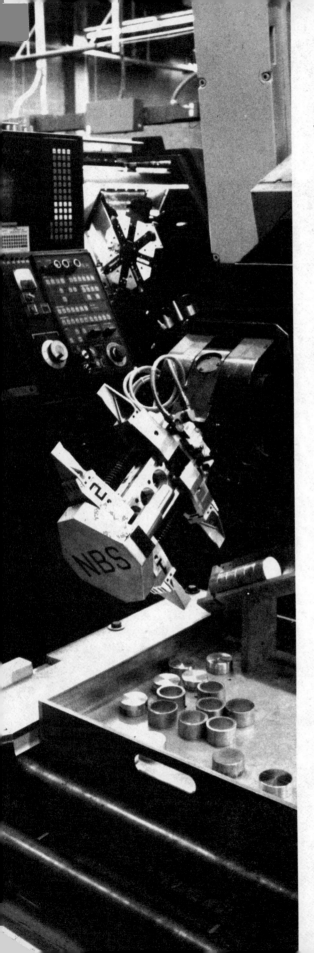

REVIEW
OF THE
YEAR

TECHNOLOGY

During 1983 the term "digital" seemed to pop up everywhere as computers and microprocessors widened their invasion to include such diverse territories as audio recording, filmmaking, and sports such as sailing. Computer technology also contributed to novel systems for automated factories and to new sensitive microscopes for delicately probing the secret nature of materials.

COMPUTING A FUTURE FACTORY

In the future, automated factories will probably look more like today's computer rooms than yesterday's dusty, noisy manufacturing plants. In such factories the computer "peripherals" will be robot arms and machine tools such as lathes instead of the printers, terminals, and other devices usually found in computer facilities. John A. Simpson, director of the Center for Manufacturing Engineering at the National Bureau of Standards (NBS), says, "Such systems will, in the future, manipulate materials much as computer installations of today manipulate information." Some of the largest U.S. manufacturers already are well on their way toward integrating robots, computers, and tools into efficient and flexible manufacturing systems. The Westinghouse Electric Corporation in Pittsburgh, Pennsylvania, for example, has an experimental assembly system that can be programmed to adapt to new product styles.

However, small companies can't afford the elaborate, integrated systems now being installed at large-scale factories. They need a way to buy new and improved machinery from various makers one piece at a time. The answer to their problem may be taking shape at a new $5 million Automated Manufacturing Research Facility in Gaithersburg, Maryland, that was first publicly displayed in 1983. There, using off-the-shelf machinery and thousands of lines of innovative computer programming, NBS researchers demonstrated a computer-engineering feat comparable with getting, say, a dozen chefs—all with widely varying abilities, individual specialties, and different native languages—to work together

Left: The experimental gripper on this industrial robot was specially designed to work with the computer-controlled turning center in the background.

NBS

to prepare a sumptuous gourmet feast. They demonstrated that computer programs could be devised to allow different equipment from a variety of manufacturers to "talk" with and control one another without requiring a common programming language or revealing trade secrets about how a particular machine works. (See "The Race to the Automatic Factory" on page 335.)

SAILING BY KEEL AND COMPUTER

Traditionally, the America's Cup race has always been a showcase for ingenious technical innovations in sailing. In the 1983 race, attention focused on the winning *Australia II*'s mysterious keel. The radical design, which seemed to make the yacht more maneuverable and faster, featured a relatively short keel from which sprouted two small fins. These winglike appendages smooth out the underwater eddies that slow a boat.

Computers also made their mark on the race. Designers used computers to help evaluate thousands of possible hull and sail shapes. They were also used on board to monitor performance, and in some cases, to aid navigation and strategy. Now many of these race-tested innovations are starting to appear in recreational sailboats.

DIGITAL ENTERTAINMENT

As the video-game boom faded and electronic arcades began losing customers, game manufacturers tried to come up with ever more exciting and involving pastimes. The sensation of the year was a game called "Dragon's Lair," in which players can see themselves hurtling down cobwebby corridors, plunging into bottomless black holes, and battling dragons with a laser sword—all to stereophonic background music. What made this creation special was the marriage of the microprocessor usually found in video games with the laser-scanned optical disk. The film images and sound effects are stored on the video disk, while the action is controlled by the microprocessor. To improve the system further, Laser Disc Computer Systems of Boston, Massachusetts, came up with a two-disk operating system to give the illusion of uninterrupted action: while one disk follows the action, the other races ahead, anticipating the player's next decision. Eventually, laser games may evolve into more educational forms— allowing a viewer to explore a country's geography, to perform surgery by trial and error, or to bring the space shuttle in for a landing.

Digital computers have also become important in the making of motion pictures. Lucasfilm Limited, in particular, has built up an extensive research and development center devoted to using computers to synthesize sounds; to control the mixing of sound tracks; and to create three-dimensional, animated models as well as to edit film and generally manage the innumerable tasks that must be coordinated to produce a film. Some of the most sophisticated and spectacular applications of computer graphics appeared in films such as *The Return of the Jedi*, *Tron*, and *Star Trek II: The Wrath of Khan*.

BETTER EARS AND EYES FOR DETAIL

Scientists have always been interested in exploring metals, semiconductors, ceramics, and organic materials atom by atom to unlock the secrets of the properties of these materials. In the past, they have been able to see individual atoms, and even make motion pictures of them, only under special conditions. Now a series of advances in microscopy is making it easier to study atoms that are tightly packed together in metals or other crystalline materials. One new instrument is the Atomic Resolution Microscope at the University of California's Lawrence Berkeley Laboratory; it is the first electron microscope capable of resolving individual atoms in solid material. (See the article "To See An Atom" on page 304.)

Taking a different approach, researchers at the IBM Corporation in Zurich, Switzerland, have come up with a powerful new technique for producing three-dimensional images of the surfaces of solids. Called "scanning tunneling microscopy," the technique depends on the ability of electrons to pass between two conducting solids separated by a vacuum. The number of electrons that get across depends strongly on the separation between the two solids. The method is so sensitive that a change in distance by the diameter of a single atom changes the current by a factor of 1,000. Using this technique, scientists can construct precise "topographic maps" of the bumpy surfaces of materials such as gold and silicon.

Finer details are also showing up in electron-microscope images of cells and complicated molecules like proteins with the use of a new technique that involves flash-freezing fresh tissue to near absolute zero ($-273.16°$ C or $-459.67°$ F). This procedure preserves a cell's fragile contents, and the cold keeps the atoms from constantly shuffling about. The result is clearer pictures and information about complex, rapid changes that a living cell may undergo.

Meanwhile, the recent development of acoustic microscopy, which uses sound waves to reveal the structure of materials in fine detail, opened up the possibility of discerning features not visible with light or electron beams. This technique makes it possible to look inside a piece of material, whether it's a living cell, a silicon chip, or a chunk of metal.

IVARS PETERSON

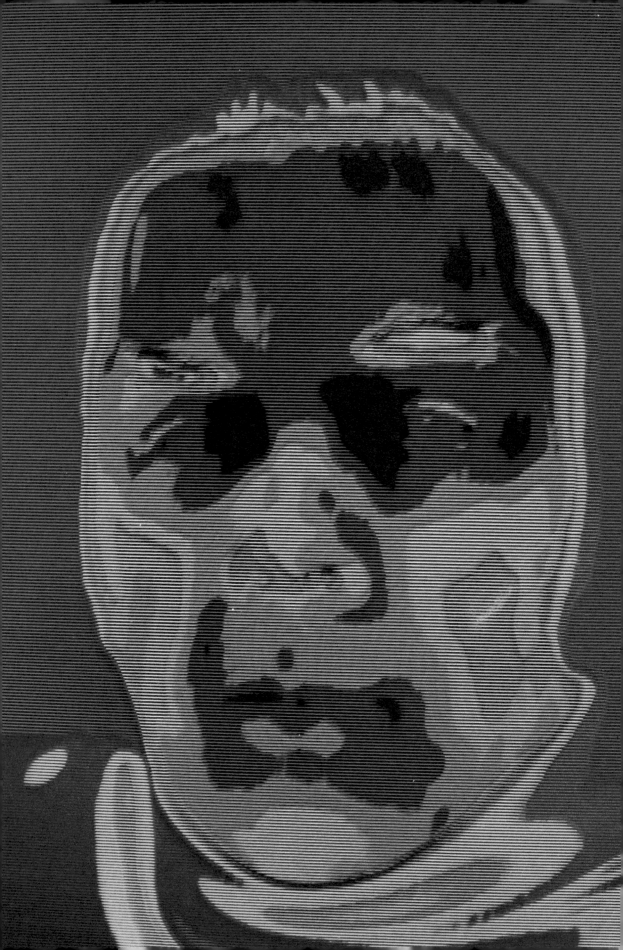

THE COLOR OF PAIN

by David Monagan

A new technology is sending shivers through the legal profession and the insurance industry, as well as rudely awakening doctors who have been too quick to dismiss pain they couldn't previously diagnose. Called thermography, it works by taking infrared pictures of heat variations in every surface nook and cranny of the human body—not to detect fever, but to draw temperature profiles of hidden internal damage. The most sophisticated electronic thermographs register body temperatures from up to 64,000 points on the torso and limbs. In the resulting composite, thermal differences around key nerves serve as a map of musculoskeletal problems that often could not be documented—or accurately identified for treatment—before.

Heat Aberrations Accompany Injury

Unlike X rays and other methods of recording strictly anatomical damage, thermography displays the abnormal heat patterns around irritated sensory nerves so that physicians can follow the physiology of pain itself throughout the body—a crucial advantage when there are no fractures or lacerations. The technology's advocates claim it eliminates much of the guesswork formerly involved in the diagnosis and treatment of whiplash and all manner of so-called soft-tissue injuries. Hence, it could benefit not only the one out of four adults who suffers at some time from often clinically vague back pain, but the millions more who are afflicted with once unprovable and unseeable aches, from muscle spasms to migraine headaches.

It was Hippocrates who observed 2,400 years ago that one side or part of the human body is likely to be warmer than the other when something is internally amiss. Painting the bodies of his patients with a slurry of wet clay, he observed that one side dried faster than the other—leaving a telltale pattern of skin temperature that seemed to coincide with injury.

A thermogram of a relatively healthy man's head shows temperature increases around his congested sinuses.

Dan McCoy/Black Star

Modern military innovators developed the infrared-sensing technology—originally to detect snipers at night by body heat; later to detect encampments, planes, and the like from spy satellites—that finally made it possible to analyze what Hippocrates was talking about.

Electronic and Contact Systems

Today's most sophisticated thermographic systems resemble television cameras, except that they use high-speed oscillating mirrors and prisms to scan for infrared radiation that is invisible to the naked eye. At the core of these units are chips of semiconductor material, .0025-centimeter (.001-inch)-square, composed of exquisitely temperature-sensitive alloys with either mercury or indium bases. When cooled to $-196°$ C ($-321°$ F) with liquid nitrogen, these detectors respond to the narrow range of infrared energy given off naturally by the human body. The display electronics amplify and process anywhere from 30,000 to 64,000 heat signals per second into televisionlike images on a video screen. The computer then arranges this information into a color scale. The result is an easily photographed image of bodily temperature variations.

Far more prevalent than these electronic systems, however, are newer and cheaper—though less precise and reliable—devices called contact thermographs. These systems consist simply of ordinary still cameras and rubber Mylar pads impregnated with heat-sensitive liquid crystals. Generally, the pads include a variety of different color-coded chemicals, each of which will react only under a very specific temperature condition. Just as does its electronic counterpart, contact thermography records all heat variations coming from beneath the skin.

Temperature Variations Abound

The resulting brilliantly silhouetted photographs—called thermograms—show that the body is a caldron of temperature variations—with as much as a six-degree [Fahrenheit] difference from nerve centers to the tips of the extremities. Usually, these variations appear on film as rippled, regular contours that are as symmetrical as the anatomy they overlap. But as Hippocrates suspected, heat aberrations—on thermograms, streaks, and leopard spots of color—invariably accompany internal injury.

Hot spots frequently appear after an injury as a result of the rush of warm blood that replenishes areas of contusions, inflamed muscles, sprains, skeletal dislocations, and the like. Ironically, however, chronic musculoskeletal problems—and some sudden muscle spasms—may often be surrounded by significant enough vascular constriction to undermine the flow of blood and produce pronounced cold spots in a damaged area. These abnormal temperature pockets—differences as slight as one-tenth of one degree Celsius—can be detected on a thermogram and turned into a map of damage beneath the skin.

An Important Diagnostic and Legal Tool

"An X ray is a two-dimensional picture of anatomy," explains Dr. Charles E. Wexler, an Encino, California, radiologist and a leading medical thermographer. "A thermogram [shows] the dynamic physiology of the sensory nerves, the missing link in the evaluation of pain."

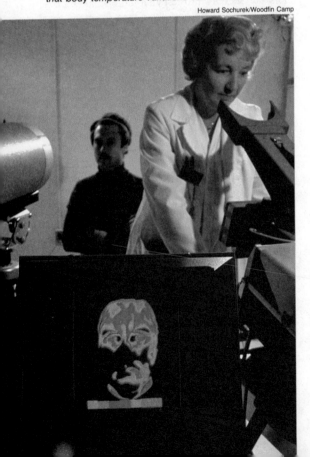

A concussion victim's thermogram demonstrates the fact that body temperature variations can indicate ill health.

Howard Sochurek/Woodfin Camp

Wexler has used thermography to chart the far-flung—and often previously ignored—peripheral neuromuscular problems that accompany chronic lower back and neck pain.

Before thermography, diagnosis of many soft-tissue injuries was a matter of say-so. Doctors could take their patients' complaints at face value and treat them accordingly or dismiss pain symptoms they couldn't objectively verify. As Dr. Paul H. Goodley, a Los Angeles, California, doctor of orthopedic medicine, puts it: "If there was nothing on the X ray, we were told to suspect the patient."

But thermography has proved a powerful tool in changing that way of thinking—in the courts as well as in the medical office. In the less than five years it has been systematically used in diagnosing soft-tissue injuries, thermography has been introduced to hundreds of courts in California, New York, New Jersey, Louisiana, Illinois, Michigan, Wisconsin, and Pennsylvania. The plaintiff with thermography on his or her side is hard to beat. "I think it's a great help, especially for the guy who's hurt but has no objective signs of injury," observes Edward B. Rood, former president of the Association of Trial Lawyers of America and senior partner in a Tampa, Florida, firm that specializes in personal-injury lawsuits. "We've used it in [dozens of] cases so far. Of course, the defendant has challenged the thermograms in every case—always unsuccessfully." Rood added that the technology had exposed four outright fakes, whose cases were dropped by his firm.

A Growing Market

Not surprisingly, given the enormous stakes of today's personal-injury lawsuits, medical thermographers have been quick to realize that they're holding a lucrative legal product. Many report that half of their diagnostic work or more is performed for disputed insurance cases—several of the leading physician thermographers even tout their services in advertisements in *Trial Magazine*. No wonder, then, that after several years of harboring suspicion and hostility toward the technology, a number of the Workmen's Compensation carriers have started buying thermographs of their own.

Sales of the sophisticated $14,000 to $100,000 electronic thermographs—made by Inframetrics of Boston, Massachusetts, and a Swedish company called AGA—have more than doubled in recent years. According to Sy Katz, a vice-president of Flexi-therm, of West-

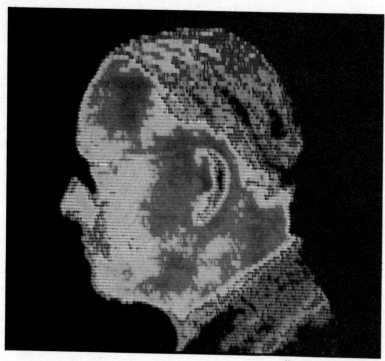

This unfortunate person's thermogram shows very little blood circulation in a frostbitten ear. Red areas indicate normal circulation.

bury, Long Island, sales of that company's $5,200 contact thermography sets have shot up 500 percent. A new competitor in the liquid-crystal contact thermograph market, BCD Products in New York, claims to have sold 500 units in its first four months. The clients for all these companies run the gamut from radiologists, chiropractors, gynecologists, neurosurgeons, and athletic trainers to veterinarians with a thoroughbred clientele.

Obstacles to Technique's Use

Still, a more widespread use of thermography has been held back by two problems. One difficulty was a lack until recently of systematic standards for determining what different thermal patterns meant or did not mean. Medically, it's obviously crucial to know, for example, the difference between heat patterns indicating chronic bursitis and those surrounding a sprain. However, medical thermographers have finally begun to agree upon a standardized key to the body's heat map.

Much of the real progress in this direction is the result of work by Dr. Wexler, a great promoter of the technology, who has performed more than 6,000 thermographic exams. Wexler found that thermograms revealed the nature of 93 percent of the musculoskeletal injuries he

looked at, as opposed to 82 percent for the nearest competitor, electromyography, or EMG.

Thermography was also handicapped by premature application. Early claims for the technology put it squarely in the middle of the National Cancer Institute's (NCI's) mass Breast Cancer Detection Demonstration Project of the mid-1970's—apparently years before the technology was ready. According to Dr. JoAnn Haberman, professor of radiology at the University of Oklahoma, diagnosticians who had no idea how the technology worked were told to try it because "it couldn't hurt." It didn't help either. By 1975 it was clear that many thermographically screened women were being told they had possible breast cancer when further probing indicated that they had none. So NCI backed away from thermography, while medical articles criticized the technology.

But Haberman and a few others pressed on. A recent study by researchers from the University of Louis Pasteur in Strasbourg, France, found that many of the so-called "false positives" generated by thermographic tests were real positives—cancers diagnosed at a stage when no other technology could verify them. Haberman's recent research with sophisticated new electronic thermographic machines supports these findings. "It's just like any other evaluative technique—if you go about it in a haphazard fashion, you'll get haphazard results. My experience is that the thermogram can show an abnormality months and up to several years before a mammogram,'' Haberman says of the standard X-ray screen for breast cancer that runs the danger of causing it. ''Thermography has just never been properly evaluated as a screening technique for breast cancer. One reason is that we didn't have the right equipment; another is that there is too much about breast thermography that has never been worked out.'' As a result, although Haberman swears by the technique as an adjunct in breast-cancer diagnosis, she worries that practitioners may be ill-prepared to use it.

Many Take Advantage of New Technology

Unfortunately, the technology has had its share of opportunists. Fabergé, the well-known cosmetic company, has been after the U.S. Food and Drug Administration (FDA) to approve a liquid crystal–like cup that women could periodically put in their bras to check for what the company says are telltale color reactions of breast cancer. The prosaically named BCD (Breast Cancer Detection) Products, meanwhile, sought approval of a similar device, which it planned to sell for $200. ''Eventually, when the technology's all worked out, when reasonable approaches have been developed, a home thermographic system could be useful for detecting temperature changes in breasts,'' says Dr. Lillian Yin, director of the FDA's division of obstetrical and gynecological devices. ''But right now, there are no devices ready for home market. A woman could easily make some mistake in using the technology and get some color change that could scare her half to death.'' At the latest word, both Fabergé and BCD are petitioning for FDA approval of their breast-cancer screening systems for use only under a doctor's prescription.

Meanwhile, thermography mills—storefront and shopping-center mass-screening operations offering ''guaranteed'' diagnosis of any pain—have begun to worry established medical thermographers. The American Thermographic Association has consequently decided to certify only M.D.s (doctors of medicine) and D.O.s (doctors of osteopathy) as professionally approved thermographers.

But this does not prevent many qualified thermographers from building lucrative second practices—usually with high-sounding names like Pain and Diagnostics Institute—around their machines. A number of physicians have farmed out a lion's share of their work to paramedics or research technicians, who, unlike X-ray technicians, need no government licensing to operate thermographic equipment. The charges for such services range from $50 for a

The normal breast at left shows symmetrical temperature variations; the abnormal pattern at right may indicate cancer.

Photos: Thermascan Inc.

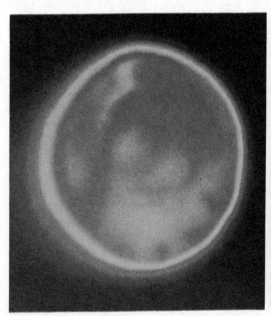 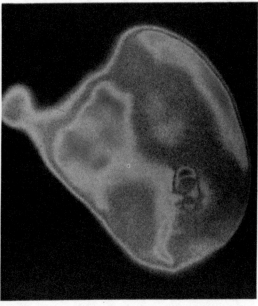

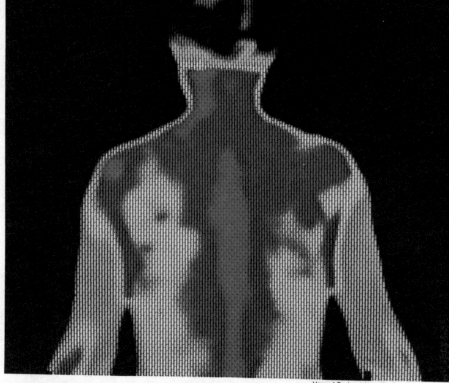

Pain sites and soft-tis-
sue injuries appear as
spots and streaks of
color in asymmetrical
patterns, as seen in this
back thermogram.

Howard Sochurek/Woodfin Camp

quick breast screening with a physical examination up to $235 for a full 55-picture work-up of
complex back problems.

Value of Prescreening Gets Mixed Reviews

Some thermographers have even tooled up to
sell their services en masse to corporations as
a means of screening prospective employees for
chronic back problems and other hidden ailments. For example, Dr. Pierre LeRoy, a New-
ark, Delaware, neurosurgeon who has set up an
entire facility specifically for thermography,
gets patient referrals from the Southeast Pennsylvania Transit Authority and the Wilmington
plants of General Motors, Chrysler, and Du
Pont. Dr. Wexler, meanwhile, has tried to persuade companies of the benefits of mass preem-
ployment screening by offering to set up free
demonstration programs.

Dr. Goodley, for one, worries that there is
potential for abuse in any mass screening procedures that elevate thermography above the physician's knowledge of individual patients. "If
somebody gets a thermogram on the day that he
slept wrong, and he's labeled as having a bad
back, and 36 hours later he's completely better—why should he carry the stigma of a test on
him that had no clinical validation?" Wexler,
however, defends prescreening as a means of
telling whether "a guy is going to be a hazard to

himself or his fellow employees." The real
question, of course, is whether the use of thermography in obviously nonclinical ways consti-
tutes invasion of privacy. One National Football
League team, for example, is reportedly making
thermographic records of injured athletes that
could be used against them at contract time.

While the diagnostic potential of thermography is unmistakable, just how and when it
should be used remain open questions. Dr. Haberman, the University of Oklahoma radiologist, says, "I wouldn't do breast [cancer]
screening without it." But she also says it's
high time the users of the technology come up
with some firm professional standards. "People
have been led to believe that all you have to do
is stick a patient in front of a machine and turn
on a button. Unfortunately, because it's nonin-
vasive, there's been a lag in properly restricting
it. In truth, it's a very complex and misunder-
stood technology" □

SELECTED READINGS

"Beating breast cancer—the CAP test (Cholesteric Analy-
    sis Profile)" by Stephen Kindel. *Forbes*, December 6,
    1982.
"A picture of pain" by C. A. Helwick. *Health*, March
    1982.
Fundamental Aspects of Medical Thermography by W. M.
    Park. British Institute of Radiology, 1980.

DISC

MAGNET

OBJECTIVE
LENS

FOCUS COIL

COLLIMATOR
LENS

Philips

PHOTO
DETECTOR

PRISM

LASER

Compact Discs

by Stephen A. Booth

A new music-playback system that promises to revolutionize sound recording has made its U.S. debut. And there are those who think it could make the LP (long-playing) record and phonograph obsolete. With its laser pickup and grooveless, saucer-sized record, the Compact Disc (CD) digital audio system signals the finale for all the snap-crackle-pop, hiss, and compressed music that fit into the "microgrooves" of today's LPs.

But don't get rid of your stereo gear yet or hold your breath waiting for your favorite artists on CD. Although the Compact Disc player will

Left: As a Compact Disc revolves on a player, a laser beam scans microscopic pits in the disc and transforms the reflections into a digital code that is then converted into superb sound. Right: A close-up of a laser beam "reading" the pits in a Compact Disc.

Index Stone International

patch into your stereo system just as a turntable does, it will be some time before the CD catalog can accommodate every musical taste.

Plastic-coated Pits

The Compact Disc, jointly developed by Japan's Sony Corporation and Holland's N.V. Philips, is stunning to behold. Its mirrorlike surface measures 12.06 centimeters (4.75 inches) in diameter—smaller than a 45-rpm (revolutions per minute) single—yet it carries up to an hour's worth of stereo music on its playable side (the other side is covered by the label).

Sandwiched between two layers of clear plastic is the shiny playing surface, and it is here that one sees what passes for grooves on a Compact Disc. Instead of the squiggles that set a stylus vibrating on an LP, the spiral tracks of the Compact Disc are composed of microscopic "pits" that are read by a laser beam. The layers of clear plastic protect these pits from scratches, dust, or fingerprints.

CD players aren't cheap. As of early 1984, the lowest-priced model in the U.S. was $499. Industry specialists predict that prices will drop significantly in the next few years; attributable, some say, to mass production and competition among CD manufacturers.

A Musical Morse Code

Besides size, price, and a laser, what sets the Compact Disc digital audio system apart from its needle-in-groove forebears? The answer lies in those microscopic pits embedded in the CD sandwich. As the laser scans the pits, it transmits a musical Morse code—a 16-digit code made up of ones and zeros, with each number signifying a musical signal of specific frequency and level—to the CD player's electronic circuitry, where the numbers are deciphered back into electrical impulses that correspond to musical wave forms. These impulses are passed along to the stereo system's amplifier, which boosts the signal in a routine manner and shuttles it off to the loudspeakers, where the electrical impulses are converted back to mechanical energy—sound waves.

It's this music-by-numbers aspect of digital recording that makes it fundamentally different.

The traditional method of recording sound—called analog recording, because it traces a mechanical replica, or "analogy," of the actual sound waves—hasn't changed much since Thomas Edison invented the cylinder player more than 100 years ago. Certainly the medium for conveying sound has evolved over the years: cylinders gave way to 78-rpm discs, noisy 78s vanished in the wake of long-playing microgroove discs spinning at 45 and 33⅓ rpm, and the more realistic-sounding stereophonic recordings made their debut along the way. With each change of format, the quality of sound improved. But each format—cylinders, 78s, and today's LPs—uses analog-recording techniques: sound waves either cause the mechanical vibrations of a stylus to cut squiggly grooves in a plasticlike substance, or they are converted to electrical impulses that cause the magnetic particles on a tape to form patterns that represent the changing wave forms of the music.

Analog Method Has Limitations

As good as it is, the analog recording and playback process is subject to limitations. These begin with the master tape. Tape hiss detracts from the amount of musical dynamic range that can be recorded—dynamic range being the difference between the softest and loudest passages in the music. Similarly, there's tape "headroom" to contend with: that is, the amount of musical energy that can be absorbed before the tape particles saturate or overload, causing the music to become distorted. The next hurdle is the cutting of the record. In real life, music doesn't fit into neatly spaced grooves, but in making records, the sound must be compressed. High-energy passages are diminished in order to shoehorn them into a groove that a stylus can track; low-energy sections are boosted to make them audible over the tape and record hiss.

There are other aspects of the analog playback system that diminish its fidelity to the original music. These include rumble (the sound picked up from the turntable's motor) and wow and flutter (changes in musical pitch caused by speed variations in a turntable or tape deck). Record warp is another liability of the analog disc, and even the record vinyl, phono cartridge, and tone arm add coloration and resonance (read "distortion") of their own. Finally, the musical wave forms inscribed in a record groove wear down with each scrape of the stylus, and the oxide particles on magnetic tape are shed with repeated passages over the recorder's tape head, capstan, and pinch roller. In fairness to the legacy of Mr. Edison, the makers of analog recordings and playback equipment have, over the years, reduced many of these handicaps down to manageable (and mostly inaudible) proportions. The epitaph for analog shouldn't be how poor it sounded, but how good. Yet the sad but inescapable fact remains that analog simply is not capable of reproducing the memory of a musical performance with perfect fidelity to the original. Enter digital audio.

Digital Eliminates Fidelity Problems

Digital recording assigns a numerical value to each nuance of sound, and because a 16-digit code is used, some 65,536 combinations are available to cover the entire range of sound audible to humans. In order to assign one of those 65,000-odd codes to a specific interval of sound, digital audio systems "sample" the music some 44,100 times a second. Put another way, each second of music is sliced into more than 44,000 "frames," and each frame has one of 65,536 codes assigned to it.

Digital's real claim to fame is its hands-off relationship with the music. Because it is a code that is recorded on the tape or disc, there is no danger of tape saturation or groove deformation. The music, therefore, need not be compressed to fit the medium. Mechanical limitations are nonexistent with the laser-read Compact Disc format. Since the laser pickup reads and transmits a code instead of trying to pick up and replicate a vibration, questions of stylus and tone-arm resonance, turntable rumble, and record warp are immaterial. Similarly, the hiss and ultimate deterioration of vinyl and tape are no longer considerations.

Compact Discs featuring jazz, rock, classical, country, and other kinds of music are now widely available.

Jane DiMenna

Compact Disc players are typically more expensive than conventional record players.

Tannenbaum/Sygma

Will CDs Fare Well in an LP World?

Should you wish to run out and buy a Compact Disc player, your current hi-fi (high-fidelity) component system will hold you in good stead as long as it is of fairly decent quality. Over time, digital audio recording will mandate the upgrading of other hi-fi components. But you can expect manufacturers to implant these changes gradually, so that by the time a reasonable amount of CD recordings are in circulation, most of the upgrading will have taken place.

The time frame for these changes is anyone's guess, because the long-range success of the Compact Disc system depends upon the widespread availability of music programming.

When the Compact Disc was introduced in autumn 1982, most of the digital recordings available were limited to classical music titles. Production capacity was insufficient: only three major record companies were issuing CDs, only five factories in the world could make them, and fewer than 100 recording studios had the necessary equipment to make digital master tapes. Because of this lack of resources, record companies devoted much of their CD production to classical music, on the assumption that the first Compact Disc buyers would be older, more affluent customers with a taste for the classics. Another reason for the early emphasis on symphonic works is this: the wide dynamic range and tonal textures of orchestral music show off the CDs' qualities best.

The software situation is much improved today. As of spring 1984, nearly a dozen CD plants were in production, and more than 125 studios were capable of making digital-master recordings. Every major record company in the world is now issuing CDs, and there is greater variety to the musical menu. Rock, jazz, pop, folk, country, ethnic, and other popular genres are available.

While the record companies and hi-fi component manufacturers continue making the very expensive investment in CD production facilities, you can rest assured that there'll be no shortage of vinyl LPs—or conventional turntables to play them on. That's because LPs and turntables are inexpensive compared to CD merchandise. And so long as production is limited, CD system manufacturing and selling costs will be higher than many people can afford or be willing to pay.

CDs Bring New Life to Old Recordings

Limited CD-making capacity means it will be some time before much of the music recorded for LPs in the pre-digital era (The Beatles, for instance) can be transferred to Compact Disc. So long as the artists remain in popular demand, you can expect most of this "older" music to make the transition to Compact Disc—provided the original "master" tapes or disc are in good condition. To prove the point, in May 1984, PolyGram Records planned to reissue some 90 vintage jazz albums on CD, some of these recorded as far back as the 1950's. In fact, some much older recordings, originally recorded monaurally (involving a single transmission path for sound recording) and issued on 78-rpm discs, might find new life on the Compact Disc format. A Swiss company named Studer-Revox has devised a process whereby digital technology is used to filter out the objectionable noise often found on old recordings. This means future generations will be able to hear once-famous artists like Arturo Toscanini and Louis Armstrong as they might have sounded in live (rather than recorded) performances □

Cellular Radio

by Duane L. Huff

Most of us regard a telephone on the dashboard or a "walkie-talkie" radio in the hand as a luxury or a status symbol rather than a necessity. But a new technology called cellular radio is now emerging to make high-quality, full-service mobile telephones and radio transceivers much more widely available. Mobile communications that have been for 20 years a curiosity will become commonplace—and soon enough a necessity for many.

Most market studies suggest that cellular-radio technology will increase the number of mobile telephone users tenfold within the first few years. Projections for the use of hand-held, portable radiotelephones based on cellular technology are equally optimistic. The result is likely to be one of the greatest changes in communications patterns since the invention of the telephone.

After years of effort to select the most efficient way to meet increasing demand for high-quality mobile communications, the Federal Communications Commission (FCC) has accepted the cellular-radio concept. Early in 1982, the agency began taking applications to operate such systems in the 90 largest U.S. markets, and it will permit service in other markets as well. No technical, legal, or regulatory uncertainties remain. Many other industrialized countries have already installed or are actively planning cellular-radio systems. Third World countries are also planning to use this technology in the absence of conventional wire or optical transmission facilities.

Efficient Use of Frequencies

Less than one-tenth of 1 percent of the 150 million vehicles now in use in the United States have mobile telephones. This is because, until now, mobile communications have been limited by a lack of radio channels. Each mobile telephone has required exclusive use of one channel in the available spectrum of frequencies. Cellular systems conquer this limitation by using low-power transmitters of limited range, so that

Top: An artist's view of overlapping cells depicts the cellular-radio concept. Below: This cellular phone system is more dependable than earlier mobile systems.

each channel can be used simultaneously in many different geographic areas called "cells." Taken together, these cells make up the total service area of a system.

This idea of frequency reuse is familiar. Hundreds of television stations across the U.S. that are out of one another's ranges, and therefore cannot interfere with one another, reuse television channels 2 through 13 in the VHF (very high frequency) band. Cellular-telephone systems reuse frequencies on a much smaller geographic scale. Instead of covering an entire service area with one transmitter with high power and an elevated antenna—the technology used for conventional mobile communications—cellular service relies on transmitters of moderate power distributed throughout a service area.

Each transmitter is only powerful enough to communicate with the radiotelephones in its "cell"—the area surrounding it. Thus, other transmitters in distant cells can use the same frequencies at the same time to communicate with mobile telephones in their cells. The only requirement is that transmitters in adjacent cells avoid using the same frequencies to keep from interfering with one another. Cellular radio's more efficient use of the frequency spectrum was the reason the FCC chose it as the standard technology for mobile public communications.

Searching for the Choicest Cell

Consider how a mobile telephone in a cellular system operates. As a vehicle in which a telephone call is in progress on a particular channel moves from one cell to another, the call is automatically transferred to the neighboring cell. There the call is conducted on different frequencies, to which both mobile and fixed transmitters and receivers are automatically assigned. This transfer should take place without the users' awareness; there is no break or perceptible difference in the communication. This function—the transfer of a conversation from one cell to another—is usually called "hand-off."

Although the system is simple in concept, it is complex in execution, demanding specialized computers and intricate sensors and controls. For example, one popular way to perform the hand-off function is to monitor the quality of

Top: Tom Norton; Bottom: Ray Stanek/
Illinois Bell Telephone Co.

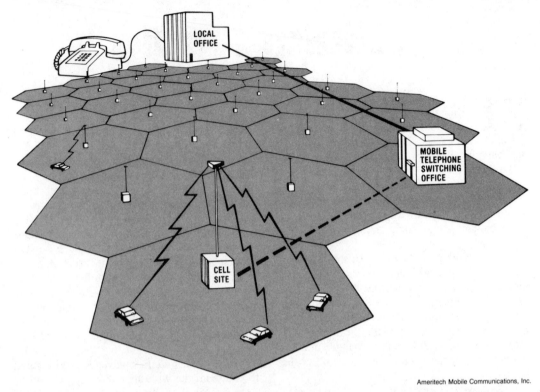

LOCAL OFFICE

MOBILE TELEPHONE SWITCHING OFFICE

CELL SITE

Ameritech Mobile Communications, Inc.

This diagram shows how a driver's call is instantly connected by radio to the nearest cell site, carried over telephone lines from the switching office to the local phone company office, and then transmitted to the recipient.

the signal for each telephone call received at a cell site. When the signal quality starts to degrade, nearby cells are automatically requested to measure the quality of the signal to determine which cell could better serve the call. When a better candidate is chosen, an idle radio channel at the new cell is selected, and a digital message on the "old" radio channel instructs the mobile equipment to tune to the new channel in the "new" cell. Control techniques have now been developed so that conversations continue without interruption. (If no new channel is available, the hand-off procedure can be briefly delayed, or the call can be transferred to an adjacent cell where transmission is adequate.)

Incoming and Outcoming Calls

When a customer's mobile equipment is turned on but not in use, it periodically scans special control channels broadcasting at each cell site in the system, selecting the strongest signal to monitor for incoming calls. When an incoming call is detected, or when the user wants to make a call, the user's equipment rescans the cells to be sure it is still working with the strongest signal (usually the nearest cell).

Calls to a mobile terminal are initiated just like normal telephone calls. A seven-digit number (ten digits if the mobile party is in a different area code) is dialed, and the conventional telephone network routes the call to the central computer of the mobile party's cellular system. From here the number is sent to all cell sites in the system, and each transmits the number of the called terminal on its control channel. When the called mobile terminal detects the incoming call, it selects the best cell with which to establish the communication and sends its identification back to the system through this cell. The system then uses a digital message to designate a frequency for the mobile terminal to use, the receiver tunes to this frequency, and the user is alerted to the incoming call by ringing.

A similar sequence is involved when a mobile user originates a call. The user first dials the desired number at a register in the mobile unit. The unit then chooses the best cell site and transmits the called number over the control channel. The main computer designates the frequency to be used to complete the call, and the mobile unit is automatically tuned to this same frequency.

All the control information between the terminal equipment and the system is redundantly coded and repeated to prevent errors. The entire exchange of control information takes place in a fraction of a second, and none is heard by the user. To the user, the sending and receiving of calls is routine—standard telephone procedure.

Cellular radio systems actually use two separate frequencies for each transmitter-receiver pair. This allows "full-duplex" service, which means that both parties in a conversation can talk at once. This contrasts with one-way systems such as citizens'-band, ship-to-shore, and private radio. In these, the same frequency is used for transmission and reception, so transmission requires pressing a "push-to-talk" button that disables the receiver.

Functions and Features

The computer-control systems are the key to efficient operation of cellular-radio communication. Computers at the mobile telephone switching office (MTSO) activate computers at the cell sites and, through these, the computers in the mobile units. Both cell sites and mobile units possess a degree of autonomy and a considerable amount of intelligence. In many cellular systems, the computer in the cell site performs

Left: A skyward view of the transmission-antenna tower of a cell-site building. Below: Computer-control systems at switching offices are highly autonomous.

most of the hand-off functions, including measuring signal quality and comparing results with other cell sites. It also performs diagnostic tests and reports trouble in any equipment at the cell site. The microprocessors at these sites must cooperate with the processors of the MTSO to form an efficient system. The microprocessors in mobile units have important error-detection functions, in addition to controlling some terminal hardware and keeping mobile units tuned to the strongest cell sites.

Special features for cellular systems are limited only by the ingenuity of the developers. These currently include options such as one- or two-digit speed calling; call-forwarding, message-waiting, and three-way calling; locking devices to prevent unauthorized use; data transmission; and automatic callback. In short, mobile telephones based on cellular technology offer essentially all the services available on conventional telephones.

How Radios Roam and Cells Divide

A service called "roaming" allows customers with mobile radio service in one cellular system to use their units in another system as they travel. Roaming works this way: When mobile equipment is taken into a new cellular system, the roaming feature identifies the equipment to the new system, which then informs the home system of the user's whereabouts. Calls coming into the home system for the user are rerouted to the correct "foreign" system. The roaming function includes computerized systems that exchange billing and payment information and permit operators to identify fraudulent users, illegal equipment, and bad credit risks. When cellular systems are fully developed, roaming—essentially complete mobile long-distance service—will be possible, with no special actions required by the user.

Though developers expect strong demand for cellular mobile services, start-up facilities can be small, with orderly growth possible as mobile traffic increases. When the traffic in one cell becomes greater than that cell can handle, the cell can be divided into smaller cells. Transmission power is reduced to avoid radio interference among the new cells, each of which serves about as many customers as the original cell.

Most start-up systems use cells with radii of 13 to 19 kilometers (8 to 12 miles). These systems can be split three times if necessary, with the minimum radius being 1.6 to 2.4 kilometers (1 to 1.5 miles). New cells can be created to meet demands for service outside the original area. In most systems, the maximum number of customers per cell, without excessive service delays, is 3,000.

This man can enjoy eating outdoors and still attend to important business calls with his hand-held cellular phone.

Motorola, Inc., Communications Sector, Schaumburg, IL

The Demand to Talk on the Move

The potential market for mobile communications was recognized soon after the invention of radio. The first major use of this potential came quickly, as radio was employed to communicate with moving vessels at sea for navigation and safety. Today the use of mobile radio has spread dramatically to include land vehicles—public safety and service vehicles and taxis, to name but three—as well as aircraft and ships. Indeed, more than 8 million mobile units are licensed for private radio service in the United States today. Over 8 million "citizens'-band" radio units are also in use in this country.

At the end of World War II, the demand for public mobile telephone service induced the Bell System to test-market the service in St. Louis, Missouri. Six frequencies were chosen, but the equipment was not sophisticated enough to prevent interference from adjacent channels. Thus, just three frequencies were actually used, so only a few customers could be served. One year later, the Bell System tested a public mobile system along the highway between New York City and Boston, Massachusetts. This system also proved troublesome because of erratic radio propagation: short-range radio was blanked out in some areas, while interfering conversations were carried for long distances.

Despite these difficulties, Bell gradually extended mobile telephone service to other markets, and by the late 1940's there were long waiting lists in several metropolitan areas. Indeed, over 25,000 people are on waiting lists for mobile telephones today, with many more would-be users not even bothering to add their names. The number of channels assigned to a typical mobile telephone system are so few that subscribers frequently cannot make or receive calls when they wish, so new customers obviously cannot be accommodated.

The First Market Test

Bell Laboratories planners were already looking forward to a more economical and efficient form of mobile telephone service when the first systems were being installed in the 1940's. The basic cellular concept of reusing frequencies was conceived as early as 1947. However, there was then no way to rapidly change the frequencies of transmitters and receivers and no techniques for managing the movement of calls among many cells.

Advances in electronic switching systems, low-cost frequency synthesizers, and high-

Ameritech Mobile Communications, Inc.

The demand to talk on the move extends to many professions and is creating a vast market for the technology.

capacity microprocessors stimulated the Bell System to suggest late in 1971 how a cellular system might be developed and operated. The FCC responded in 1974 by allocating a substantial block of frequencies and soliciting proposals for tests that would prove the feasibility of cellular systems. A year later, American Telephone and Telegraph Company (AT&T) applied to operate a trial cellular system in Chicago, Illinois. In March 1977, the FCC granted a license, and the first such service began in late 1978, growing quickly to its limit of 2,000 customers.

The Chicago system provided the first market test of this new technology. There were 136 voice channels spread across 10 cells, making possible "full-duplex" service—with few busy signals—to a 5,440-square-kilometer (2,100-square-mile) metropolitan area. Potential customers, randomly selected from lists of businesses in the Chicago area, were reached by direct mail, telephone, and personal visits. One out of eight companies contacted subscribed to the service, and most requested more than one mobile unit. Construction companies were especially receptive to mobile telephone service. Least receptive were food retailers, whose customers generally come to them.

Almost all customers cited time saved and convenience as major benefits of cellular radio. Many said that productivity increased 20 to 30 percent. An executive of a fast-food chain predicted his company could save 3.7 million liters (1 million gallons) of gasoline a year if all service vehicles were equipped with cellular radios. A trucking company attributed a 30 percent increase in business and a 15,140-liter (4,000-gallon) fuel saving to cellular service. A real estate executive claimed a 50 percent increase in productivity, and a representative of a waste-disposal company reduced his automobile mileage by 20 percent. Many executives said they extended their working days by using their telephones while commuting or traveling.

The Beginning of the Cellular Age

With testing complete, widespread use of cellular technology is beginning in the United States. Applications for two competing systems in each of the largest 90 cities in the U.S. have been filed with the FCC, and more than one-third have been granted, with systems construction underway.

Consumer rates for the new telephone service and for leasing the mobile equipment are being established by the marketplace. Prices will vary widely around the country, depending on individual companies' plans for recovering their large investments. However, the prices tested in Chicago in 1978–79 may typify what many companies will start off with. Basic service (a telephone number and connection to the system) with 120 minutes of free usage cost $25 a month, with "overtime" usage charged at 25¢ per minute. Lease of the mobile terminal cost either $45 or $60 per month, depending on the type of set. Thus, the minimum bill was either $70 or $85 per month. Most customers used much more than the 120 minutes of free time, so that the average bill was double the minimum—over $150.

Most experts predict that minimum costs for cellular service will run about $150 per month during the first few years of operation.

As the number of users increases, the cost of mobile equipment and services will decrease. If the FCC allocates more frequencies to cellular radio, construction of new cell sites can be delayed, decreasing costs further.

Cellular systems are also being developed overseas. The Nordic Mobile Telephone System, which began operating in 1981 in the four Scandinavian countries, had over 40,000 subscribers in the fall of 1983. The system was originally designed to accommodate only 96,000 subscribers, and expansion to a capacity of 200,000 is now being planned.

A cellular system designed by Nippon Telephone and Telegraph, in operation in Tokyo, Japan, for a number of years, has recently been extended to Osaka and Kyoto. The system currently serves more than 10,000 subscribers, with over 100,000 projected by 1987. Smaller systems are in use in Mexico, Qatar, the United Arab Emirates, Australia, Canada, Austria, and Singapore.

Advances on the Way

Meanwhile, even while cellular radio is being commercialized, the technology on which it is based is by no means standing still. Improvements such as digitized transmission will reduce the amount of the radio spectrum required for a single call, thus increasing the number of calls that can be processed in a cell. Mobile telephone equipment will shrink in size, and data terminals and printers will be available. Public telephones will be installed in trains, buses, and taxis; indeed, trials of public telephones in commercial airliners are already under way. Cellular telephone services will soon be coupled with answering and message services, encryption and scrambling for privacy, dictation services, data transmission, alarm calls, automatic callback, and all the other auxiliaries now available on conventional telephones.

Cellular services will be available in all major U.S. markets by 1986–1988. Thus, this technology will form the basis of a competitive marketplace, with unlimited opportunity to provide truly personal telecommunications to users on the move □

SELECTED READINGS

"Gold rush at the FCC" by John W. Dizard. *Fortune*, July 12, 1982.
"Car telephones: cellular technology promises more channels" by Danny Goodman. *Radio-Electronics*, February 1982.

Tomas Sennett

The Race to the Automatic Factory

by Gene Bylinsky

The armaments for the next industrial revolution are at hand. During the past few years, machine-tool makers—many in the U.S. and even more in Japan—have begun to supply so-called flexible manufacturing systems (FMS) that herald something very close to the workerless factory. The repercussions of the new technology go well beyond the predictable improvements it brings to productivity.

Flexible manufacturing systems complete a process of factory automation that began back in the 1950's. First came numerically controlled machine tools that performed their operations automatically according to coded instructions on paper or Mylar tape. Then came computer-aided design and computer-aided manufacturing (CAD/CAM), which replaced the drafting board with the CRT (cathode ray tube) screen and the numerical control tape with the computer.

New Systems Costly But Yield Big Savings

The new systems integrate all these elements. They consist of computer-controlled machining centers that sculpt complicated metal parts at high speed and with great reliability, robots that handle the parts, and remotely guided carts that

Above: A supervisor scans Deere & Company's giant flexible-automation tractor-assembly plant in Iowa.

deliver materials. The components are linked by electronic controls that dictate what will happen at each stage of the manufacturing sequence, even automatically replacing worn-out or broken drill bits and other implements.

Measured against some of the machinery they replace, flexible manufacturing systems seem expensive. A full-scale system, encompassing computer controls, five or more machining centers, and the accompanying transfer robots, can cost $25 million. Even a rudimentary system built around a single machine tool—say, a computer-controlled turning center—might cost about $325,000, while a conventional numerically controlled turning tool would cost only about $175,000.

But the direct comparison is a poor guide to the economies that flexible automation offers, even taking into account the phenomenal productivity gains and asset-utilization rates that come with virtually unmanned round-the-clock operation. Because an FMS can be instantly reprogrammed to make new parts or products, a single system can replace several different conventional machining lines, yielding huge savings in capital investment and plant size.

Economy of Scope

Flexible automation's greatest potential for radical change lies in its capacity to manufacture goods cheaply in small volumes. Since the era of Henry Ford, the unchallenged low-cost production system has been Detroit-style ''hard'' automation that stamps out look-alike parts in huge volume. There is little flexibility in hard automation's transfer lines, which get their name from the transfer of the product being worked on via a conveyor from one metalworking machine to another. But such mass production is shrinking in importance compared with ''batch production'' in lots of anywhere from several thousand to one.

Seventy-five percent of all machined parts today are produced in batches of 50 or fewer. Many assembled products, ranging from airplanes and tractors to office desks and large computers, are also made in batches. Even such stalwarts of inflexible mass production as the automakers are developing systems to produce more low-volume models to cater to small market segments.

In the past, batch manufacturing required machines dedicated to a single task. These machines had to be either rebuilt or replaced at the time of product change. Flexible manufacturing brings a degree of diversity to manufacturing never before available. Different products can be made on the same line at will. General Electric (GE), for instance, uses flexible automation to make 2,000 different versions of its basic

This new automatic machining system at a General Electric plant makes frames for traction motors.

General Electric

A robot rapidly seizes an electric-motor housing from a machining line at Fanuc Ltd.'s Fuji complex in Japan.

electric meter at its Somersworth, New Hampshire, plant.

The strategic implications for the manufacturer are truly staggering. Under hard automation the greatest economies were realized only at the most massive scales. But flexible automation makes similar economies available at a wide range of scales. A flexible automation system can turn out a small batch or even a single copy of a product as efficiently as a production line designed to turn out a million identical items. Enthusiasts of flexible automation refer to this capability as "economy of scope."

Japan Quick to Implement FMS

Flexible manufacturing systems were developed in the U.S. more than 10 years ago by Cincinnati Milacron, Kearney & Trecker, and White Consolidated. The U.S. remains a world leader in the technology: the major machine-tool builders are being joined by new suppliers with great financial resources and technical abilities, such as GE, Westinghouse, and Bendix. The most unusual new venture is General Motors' (GM's) linkup with Fanuc Ltd., Japan's leading robot maker, to form a new company—GMF Robotics. The joint venture has brought together GM's considerable capabilities in design and software and Fanuc's expertise in building and applying robot systems.

However, most of the action in flexible automation is now in Japan, and both American and European manufacturers will soon start feeling the pressure. Like many other manufacturing technologies conceived in the U.S.— among them numerically controlled machine tools and industrial robots—the FMS was greeted with a yawn by U.S. manufacturers. The Japanese have become the implementers par excellence of this new type of factory automation not because they are great technical innovators, which they admit they are not, but because they have moved fast in putting the new systems into their factories.

A Few Model Systems

A visitor to Japan these days finds the new manufacturing systems turning out parts for machine tools in Nagoya, electric motors near Mount Fuji, diesel cylinder blocks in Niigata, and many other products elsewhere. In most cases these plants run on three shifts. During the day skeleton crews work with the machines. At night the robots and the machines work alone.

In Fanuc Ltd.'s cavernous buildings near Mount Fuji, automatic machining centers and robots typically toil unattended through the night, with only subdued blue warning lights flashing as unmanned delivery carts move like ghostly messengers through the eerie semidarkness. This plant, one of two in the Fuji complex, makes parts for robots and machine tools (which are assembled manually, however). The machining operation is supervised at night by a single controller, who watches the machines on closed-circuit television. If something goes wrong, he can shut down that particular part of the operation and reroute the work around it.

The total cost of the plant was about $32 million, including the cost of 30 machining cells, which consist of computer-controlled machine tools loaded and unloaded by robots, along with materials-handling robots, monitors, and a programmable controller to orchestrate the operation. Fanuc estimates that it probably would have needed 10 times the capital investment for the same output with conventional manufacturing. It also would have needed 10 times its labor force of about 100. In this plant one employee supervises 10 machining cells; the other workers act as maintenance men and perform assembly. All in all, the plant is about five times as productive as its conventional counterpart would be.

But the most astonishing Japanese automated factory was started up by Yamazaki about 32 kilometers (20 miles) from its headquarters near Nagoya. The new plant's 65 computer-controlled machine tools and 34 robots are linked via a fiber-optic cable with the computerized design center back in headquarters. From there the flexible factory can be directed to manufacture the required types of parts—as well as to make the tools and fixtures to produce the parts—by entering into the computer's memory the names of various machine-tool models scheduled to be produced and then pressing a few buttons to get production going. The Yamazaki plant is the world's first automated factory to be run by telephone from headquarters.

The plant has workers, to be sure: 215 employees helping produce what would take 2,500 in a conventional factory. At maximum capacity the plant will be able to turn out about $230 million worth of machine tools a year. But production is so organized that sales can be reduced to $80 million a year, if need be, without laying off workers. The Yamazaki plant illustrates yet another aspect of economy of scope: with flexible automation, a manufacturer can economically shrink production capacity to match lower market demand.

Though Japanese machine-tool makers are the most ambitious installers of flexible automation, they are by no means alone. FMS is spreading throughout Japanese manufacturing, with Panasonic, Mitsubishi, and other consumer and industrial goods producers installing the new systems.

Automation in Iowa

Yet the Japanese do nothing that Westerners can't—marvelously efficient factories using the latest automated equipment exist in the United States and Europe. For example, Deere & Company's giant new tractor assembly plant in Waterloo, Iowa, was restructured from a gigantic, somewhat chaotic job shop into a world-class producer. Deere has poured $500 million into the complex. Chassis and engines received from

Computerized machine tools and robots at the Yamazaki plant in Japan are run by telephone from corporate headquarters.

Richard Kalvar/Magnum

sister plants are joined with tractor cabs and bodies made at Waterloo into gleaming mechanical behemoths. Almost all the materials handling at Waterloo is under computer control. Each part of subassembly—engine, transmission, wheels, and so on—is automatically assigned to a specific customized tractor ordered by a dealer; it is retrieved from storage and delivered automatically to the assembly line just when it is needed. Putting this "just-in-time" system to work, Deere has cut inventory in some areas by as much as 50 percent, saving millions of dollars.

Flexible automation allows Deere to build a tractor at least twice as fast as before. And it has given the company a new agility: Deere can now compete successfully not only against other big manufacturers but also against "shortliners" that make only one farm implement in higher volumes.

U.S. Manufacturers Reap Rewards of FMS

But if manufacturers think that they have to pour hundreds of millions of dollars into flexible automation to reap its rewards, they are mistaken. They can begin by acquiring smaller machining centers to modernize portions of their operations.

GE, Ford, and GM are among the manufacturers that have successfully revitalized old plants by installing new machinery. One ancient GE plant—the Erie, Pennsylvania, locomotive facility—is being transformed with a $300-million investment into an ultramodern automated factory—inside if not on the outside. Building a batch of locomotive frames formerly took about 70 skilled machine operators 16 days; the newly automated factory will turn out these frames in a day—untouched by human hands. The displaced workers are being retrained for other, more sophisticated jobs. As a general matter, in fact, flexible automation threatens employment less than might be supposed. The U.S. faces a shortage of skilled machinists for the rest of the decade, and automation of assembly, where semiskilled jobs predominate, will proceed much more slowly than automation of machining.

GM is also advancing. In October 1982 it installed its first flexible automation system, an Italian-built Comau system with three machining centers, at the Chevrolet Gear and Axle Division in Detroit, Michigan. Almost immediately GM discovered just how valuable the new manufacturing flexibility can be. When an out-

General Electric

A technician reprograms a General Electric flexible manufacturing system to alter production procedures.

side supplier failed to deliver a front-axle component up to quality standards, GM brought the job in-house. It designed and built the tooling for the component on the FMS in 10 weeks—a job that would have taken up to a year.

A Growing Interest

Awareness is spreading among U.S. manufacturers that time is running short to reorganize production processes and begin investment programs for new technology. Surveys show growing interest in new machine-tool purchases. Machine-tool builders, still shocked by recession, are cautious. But the new high-technology entrants are optimistic, indeed. They expect sales of accessories alone—robots, computer controls, and materials-handling systems—to soar to $30 billion worldwide by 1990, compared with only $4 billion in 1982. That would be great news for almost everyone—toolmakers, their stockholders, manufacturers, and the U.S. economy as a whole □

HERE COMES THE SMART CARD

by Martin Mayer

Right: Soldiers in the U.S. army must now carry smart-card identification to use post facilities at Fort Lee, Virginia. Each smart-card is embedded with a tiny microchip, as enlarged below.

Top: Shepard Sherbell/ Picture Group
Below: Courtesy Intel Corporation

Among the technological wonders of the 1970's was the magnetic-stripe plastic card. It allowed merchants to verify a cardholder's credit status instantly, and banks to replace tellers with cash-giving machines. Now the mag-stripe card has a challenger with a built-in brain: the smart card or something like it could change the way the world banks, pays bills, keeps records, and fights fraud.

The First Promoters

Adding a microchip to a card makes it possible to store data—such as the details of a purchase or banking transaction—right on the card, to be read or printed out later. The United States Defense Department has bought microchip cards, and devices to read them, as part of its search for improved personnel identification procedures. It has been said that the National Security Agency is doing something mysterious with the smart card as a way to control access to data. State and city governments are interested in microchip cards to keep records and reduce fraud in welfare and Medicaid programs.

Commercially, the companies most likely to promote a chip card are not banks—which have huge investments in present credit-card and check-processing systems—but those mulling plans to move in on the banks' territory. These include American Express, J. C. Penney (which has acquired the first U.S. commercial experiment with chip cards, a home-banking pilot project in North Dakota), and General Electric (which services a system of checkless, cashless, creditless payments at several gas stations in Phoenix, Arizona).

France Develops New Technology

Unlike mag-stripe cards, which have become a favorite of thieves and forgers, chip cards can be read only with exotic equipment and are all but impossible to forge. "Mag stripe is World War II technology," says Jerome Svigals, manager of growth planning for International Business Machines (IBM), and chairman of the working group on smart cards of the American National Standards Institute. IBM's first patents on a chip-bearing card go back to 1970. But the first marketable chip card—a *carte à mémoire,* or memory card—was developed in France in 1974 by Roland C. Moreno, then 29, a journalist and self-taught tinkerer.

At the start the chip card was distinctly a tinkerer's technology—a solution, as is still said of it, looking for a problem. Moreno conceived it mostly as a substitute for the magnetic stripe on a bank card, but more secure, because the password registered on the chip couldn't be copied. Coupled with an electronic cash register to capture the data about the customer and the transaction—and then with a communications device to report the transaction to a bank and to an automated clearinghouse at the end of the day—the chip card could substitute for checks at the point of sale. This is important for France, where checks are costly to process and increasingly popular even for small purchases.

Moreno persuaded a handful of smaller French banks that a microchip card would enable them to offer retail services comparable to those of bigger banks, with much smaller upfront costs and greater flexibility. But the sale he made that mattered most was to the state-owned PTT (the postal and telecommunications systems). PTT was up to its eyeballs in videotex technology and was projecting a future in which every home would have a French-made Minitel terminal with screen and keyboard. This setup would allow the householder to tap various computer data bases to be transmitted by PTT (starting with the telephone directory, which would no longer be printed). (A videotext system allows you to connect your television, via the telephone line, to data bases in order to send and receive information.)

With Moreno's smart card as a secure device for granting customers access to the system and as a personal record-keeping tool, the French could proceed more rapidly to offer mail-order purchasing, home banking, airline ticketing, electronic mail, and other services through Minitel terminals. PTT also saw in the smart card a way to replace coin-operated public telephones and simplify billing.

Interest Grows

With PTT on board, it was no great trick for Moreno to interest CII-Honeywell Bull, the Franco-American venture encouraged by Charles de Gaulle (the late president of France) in his search for a plausible French challenge to IBM. Honeywell Bull saw that Moreno's memory card could be made intelligent by adding a microprocessor chip—a tiny computer—to the memory chip. In 1976 Honeywell Bull licensed the Moreno patents, blended in some patents of its own, and began to develop CP8, which it calls a "portable computer for the Eighties."

Others followed. The French government, as part of its plan for a high-tech future, set up experiments involving all French banks and the smart-card devices of three major manufacturers. Honeywell Bull had its CP8; Flonic, a division of the petroleum equipment–maker Schlumberger, had a less programmable card and a more complicated reader; and Philips of Holland had a card with two chips. For the pilot projects, the manufacturers supplied cards, readers, and registers—120,000 cards and 650 terminals—to be used in shops and supermarkets in three test markets: downtown Lyons, Blois, and Caen. Because French shoppers are accustomed to a three-day float (transfer period) on their checks, the smart-card systems were programmed to debit the customer's account three days after a purchase.

U.S. Explores Potential

The smart card came to public attention in the U.S. through the enthusiasm of Arlen Richard Lessin, a young communications consultant, who came upon it in an obscure booth at a trade show in Los Angeles, California, in late 1980. He then began talking it up as something that was going to revolutionize banking.

But work was already being done on chip cards in U.S. corporate labs. IBM knew all about Moreno's invention through its French subsidiary. Honeywell Bull was part American from the start. American Telephone and Telegraph Company (AT&T) was up on the activities of the French PTT, and Bell Laboratories was fooling around with a chip card. Thomas H. Wood, Bank of America vice-president for financial services planning and research, remembers he began tracking the smart card in 1979.

Probably the most thorough exploration of microchip cards in the U.S. was done in 1982 by Bank of America, Intel, and Malco Plastics, in cooperation with each other. A Maryland subsidiary of Britain's Thorn-EMI, Malco is by far the largest U.S. producer of mag-stripe cards, with an output of 200 million a year. "Smart card," says Malco's president, Larry Linden, explaining his company's interest, "represents an incredible opportunity for the first guy who learns to make it cheaply, and an unbelievable threat for everybody else." The decision by all three companies was: not yet— let someone else solve the problems of reliability, manufacturability, and durability. All three keep up with the state of the art, attend technical meetings, and wait in the wings.

Yet people in the U.S. who run into smart cards for the first time immediately begin dramatizing what can be done with them. Because the microprocessor chip can generate its own code to communicate with computers, the smart card offers a highly secure system for access not only to banking transactions but to data bases and buildings.

In mid-May 1983 at Fort Lee, Virginia, the Defense Department personnel policy office began using a Philips smart card, imported from France, to test Rapids (the Realtime Automated Personnel ID System) as an improvement on Deers (the Defense Enrollment Eligibility Reporting System). Some 2,000 to 3,000 soldiers and their dependents were issued free smart cards as their basic ID for entry to base hospitals, base stores, recreation facilities, and so on.

Tracking Transactions

Smart cards were first put to use in the U.S. in 1982 for encryption purposes in an experiment in home banking using French videotex terminals. First Bank System of Minneapolis, Minnesota, put Minitel terminals in 250 farmhouses near Fargo, North Dakota. Farmers were living with party-line telephones. "They were concerned about security," says Stuart C. MacIntire, who ran this pilot program for the bank. He has since joined J. C. Penney, which acquired the home-banking system in February 1983. To ensure the privacy of the banking messages, First Bank System bought the Honeywell CP8 card and programmed it to encode communications between the farmers' terminals and the bank's computer. As a bonus, the memory chip gave the farmer an electronic ledger of his transactions, which he could flash up on his screen by inserting the card.

The ability of smart cards to keep a journal of transactions may turn out to be significant for purposes far removed from home banking. At the New York Stock Exchange, floor brokers write slips of paper saying they have bought so many shares from, or sold so many to, such and such a broker for this or that price, and then everybody argues later about who made what mistake. In a brave new world, brokers who have made a deal could insert their smart cards into a writer-reader with two activating buttons. The details of the sale—stock, price, quantity—would appear on a display screen. When both brokers hit their buttons the sale would be registered on their smart cards. If there was any dispute later, the data could be printed off the card. "It feeds into the development of an audit trail,'' says Erik Steiner, head of the product analysis laboratory at the Exchange, very approvingly. He adds, ''It's not on our list of things currently in progress, but it's on the list of 'may's.' ''

Great Potential—Why the Delay?

Some of the paper the smart card could eliminate is fraudulent paper. ''I can't imagine,'' the president of a Federal Reserve Bank said conversationally, ''why anyone still counterfeits currency. The banks look at currency. But nobody ever looks at food stamps. There are brokers in every slum who will give cash for food stamps.'' A chip card that can be loaded with an initial credit that works its way down as purchases are made would seem a natural match for the food stamp program. But in June 1983 the Department of Agriculture, looking to substitute some sort of credit card for food stamps, decided to stick with the familiar mag-stripe for a three-year test.

In recent years the U.S. has suffered from a growing flood of fraudulent documents. Counterfeit drivers' licenses, Social Security cards, foreign nationals' green cards—such items are available from your friendly street-corner, un-

derground economist about as easily as illegal drugs. A chip card is almost impossible to forge and can be activated only with a code number. A single card could serve as everything from birth certificate to medical-insurance record, driver's license to bank-account number. Each section of the card can be kept private by the microprocessor from any but specially authorized and equipped readers. If the card could be equipped with the latest in memory chips, there would be capacity left over to record purchases and payments.

What's holding it up? In part, the problem is the incompatibility of the three French card systems and a disagreement about standards between the French and the potential licensees and competitors elsewhere. The French manufacturers agree on the geography of the card, with the chip in the upper-left-hand corner, and on the location of the electrical contact points. But the Germans have been promoting a card with the contacts on the edge rather than on the face. The Americans feel they don't have enough information. In mid-May 1983 the French urged at a meeting of the International Standard Organization in California that standards for the placement of the chip and its electrical connec-

The ability of smart cards to keep records of transactions allows this man to do banking right in his own home.

Richard Kalvar/Magnum

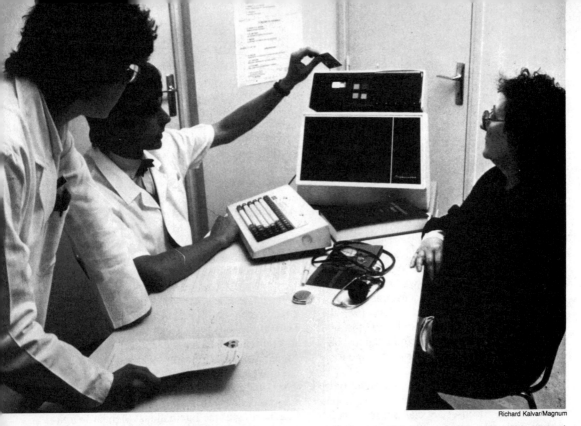

Richard Kalvar/Magnum

Doctors can review and update this patient's complete medical history, which is programmed onto a smart card.

tions be established right away. Jerome Svigals, the IBM expert who is chairman of the American working group on smart-card standards, wanted more time. His schedule called for standards to be adopted in steps—in late 1983 for the card, late 1984 for the electrical contacts, and 1985 for the information-processing systems.

It's not a schedule the French like. By 1985, says Jean Boggio, permanent representative to the International Association for Microcircuit Cards, an organization of more than 100 banks in 11 countries, "IBM will be ready to manufacture."

Memory Improvements on the Way

Beneath all this smoky argument is the real fire that the existing smart card is just not good enough. The typical 8K (8,192 bytes of memory) card in use in the Blois, Caen, and Lyons pilot programs contains a PROM (programmable read-only memory) chip that cannot be erased. (The symbol K represents 1,024. For example, 64K bytes of memory equals 64 × 1,024, or 65,536 bytes.) Even with a bare minimal entry for each transaction—nothing but the amount—there is room for at most 200 transactions. Discussing the possible use of a chip card at the New York Stock Exchange, Erik Steiner noted that an 8K PROM card wouldn't get a floor broker through a full day's work.

Two well-marked paths lead out of this corner: the memory chip can be made larger, or the memory can be made erasable. "I'm sure in five years we will have a million bits," says Jean Boggio. His countrymen do not agree with him. Hervé Nora, who runs the CP8 division of Honeywell Bull, says that 16K is no problem, but "over 16K there is a problem of size"—that is, the chip assembly becomes too big to fit conveniently on a card only 0.08 centimeter (0.03 inch) thick.

The Americans are more optimistic. A putative U.S. competitor called Smart Card Systems is using a patented process for bonding a microchip on a very thin film to produce a 64K card. Intel says that its 256K E (for erasable) PROM chip, already in production, can be shaved to fit into a bank card.

Whether by enlarged chip capacity or by programs that rewrite the memory, smart cards will soon be usable for many more transactions. U.S. banks, like French ones, may well be issuing cards that have both a chip inside and a magnetic stripe on the surface. But the banks will come to the smart card as reluctant followers, moving only in fear of losing their grip on the payments system, rather than as pioneers.

Banking Industry Holds Back

In 1981 Chase Manhattan moved prototype smart-card equipment to a closet outside the office of Robert Reffelt, the bank's bearded chief of advanced systems technology. The equipment is still there, looking rather like an exhibit in a museum, and the second generation of cards and readers has never emerged from Chase's laboratory. "The concept behind the smart card is valid," Reffelt says, "but we have so much momentum behind the technologies we have already adopted."

The banking industry cant is that the chip card is an "off-line" device—not continuously connected to a central computer—and the U.S. is an "on-line" society, in which transactions data move immediately to a central computer. But with the exception of the very high value items (average: $2 million) moved by the Federal Reserve's Fedwire, transactions in the U.S. banking system are not on-line and probably never will be. The books on interbank transactions are balanced once a day, and banks bring their accounts up to date only once, in the middle of the night. A merchant accepting smart cards forwards his point-of-sale information once a day to the bank, which meshes ideally with the way the banking system really operates.

"Banks want to do a credit check on every transaction, to make sure it's a good card," says David Aaronson, who designed the system used in General Electric's payments-processing venture. "That's the only reason you have to be on-line. It's crazy—but it's a psychology with them. Banks have a total lack of trust in the general public."

Reffelt of Chase would like to see the smart-card equipment come out of his closet. Bank executives looking at their on-line systems, he says, don't see "the communications cost, the people in the verification center, the computer cost, the terminal cost." By protecting their investments in an older technology, the banks could be in danger of losing their business.

Soon to Be Widespread

The smart card is in a sense the ultimate redistribution of processing chores, from a giant mainframe computer in a bank to millions of

Richard Kalvar/Magnum

During a purchase, money moves electronically from a customer's smart card to the bookstore's cash register.

cards in millions of wallets. As such it could make a significant difference in how people in the U.S. transact business, and in the efficiency of the payments system, that gigantic public utility that may cost as much as $40 billion a year to maintain. "In [a few] years," says Paul Finch, vice-president for systems research and development of Valley National Bank in Phoenix, Arizona, taking a prototype smart card from his wallet and throwing it on his desk, "everybody's going to be carrying one of these." If so, they'll be carrying a lot less of other things: government and private ID cards, credit cards, checks, and cash□

WILDLIFE

REVIEW
OF THE
YEAR

WILDLIFE

The wildlife scene in 1983 was mixed—with encouraging signs amid some persistent problems. Throughout, it became increasingly clear to scientists and interested observers alike that people's activity has a powerful effect on the survival and adaptability of many more species than ever before believed.

WILDLIFE COMEBACK

When people act to nurture a wildlife species, it can not only be saved, it can thrive. The Endangered Species Act became law in 1973, and the consensus is that the Act has been a mighty force for wildlife—not only for species actually listed as "endangered" but also for other plants and animals across the United States.

Prodded by the law, developers and government agencies have taken to heart the concept of species preservation. Their consideration of the welfare of wildlife early in the planning stages of highways, dams, airports, subdivisions, and other large projects has more than fulfilled the expectations of those who backed the Endangered Species Act. And this pervasive interest in preserving living space for wildlife spawns other efforts to turn the tide for beleaguered species.

There are several examples of good news on the wildlife scene:
• Elephant seals, down to about 20 in 1892, are now so plentiful (65,000) that some are moving from their preferred offshore islands to establish themselves on mainland beaches.
• Alligator populations in some prime Texas habitats have doubled since 1979, and as a result, the huge reptile has been taken off the state's endangered species list.
• The eastern brown pelican has recovered so well since the 1972 ban on the use of the pesticide DDT that its numbers have doubled in many parts of its range, and it is a candidate for "delisting."
• Musk-ox herds in Alaska now contain 1,500 animals, up from the several dozen that were transplanted from Greenland after the native herds had been killed off.

A mother moose keeps careful watch over her calf. After a 100-year absence, moose have returned to Maine.

Stephen J. Krasemann/DRK Photo

• Moose are also a success story. They have returned to Maine after a 100-year absence and are protected even though they are not on the endangered-species list.

Even a creature thought doomed—the California condor, now down to 19 individuals—is showing promising signs of survival. It has been successfully hatched in captivity, and this is often the first step toward successful repopulation of endangered birds in the wild. Captive-breeding programs have already succeeded on a limited basis with bald eagles, peregrine falcons, and whooping cranes (150 today, up from 35 in the 1930's), and are part of the reason the populations of these birds in the wild are growing. ■ While a lot of attention is rightfully focused on endangered species, it should not be overlooked that many native game species are also thriving, especially deer, doves, and turkeys.

Besides the Endangered Species Act, other recent programs have had a salutary impact on wildlife populations. More than 20 states allow taxpayers to earmark all or part of their tax refunds for wildlife conservation, and several states earmark a portion of sales-tax receipts for wildlife programs. Some states give farmers monetary incentives to take cropland out of production and provide much-needed habitat for wildlife. State activities against poachers are also helping wildlife, especially those species that are most vulnerable to killing by poachers, such as birds of prey, deer, and trophy animals—for example, bighorn sheep. In 1977 New Mexico started offering cash rewards for information leading to the conviction of poachers, and the program has been so successful that it has been adopted by more than 30 other states.

BUT STILL SOME BAD NEWS

The year did not bring all good news, however. Only 20 animals are left in the last herd of caribou in the Lower 48 states, and the mammal was put on the endangered-species list. ■ The blue pike, once plentiful in the Great Lakes, was declared extinct—a victim of pollution and overfishing. ■ The Florida panther is under increasing pressure from human encroachment, and the remaining several dozen individuals may not perpetuate the species to the end of the century. ■ And grizzly bears in the Lower 48 also seem to be in a no-win contest with people's need for more space for living and recreating in mountainous areas of the western U.S.

Many songbird species are also diminishing. About 250 of the 650 species that breed in the United States winter in Central America and South America. There, deforestation and development are destroying many habitats of the songbirds and of migratory shorebirds. When hillsides in Colombia, Ecuador, and Bolivia erode, there just are not going to be as many warblers and other songbirds returning to the United States each spring.

EFFECTS OF POLLUTION

Besides affecting the air we breathe and the water we drink, pollution kills a lot of animals. Industrial chemicals, pesticides, and heavy metals contaminate the meat of panfish all over the country. ■ Several million ducks and geese die each year from lead poisoning after eating pellets from waterfowlers' shotguns. ■ And the latest research shows that bald eagles—majestic carrion-eaters just recovering from pesticide poisoning—are now dying from eating lead-poisoned waterfowl.

The harmful effects of acid rain are also becoming more recognized. Acid rain is not only poisoning lakes across the country, but it is also leaching aluminum from the soil. Researchers have now found mountain streams with aluminum concentrations high enough to kill aquatic plants and fish.

LOSS OF HABITAT

Just as insidious and damaging to wildlife is the less heralded loss of wildlife habitat. Each year over 200,000 hectares (500,000 acres) of wetlands in the United States are drained or filled so that now only one-half of the wetlands areas that once existed in the country remain. In the minds of many, wetlands are merely swampy breeding grounds for mosquitoes. They are, in fact, key players in recharging water systems, protecting against floods, and purifying water as well as the nurseries of a wide variety of wildlife—from economically valuable shellfish to many prized sport fish. Although some legislators recognize the value of wetlands, others are faced with public pressure for development and are too often willing to let go of a wetland here and a wetland there that do not seem critical enough to save by themselves.

EFFECTS OF WEATHER

Weather continues to be a powerful factor in wildlife survival. Spring flooding, summer droughts, and fierce autumn storms combined to kill millions of animals. The thing to remember, however, is that weather devastations of wildlife populations are temporary. The animals that do survive are the strongest and fittest, and they almost always start populations back on their way to the maximum the habitat can carry.

One increasingly popular activity—the backyard feeding station—is circumventing normal weather-induced wildlife habits. Many birds, finding it easier to find food in winter, are cutting short their annual flight to warmer climates and expanding their winter ranges.

BOB STROHM

Learning to Play,

Playing to Learn

© Helen Williams/Photo Researchers,Inc.

by Barbara Ford

On the snowy banks of Wisconsin's Tomahawk River, a river otter paws another otter's face. Fiercely, the two animals wrestle, biting heads and necks. Then, otters sitting nearby join in the fray, pawing, biting, and rolling on the bank and into the water.

But curiously, amid the strife, the otters remain absolutely quiet. The only sound is the splash of the river. For scientists, that silence is the tip-off; it means the wrestling otters are playing.

"In free-ranging otters, play wrestling is the most common social activity," says Annamarie Beckel, who is studying river otters along the Tomahawk. "Even older animals wrestle a great deal." Beckel, a behavioral scientist at the University of Wisconsin, notes that although otters play in silence, they emit ear-piercing screams when they fight.

A Way of Learning

More and more, Beckel and other scientists who study animal behavior are finding that young mammals spend at least part of their waking hours playing games. In fact, social play is such an important part of many young mammals' lives that most have developed signals that clearly say: "Hey—let's play!" Apes grin or look through their legs, hippopotamuses splash water, harbor seals rest their heads on another seal's body, pandas somersault.

Young chimpanzees and some other pri-

While searching for fish, a pair of young bears takes time out for a playful wrestle.

mates spend more than half their waking hours playing games that range from fighting for fun to running and chasing each other. Captive dolphins will play with almost any objects tossed into their tanks. There's good evidence that birds play games, too.

What's it all about? Behavioral scientists suggest several explanations: exercise, practice for adult activities, social interaction, and assessment of other animals. Most of the reasons can be lumped together in a single word: learning. "Play helps animals become better monkeys or wolves or whatever creatures they are," says Frank E. Poirier, an anthropologist at Ohio State University.

Young animals are the most frequent players, and they are the ones most in need of motor, cognitive, and social skills. "Those species who need to learn the most, play the most," says Poirier. "In the higher primates, including humans, most if not all of the behavior is learned behavior. So they have to play a good deal in order to learn the behaviors they'll need in later life." The playful primates, he notes, also have the largest brains among mammals in proportion to their size so they can handle the cognitive processes needed for extensive play.

Play Behavior Varies

Recognizing play is easier than defining it, even for experts. "In my discussions with fellow behaviorists, it has sometimes taken 20 minutes to agree on what is meant by play," notes Dietland Müller-Schwarze, an animal behaviorist at the State University of New York at Syracuse. However, most scientists who study play agree on some indicators. For instance, when a lamb gambols—leaps into the air and simultaneously twists its body—it is exaggerating an action and doing it without any apparent goals. Both are elements of play. When animals interact and yet control aggressive actions, it is play, too. Animals rarely get hurt in play. Steven M. Herrero, a behavioral scientist at the University of Calgary in Alberta, Canada, has observed playing among young black bears, both in the wild and in captivity. He has never seen one bear injure another—even though the bears typically swat, bite, and claw each other. "They close their mouths, but they don't clamp down," says Herrero. "What they do is angle the paw so only the pad, not the claws, hits the other animal, and they withhold their strength." (Bear claws, unlike cat claws, are not retractable.)

Some animals play only by themselves. Two researchers with the University College of Wales, Trevor B. Poole and Jane Fish, have found that the house mouse will run and jump by itself. The Norway rat will, too. But the rats will also wrestle and box other rats, and if no rats are around, they will approach mice. The mice, however, will not even play with other mice.

Most mammals play with other members of their species, and many will approach another species for play, particularly if no partner of their own kind is available. One young chimpanzee observed by Jane Goodall in Africa regularly played with a baboon. Both animals were about the same size and age.

To humans, the most familiar signal an animal uses to invite play is probably the dog's "bow," in which the dog drops down on its forelegs. Marc Bekoff, a biologist with the University of Colorado, has seen the same signal in the coyote and the wolf, close relatives of the dog. Among coyotes, play is almost always preceded by a sign that an animal wants to play. Bekoff saw one female coyote try 39 times to engage another coyote in play. Without the right signal, she failed 39 times. On another occasion when she did make a bow, another coyote joined her in play.

Coyotes tend to be more aggressive and less playful than either dogs or wolves. Bekoff believes that the standoffish coyotes were simply afraid to enter into the proposed play without receiving some clear sign that it would not be dangerous. "In species in which aggression can easily occur, the play signal can be very important," he says.

Strongest Evidence From Controlled Studies

Some of the most important evidence that animals play to learn comes not from the wild, but from the laboratory. In the 1960's, Harry F. Harlow and a team at the Wisconsin Regional Primate Center at the University of Wisconsin raised rhesus monkeys. Some were kept with their mothers but were not allowed to play with other young monkeys. Others were given a substitute "mother"—a cloth-covered wire form—but were allowed to play with monkeys their own age. When the two groups became adults, the monkeys raised with the opportunity to play were able to get along with other monkeys. The animals that hadn't played were aggressive, nervous, and fearful.

Do animals learn in the wild through play, too? There is indirect evidence that they do.

Researchers find that primates play a variety of games, many similar to human games. Chimpanzees play tag, follow-the-leader, hide-and-seek, king-of-the-castle, and tug-of-war. They also box, wrestle, tickle, and play the most complex games with objects in the animal world. Goodall watched young chimpanzees in Africa tear branches off trees and use them to hit each other playfully.

Not all primates are playful, however. "I saw hardly any play in the langur monkeys I observed in Asia," says Poirier.

The environment clearly can have a strong influence. Sociologists John D. Baldwin and Janice J. Baldwin of the University of California at Santa Barbara found that squirrel monkeys in a naturalistic zoo, Monkey Jungle near Miami, Florida, played more than those in wild areas in Central and South America. At Monkey Jungle, food is provided. In the wild the monkeys spent almost all their time looking for food.

Vervet Games Divided by Sex

Choice of games can break down along sex lines, researchers are finding. They view it as more evidence that play is a learning experience. Jane L. Lancaster, a biological anthropologist from the University of Oklahoma who has studied vervet monkeys in Zambia, saw males fight and chase throughout their youth while females played another kind of game. "The young females are just obsessed with the newborn infants," says Lancaster. "The little males, however, aren't interested at all."

Vervet mothers allow young females to handle their infants, according to Lancaster, and animals as young as nine months—some still not weaned—stagger around trying to carry the infants. At first, Lancaster notes, all the young females are very awkward with the infants, but as time passes, they become more adept. In time, they are able to carry the baby in the correct position—that is, slung under the body

Top: Two orangutans frolic in the treetops. Bottom: Fox kits roughhouse at the entrance to their den.

© George Holton/Photo Researchers, Inc.

Young female vervet monkeys develop maternal skills by handling and learning to care for infants that belong to mothers in their group.

with the head near the nipples—and pacify it if it begins to cry or struggle. Lancaster compares this behavior to girls playing with babies. "Maternal behavior, like all other important behaviors, involves not just motivation but skill and practice," she says. "This early play allows the young to get the feel of an infant and how to handle it."

Play fighting allows young males to learn useful skills, too, Lancaster notes. "In many monkey societies, including vervets, females inherit their social position directly from their mother. So there's virtually no achievement possible for them. But most males leave the social system they grew up in and migrate to another social group at puberty. Their rank in

the new group depends very much on their ability to dominate, part of which has to do with fighting. The ability to become a successful fighter is very important to the male."

Not all animals divide play so neatly by sex. Roger L. Gentry, an animal behaviorist with the National Marine Mammal Laboratory, studies Steller sea lions and northern fur seals. "On the Pribilof Islands last summer, I saw female pups doing perfectly good blocking of pups smaller than themselves, just the way adult males do to females," he says. One explanation for this behavior, he suggests, is that in adulthood, eared seals occasionally switch sex roles. Gentry has seen two-week-old Steller sea lion pups defend turf the way adult males do. The pups rush toward each other, throw themselves on their bellies, hold their rear flippers back, and open their mouths all in an effort to protect territory that is only 1 to 2 meters (3 to 6 feet) wide.

Not Just for Youngsters

Adults play, too. Among the river otters Beckel watches along Wisconsin's Tomahawk River is an adult female who starts wrestling matches with her mate. The two play year round. Adult primates play, but their games are infrequent and low-key. In Puerto Rico, Judith A. Breuggeman, an animal behaviorist with Busch Gardens in Tampa, Florida, saw a female adult rhesus monkey hold up her hand and move it in circles while an infant wrestled with the hand. Breuggeman thinks such play is aimed at "redirecting the behavior of a young animal."

One common form of play she saw in adults occurred when mothers were trying to wean their young. One particularly determined female about a year old persisted in trying to nurse from her mother even though the youngster was well past the age of weaning. In an apparent effort at distraction, the mother would begin wrestling and chasing games every time the youngster made a move toward the nipple. Breuggeman reported that she never saw the mother play with her offspring under any other conditions.

On many occasions, Breuggeman discovered, adults would not join in the play of youngsters but would watch them, particularly if they were engaged in a very active game. One of these occasions occurred when some young monkeys were playing a follow-the-leader game of jumping off a tree branch into a temporary pond that had formed during the rainy season.

"They acted like kids on a diving board," says Breuggeman. Adults gathered around to watch, appearing fascinated by the leaping and splashing, but no older monkeys used the "diving board."

Birds Do It, Too

Most behavioral research on animal play has been directed at mammals for a very good reason: mammals are the most playful animals. But there is solid evidence that some birds play games. Dietland Müller-Schwarze watched Adélie penguins in Antarctica playing a curious sort of game. In the game, young birds gather on the beach in large numbers just before they leave the mainland for the ocean. Suddenly one bird will start to run, holding its flippers out. It stops, turns abruptly, and starts off in the other direction. Other birds copy the behavior. The birds also climb on blocks of ice on the beach. "What strikes you about this kind of play is that it is a sort of preadaptation to later needs," says Müller-Schwarze. "Penguins are the prey of the leopard seal once they are in the ocean, so they climb ice floes and run to escape them."

Parrots and ravens—both noted for vocal learning ability and relatively large brains—play more complicated games. Robert Fagen, an animal behaviorist with the University of Alaska, watched five ravens in the White Mountains of New Hampshire. The birds dove at one another and chased each other in a straight line, repeating these behaviors the length of a ridge. He was close enough to see that the birds frequently changed roles, one of the characteristics of social play.

Some of the researchers who study animal play believe the behavior not only benefits the individual but also the social group—and possibly the species as well. Play encourages flexibility and creativity, they argue, leading to innovations in behavior. If the innovations are learned by the animal's kin, a family group benefits. Eventually, a population or a species may learn the new behavior and become better able to cope with the environment. "A species' survival may ultimately depend upon allowing youngsters playful experimentation and environmental manipulation," says Frank Poirier.

This hypothesis, like most others about play, remains to be tested. The investigation of animal play is barely getting under way. Says Müller-Schwarze: "It's an exciting period for play research. We're on the threshold" □

One game that some immature Adélie penguins play may teach them techniques to escape from sea predators.

WOLVERINES

© Jack Couffler/Bruce Coleman, Inc.

by Maurice Hornocker

The wolverine has a public relations problem. From the 16th century to the 20th, the largest land-dwelling member of the weasel family has had the same image: a ravenous monster, a terror to all other beasts, a bitter enemy of most people. The wolverine, wrote naturalist Ernest Thompson Seton in the early 1900's, is a "whirling, shaggy mass with gleaming teeth and eyes, hot-breathed and ferocious," that chases coyotes, wolves, mountain lions, grizzlies, and black bears. Another writer once claimed that a wolverine killed a polar bear—an amazing feat indeed for an animal that seldom weighs more than 18 kilograms (40 pounds).

Separating Fact from Fiction

Accounts such as these have given the wolverine an almost supernatural character. This is somewhat understandable. The wolverine lives only in remote, inhospitable areas, and it is seen rarely by humans. Its strength and sometimes belligerent nature only add to its mysterious aura. The wolverine is a scavenger, and in Alaska it traditionally raided Eskimo animal traps and food caches often enough to earn the name *Kee-wa-har-kess*—the Evil One. Early European settlers in North America had other nicknames for the creature: Indian devil, glutton, devil bear, skunk bear. And because no scientist had ever objectively studied the animal, myths about the wolverine persisted.

Recently, however, new research conducted by a team of biologists has dispelled many of these myths. Our group intensively studied a population of wolverines in northwestern Montana from 1972 to 1977, and we have just completed our analysis of the results. Our effort was the first long-term research project on wolverines, using radio telemetry and other modern techniques. We learned that the animals are tremendously strong for their size, but that most popular accounts of this strength are vastly exaggerated. We learned that wolverines are sometimes truculent in their behavior, but most of the time they are shy and secretive. We learned that wolverines scavenge for food, carrion being an important part of their diet in winter, but this scavenging life-style is not a great deal different from some other species, such as bears. We learned that wolverines hunt other animals, but not anywhere near the extent many popular accounts claim.

The Montana Study Team

My interest in the wolverine began more than 20 years ago when I was a wildlife management student at the University of Montana. One of my professors, Phil Wright, was an expert on the mustelidae, the weasel family. He told me he believed the only healthy population of wolverines in the lower 48 states was in northwestern Montana. He said he thought research should be done there, but that studying wolverines would be difficult because the animals live in such remote wilderness, travel great distances, and are elusive.

The same species of wolverine lived historically in northern latitudes of both the Old and New Worlds. It has disappeared from much of that territory largely because of overtrapping and poisoning by people and loss of wild places. By the 1920's, one remnant population remained in northwestern Montana. This population, according to Wright, hung on and, with help from Canadian immigrants, actually increased between 1940 and 1960. What researchers consider to be ''good'' populations of wolverines occur today in western Canada and Alaska.

In the early 1970's, fresh from a study of mountain lions, I decided to look into the Montana wolverines. In the summer of 1972, I traveled to the Flathead country of northwestern Montana. I first sought out Ray Belston, an experienced fur trapper, who had contacted me

An extremely powerful animal armed with long, sharp claws, the wolverine is an adept climber.

years earlier volunteering information on wolverines. He had successfully trapped live wolverines for sale to zoos and animal parks.

Ray's home was located in the pines at the base of Columbia Mountain, a towering peak overlooking the Flathead Valley. Ray had trapped the South Fork of the Flathead and the Bob Marshall Wilderness for years. He had caught wolverines and knew what would be involved in studying them.

"It will be rough," he said. "Wolverines are tougher than any other animal. They eat steel, you know." I looked up, smiling at the remark. Then I saw he was serious. He related how wolverines he had captured had escaped metal cages by literally chewing through the metal. I mentally recalled hyenas in Africa chewing through our heavy metal leopard traps, from the outside, to get at baits we had placed inside for leopards. Can wolverines be that powerful, I wondered? Little did I realize what we were about to find out.

My next stop was to visit Dick Weckwerth, the area biologist for the Montana Department of Fish, Wildlife, and Parks. Dick, an old friend from my University of Montana days, was enthusiastic about the project. So was Dave Minister of the U.S. Forest Service. Dave arranged for funding from the Flathead National Forest, and with additional financial support, we launched an exploratory project that winter.

Ray had agreed to trap for us, and Ted Bailey, one of my former students, worked with him. They captured and marked four wolverines during a three-month period, and learned a great deal about what to do and what not to do. On the strength of their experience, we launched a full-scale project in the winter of 1973–74. We were funded by the National Science Foundation and several private groups, including the National Wildlife Federation.

Two longtime associates, both former students and experienced backcountry researchers, joined the project. Howard Hash had worked on elk in the Idaho backcountry and was an experienced mountain pilot. Gary Koehler had worked on various wilderness projects. Ray agreed to work another winter trapping, and later, Pedro Ramirez was to join us. Phil Wright agreed to analyze our biological material.

A Landmark Effort

Our project was designed to study a free-ranging population of wolverines by observing marked individuals—something that no other researchers had done. We set out to study the "dynamics" of the population: size, sex and age makeup, reproduction, behavior, and food habits. We also wanted to determine the movements and habitat use of individuals.

We worked along 160 kilometers (100 miles) of the South Fork of the Flathead River, south of Glacier National Park, in some of the most rugged country in the United States. Metal boxlike traps were baited with meat to capture the animals. Each wolverine captured was tranquilized to enable us to examine, weigh, measure, and mark it. We also placed a flexible collar containing a small radio transmitter around the neck of each wolverine. This small transmitter emitted a signal that we could pick up on our receivers as far as 16 kilometers (10 miles) away. In such rugged, mountainous country, it was impossible to keep track of animals on foot. Howard Hash and his airplane solved the problem by flying the area and tracking the radio signals.

We operated three different traplines within the 160-kilometer length of river drainage. One was near the mouth of the river, or "lower end," and was about 30 kilometers (18 miles) in length; another, from Spotted Bear, 80 kilometers (50 miles) upriver, was 50 kilometers (31 miles) long. Both of these lines were along logging or Forest Service roads and were traveled by snowmobiles. A third trapline, 40 kilometers (25 miles) long, was operated within the Bob Marshall Wilderness Area, in the upper reaches of the South Fork. Gary Koehler and his brother, Tim, ran this line entirely on snowshoes and skis. We normally trapped every day from the first of December to mid-April, depending on weather and snow conditions.

During the five years of the study, we traveled—by motor vehicle, snowmobile, skis, and snowshoes—a total of 60,670 kilometers (37,700 miles) tending these traplines. In chest-deep snow we had to couple two snowmobiles in tandem in order to power through.

The Koehler brothers' task in the Bob Marshall Wilderness was even more strenuous and demanding. They covered 19 to 32 kilometers (12 to 20 miles) daily on skis. More often than not, snow was either too deep or too sticky for easy travel. It was either bitterly cold or there was a blizzard howling.

Significant Travelers

Despite such conditions, we captured and marked 24 wolverines in the five years: 11

Right: A wolverine is fitted with a collar containing a radio transmitter that will allow scientists to keep tabs on where the animal goes. Below: By tracking wolverines for long distances, the scientists learned much about their behavior.

males and 13 females. Ten were recaptured 74 times. We placed radios on 20 individuals. We estimated a minimum of 20 wolverines stayed year round in our 1,300-square-kilometer (500-square-mile) study area based on our capture data, radiotelemetry, and observations of trails in the snow. We found that wolverines travel far more widely in winter than other species we had studied. Geographic barriers like mountain ranges and large rivers confine most species, especially in winter, but wolverines crossed the highest ranges in midwinter. Wolverines have very large feet and, in winter, these become grown over with hair, making perfect snowshoes.

The average yearly range of males was about 430 square kilometers (165 square miles), and for females 310 square kilometers (120 square miles)—a remarkable area for an animal this size. For the most part, wolverines remained within this large area, but from time to time both males and females traveled great distances and remained away for as long as 30 days. They always returned to the home area, though. Males traveled longer distances than females; the greatest distance traveled in a three-day period was 64 kilometers (40 miles) for males and 39 kilometers (24 miles) for females. These are straight-line distances—the actual kilometers traveled in such rugged country had to be much more. During a longer period, both radio-collared males and females traveled more than 160 kilometers (100 miles). In addition, markers from young males—obviously leaving our study area permanently for ranges elsewhere—were returned to us by hunters and trappers who found them as far as 160 kilometers from our study area. In light of these findings, there is no question that wolverines rank alongside the other major travelers among mammals.

Wolverines must travel far to find food. A scavenging life-style dictates long seasonal movements and a large home area. An inefficient hunter, the wolverine must depend on carrion, particularly in winter. It has learned to rob traplines, cabins, and food caches in some areas. However, none of our camps or cabins were bothered, and none of the people we contacted had experienced problems. Some sportsmen did report seeing wolverines carrying elk or deer bones, and trappers noted occasional losses to wolverines. But in Montana, such losses are insignificant.

Encounters Avoided

Wolverines were not territorial in our area and made no attempt to defend their home ranges, which overlapped between animals of the same and opposite sex. They scent-marked trees and shrubs and noted their presence by biting on trees, but apparently they did so to maintain a spacing in time, not area. It does not appear that strict territorialism would be an advantage to a species often dependent on carrion.

There are many advantages to a species living under such an orderly system, and it probably represents the ultimate in wolverine behavior. Each individual knows its area well, making the process of survival much easier. Our

Although wolverines eat a variety of foods, carrion is a vital part of their diet, especially during the winter.

© Tom McHugh/Photo Researchers, Inc.

Nicknamed the "devil bear" by early European settlers in North America, the wolverine is normally shy and secretive.

population in northwest Montana was not territorial, probably because of high mortality—too many residents were being removed by trappers for such a system to operate. Obviously, these trapped animals were replaced naturally—the population level remained the same—but the "excess" mortality caused instability within the population. Individuals simply did not have time to establish any orderly spacing system before they were removed and then replaced by new wolverines.

Our study area included other carnivores such as grizzly and black bears, cougars, coyotes, lynx, fishers, bobcats, weasels, and foxes, and we were interested to learn how the wolverine "fit in" with this community of predators. According to popular belief, the wolverine will chase everything out of its path. We found that this was not the case. The radio-collared wolverines avoided the larger, dangerous carnivores. Wolverines are, however, capable of displacing smaller scavengers in a direct confrontation, but not black and grizzly bears, two species with feeding habits similar to that of the wolverine.

In addition, there is no question that a mature mountain lion could kill a mature wolverine if it were hungry and the opportunity presented itself. After all, a big male wolverine weighs in the neighborhood of 18 kilograms (40 pounds); a big male cougar weighs about 80 kilograms (175 pounds). The wolverine can be very aggressive, and no doubt it does occasionally frighten other carnivores, even inexperienced young cougars and grizzlies. But it simply is no match for a mature individual of the bigger, powerful predators.

Tremendously Strong

Even so, people are the number one enemies of wolverines. Of the 18 known wolverine mortalities during our five-year study, 15 were caused by humans. Contrary to popular belief, the animals are highly susceptible to trapping because they travel widely and are readily attracted to baits. Many of the adults we captured were missing one or more toes and had broken teeth—possibly the result of being caught in traps. Wolverines are so powerful that they can and do pull their toes off to escape. Experienced trappers have no difficulty in catching wolverines in steel traps; holding them is a different story.

During our study, we saw evidence of the wolverine's tremendous strength. In the winter of 1973–74, we had tracked a big male several kilometers in fresh snow to a baited trap where he obviously had been caught. We found the steel jaws of the trap pulled from the frame—an incredible feat of strength.

On other occasions, wolverines escaped from the live traps Gary and Tim Koehler were tending in the Bob Marshall Wilderness Area. The two researchers lived in the "Bob" from early December until mid-April each year. One February morning, we heard Gary's voice crackling over our radio: "They're chewing out of the traps!" He related how two different wolverines had literally chewed through the heavy chain-link fencing wire we had used in constructing the traps. However, the wolverines were not actually "eating" the wire but were grasping it in their jaws and eventually breaking it.

New Insights

Wolverines in our area ate a variety of different food items, but carrion was vital to them in winter. They possess an extremely keen sense of smell and can locate food under deep snow. Different individuals during the course of our study came directly to our baits. By backtracking, we learned that some wolverines came directly to the baits from distances of more than 3 kilometers (2 miles).

We also tracked wolverines through the snow for many kilometers, and they appeared to wander around looking for something to eat rather than to prey on another animal. We saw no evidence that wolverines killed game animals, nor did it appear they even tried. This

Even in the most rugged mountainous terrain, wolverines are active travelers all through the year.

contrasts with observations in Scandinavia, where wolverines prey on reindeer and, in some areas, on domestic sheep. We did see where wolverines killed small mammals such as ground squirrels. In summer they ate a variety of foods including wild berries.

The biggest disappointment in our work was our failure to capture intact family groups of the animals. Young wolverines are born in our area in late winter and early spring. By radio-tracking different pregnant females, we knew when they denned and had young—they all confined themselves to one small area for several weeks. We were reluctant to disturb the den, fearing the female might abandon her offspring.

Despite that setback, some important factors in wolverine ecology have become evident as a result of our work. We now know that wilderness or remote country where human activity is limited is essential to wolverine populations. Our study area is used by humans during the summer and fall months, but in winter and early spring it is "snowbound." The whole area is bordered by rugged, relatively inaccessible mountains, and the wolverines moved to these high, cooler areas in summer when people began to move into the region.

Our radio tracking showed that individual wolverines travel great distances, often crossing major mountain ranges in a relatively short period of time. This can give a false impression of abundance. In fact, there never will be, under any conditions, large numbers of wolverines in the lower 48 states. They just naturally are a very scarce animal. But with stringent regulations now in effect and with adequate wilderness habitat, wolverine populations should be secure in Montana. And there is evidence wolverines are reappearing in a number of other western states where they formerly ranged, including Washington, Oregon, Wyoming, Colorado, and California. So, in general, the future looks bright for this remarkable member of the weasel family.

From our study, we gained tremendous respect for the animal's tenacity, determination, strength, and endurance. Flying over the highest snowswept peaks of our study area in midwinter and seeing a set of wolverine tracks going straight up and over, Howard Hash and I could only look at each other and marvel. No other animal I know of is capable of that. It's small wonder the Eskimos and Indians regarded it as supernatural □

Michael Habicht/Animals Animals

Star-nosed Moles

by Terry L. Yates

Millions of years before people walked the surface of the earth, moles were tunneling beneath it. Today, the meandering surface runways made by these animals in their constant search for food are a familiar sight to many Americans, and yet few people have more than a vague understanding of the creatures that inhabit them. Of the many mole species living today, none is more bizarre than the semiaquatic star-nosed mole.

A Unique Nose

The star-nosed mole gets its name from a ring of 22 fleshy appendages on the end of its nose. No other mammal has such a structure. Each tentacle contains highly sensitive tactile organs, called Eimer organs, which function as mechanoreceptors (neural organs that respond to mechanical stimuli). This creature makes great efforts to keep its nose clean, even occasionally dunking the nose in water to shake off dirt, and

its nose is constantly in motion. Star-nosed moles often test potential food items by exploring them with their noses. With the aid of this remarkable structure and countless vibrissae (whiskers) on their faces, hands, feet, and tails, the moles feel their way through bogs and marshes. Their tiny eyes provide little assistance in this task because they are useless except for light detection.

The star-nosed mole occurs throughout much of the northeastern United States and eastern Canada. In the northern part of its range, it is found from Manitoba and Minnesota to as far northeast as Labrador and Nova Scotia. The species ranges southwestward through much of Wisconsin, northern Indiana, and Ohio; along the Atlantic coast as far south as southeastern Georgia; and in the Appalachian Mountains to eastern Tennessee and western North Carolina. The species becomes rare in the more southern portions of its range but is often abundant in northern areas where it is free from competition with other mole species. Although most of its relatives are solitary species, the star-nosed mole appears to be gregarious or colonial.

Adaptations for a Semiaquatic Life

The star-nosed mole is the only species of mole that is semiaquatic. Part of its time is spent burrowing beneath the ground, and part is spent in the water. Its tunnels are often found near marshy areas of streams and frequently open directly into water. In winter, star-nosed moles often burrow beneath the ice in frozen lakes and streams. They are frequently captured in minnow and muskrat traps set beneath the surface of the water.

Probably as a result of this penchant for the aquatic, the star-nosed mole has a number of features not shared by other moles. Its fur, which ranges in color from black to brown, is longer and coarser than that of most other mole species and sheds water easily. Its hind feet are not webbed, but they are longer and wider than those of other species and obviously useful for swimming. Even the star-nosed mole's tail seems to be adapted for life in the water. It is considerably longer than those of other fossorial (underground) moles and may serve as a rudder. During winter and spring the tail swells up with an increased deposition of fat. Although its exact function is not known, the extra fat may provide a reservoir of energy during the breeding season.

Efficient Burrowers

If these features dramatically differentiate the star-nosed mole from other North American moles, its method of burrowing clearly links it to the others and removes any doubt that it is part of the same family—Talpidae. Most burrowing mammals, such as pocket gophers, hold their forefeet beneath their bodies when they dig; moles dig with their forelimbs held to the side. They can do this because their pectoral

Dwight R. Kuhn/Bruce Coleman

Ever hungry, star-nosed moles spend most of their time searching for food. With such poor vision, the mole uses its sensitive, tentacled nose to help it locate prey.

Massive muscles and powerful claws make the star-nosed mole an efficient burrower. The creature's shallow tunnels are often visible on lawns, golf courses, and pastures.

Dwight R. Kuhn

girdles are drastically modified and their pelvic girdles, by comparison, are relatively narrow and unmodified. One of the most striking features of the mole pectoral girdle is a joint between the humerus and clavicle. In most mammals, this joint is between the clavicle and scapula. The star-nosed mole, like other moles, has a long scapula, which articulates directly with the humerus. The humerus is a massive rectangular bone, very different from that of most other mammals, and it provides a large surface area to which the well-developed digging musculature is attached.

The massive muscles used by moles for digging tend to obscure the neck and cause the forepaws to be rotated parallel to the body. Mole forepaws are usually spade-shaped, and the palms are as broad as they are long. The fingers have powerful claws, and the first four fingers have three flat, triangular flaps on the outer edges. The result of all of these modifications has been that moles have an increased mechanical advantage and efficiency in burrowing, even though they are not bigger than other burrowing animals.

Two Types of Tunnels

Star-nosed moles build tunnels similar to those of other mole species but typically construct them in poorly drained soils near marshes or streams, areas that other moles would shun. As such, their tunneling usually does not bring them into conflict with humans. The star-nosed mole generally constructs two types of tunnels,

one shallow and one deeper. The shallow kind, which is usually dug in great numbers, is the result of the animal's constant search for food. This type of tunnel is familiar to many Americans because it is usually only centimeters below the ground and forms a visible ridge of dirt, often extending across pastures, lawns, and golf courses. Most moles are pretty particular about their tunnels and won't tolerate even the slightest opening, but star-nosed moles are less fastidious that way. In extremely wet areas I have found active tunnels that were almost entirely exposed—they were little more than deep runways covered with grass.

At the beginning of my research on star-nosed moles, more than 10 years ago, I found that they were frequently rebuilding the shallow tunnels in the same spots year after year, even though the tunnels were often completely destroyed by rain and other natural forces. In an effort to determine how a blind animal could perform such a feat, I selectively destroyed sections of tunnel and marked precisely where each had been. In every case the moles were able to rebuild the old tunnel system. Examining these areas further, I discovered that the soil was less compacted in areas where old tunnels had been and realized that the animals were simply following, by touch, the path of least resistance. Since that time I have been able to divert individuals a short distance from their normal tunnels by artificially compacting certain areas of soil. This tactic has proved crucial to our ability to capture live moles for research.

The other kind of tunnel the star-nosed mole constructs is deeper, perhaps a few meters below the surface, and more permanent. This system of tunnels is used for resting, rearing young, and foraging for food when the surface of the ground is frozen during the winter. These deep tunnels form the classic molehills out of which mountains are made. The moles can't build them simply by pushing into the soil, as they do with the surface tunnels. To build the deeper tunnels, the moles have to excavate, bringing soil to the surface and depositing it in a mound.

Breeding and Feeding

Star-nosed moles are not picky about their surface tunnels, but they are very choosy about where they dig their nests and pick areas that are above high-water mark and near abundant food. Usually they find some natural rise and create the nest by enlarging a section of tunnel and filling it with dry grass and leaves. Gardeners occasionally unearth nests because star-nosed moles sometimes choose manure piles or compost heaps as nesting places.

The moles breed once a year, and peak breeding time varies geographically. Litters range from three to seven young; they have been recorded from late March to early August. Although baby moles are naked at birth, the feet are well formed, the vibrissae on the snout are 3 to 6 millimeters (0.1 to 0.2 inch) long, and the star is evident although enclosed in a thin membrane. The young develop rapidly and apparently can leave the nest when they are about four weeks old.

One cannot help but be impressed with the amount of food these moles eat daily. They have voracious appetites and consume 50 percent or more of their weight every day. I kept three of them alive in the laboratory for two days, and they consumed $17 worth of earthworms. Their diet appears to vary by locality, but these animals seem to eat certain invertebrates and little else; they show little interest in vegetable matter. In many areas they seem to covet white grubs and earthworms, but those living near large bodies of water prefer aquatic annelids (worms) and insects. These feeding habits probably reflect what is available.

Unlike other kinds of moles, the star-nosed mole is semi-aquatic. Here, after capturing a large worm, the mole will swim to shore to feast on its prey.

Dwight R. Kuhn

sity has suggested that large fish may occasionally grab a star-nosed mole, and house cats take large numbers.

Evolutionary Split Occurred Long Ago

My general interest in the natural history and biology of this unique mammal has led me more recently to wonder about its evolution. The star-nosed mole's pectoral girdle shows it to be related to other moles; but what, I wondered, had produced the extensive morphological and presumably genetic divergence of this species? Either it had not shared a common ancestor with other mole species for an extremely long time, or it had evolved at a more rapid rate than the other North American species.

Based on extensive studies I conducted and evidence from fossil records, I estimate that star-nosed moles diverged from other moles perhaps 30 million years ago, and somewhere in Eurasia. Ancestors of the other fossorial moles appeared in North America during the Miocene about 25 million years ago, long after their first appearance in Europe. Thus, the arrival of star-nosed moles in North America appears to be the result of an invasion separate from the one that brought other North American species here from the Old World. From a geological standpoint, their arrival may be relatively recent. Fossils of star-nosed moles dating from the middle or late Pliocene, 3 or 4 million years ago, have recently been discovered in Poland, whereas the oldest North American fossils of this genus date back 700,000 years. What all of this strongly suggests is that star-nosed moles didn't come to North America until the late Pliocene or early Pleistocene, when they crossed via the Bering Strait land bridge.

We don't know why they became extinct in the Old World. Perhaps they were unable to compete successfully with the fully aquatic desmans (Old World water moles) or were unable to adapt to changing climatic conditions or other factors. Whatever the reason, star-nosed moles obviously have lived, and evolved, apart from other moles in their family for a long time. Virtually untroubled by competition in the New World, these immigrants have managed to settle down comfortably in the unlikeliest of mole habitats □

Whatever the food source, the mole's high metabolic rate, coupled with the large amounts of energy needed for tunnel construction, makes it constantly hungry. Deprived of food, star-nosed moles will starve to death in a matter of hours. I have also found that if temperatures drop very low and food is in short supply, these mammals have difficulty regulating their body temperatures. That such a creature exists so far north, and even thrives there, is amazing. The moles are active year-round and appear to spend more time in the water in winter than they do in the summer.

They are active day and night and spend more time on the surface of the ground than most mole species. I have frequently trapped this species in above-ground runway traps set for small rodents or under overhanging banks or in containers buried in the ground. They are bold creatures, which would be admirable if they weren't so blind. As it is, their incautiousness makes them prey to a great many animals, including great horned owls, screech owls, red-tailed hawks, foxes, skunks, weasels, and snakes. William J. Hamilton of Cornell University

The Fish-in-the-BOX

by Ronald Thresher

The Japanese call it *shimaumasizume*. Nearly 20 centimeters (8 inches) long when mature, it is as supple as a rock. Indeed, a diver could hardly be faulted for mistaking the thornback cowfish for a stone rather than a short, squat, lumpy fish.

Even when it moves, the animal is hardly striking. Drab brown mottled with olive, it spends most daylight hours plodding from place to place, peering into crevices and around rocks in search of worms and crustaceans. In the flashing beauty, sweeping submarine vistas and piscine elegance of the coral reef, the cowfish is one of nature's most comical oddities.

Armadillos of the Sea

And yet, the thornback cowfish and its relatives are a succesful lot. The family, Ostraciidae, encompasses only about two dozen forms, but they inhabit all shallow tropical seas—ubiquitous though low-key elements of the coral-reef fish community. Their success derives mainly from their peculiar anatomy.

Cowfishes and trunkfishes—the common names most frequently given the creatures—are encased in bony, boxlike coverings rather than flexible, scaly fish skin. The rigid skeleton is pierced only for gills, eyes, mouth, anus, and fins. Cowfishes are the marine analogs of pangolins and armadillos—animals that have traded grace and speed for the benefit of having armor-plated security.

But cowfishes have taken this anomaly further than their terrestrial counterparts. The fishes' armor is rarely smooth. In most species, sharp, permanently erected spines protrude

Above and opposite: R.E. Thresher

Above: A female cowfish before the mating season. Opposite: A male cowfish displays brilliant courting colors.

somewhere on the body. Several forms have a pair of spines projecting from the top of the head, just above the eyes, reminiscent of cow horns. Hence the common name.

Shells and spines are not cowfishes' sole means of defense, however. Numerous minute glands that secrete a noxious-tasting chemical—ostracitoxin—are scattered over the skin. Even in small doses oxtracitoxin is highly poisonous to most marine animals. Any piscivore (fish-eater) foolish enough to attack a cowfish deserves its misery. One could hardly design a less palatable package of flesh.

Interested Observer Gathers Data

Until recently, little was known about the behavior of cowfishes. Diving biologists around the world had noted their curious shapes and slow-paddling swimming style but had gathered no data on their social patterns or methods of reproduction. This dearth of information is understandable, given the often bewildering variety of fishes on the reef and the scarcity of people trained to study them. It wasn't until the mid-1970's that someone took cowfishes seriously, and he is a biologist by vocation.

Jack T. Moyer is Director of the Tatsuo Tanaka Memorial Biological Station in Japan and watches birds and fishes in his spare time. In 1976 he started observing a population of the thornback cowfish, *Lactoria fornasini*, living just off the volcanic island of Miyake-jima. Lacking academic pressure to produce exciting scientific papers, Jack takes the old-fashioned approach and chooses his subjects because they interest him. He watches them for hours at a time to learn as much as possible. It is not uncommon for him to make three or four dives a day for weeks on end, especially during summer vacation. Moyer's information on cowfish behavior is unique, and since I too had observed reproductive biology in marine fishes for several years, I was eager to work with Jack.

Cowfish Society

Three species of cowfishes live off Miyake-jima, but only the thornback is abundant. It lives close to shore in shallow water, so we could easily trace its social organization.

Moreover, the animal is relatively placid, which doubtless reflects its immunity from predators. A diver blowing bubbles and thrashing across the reef must be an imposing sight to a 20-centimeter (8-inch)-long fish. It is not surprising that most sea creatures flee, at least initially. The thornback, however, blithely goes about its business, unconcerned about noisy apparitions following it.

Cowfish society, like that of many other fishes, is based on small harems dominated and defended by one male—defended against other males, that is. Throughout the day each female swims ponderously about her home range—an area of large volcanic rocks covered with thick algae—searching out delectables. Few other fishes bother her, but when pressed, she simply disappears under a rock or coral outcrop and waits for the nuisance to go away.

Not quite so for the male. He too eats during the day, but mainly on the run (or swim). Moving from female to female, he spends a short time with each, confirming their presence and watching for rivals that might threaten his exclusive rights to a mate. The similarities between the sexes (males are only slightly larger than females and color patterns are the same) complicate his patrol. At first glance, a male doesn't know an intruder's sex and hence must be a little unsure of the proper reaction. Still, during the day, life for the male is not hectic. Territorial borders are stable, females munch contentedly, and all is quiet.

The Spawning Transformation

Things change at dusk. Along with most of the other reef fishes, the thornback spawns nearly every evening from April to October. These hours are critical to a male. He must find each of his mates and, if they are ready, spawn with them. It is also the time when rival males will most likely invade his territory to try to steal a spawning.

The problem of locating both mates and rivals is magnified by fading light, rendering the brownish fishes difficult to see. Perhaps this is why the spawning male undergoes a transformation. Over his drab daytime hue he develops electric-blue lines on his head and body, advertising to females his presence and willingness to spawn, as well as possibly warning off other males.

In this courting livery the male dashes about his territory, paddling paired fins at full speed to propel his clumsy shape. He scoots across clearings and up to the peaks of small boulders. There he hesitates, peers over the edge, and slowly scans the terrain like an Indian looking for the cavalry. Nothing! He dashes over the top and up the next boulder to search again. He keeps up this stop-and-go hunting until he finds either a female or a rival.

Each female, meanwhile, signals her readiness to spawn by swimming to a preferred rock, night after night, and waiting for her mate. When the male finally sees a member of his harem at the rendezvous site, he rushes up to her, colors flashing, and races around her in tight circles. The female's colors also brighten, but they are never quite as brilliant as the male's mating colors.

A male cowfish tags behind a female as the pair ascends from the deep to shallower waters to spawn.

R.E. Thresher

The pair immediately begin their spawning ascent. With the female leading, they slowly swim upward in a long, sloping incline. A meter or so below the surface, the female halts and the male comes up beside her. Abruptly the pair face in opposite directions; only their tightly curled tails remain close together.

Rigidly holding this position, the male begins to hum—a soft, low-pitched thrumming that even a diver, hovering quietly, can hear. The male hums for a few seconds, then stops. Just as he ceases, the pair spawn, shedding eggs and milt (secretion from the reproductive glands of male fishes) in a milky cloud between them. Buoyant round eggs tumble into the water column, are fertilized, and quickly disperse in the currents to begin their development. The adults remain in place a few moments more, then dash to the bottom. The female heads for cover, seeking a place to spend the night; the male searches for another partner. With as many as five females in his territory, he has a busy evening ahead.

Should the male spy a rival rather than a waiting female, his behavior changes drastically. Seeing the intruder, who is also bedecked in spawning colors, the dominant male rushes furiously. The invader turns off his brilliant colors and flees across the bottom, weaving around the rocks with the first male close behind. The chase continues until the intruder either leaves the territory or eludes his pursuer. There's not much a dominant male could do even if he caught a rival, short of ramming it a few times and bumping it out of his domain. The fish's hard skeleton makes severe damage unlikely.

Mating Habits Unlike Other Reef Fishes

The thornback cowfish's social system and reproduction differ from those of other reef fishes. Most other reef fishes have harems, spawn at dusk, and ascend from the bottom in pairs before spawning (by shedding eggs in a watery column, they maximize dispersal and minimize losses). But the thornback's spawning behavior differs in two respects—both consistent with speculated effects of predation. It rises much higher to spawn than do other similarly sized fishes. Typical ocean-dwelling, egg-producing fishes of the same size as cowfishes will ascend only between 0.6 and 1.8 meters (2 and 6 feet) before shedding their eggs. Scientists believe this height is constrained by the adults' vulnerability while hanging in midwater—easy targets for creatures that hunt at dusk. Yet the thorn-

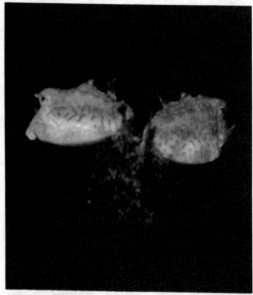

Jack T. Moyer

After a female sheds her eggs and the male fertilizes them with milt, they are dispersed by ocean currents.

back has been recorded spawning as high as 12 meters (40 feet), about 10 times the distance of vulnerable species. Jack and I saw them ascend to at least 4.6 to 6 meters (15 or 20 feet) from the bottom.

Furthermore, during their ascent, a pair of thornback cowfishes swim toward the surface for several minutes. Rather than hurriedly spawning, they hover and the male hums noisily, announcing their presence to any hunter that might come along. Compare this with a similar-sized parrotfish, a delectable species, which ascends at nearly 40 kilometers (25 miles) per hour—an incredible speed for such a small animal. Many fishes spawn so quickly that it is difficult to get pictures of them. Certainly if a fish is eminently edible, it would try to minimize its exposure time. In these respects at least, the cowfish's armor plating and noxiousness seem to allow greater liberties in courtship and mating.

Observations like these provide biologists with insight into the mechanisms that underlie animal behavior and ecology. Yet there is still so much we don't know about the inhabitants of the coral seas. Many species haven't even been formally described. Jack Moyer's work with the thornback cowfish may reveal more secrets about the strange, little creature and of the dynamic reef community as well□

Blueprint
for
Bluebirds

by Eileen Lanzafama

During a fit of spring cleaning a few years ago, Cheryl Dodenhoff found an old wooden birdhouse buried in the back of a closet. The Long Valley, New Jersey, homeowner dusted it off and nailed it to a post in her backyard. Then she forgot about it—that is, until a bird-watching friend visited her home a few weeks later. "My God!" her friend exclaimed, looking out a window. "You've got a bluebird!"

Decline Prompts Concern

A generation ago, the presence of an eastern bluebird in Dodenhoff's backyard would hardly have produced such a response. As recently as the 1940's, the bird was commonly seen in the fields and yards of farms and suburban communities throughout much of the eastern half of the country. After World War II, however, the situation began to change. The eastern bluebird, plagued by a series of harsh winters, habitat loss, and increasing competition with more-aggressive birds for available nesting sites, began to disappear. "When I was a boy, bluebirds were almost as common as robins," recalls Lawrence Zeleny, a retired biochemist from Maryland. "Now, most people under 30 have never seen one."

Experts disagree over the extent of the bluebird's decline. Some people believe their numbers have been reduced by as much as 90 percent during the past half-century; others think that some of the birds have merely become less visible, shunning heavily populated areas in favor of more-remote habitats. No one knows for sure. However, one thing is certain: the little bird has been sorely missed. In recent years, its absence from traditional nesting areas has resulted in an almost unprecedented voluntary effort by hundreds of concerned citizens.

Some people, like Cheryl Dodenhoff, who has since put up more nesting boxes in her backyard, are attempting to entice the birds to their property. Others, like Lawrence Zeleny, one of the founders of the North American Bluebird Society, have erected dozens of birdhouses to help the birds return to their former haunts.

"Placed in the right environment, nesting boxes do attract bluebirds, and thus can help build up local populations," says Eugene Morton, a biologist at the Smithsonian Institution in Washington, D.C. Adds Cornell University researcher Charles Smith: "If nothing else, these people are drawing important attention to the plight of the bluebird."

Three Separate Species

A member of the thrush family, the bluebird ranges only in North America, where it is divided into three distinct species. In each case, the males are more colorful than the females. The male of the eastern species sports a bright blue back, blue tail feathers, and a russet breast. It breeds in the United States from the Rocky Mountains to the Atlantic Ocean. The western bluebird, marked by brownish feathers on parts of its back, lives from the West Coast to the Rockies. The mountain bluebird, all blue except for a whitish belly, spreads as far north as Alaska. All three range into Mexico.

Of the three, the eastern bluebird has suffered the most in recent years as its rural farm habitat has given way to highways, shopping centers, and subdivisions. The hollowed-out trees where it builds its nests have become scarcer, and the rotten wooden fence posts it also inhabited have been replaced in many areas by metal.

Nesting Sites in Great Demand

To make matters worse, the natural nesting sites that remain are often taken over by the European starling or the house sparrow, two aggressive, hardy birds imported from England in the late 1800's. "The biggest problem bluebirds face is this competition over housing," says Zeleny. "These tough immigrants frequently usurp the bluebird nesting boxes, throwing out or attacking the adult bird. If the bluebirds are nesting, the starlings will sometimes drive them out and possibly kill their nestlings."

In the northern part of their range, most bluebirds migrate south in loose flocks during the winter to areas where their food supply of insects and berries is abundant. Others tend to remain throughout the year where they breed. During winter, the birds live in flocks. Some birders have seen more than a dozen of the creatures enter a single roosting box at once, and then huddle together through the night to keep warm. But in late winter, the birds turn to domestic affairs. After the birds pair off, they select a likely nesting site. During the next four to five days, the female builds a nest—usually a loosely constructed cup of grasses—while the male remains in the area.

Bluebirds usually produce two or three broods a year, laying three to eight eggs per clutch. Following a two-week incubation period completed solely by the female, the eggs usually hatch on the same day. Then, for the adults,

A cluster of adult bluebirds huddles together for warmth during a cold spell. Many bluebirds do not migrate for the winter.

it's an almost continuous hunt for food, as their hungry youngsters require feeding about once every 20 minutes. After about 18 days, the young birds leave the nest, flying to a nearby tree or perch. Within a few days, the father takes over caring for the youngsters while the mother prepares the nest for the next brood. After the fledglings can fend for themselves, they sometimes stay around to help the parents feed the next brood.

People Lend a Hand

With so many young growing so quickly, it's not surprising that bluebirds need plenty of nesting sites to flourish. "They're naturally curious," says Morton. "They like to investigate dark holes that might lead to a nesting cavity. Artificial nesting boxes provide a strong stimulus: a dark hole in light-colored wood."

People have been supplying such boxes for more than 150 years in this country. "In fact," notes Zeleny, "Thoreau's diary speaks of several bluebirds that nested in his boxes." Ze-

leny's own efforts span seven decades. "When I was about 14, I began trying to entice bluebirds into my homemade birdhouses," he says. Today he maintains a 60-box bluebird trail that he established in 1967 near his home in Maryland. Such a trail is made up of strategically spaced, well-monitored nesting boxes with holes 4 centimeters (1.5 inches) in diameter to keep out starlings. Zeleny spends about five hours a week during the nesting season monitoring his boxes, keeping meticulous records of eggs hatched and birds fledged. He believes that about 2,500 young birds have fledged on his trail since it was first put up.

The bluebird society Zeleny helped found in 1979 has since grown to more than 4,000 members. Camp Fire, Incorporated, a youth organization, has also taken to building nesting trails. In the past few years, more than 10,000 children have been awarded badges for bluebird conservation. The group recently helped establish a 112-box trail at the Air Force Academy in Colorado. In 1983, 46 were occupied.

The main thrust of the bluebird revival, though, has come from individuals, many of them retired people who grew up when bluebirds were abundant. Around Tupelo, Mississippi, for instance, people are seeing more of the birds these days, thanks to the efforts of Gale Carr, a retired dairy farmer. In 1964 Carr was surveying a site for a new home when he spied a bluebird in a nearby tree. It was the first one Carr had seen in some time, and it inspired him. From then until a few years ago when he became disabled, Carr made more than 6,500 nesting boxes, which he gave away to people in Tupelo and some 20 other cities. "The boxes really attract the birds," says Carr. "Today, there are a great many bluebirds in this area."

Above: Bluebirds typically produce two or three broods each year and lay three to eight eggs per clutch. Left: Adult bluebirds almost constantly hunt for food to keep their ever-hungry youngsters well nourished.

A Teacher's Special Program

Carr's northern counterpart may be Junius Birchard, a modern-day Johnny Appleseed from Hackettstown, New Jersey, who travels through parts of his state dotting the landscape with nesting boxes and getting other people to do the same. Since 1978 the spry former teacher has been visiting area schools and local clubs, preaching bluebird conservation.

"As a boy in Pennsylvania, I'd go camping along the Allegheny, and there'd be bluebirds everywhere," he says. In 1977 Birchard read an article about the eastern bluebird's decline. It spurred him to action. "At the time," he recalls, "I didn't even know where to begin to find a bluebird."

Before long, Birchard thought of a way to combine his love for teaching with his new mission to help bluebirds: "I arranged to teach a group of second-graders about bluebirds by helping them build nesting boxes," he says. "It fit in beautifully with their curriculum, which called for using tools and also learning about various aspects of the environment." With Birchard's help, the students built some 50 nesting boxes, and the following season, they recorded a successful nesting.

Word of his program spread quickly, and, he says, "the thing just mushroomed." As more and more groups began asking for his program, Birchard developed nesting-box kits that people could easily assemble themselves. Originally, he cut the lumber for the kits himself, but soon enlisted the aid of students at two nearby schools. As of early 1984, Birchard had sold more than 6,000 of his nesting boxes. They sell for $3 each, and he turns all profits into the purchase of more lumber for boxes. So far, he's spent more than $10,000 of his own money on the project. Each of his bluebird boxes is stamped with an emphatic plea: "Please do not let the very destructive English sparrow breed in this birdhouse."

Bluebird Admirers Everywhere

Another determined self-starter, Laurance Sawyer of Ringgold, Georgia, has been building and delivering nesting boxes to area residents since he retired from a local bakery in 1975. Sawyer fashions his boxes out of hollowed-out logs and sells them at a minimum fee. "Many of the older people who want a box don't have the tools or the strength to put them up properly," he says. "Some of them simply need the therapeutic value provided by offering care to a living creature."

Jack Finch, a retired tobacco farmer from Bailey, North Carolina, channels his energy in

With his feathers puffed out to help ward off the cold, an adult male bluebird perches on a fence post near his box.

another direction: he invents gadgets to keep predators away from the nests. After maintaining a bluebird trail of more than 2,000 boxes for several years, Finch realized that he couldn't take care of the project properly. He took down the boxes and replaced them with 360 new birdhouses perched atop 3-meter (10-foot) pipes. Finch, who had also invented a device to keep snakes out of the boxes, created the new design to stop flying squirrels from nesting in the bluebird homes. "It works," he says.

Mixed Reviews About Effects of Aid

Despite such success, some scientists question whether it's wise to attempt to restore large numbers of bluebirds to small areas. T. David Pitts, a researcher at the University of Tennessee at Martin, suggests that by doing so, people may be inadvertently exposing the birds to some of the same problems that wiped out populations of the creatures in certain areas in the first place. "Sooner or later," he says, "we're going to be hit with a severe winter similar to those of the late 1970's, when large numbers of the birds died. In cold weather, food supplies are limited. Perhaps if the species were more widely dispersed, a larger percentage of the birds could get enough food to survive."

Ornithologist and renowned bird artist Roger Tory Peterson disagrees. Density, he says, is not a problem for bluebirds. "The birds will sort it out," observes Peterson. "They'll space themselves. When it gets too crowded, some of them will move to the periphery."

"Undoubtedly, when you entice birds into suburban situations, as many people are now doing, you increase the risks of mortality," adds Craig Tufts, chief naturalist with the National Wildlife Federation. "We're finding, though, that the number of birds is increasing far faster than the number that is being lost to predation and disease."

A Definite Revival

"Can a nesting-box program really help safeguard a species?" says Eugene Morton. "The wood duck was nearly extinct in the 1930's. But through a similar citizen effort, it was protected with nest houses. Now, the wood duck is very common and no longer dependent on humans for its survival."

Though no accurate population figures are available for bluebirds today, scientists at the U.S. Fish and Wildlife Service Research Center in Laurel, Maryland, are beginning to see some

Bluebird lover Lawrence Zeleny checks one of 60 nesting boxes he put up many years ago in Maryland.

trends. "Bluebirds appear to be increasing in some local areas," observes biologist Chandler Robbins.

Whether or not they ever come back in vast numbers, bluebirds are indeed returning. Their revival should be delightful, both for those people who remember the little songbird from years past, and for those who will have the chance to see one for the first time□

SELECTED READINGS

"Science helps the bluebird" by Joan Arehart-Treichel. *Science News*, June 11, 1983.
"Bring back our state bird . . ." by James Mordovancey. *The Conservationist*, March–April 1982.
"Give the bluebird a housing hand" by Mary Durant. *Audubon*, March, 1981.

Editor's Note

For more information about nesting boxes, write to the North American Bluebird Society, P.O. Box 6295, Silver Spring, Maryland 20906-0295.

IN MEMORIAM

AHLQUIST, RAYMOND (68), U.S. pharmacologist whose study of how adrenalinelike hormones affect the heart led to the development of propranolol (Inderal), a drug now widely used to treat heart disorders and hypertension. He (and James W. Black) was awarded the Albert Lasker Clinical Medical Research Award in 1976; d. Augusta, GA, April 15.

BLOCH, FELIX (77), Swiss-born professor emeritus of physics at Stanford University, who shared (with E.M. Purcell) the 1952 Nobel Prize in Physics for the discovery of the nuclear-induction method of studying atomic nuclei, a discovery that later led to the development of the nuclear magnetic resonance (NMR) diagnostic device; d. Zurich, Switzerland, Sept. 10.

BOK, BART J. (77), Dutch-born astronomer and leading authority on the Milky Way. He was professor of astronomy at Harvard University (1933–57), director of the Mount Stomlo Observatory at the Australian National University (1957–66), and director of the Steward Observatory of the University of Arizona (1966–70) and professor emeritus of astronomy there. His classic work, *The Milky Way* (1941), was written with his wife Priscilla; d. Tucson, AZ, Aug. 5.

BOYD, WILLIAM C. (79), U.S. biochemist whose early work on blood led to modern immunology. He detected new blood types, found ways of making fine distinctions within blood types, and provided evidence concerning the relationship of blood groupings and race. He served on the faculty of the Boston University School of Medicine as professor of biochemistry and later (1948–68) as the school's first professor of immunochemistry. He wrote texts on immunology and anthropology, including *Genetics and the Races of Man* (1950); d. Falmouth, MA, Feb. 19.

CATTELL, McKEEN (91), U.S. physician and professor emeritus of pharmacology at Cornell University Medical School; pioneer in research on how the powerful stimulant digitalis affects heart muscle; d. Beacon, NY, Feb. 8.

CHIPMAN, JOHN (86), U.S. chemist, professor emeritus and former head of the department of metallurgy of the Massachusetts Institute of Technology; world renowned for research on steel; member of the Manhattan Project team that developed the first atomic bomb; d. Winchester, MA, May 14.

CLAUDE, ALBERT (84), Belgium-born cell biologist who shared (with Christian René de Duve and George Emil Palade) the 1974 Nobel Prize in Physiology or Medicine for work on the structure and function of parts of the cell, much of it done at what is now the Rockefeller University in New York City. The first to apply the electron microscope to the study of cells, he discovered several cell organelles, including the mitochondrion, and was the first to provide a detailed view of cell anatomy; he also studied cancer viruses; d. Brussels, Belgium, May 22.

CROHN, BURRILL B. (99), U.S. gastroenterologist whose work led to major advances in the identification and understanding of ileitis (an inflammation of the intestinal tract), now known as Crohn's disease. He was long associated with Mount Sinai Medical Center in New York City and was the author of several books on gastroenterology; d. New Milford, CT, June 29.

DAVIS, ALLISON (81), U.S. psychologist and social anthropologist, and professor emeritus of education at the University of Chicago, who did pioneering studies on the influence of social and economic factors in the education of poor children and who criticized intelligence testing as being culturally biased. He was the author of several books, including *Deep South* and *Social-Class Influences Upon Learning*; d. Chicago, IL, Nov. 21.

DAYHOFF, MARGARET OAKLEY (57), U.S. biochemist who developed compilations of protein structures that are much used in genetic engineering and medical research; she also developed evolutionary "trees" based on correlations between organisms and proteins; d. Silver Spring, MD, Feb. 5.

DEBUS, KURT HEINRICH (74), German-born engineer who helped develop modern rocketry. After working on the U.S. Redstone Ballistic program developing launch vehicles, he became director of operations in 1952 at what was to become the Kennedy Space Center, overseeing the launch of the first U.S. earth satellite, manned space flight, and landing on the moon; d. Cocoa, FL, Oct. 10.

FULLER, R. BUCKMINSTER (87), U.S. inventor and futurist best known as the father of the geodesic dome. An original thinker long dismissed by the scientific and architectural establishment, he was an eternal optimist, convinced that through rational planning and the application of simple principles of geometry and design, "technology could save the world." Among his many inventions—he held more than 2,000 patents—was the prefabricated Dymaxion House (1927), a radical break with conventional architecture; Dymaxion three-wheeled cars (1933–35); and the Dymaxion Airocean World Map (1943). In his later years he traveled about the world preaching his gospel of rational planning for life on "Spaceship Earth"; d. Los Angeles, CA, July 1.

GERSHON, RICHARD K. (50), U.S. physician; professor of pathology, immunology, and biology at Yale University School of Medicine, who was a leader in research on the immune system and also contributed to the understanding of viral hepatitis and tumor biology; d. New Haven, CT, July 11.

GODOWSKY, LEOPOLD, 2d (82), U.S. inventor and violinist, who developed (with Leopold Mannes) the Kodachrome color photography process; d. New York, NY, Feb. 18.

HARRIS, ROBERT S. (79), U.S. nutritionist; professor emeritus of nutritional biochemistry at Massachusetts Institute of Technology; known for his studies of the absorption of nutrients and the roles of vitamins and minerals in metabolism; d. Wellesley, MA, Dec. 24.

HARTLINE, H. KEFFER (79), U.S. biophysicist who shared (with George Wald and Ragnar Granit) the 1967 Nobel Prize in Physiology or Medicine for work on the electrical nature of vision. He was professor of biophysics at Johns Hopkins University (1949–53) and at Rockefeller University (1953–74); d. Fallston, MD, March 17.

HEIDELBERGER, CHARLES (62), U.S. organic chemist who explained some of the chemical processes of malignancy and developed a drug—5 fluorouracil—that is widely used against some cancers. After many years at the University of Wisconsin's McArdle Laboratory for Cancer Research, he became director of basic research at the University of Southern California's Comprehensive Cancer Center in Pasadena in 1976; d. Pasadena, CA, Jan. 18.

HILDEBRAND, JOEL H. (101), U.S. physical chemist whose renowned career as a researcher and teacher spanned the birth and development of modern chemistry. He contributed to knowledge of the behavior of liquids and solutions; wrote *Principles of Chemistry* (1918), a standard text for many years; and taught at the University of California at Berkeley since 1913, becoming one of the most revered faculty members; d. Kensington, CA, April 30.

JACOBSON, EDMUND (94), U.S. physician, a specialist in tension control, who introduced an approach to illnesses now termed psychosomatic, developed relaxation treatments for various anxiety-caused disorders, and established methods that formed the basis for natural-childbirth techniques; d. Chicago, IL, Jan. 7.

KAHN, HERMAN (61), U.S. scientist, futurist, and thinker on nuclear strategy; founder and director of research of the Hudson Institute, a private center for national security and public policy research; best known for his writings on nuclear war, its likelihood and probable consequences. Starting his career as a physicist and mathematician, he developed a system that reduced the time needed to solve mathematical problems before the computer age. Then his interests widened, and he did research and wrote on many topics, including armament, educational programs, conditions in Central and South America, and environmental issues; d. Chappaqua, NY, July 7.

KUNKEL, HENRY G. (67), U.S. physician and leader in immunology research, who developed a technique for distinguishing antibodies that became an important tool in cellular immunology and who contributed to the understanding of immune-related diseases (e.g., rheumatoid arthritis); he was associated with Rockefeller University since 1945; d. Rochester, MN, Dec. 14.

MELICOW, MEYER M. (88), Russian-born physician, Given Professor Emeritus of Uropathology Research at Columbia University's College of Physicians and Surgeons; an expert on cancer of the urinary tract; d. New York, NY, June 3.

NACHMANSOHN, DAVID (84), Russian-born, German-educated biochemist who made important contributions to the identification and isolation of chemicals involved in the transmission of nerve impulses. He joined Columbia University in 1947, taught for many years, and retired as professor emeritus of biochemistry in the neurology department; d. New York City, NY, Nov. 2.

NEWELL, HOMER E., JR. (68), U.S. mathematician and space scientist primarily responsible for organizing early U.S. space efforts. At the Naval Research Laboratory (1947–58), he was responsible for upper-atmosphere research with rockets. Later, at the National Aeronautics and Space Administration (1958), he organized research efforts and held several posts, retiring in 1973 as associate director; d. Arlington, VA, July 18.

NOTESTEIN, FRANK W. (80), U.S. research demographer who helped establish and guide efforts in population control. He was professor of demography at Princeton University, established the Office of Population Research at Princeton in 1936, and served as first director of the population division of the United Nations (1946–48) and as president of the New York City-based Population Council (1959–68); d. Langhorne, PA, Feb. 19.

REICHELDERFER, FRANCIS WILTON (87), U.S. meteorologist who headed the U.S. Weather Bureau for 25 years and ushered in the use of space and computer technology in forecasting. After 20 years as an aviator and meteorologist in the Navy, he took over as head of the Weather Bureau in 1938. There he adopted the air-mass theory of forecasting with attention given to the upper atmosphere. He added to the Bureau's services such aids as crop forecasts for farmers, frost warnings for fruit growers, hurricane warnings, marine weather reports, and reports on river and flood conditions. He also helped found the U.N. World Meteorological Organization and served as its first president (1951); d. Washington, D.C., Jan. 25.

ROWE, WALLACE PRESCOTT (57), U.S. virologist and cancer researcher; chief of the laboratory of viral diseases at the National Institute of Allergy and Infectious Diseases. His research helped to show the relationship between viruses and cells, and specifically, the actions that produce cancer; d. Baltimore, MD, July 4.

SHEAR, MURRAY J. (83), U.S. biochemist and pharmacologist sometimes called "the father of cancer chemotherapy" for his pioneering work on the origins of cancers and the isolation and purification of substances capable of destroying cancer. He also contributed to the development of a vaccine against typhus and studied the relationship between cancer and air pollution. He was associated with the National Cancer Institute for more than 30 years, serving as head of the laboratory for chemical pharmacology (1951–64) and later as a special adviser; d. Bethesda, MD, Sept. 27.

SPIEGELMAN, SOL (68), U.S. microbiologist who made important contributions to genetics, virology, and cancer studies. His work on DNA (deoxyribonucleic acid, the hereditary material) helped provide the basis for current recombinant DNA technology; his study of mutations in bacteria led to an appreciation of bacteria as ideal models for studies of cell genetics; and his work on tumors of viral origin advanced the understanding of the molecular basis of cancer. He was professor of human genetics and development at Columbia University's College of Physicians and Surgeons and director of the Institute of Cancer Research (1969–75) and later held the post of University Professor; d. New York City, NY, Jan. 21.

STEINBERG, ISRAEL (80), U.S. physician and heart specialist who developed (with G. P. Robb) the technique of angiocardiography, an important aid in the diagnosis of heart and blood vessel disorders; d. Phoenix, AZ, Feb. 15.

SULZBERGER, MARION BALDUR (88), U.S. physician who helped establish the modern field of dermatology. He studied contact allergies, dermatologic immunity, and the use of steroids for skin diseases and served as chairman of the department of dermatology at New York University School of Medicine (1949–60); d. San Francisco, CA, Nov. 24.

TAFT, HORACE DWIGHT (57), U.S. physicist and leader in research in high-energy physics and the use of the computer in that area, who was one of a team that discovered the last predicted antimatter particle (1962). He was associated with Yale since 1956, serving as dean of Yale College (1971–79) and later as professor of physics; d. New Haven, CT, Feb. 12.

THOMPSON, FREDERICK ROECK (75), U.S. orthopedic surgeon who developed a metal device (the Thompson Vitallium Hip Prosthesis) that is still used in the treatment of hip fractures and was the forerunner of the full artificial hip. He was director and chief of the orthopedic staff at St. Luke's Medical Center in New York City (1961–72); d. New York City, NY, April 13.

VINOGRADOV, IVAN M. (92), Soviet mathematician, director of the Institute for Mathematics in Moscow since 1932; he contributed to number theory and was the recipient of numerous Soviet awards and an honorary member of international science societies; d. Moscow, USSR, March 20.

WEITZMAN, ELLIOT D. (54), U.S. physician and leader in the study of sleep. He was the founder of the Institute of Chronobiology in White Plains and professor of neurology at Cornell University Medical Center; former chief of neurology and director of the Laboratory for Human Chronophysiology and of the Sleep-Wake Disorders Center at Montefiore Medical Center in the Bronx; and professor of neurology and neurosciences at Albert Einstein College of Medicine, all in New York; d. New York City, NY, June 13.

WESTCOTT, CYNTHIA (84), U.S. plant pathologist and rose expert who worked as a plant doctor in New York area gardens and wrote widely on rose growing and plant diseases. She was the recipient of numerous awards from horticultural and rose societies and served as an active member of many scientific societies; d. Tarrytown, NY, March 22.

WILLIAMS, CLARKE (80), U.S. nuclear physicist whose study of neutron physics, particularly the gaseous diffusion method of separating the uranium-235 isotope, contributed to the development of the atomic bomb. He was long associated with the Brookhaven National Laboratory at Upton, Long Island, NY, as chairman of the department of nuclear engineering (1952–62), as deputy director (1962–67), and finally as deputy director emeritus; d. Bellport, NY, March 15.

WILSON, CARROLL L. (72), U.S. science and energy expert, the first general manager of the Atomic Energy Commission (1946–50); founder and director of a "Workshop on Alternative Energy Strategies" (1974) that assessed energy options for Europe, North America, and Japan; executive director of the United Nations Environment Program on the limits of global ability and energy; and recently the author of an optimistic report on the role of coal in future energy needs (1980); d. Providence, RI, Jan. 13.

WINSCHE, WARREN E. (66), U.S. energy engineer, deputy director of the Brookhaven National Laboratory at Upton, Long Island, NY, where he directed programs for nuclear-reactor safety and nuclear-materials handling, and atmospheric and coastal studies. He earlier worked at E. I. du Pont de Nemours & Company and served as chairman of the nuclear-engineering department (1962–75) and associate director for energy (1975–79) at Brookhaven; d. Upton, NY, June 19.

INDEX

Parkinson's disease (med.) 301
Parks, National see National parks
Particle, Elementary (phys.) 2
  intermediate vector bosons 282
Peaceful nuclear explosives 160–165
Peking, China 253
Pelican (zool.) 348
Pendulum, Foucault 295; illus. 292
Penguin (zool.) 355
Pentagoet, Fort (archaeo. site) 253
Permafrost (soil) 140
Pesticides (chem.) 193
Petroleum 158
  nuclear explosion technology 164
  offshore oil leasing 189
Pharmacology see Drugs
Phase space (phys.) 287
Phosphorus (elem.) 283
Photochromic materials (chem.) 118
Photography 379
  holography 308
  underwater 54
Physiatry (med.) 236–243
Physics 282, 379, 380
  Atomic Resolution Microscope 304
  chaotic dynamics 285
  Coriolis effect 292
  Nobel prize 311
Physiology see Body, Human; Health
  and disease
Physiology or medicine, Nobel prize
  in 93–95
Pike (fish) 349
Piltdown Man 267–273
Pioneer mission (space probe) 2
Pitch (voice) 57, 58
Planetary nebula (astron.) 3
Planets (astron.)
  Coriolis effect on geologic features
  298
  See also names of planets
Plants see Botany
Plastic pollution at sea 196–203
Plate tectonics (geol.) 131, 135
Play, Animal 350–355
Plowshare, Project (U.S.) 161
Pollination (bot.) 193
Pollution
  effects on wildlife 349
  peaceful nuclear explosions 162
  plastic pollution at sea 196
  poisonous waste pollution 189
Polylysine (synthetic protein) 120
Polymer (chem.) 118
Polynomial equation (math.) 99
Population 252, 380
Potato (bot.) 65; illus. 64
PowerPad (computer tech.) 99
Precession (phys.) 264
Predation (zool.) 65
Prehistoric man 253
  Piltdown Man 267
Pressurized fluidized-bed
  combustion unit 171
Primate (zool.) 350, 352
Programming, Computer 98
Project Daedalus study 274–279
Project Plowshare (U.S.) 161
PROM (programmable read-only
  memory chip) (tech.) 344
Pronghorn antelope (zool.), illus. 209
Propranolol (drug) 379
Protein (biochem.) 379
  aequorin 302
  influenza virus proteins 231
  molecular computer 120
  origin of life 65
  tissue plasminogen activator 219
Proton (phys.) 282, 283
Proton-precession magnetometer
  264
Psychobiology 66–73
Psychology see Behavioral sciences
Psychology profiling of criminals
  40–47
Psychosomatic illness (med.) 72
Pygmy chimpanzee (zool.) 253

R

Radar (electron.)
  Doppler weather radar 142
  glacier surveys 152
Radiation see Infrared radiation; Light;
  X rays
Radio (electron.)
  cellular radio 329
  wolverine tracking 358; illus. 359
Radiocarbon dating 129
Radioactive dating 129
Radioactive fallout 162, 163
Radioactive waste disposal 159, 164
Radioactivity, Two-proton decay
  (phys.) 283
Rain forest 190–195
Random fractals (math.) 102
Randomness (phys.) 284–291
Random sampling (math.) 263
Rapid eye movement sleep 33
Rat (zool.) 352
Raven (zool.) 355
Reclamation of land 205
Recombinant DNA (biol.) 65, 88
Recording, Sound 324–327
Red blood cell (physiol.) 224
Red dwarf (astron.) 275, 276, 278, 279
Red giant (astron.) 10
Rehabilitation medicine 236–243
Reichelderfer, Francis Wilton (Amer.
  meteorol.) 380
Reproduction (biol.)
  genetically engineered cattle 86
  thornback cowfish 370; illus. 371
Rhesus monkey (zool.) 352, 354
Rhinoceros (zool.) 52
Ribonucleic acid (RNA) (biochem.) 65
  influenza genes 231, 233
Richter scale (seismol.) 128
Ride, Sally K. (Amer. astro.) 4
Rio Grande Rise (geol.) 136
Rivers
  Brazil's hydroelectric potential 174
  Coriolis effect 296
RNA see Ribonucleic acid
Roaming (cellular radio) 332
Robots (mach.)
  aid systems for the handicapped
  239; illus. 240
  automated factories 316, 335
  Daedalus project study 274
Rock (geol.) 129
Rockets 379, 380
  Ariane 5
Rodent (zool.)
  pacas 195
  play 352
Rose (bot.) 380
Rowe, Wallace Prescott (Amer.
  virologist) 380
Rubber 283
Ruckelshaus, William (Amer. pub.
  offi.) 188
Running, Compulsive (psych.) 33

S

Saber-toothed cat (paleon.) 129
Sage grouse (zool.), illus. 207
Saint Helens, Mount (volcano) 129
Salmon (zool.) 212
Salton Sea geothermal-electric
  project 159; illus. 158
Samoan society (anthro.) 253
Santa Catalina mission
  archaeological expedition 260–
  266
Satellites, Artificial
  communications satellites 4, 5
  Exosat 5
  InfraRed Astronomical Satellite 3, 5,
  6
  seafloor gravity mapping 130, 131,
  136; illus. 135

Saturn (planet) 3; illus. 26
Scanning tunneling microscopy 317
Schizophrenia (psych.) 33
Sea see Ocean
Seafloor gravity mapping 130–136
Seal (zool.)
  effects of pollution on 198, 200
  elephant seal 348
  influenza 229
Sea lion (zool.) 354; illus. 201
Seasat (artificial satellite) 131, 136;
  illus. 135
Seismic surveying (geol.) 163
Seismology see Earthquake
Semiconductor (phys.) 319
Senile macular degeneration (med.)
  246
Sex differences
  animal play 352
  mathematical ability 99
Shale oil 159, 164
Shark (zool.) 49
Shaving, Preoperative 221
Shear, Murray J. (Amer. biochem.)
  380
Shelters for the homeless 35, 39
Shock wave lithotripsy,
  Extra-corporeal (med.) 220
Shrimp (zool.) 210, 214
Silicon chip (computers) 99, 115
Skin fungus (bot.), illus. 216
Sleep research 33, 380
Slime mold (zool.) 291
Sloth (zool.) 192; illus. 194
Smart card (microchip card) 340–345
Snow (meteorol.) 138, 140, 141
Software (computers) 98
Solar energy 159
Solar flare (astron.) 2
Solar system (astron.) 2
  Coriolis effect 298
Son of Sam (Amer. criminal) 44; illus.
  45
Sound recording 324–327
Sound wave (phys.)
  acoustic microscopy 317
  seismic surveying 163
Soyuz (spacecraft) 5
Space art 24–29
Space exploration and travel 2, 379,
  380
  Daedalus project study 274
  Halley's comet probes 23
  Space Camp 13
Spacelab 4
Space shuttle 4; illus. 5
  ice damage 141
  Space Camp simulation 16
Speech development 56–61
Spider (zool.) 65
Spiegelman, Sol (Amer. microbiol.)
  380
Spinal cord injury 236, 239 fol.
Squirrel monkey (zool.) 352
Star-nosed mole (zool.) 363–367
Stars (astron.) 3
  age and collapse 311
  Barnard's star 275, 276, 278, 279
  InfraRed Astronomical Satellite
  discoveries 8, 12
Statistics, Baseball (math.) 108
Steinberg, Israel (Amer. phy.) 380
Strabismus (med.) 301
Strange attractor (phys.) 286 fol.
Streptokinase (enzyme) 219
Stress (biol., psych.) 72
Strong force (phys.) 282
Stypoldione (drug) 301
Subatomic particles see Particle,
  Elementary
Suicide 73
Sulfur dioxide (chem.) 2, 3, 189
Sulfuric acid (chem.) 128, 189
Sulzberger, Marion Baldur (Amer.
  dermatol.) 380
Sun (astron.) 2
  comets 19, 20

CONTRIBUTORS

TOM ALEXANDER, Board of Editors, *Fortune*
PSYCHOBIOLOGY

LAWRENCE K. ALTMAN, M.D., Medical reporter, *The New York Times*
REVIEW OF THE YEAR: HEALTH AND DISEASE

VIC BANKS, Free-lance science writer
THE ITAIPU DAM

MARCIA BARTUSIAK, Free-lance science writer with a master's degree in physics
SEAFLOOR PANORAMA

MARY BATTEN, Free-lance science writer
TROPICAL FORESTS IN TROUBLE

BETH BIRDSONG, Free-lance writer specializing in child-development and language topics
MOTHERESE

STEPHEN A. BOOTH, Senior editor, *Audio Times*
COMPACT DISCS

HARRIS BROTMAN, Free-lance writer with a doctorate in genetics
ENGINEERING THE BIRTH OF CATTLE

MALCOLM W. BROWNE, Senior editor, *Discover* magazine
PEACEFUL NUCLEAR EXPLOSIVES

GENE BYLINSKY, Associate editor, *Fortune* magazine
THE RACE TO THE AUTOMATIC FACTORY

ANDREW CHAIKIN, Assistant editor, *Sky & Telescope*
ASTROARTISTS

DANIEL JACK CHASAN, Columnist for *Pacific Northwest* magazine
COLUMBIA: A GLACIER IN RETREAT

LAURENCE CHERRY, Free-lance writer specializing in medicine
co-author NEW HOPE FOR THE DISABLED

RONA CHERRY, Executive editor of *Glamour* magazine
co-author of NEW HOPE FOR THE DISABLED

FELICIA C. COLEMAN, Professor at Cornell University's Shoals Marine Laboratory; free-lance writer and author
co-author PLASTICS AT SEA

WILLIAM J. CROMIE, Executive Director of the Council for the Advancement of Science Writing
TO SEE AN ATOM
AQUACULTURE

BARBARA FORD, Free-lance writer
LEARNING TO PLAY, PLAYING TO LEARN

ROBERT HADDON, Member of the technical staff in the chemical physics research department at Bell Laboratories
co-author THE ORGANIC COMPUTER

DAVID HALBERSTAM, Pulitzer Prize-winning journalist; author of *The Best and the Brightest* and *The Powers That Be*
BONSAI

DIANNE HALES, Contributing editor, *American Health*; author *The Complete Book of Sleep*
co-author AEROBICS: HOW MUCH IS ENOUGH?

ROBERT E. HALES, M.D., Psychiatrist with the U.S. Army Medical Corps at Fort Worth, Texas
co-author AEROBICS: HOW MUCH IS ENOUGH?

STEPHEN S. HALL, Free-lance writer
THE FLU

KATHERINE HARAMUNDANIS, Research Associate, Smithsonian Astrophysical Observatory; co-author of *An Introduction to Astronomy*
REVIEW OF THE YEAR: ASTRONOMY

WRAY HERBERT, Behavioral Sciences editor, *Science News*
REVIEW OF THE YEAR: BEHAVIORAL SCIENCES

GLADWIN HILL, Free-lance writer; former Environment editor of *The New York Times*
REVIEW OF THE YEAR: THE ENVIRONMENT

JUDITH HOOPER, Free-lance writer
CONNOISSEURS OF CHAOS

MAURICE HORNOCKER, Leader of the Cooperative Wildlife Research Unit at the University of Idaho; one of the foremost wildlife scientists in the U.S.
WOLVERINES

DUANE L. HUFF, Vice-President of technical services for Advanced Mobile Phone Service, Incorporated
CELLULAR RADIO

JEFFREY KLUGER, Assistant editor, *Science Digest*
SPACE CAMP

MARC KUSINITZ, Associate editor, *Science World*
REVIEW OF THE YEAR: PHYSICAL SCIENCES

ANNE LABASTILLE, Authority on Central American wildlife and wild lands; Ph.D. wildlife ecologist; author of *Women and Wilderness* and *Woodswoman*
EIGHT WOMEN IN THE WILD

ANGELO LAMOLA, Head of the molecular biophysics research department at Bell Laboratories
co-author THE ORGANIC COMPUTER

JOHN LANGONE, Staff writer, *Discover* magazine
CHEMICAL TREASURES FROM THE SEA

NEW WOOD ARCHITECTURE

Published in North America by
Yale University Press
P.O. Box 209040
New Haven, CT 06520-9040
U.S.A.

First published in Great Britain in 2005
by Laurence King Publishing Ltd,
London

Text © 2005 Ruth Slavid
This book was designed and produced
by Laurence King Publishing Ltd

Library of Congress Control Number:
2004113112

ISBN: 0-300-10794-3

Designed by Mark Vernon-Jones

Printed in China

NEW WOOD ARCHITECTURE

RUTH SLAVID

Yale University Press

CONTENTS

Is timber a contemporary material? The projects in this book show that the answer must be a resounding yes, but it is a question that would have received a much more tentative answer only a couple of decades ago. Steeped in tradition – the oldest known timber buildings are log houses found in Poland, dating from about 700 BC – timber is a material that has scarcely seemed at the forefront of technology in recent times.

The heroic age of the Industrial Revolution was concerned with showing how structures could be made larger than ever before. As the potential of cast iron, and, later, of steel, came to be understood better, and as reinforced concrete was developed, timber seemed like the poor relation – the traditional material that was all right for a vernacular building, cheap housing or a small footbridge, but was not a player in the big league.

Two factors have led to change. One is the advance of the environmental movement and a concomitant growth

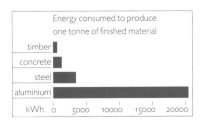

Energy consumed to produce
one tonne of finished material

timber		
concrete		
steel		
aluminium		
kWh 0	5000 10000 15000	20000

of interest in traditional methods of construction. Timber is a renewable material and, given sustainable forestry techniques, an environmentally acceptable one. Wood absorbs carbon dioxide as it grows, and the gas remains locked up during the wood's lifespan in a building and during any subsequent reuse. It also scores well in terms of 'embodied' energy, the amount of energy needed to produce a unit of timber, compared to other structural materials. For example, it takes 750MJ of energy to create a cubic metre (35 cubic feet) of rough-sawn timber, compared to 266,000MJ for a cubic metre of steel and 1,100, 000MJ for a cubic metre of aluminium. Timber is, of course, less dense than the other materials, but make the comparison on weight and it is still impressive: 1.5MJ per kilogram (2¼ pounds) of rough-sawn timber, 35MJ per kilogram of steel and 435MJ per kilogram of aluminium.

At the same time, timber has become much more engineered than it used to be, with the development of glue-laminated material, the understanding of stressed-skin structures, and the ability to use almost every part of a tree, including the sweepings that go into chipboard. This is partly a measure of the ingenuity of manufacturers, encouraged by imaginative architects and engineers, and partly a response to changing circumstances.

A DIFFERENT MATERIAL

Timber is not the same as it once was. Just as the agricultural revolution changed the nature of crops and farm animals – so that, although the word 'pig' was commonly understood in the Middle Ages and is so today, the animal has changed out of all recognition in the intervening centuries – so timber has altered as well. Originally, timber came from virgin forest, a practice that did not end until Asia and South America were persuaded to stop depleting their resources in a way that Europe had done centuries earlier. Nearly all timber now comes from managed second- or third-growth forest. Silviculture is a complex discipline, which attempts to balance productivity with environmental concerns, ranging from the preservation of natural fauna to the protection of watercourses, but there is always pressure for relatively rapid commercial returns.

In America, for example, it was believed until the 1920s that only old-growth timber from virgin forest was suitable for structural use. Increased understanding of the way that timber grows and should be cut means that this is no longer the case. But using secondary growth means that there will be a lot more juvenile timber. In most cases, timber from trees less than about 30 years old is considerably softer than that from mature trees. It is also smaller, adding to the cost of large structural elements or even of the kind of wide floorboards that were common in houses until the end of the nineteenth century. Another consideration is that production is seasonal. This may not be immediately apparent if you are buying standard timber harvested and stored in large quantities. But manufacturers of specialist veneers, for example, find that architects show little appreciation of the fact that all harvesting takes place during the dormant period in autumn, and that, by the following summer, stocks may be running low.

One solution to these problems is to use specialist cutting and grading, which ensures that architects get exactly what they want, but manufactured products come much closer than 'raw' timber to offering the consistency of properties found in steel and concrete, while offering the moral satisfaction of cutting down on waste. However, the environmental purist is likely to be concerned about the nature of the glues and resins that bind these products, and there are implications for recycling.

STICKING TOGETHER

Of all engineered solutions, probably the most influential is 'glulam', short for glued laminated timber. This is a structural timber product manufactured

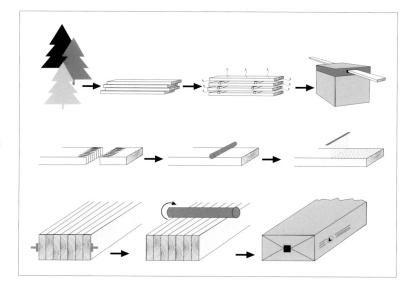

by gluing together individual flat pieces of wood under controlled conditions. The timber laminates are strength-graded before fabrication. The member can be straight or curved, and can be made with a variable section to satisfy structural requirements.

Early structures built using glue-laminated timber included a lantern roof, with laminated, curved support beams, at the Old Rusholme Chapel in Manchester (1827), which survived until it was demolished in 1962, and the assembly hall of King Edward College in Southampton (1860), but these projects were rarities. The technology derived from earlier work in Germany, where Carl Friedrich von Wiebeking had used

laminated timber in the construction of road bridges. The normal technique for securing those elements was with bolts, although on a bridge at Altenmarkt the laminates were glued.

The French developed the idea of using curved laminated timbers, with Armand Rose Emy applying his ideas to a roof of horizontally laminated timber arches in a trial at Marac, near Bayonne, in 1825. But this was something of a false dawn. The development of wrought iron in 1850 squeezed out the need for large-span timber structures, which, in any case, experienced problems with rot and with the glues that were used. The next growth in importance of glulam occurred when Otto Hetzer, a scientist

LEFT
One of the earliest uses of glulam in the UK was on the Waterloo entrance arches built for the Festival of Britain in 1951, and subsequently demolished.

BELOW
Architects Pringle Richards Sharratt created parabolic arches, reminiscent of the Waterloo entrance, for their Sheffield Winter Gardens, completed in 2002.

by Pringle Richards Sharratt. This building has structural arches shaped as parabolas, so that the line of action of the forces lies within the arches when they carry the building's self-weight. The laminates are made from larch, which is relatively hard, minimizing minor damage where it is exposed in the public space, and durable enough to last for a long time if it is detailed to ensure that it does not remain wet. Sheffield Winter Gardens is an example of a project where glulam not only supplies a structurally intelligent solution, but also forms one of the main visual elements of the project. It is a harmonious solution for a space populated by plants.

OTHER ENGINEERED FORMS

A newer type of parallel laminate product is known as laminated veneer lumber (LVL). Thin sheets of wood are peeled from the log (in a similar manner to, but more thickly than, the veneers used in making plywood) and cut into shorter lengths, which are overlapped and glued together to provide the required thickness, and then cut into structural-sized sections. With all the grains running parallel, LVL has the same directional properties as sawn timber, but is more uniform. It is possible to make unjointed lengths of up to 26 metres (85 feet) in this way. Although it may be stronger than glulam, LVL is

at a factory in Weimar, registered a patent in 1906 for glulam structures using casein glue. Casein, which is derived from cows' milk, is not moisture-resistant, however, and the real success of glulam depended on the development of modern glues, such as resorcinol and melamine formaldehyde, which resist both water and heat.

In the UK, for example, the use of glulam dates back to the Festival of Britain in 1951, when it was used for the parabolic entrance arches. In the 1950s, shapes that became fashionable, such as conoids and hyperbolic paraboloids, were built of glulam. The material tended to be used in such buildings as churches, where its aesthetic seemed

appropriate, or in swimming pools and ice-hockey stadiums, where the humid atmosphere could be inimical to structural steel and to the reinforcing steel in concrete.

In the 1970s, glulam became more of a commodity material, as techniques for producing curved beams improved, and high-volume plants were built in many northern European countries to produce straight beams in a wide choice of standard section sizes. As a result, consumption of glulam in the UK doubled between 1985 and 1995.

Recent projects making exemplary use of glulam include the Sheffield Winter Gardens in northern England, designed

commonly seen as a less attractive material, and is therefore less likely to be used as a central architectural feature.

LVL was used by British architect Studio E, with engineer Techniker, to create the panels on a curved pod for an experimental educational building in London. Shigeru Ban used it to create a children's nursery in Odate, Japan, (see page 172), but on his Atsushi Imai Memorial Gymnasium in the same town (page 180) he used laminated strand lumber (LSL). Another material made from veneer sheets is parallel strand lumber (PSL). The sheets are cut into small strips, dried and, after spraying

with adhesive, formed into a billet, which is dried again. This allows the formation of strong, relatively deep structural members that can be used for beams or columns.

BOARD GAMES
There is a wide variety of boards available for cladding, flooring and other

applications. Midway between a board and a structural element is laminated strand lumber (LSL), which can be up to 120 cm (48 inches) long, up to 19 cm (8 inches) wide and 13 cm (5 inches) deep. It is made by cutting logs into short 'bolts' that go through a rotary slicer, making them into strands about 30 cm (11 inches) long. The dried strands are formed into a mat, so that they are parallel to each other, and the mat is bound with adhesive, pressed and dried, before being cut into smaller units.

The best-known board is plywood, a product so common that one tends not to think about how it is made. (Its origins are very old. The sarcophagi of Egyptian mummies were made from a form of plywood.) In fact, the manufacturing method is fascinating and far more closely controlled than might be imagined. Plywood is made by a veneer process, meaning that it is made from peeling a log, but the veneers are much thinner than in the manufacture of LVL.

Modern production began with the introduction of the first veneer lathes in the USA in the mid-nineteenth century.

To make plywood, the veneers are laid so that the grain of each layer is at right angles to that of the layer below, and the two outermost layers are placed in such a way that their grains run in the same direction. A special resin adhesive is used to bond the boards, which are then pressed together under heat. After drying, they are trimmed, then surface defects, such as knotholes, are repaired and the plywood is sanded. Different grades of plywood are determined by the quality of the surface. Birch-faced

plywood – available from Finland, Russia and Latvia – also comes in various grades, determined by whether birch veneers are also used for the internal layers. The more birch there is, the stronger the material will be. Whether or not a plywood is suitable to be used externally depends largely on the glue that has been used in its construction.

(Phenol formaldehyde is fine out of doors, but urea formaldehyde is not.)

The toughest of all plywoods is marine plywood, made from durable hardwoods. Many architects specify marine plywood when it is not really necessary. As a result, there are two types of marine plywood. The type aimed specifically at the construction industry is not, despite its name, deemed suitable for marine use.

Plywood is defined as a layered composite, but there are also more complex kinds, with names such as blockboard and laminboard, which have timber-strip cores and a stiffer surface, often of plywood. These composites offer a way of using lower-quality elements at the centre of a board, while achieving good qualities and attractive properties.

Particle composites come in three categories: particleboards, oriented strand boards and structural particleboard. Particleboards can be of wood chipboard, flaxboard or cement-bonded particleboard. Chipboard, a common and inexpensive material, is made by breaking up wood mechanically into particles, which are graded, dried, blended with glue and formed into mats. The mats are pressed at high temperature and dried to provide a panel that can be trimmed

RIGHT
The McDonald's headquarters in Helsinki, Finland, designed by Heikkinen-Komonen Architects and completed in 1997, was one of the first buildings to be clad in Thermowood, softwood that becomes harder and more durable as a result of heat treatment.

and sanded to size. Again, there are a number of grades available, distinguished by the particle size and the glue used.

Flaxboard is not a timber product at all. It uses flax waste, rather than wood particles, and has a lower specification than the other particleboards. Cement-bonded particleboard uses cement, rather than glue, to bind wood particles, which form only 20 to 30 per cent by weight of the total. It is about twice as dense as plywood, but offers advantages in wet conditions, in a fire or when under fungal or insect attack. It also has good sound absorbency.

Oriented strand boards are intended for external use. They are similar in both price and performance to softwood ply. Softwood strands, at least twice as long as they are thick, are used to make the boards. The strands are in three layers, with those in the top and bottom layers oriented roughly parallel to the length of the board and those in the middle layer at right angles. Exterior-grade phenolic resins are used as a binder.

Fibreboards also come in a variety of forms. The best-known are hardboard and MDF (medium density fibreboard), the latter beloved of the makers of inexpensive furniture and participants in television makeover shows. All the boards mix wood fibres with a binder, by either a wet or a dry process. MDF is made by a dry process that gives it a very smooth finish. It is fairly dense, although it will creep under load with time; and, although it is easily cut, the process produces large quantities of dust.

HARD TIMES

Composite materials make it possible to get more out of relatively low-grade timber. Another approach is to improve the properties of the timber itself, enabling the use of relatively inexpensive softwoods where, previously, hardwoods would have been specified. There are two main techniques, one that 'cooks' the timber and another that impregnates it with a binder. Both tackle the weak points of timber – susceptibility to rotting and lack of dimensional stability – as well as improving its mechanical properties.

Water is the enemy of timber. Damp wood is susceptible to attack by fungi and insects. In addition, timber will swell when damp and shrink when it dries out, a cycle that can lead to cracking or to problems with adjacent materials. The rule of thumb is that hardwoods perform better than softwoods, and that some very dense (and very expensive) hardwoods perform very well indeed. In many applications, the solution has been to use softwoods treated with a preservative. The most commonly used preservative is copper chrome arsenate (CCA), which, although highly effective, is toxic to wildlife, and the ingredients are unpleasant for humans, too. Many people argue that, used properly, CCA is perfectly safe, but even its proponents agree that waste timber treated with CCA should not be used for domestic fires or barbecues. Concern about CCA arose particularly because of its widespread use on playground equipment, and some countries, including Germany and Austria, have placed severe restrictions on its use.

Environmentalists see boron as the most acceptable alternative to CCA. But a treatment introduced by Finnish company FinnForest actually removes the need for chemicals. Called Thermowood, it involves heating timber for up to 25 hours at a temperature of 190–240°C (375–475°F) in an oxygen-free environment, so that it doesn't catch fire. During heating, the moisture content drops dramatically and the wood undergoes an irreversible change, becoming harder, more stable and darker in colour, with a moisture content of only about 8 per cent. There is a similar process available in France, and one in the Netherlands called PLATO. The only disadvantage, apart from the energy used in heating, is that the bending strength of the wood reduces slightly and it becomes more brittle. This can make fixing difficult unless, for example, holes are pre-drilled. Pioneered in Finland, and first used on such projects as the McDonald's

BOTTOM AND BELOW
Careful detailing,
such as the use of
overhanging roofs and
of plinths, has allowed
architects van
Heyningen & Haward
to specify softwood
cladding on buildings
such as the Sutton
Hoo visitor centre in
Suffolk, England. Timber
is also used extensively
in the interior.

headquarters in Helsinki, Thermowood has been specified on a number of residential projects in Denmark. In the UK, one of its first commercial uses was on the Beehive shopping centre in Cambridge, designed by Benoy.

The alternative is a process offered by Indurite, a company founded in New Zealand but now based in London, which involves impregnating timber with a starch-based, non-toxic food derivative and then drying it at relatively low temperatures. The starch reacts with the timber, increasing its hardness. Dyes can also be included in the process, changing the colour of the

wood. According to Indurite, durability and dimensional stability improve greatly. Originally used on New Zealand softwoods, the process is now being extended to other timbers. The Building Research Establishment in England has been awarded a grant from the Forestry Commission and Scottish Enterprise to evaluate the performance of the Indurite treatment on UK plantation species. Indurite also has an office in Seattle, Washington, from where it is exploring the possibility of introducing the process to the USA. Until now, the focus has been on flooring and some specially designed furniture, but this may be extended to cladding.

BETTER BY DESIGN

Improved design has made it possible for softwoods to be used in applications not previously considered appropriate. Architect Chris Wilderspin of van Heyningen & Haward, who has worked on a number of buildings that use softwood cladding, advocates the use of preservatives, but likes them to be environmentally non-aggressive. 'My practice's timber-clad buildings, such as the Sutton Hoo visitor centre in Suffolk, and the Gateway to the White Cliffs visitor centre in Dover, benefit from generous roof overhangs and proper plinths to lift the cladding away from the ground,' says Wilderspin. Softwood cladding should be used as a rainscreen, he argues, with adequate ventilation

behind it. Even with good protection, there will be some weather-induced changes to dimensions, which the design must accommodate.

WASTE NOT

It is environmentally desirable to keep waste to a minimum. Although lower-grade timber can be used in items such as particleboards, it would be marvellous if the material could find more direct applications. This was the thinking behind the work of architects Ahrends Burton and Koralek, and Edward Cullinan Architects, at Hooke Park, a furniture-making college in England. Working with engineer Buro Happold on a number of projects, they developed ways of building structures

BOTTOM
British architect Benoy
used Thermowood on
its Multiyork furniture
store in Cambridge,
England.

BELOW
At Hooke Park in the
south of England,
Edward Cullinan
Architects pioneered
the use of roundwood
thinnings.

with unseasoned roundwood, the timber from the small trees thinned out from a plantation. Although the aesthetic is slightly folksy, the engineering is not. For example, the professionals developed new steel joints for the timber that would accommodate both its mechanical properties and the shrinkage that would occur as it aged.

A research project funded by the European Community on the use of roundwood concluded that its main application was likely to be in rural areas, particularly for holiday and leisure use, in buildings such as cottages and footbridges. Austria and Finland, in particular, were identified as potential markets. In Finland, for example, 1.8 million cubic metres (64 million cubic feet) of roundwood of 8 to 15 cm (3 to 6 inches) in diameter are produced every year. At present, most of this is used for paper-industry pulp or firewood. Part of the task, the researchers acknowledge, is changing attitudes, but they are also tackling such issues as harvesting, jointing and structural details. Purdue University in Indiana is also carrying out research on the topic, producing prototype frames and details, as well as recommending that the material could find uses in furniture for schools.

GOING GREEN

Another way of reducing the amount of intervention is to use a material such as 'green' oak for either framing or cladding. Green oak is oak that has been stored for 12 to 18 months, allowing it to lose the tensions set up by the felling process. The timber still has a high moisture content, which makes it easier to work than seasoned timber, but the architect must take account of the shrinkage that will occur over the first few years of the building's life.

DEALING WITH STRESS

The processes described above concern the use of timber that is as near to its natural state as possible. But it is equally valid to push the properties of timber to the limit, either by sophisticated engineering or by pairing it with other materials. This extends its performance, allowing it to replace another material or saving on total material quantities, or both.

One common approach is to use stressed-skin panels, composite members in which the framing and the sheet materials are designed to act together for greater efficiency. It was first developed for aeroplane construction, on such planes as de Havilland's Mosquito in the Second World War and Howard Hughes's doomed giant, the Spruce Goose, in 1947. As with glulam, the technique was dependent on the availability of a new generation of glues, with high-strength options based on

BELOW LEFT
Canadian architect
Bing Thom used
stressed-skin
construction for the
roof of the Lo
Residence in
Vancouver, Canada.

BELOW
Konrad Frey used
stressed-skin ply in
box sections for a roof
that spans 16 metres
(52 feet) on his
prototype solar-
production building at
Hartberg in Austria.

resorcinol formaldehyde replacing the traditional casein glues. From aeroplanes, the technology migrated to floors and, later, to buildings. American architect Bruce Goff was one of the first to recognize its potential. Working with engineer J. Palmer Boggs, he designed undercut walls and coffin-shaped profiles of stressed-skin plywood in the geometrically extraordinary house he created for his patron, Joe Price, in 1956. Some recent examples of the technique include elegantly understated buildings for the Ecopark in Hartberg, Austria, by Konrad Frey, and the much more exuberant stressed-skin roof of the Lo Residence in Vancouver, by Canadian architect Bing Thom, which combines plywood with zinc cladding.

ON A PLATE

Another flexible technique involves the use of lamellar roofs, where a number of small regular elements are built up to form a plate that acts structurally in a continuous way. It makes possible a building of great technical efficiency and often of considerable beauty. The technique has been used successfully at Hounslow East station in London (see page 118) and also at an office headquarters in southern England, where Anglo-American architect Gensler created four gridshell roofs that define the form of the buildings.

As with all truly three-dimensional modern design processes, the successful and efficient execution of such

BELOW LEFT AND
BELOW
Gensler designed four
gridshell roofs for this
financial services
headquarters in
southern England.

structures depends on sophisticated computer design techniques, as well as on the imagination of architects and the clear thinking of engineers. Already, it is possible to perform routinely calculations that once would have been impossible. As computing power continues to improve, we should see an increasing and increasingly sophisticated number of these natural-seeming forms.

MIX AND MATCH

Now that timber is no longer the preserve of the purist, imaginative designers often use it in combination with another material, most often steel, to take advantage of the properties of both. At the Swiss pavilion for Expo 2000 in Hanover, for example (see page

56), Peter Zumthor used steel tension rods to give stability to the stacks of unseasoned timber that formed the walls.

Another approach is to use flitched timber, where steel plates are sandwiched between timber elements. The steel element provides strength and rigidity, with the thicker timber elements preventing buckling of the steel plate. In this way, the structure makes very efficient use of both elements. Sharples Holden and Pasquarelli, working with engineer Buro Happold, used this technique on the house for a carousel at Greenport Village, New York (page 196), in order to make the structure as slender as possible.

It is also possible to make composite beams by combining different types of timber products. Typically, a webbed beam will use solid timber or structural timber composites for the top and bottom flanges, and plywood or other board material, such as oriented strand board, for the web. The flanges resist the tensile and compressive forces, and the web resists the shear.

HOW DOES IT LOOK?

Given the wealth of ways in which timber can be used, is wood architecture moving in a particular aesthetic direction? The simple answer is no. As more architects embrace the use of timber, so the diversity of ways in which it is used increases.

Some architects take a deliberately traditional approach. You will have no difficulty in buying an off-the-peg log cabin if that is what you want, a traditional structure that comes complete with electric sockets, sewage outlets and doubtless a place to position a satellite dish. This sort of lowest-common-denominator architecture is often far less offensive than a lot of the speculative housing that goes up in brick and block, or in concrete and render.

But even the log cabin has been embraced as an idea that can be brought up to date. On the outskirts of Moscow, the traditional *dacha*, or wooden chalet, has been recreated on a Brobdingnagian scale to contain a

BELOW
At first sight, this log-built house outside Moscow by First Architectural Studio/ARP Studio looks relatively traditional …

LEFT
… but internally, the illusion disappears rapidly, as the architect delights in the contrast between the conventional material and the high-tech elements.

treasure-house of historic fittings and furnishings for sale, known simply as 'Dacha'. More radically, Moscow practices Savinkin/Kuz'min and First Architectural Studio/ARP Studio have designed out-of-town homes that use stacked logs as a major element, but juxtapose them with concrete and steel. There is something shocking about the modernist cliché of a bright-blue swimming pool crossed by a steel and glass bridge when it sits in a log building.

Russia, which has more forest than any other country, also has a tradition of workmen skilled with axes, capable of adapting any element in situ to make it fit. This craft approach still survives in many countries and is used on the most surprising buildings, for instance on Foster's Chesa Futura building in Switzerland (see page 70), where the international high priest of high-tech has used local craftsmen to cut larch shingles in a traditional manner. Such pragmatism was probably behind the decision of a trio of young London-based architects, Silvia Ullmayer, Annalie Riches and Barti Garibaldo, to choose engineered timber as the structural material when they experimented with the concept of designing, and partly constructing themselves, three linked houses for their own occupation.

TRANSGRESSING TRADITION
Some of the projects in this book are the opposite of pragmatic. Architects and engineers have pushed timber to its limits to create tours de force of structural engineering, elegant structures that also leave observers muttering, 'I didn't realize you could do that.' Yet there is a kind of transgression here. Yes, timber can perform magnificently and be used in more slender forms than anybody once imagined; it can span great distances and adopt graceful curves; it can be designed so that the traditional enemies of fire and damp and rot are virtually banished. Timber, so often associated with tradition, with the vernacular, can be a thoroughly modern material.

It can, but it isn't exactly easy. One architect has calculated that there is a substantial cost premium in choosing to use wood rather than steel or concrete, simply because of the extra work required by the designer. Steel and concrete are the materials that evolved to satisfy the ambitions of today's ever more demanding designers – and they still have the edge. We can now construct six-storey buildings in timber, but we will never see a 40-storey skyscraper with a timber frame. There are some magnificent timber footbridges, but we will never see a timber suspension bridge carrying six lanes of traffic over a river estuary. Timber may work well as a material for railway stations, but it will not be used to form the tunnels that carry those railway lines underground.

and pushing the boundaries sets up an opposition to that idea, a tension that balances tradition with invention and makes timber one of the most exciting and modern materials with which to work.

British architect Bernard Stilwell admits that he used timber as a Trojan horse when he designed a library building in March, Cambridgeshire. 'If we are doing some fairly serious architectural things with severe geometry,' he says, 'timber avoids people's feeling that the shape is too harsh, and hence trying to soften it up. It is something they can empathize with.'

By that definition, some of the buildings in this book are also Trojan horses. And all of them are surprising. Each is an ambitious building in its own way, and because it marries ambition with a material that people know through historical association and everyday familiarity, it confounds the users' expectations. The best timber buildings convey a sense of newness and unfamiliarity, while remaining almost incapable of being ugly or alienating. That is what unites the buildings in this book, whether grown in place from a group of trees or engineered with the most advanced CAD tools available.

At its limits, timber will never be able to do what steel and concrete can do. All the efforts of designers, all the clever calculations, the manufacturing technology and the tight tolerances of construction can only help to bring it closer to the performance of rival materials. It will never surpass them. So why bother?

WARM FEELINGS

For some (probably most) architects working with timber, part of the justification for its use is environmental. But there is also the issue of appearance. Timber is warm, non-threatening, tactile. Even the grey to which exposed cedar weathers is a warmer grey than the grey-white

of concrete or the blue-white of untreated steel. In all buildings that use timber, with the strange exception of the timber-framed house in places with a tradition of masonry construction, you see the timber that is used in the building. It gives a human scale; it is a material that most people feel happy with, which they like to touch and, often, smell.

Architects designing in steel and concrete can fall into the trap of making a building seem too impersonal, too machine-like, too large and cold. With timber, the danger is the opposite – a building can seem overcrafted, too *heimlich*, too folksy. But an architect who starts playing with technology

What could be more natural than a timber building? Despite the engineering expertise, the careful fabrication, the timber treatments, the glues that hold manufactured timber together, wood is still acknowledged and valued as a natural material. For those who want to build in beautiful spots, it seems, dare one say, the natural choice.

It certainly was for Steven Holl when he designed the Y House in the Catskill mountains of New York State, and for Cutler Anderson with the Reeve Residence on Lopez Island off the coast of Seattle (see pages 40 and 44). Both these houses occupy lovely sites and are oriented to make the most of stunning views. Australian architects, such as Sean Godsell and Ken Latona, have followed the same strategy, establishing what seems to be a new national tradition; timber has also been used in such projects as Brian MacKay-Lyons's Danielson House in Nova Scotia, Canada.

The examples mentioned are all relatively large houses that offer the opportunity of gracious living, but there is another approach that harks back to the idea of a simple wooden hut. If you want to get away from it all – and many new timber houses tend to be weekend homes, rather than full-time residences – then surely it would be nice to divest

yourself of a lot of the paraphernalia of everyday life. Roberto Briccola's simple, but sophisticated, wooden refuge in Switzerland draws on the history of wooden chalets and log cabins in northern Europe (see page 24). In Denmark, Henning Larsen has created an artist's house in Zealand of similarly breathtaking simplicity; in Finland, Olavi Koponen has designed a house on an island that was not only built by hand, but also had many of its materials delivered by horse and cart. Nor does this philosophy need to be restricted to one-off buildings for the well-heeled. At Paintrock Camp in Wyoming, Charles Rose Architects have designed an education centre for city children, dedicated to ensuring that they are as close as possible to nature (see page 34). The project elevates them above the ground on a series of timber platforms and lets them sleep in buildings that can open up to the natural surroundings as much as the weather allows.

Being in touch with nature involves more than lovely views and minimal visual intrusion on the landscape. Another issue is using wood in a way that responds to how trees grow. At the Weald and Downland Museum in southern England, Edward Cullinan Architects have collaborated with one of the most advanced engineering practices to use green oak (that is, oak

that has not been dried) to construct a gridshell and clad it in locally sourced timber (see page 28). This follows earlier projects that made extensive use not just of green oak, but also of timber thinnings, trees usually discarded as part of the cultivation process.

Such projects incorporate timber that has received minimal treatment after being cut down. But the real purist may ask why it is necessary to cut down trees at all. For example, in Germany, Marcel Kalberer has brought a strong architectural sensibility to a project that otherwise might seem to be on the lunatic fringe: training trees to grow into 'living' structures (see page 20). Although hardly likely to become a mainstream design method – shelter is incomplete and seasonal – Kalberer's work demonstrates that it is possible to adopt an extremely fundamental approach to the use of timber and still create something beautiful.

VEGETAL BUILDING

Auerstedt, Germany | Sanfte Strukturen | 1998–2001

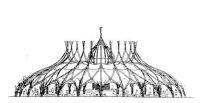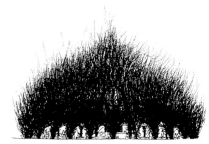

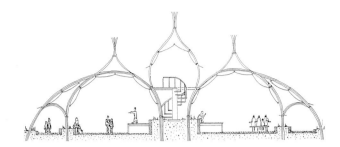

OPPOSITE

TOP   Marcel Kalberer visualized the way that the Auerworld palace would evolve through growth, from shortly after completion in 1998 (far left) up until 2012 (far right).

BOTTOM   Leaves are starting to obscure the structure in summer.

THIS PAGE

TOP   Section through the Auerworld palace.

TOP RIGHT   The flexibility of willow whips allows them to adopt the architect's chosen curves.

ABOVE   The bare structure in winter.

Good buildings are often seen to improve as they mature – but this can never have been so literally the case as in the willow constructions of Marcel Kalberer, a Swiss-born architect who works in Germany. Kalberer's buildings literally grow into place, since they are living both at the time of construction and throughout their lives. Whereas the structure is evident from the beginning, the 'cladding' takes time to appear – and, indeed, disappears each winter. Kalberer exploits the extraordinary properties of willow, one of the easiest woods to grow. It can be harvested in winter as whips (long, slender branches) that are then put in the ground and will most probably take root and start to grow. Their flexibility means that the whips can be tied over frames or simply fastened together to form shapes.

Kalberer did not invent the idea of living construction. Popular in the UK and Scandinavia, and now being adopted in parts of the USA, it is often used to create living fences and children's play structures. At Grosvenor County primary school in Staffordshire, England, in another recreation of a traditional art, it has been used to make a maze, appropriately in the form of a Staffordshire knot. There are also designers who create living furniture, either using such materials as red alder to train a bush into a shape and then harvest it in finished form, or making garden furniture in willow that can then take root in a new location. What makes Kalberer special is that he brings an architect's sensibility and an interesting set of cultural references to his living buildings. He uses ideas drawn from Mesopotamian architecture to design his green buildings, creating interlinked systems of arches with pointed tops that seem part-Gothic and part-oriental.

Kalberer's first major construction was the Auerworld palace at Auerstedt near Weimar, built by 300 volunteers from all over the world in the spring of 1998. On this and subsequent projects, he has worked with a group of builders and constructors known as Sanfte Strukturen. The palace now serves as a focus for community events and has attracted more than 80,000 visitors to a very ordinary location, in particular to attend moon festivals. Kalberer has photographs of the festivals that show naked people frolicking among the greenery – an indication of the kind of Arcadian innocence mixed with mysticism for which he strives. Since this is the oldest of Kalberer's major willow constructions, the Auerworld palace is also the one that has filled in the most, with the growth of secondary and tertiary branches. Indeed, he has drawings of how he expects it to look up until 2012, when its shaggy, overgrown appearance almost entirely obscures the original structure. Like all

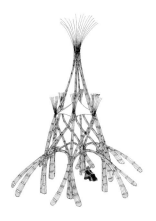

buildings, willow constructions require maintenance: twice a year, unrestrained branches must be tied in and any undesirable or dead growth trimmed.

Since Auerstedt, Kalberer has moved on to even more ambitious projects. Whereas the Auerworld palace has a circular plan, the Weidendom cathedral in Rostock has an almost conventional, cruciform church plan, with a nave and an apse. In fact, Kalberer's constructions are too ambitious to be constructed entirely of willow. Look closely at Auerworld and you will see some slender, steel circumferential elements, plus a crowning item, reminiscent of the spikes seen on the top of German dress helmets of the First World War. At the Weidendom, preformed steel tubing was used to support between 700 and 900 cubic metres (24,700 and 31,800 cubic feet) of willow, formed into arches up to 9 metres (30 feet) across. Huge numbers of volunteers also worked on this, and 600 of them completed the construction in two months, with no mechanical assistance. Ecumenical Christian and Jewish services were held in the building during the Rostock garden festival in 2003. To improve the level of shelter, tents were suspended within parts of the structure.

Kalberer employed a similar idea – of suspending a tent within a willow structure – at the Boo1 City of Tomorrow exhibition in Malmö, Sweden, which used a newly created city district on reclaimed industrial land to demonstrate sustainable approaches to building and city planning. Kalberer's pavilion, a single, tall Gothic element, formed the entrance to a series of 25 'secret gardens', set, appropriately, within a willow forest. But it also served as the shelter for Sweden's Crown Princess Victoria when she formally opened the exhibition – hence its name of Victoria Pavilion. Built only weeks before the start of the show, the pavilion was irrigated intensively, so that it started to grow almost at once.

Victoria Pavilion was a very temporary structure. Some of Kalberer's other shelters can expect a longer developmental period and lifespan, although none is likely to endure into the twenty-second century. But that is not the point. Providing some kind of shelter in summer, they are a means of getting in touch with nature and, one suspects, a whole raft of semi-mystical ideas. Once their time is past, through redundancy, collapse or death, they can (give or take a few bits of steel) simply rot down into the ground again.

THIS PAGE

TOP LEFT AND RIGHT   The Victoria Pavilion was built in 2001 as an entrance feature for the Boo1 City of Tomorrow exhibition in Malmö, Sweden, and to shelter Crown Princess Victoria at the opening.

ABOVE   The Weidendom cathedral was completed in 2001.

OPPOSITE

TOP LEFT   Like all Kalberer's work, the Weidendom draws on forms from Mesopotamian architecture.

TOP RIGHT   The plan of the cathedral, however, is close to that of a conventional Gothic church.

BOTTOM   Tents were suspended within the cathedral structure to provide extra shelter during the Rostock garden festival in 2003.

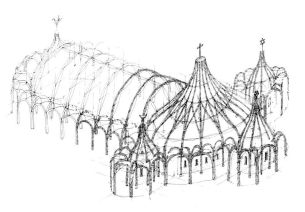

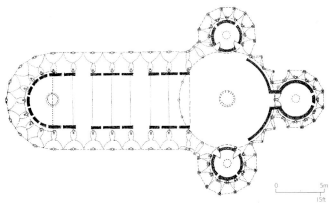

SHELTER

Campo de Vallemaggia, Ticino, Switzerland | Roberto Briccola | 2001

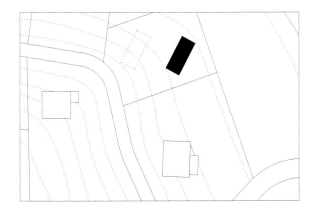

OPPOSITE

TOP — The deliberately asymmetric positioning of the openings in the four façades gives articulation to this simple building.

BOTTOM — (and this page above) Four concrete columns support the building, minimizing its interference with the ground beneath.

THIS PAGE

TOP LEFT — The shelter is sited at an angle to the contours, reducing the change in ground level and taking maximum advantage of the views.

TOP RIGHT — In contrast to the nearby log-built chalets, Briccola's design draws on the traditions of the grain barn.

BOTTOM RIGHT — External cladding consists of tongued-and-grooved larch planks.

Who wouldn't like a weekend house in the Ticino region of Switzerland? It offers pristine Alpine scenery, coupled with an Italian flavour to the language and the climate, which is warmer than elsewhere in Switzerland. The flora, a combination of Alpine and Mediterranean species, is one of the richest and most interesting in Europe. But building there carries heavy responsibilities. This is not a region devoid of architectural heritage – Mario Botta is one of its native sons – but in the most glorious places, precisely those places that might be desirable to spend a weekend, any kind of construction can feel like sacrilege.

Roberto Briccola has tackled the problem by designing a house that takes to an extreme the concept of 'touching the ground lightly' and is such an exquisitely crafted object that all but the most curmudgeonly of critics would see it as an enhancement of the landscape. Its smallness and austerity doubtless make the owners feel innocent of any charges of city sybaritism. Essentially a timber box measuring 48 square metres (520 square feet), the house sits level on four concrete columns of different heights to accommodate the slope of the site. The idea is that the meadowland beneath the house can continue to flourish, but its resilience when deprived of much of its sunlight and rain remains to be seen.

Although there are some traditional log-built chalet houses within sight, Briccola has drawn on a different tradition, that of the grain barn – which, because it is much more austere, has a greater appeal to contemporary sensibilities. Entrance is from the higher, northern side, with a metal-sheltered porch forming the only extrusion from the rectangular box shape. Internally, there are effectively two open spaces, stacked one above the other and joined by a spiral staircase in the northeast corner. The ground floor is a kitchen, dining and living area, with one small window along the west side, where the kitchen units are, and a larger one on the east to allow views out from the dining table. The whole of the southern end opens onto a balcony, sheltered by the floor above. In this way, the building takes advantage of

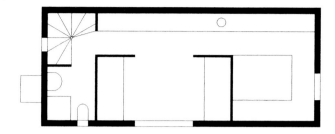

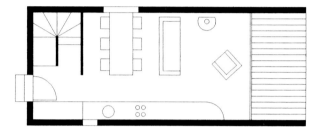

the best views, those to the south and east. The upper floor has a bathroom tucked into its northwest corner, a wardrobe running along the eastern wall and two sets of built-in bunks defining the spaces of two bedrooms. Windows are on the west and south sides, plus a north-facing window above the stair.

One of the pleasures of a building as small as this is that the architect solves the jigsaw puzzle of where to put the internal elements and these then remain fixed for all time. There is no room to permit flexibility and rearrangements, so the windows can be placed precisely to suit the use. This results in a pleasing,

well-considered asymmetry on the perfectly rectangular façades.

The structural frame is of pine, with the exterior post-ventilated cladding in horizontal tongued-and-grooved larch planks. Internal cladding is of plywood for the roof, ceilings and floor. To ensure that the building's austerity does not tip over into discomfort, a reassuring mineral-wool insulation, 14 cm (5½ inches) thick, has been set into the walls, ceilings and floor. The weekend house at Campo de Vallemaggia is indeed a timber box, but one that, given the wonderful views that tempt people both to look out and to venture out, should never feel claustrophobic.

THIS PAGE

ABOVE LEFT   The simple rectangular form makes a pleasing contrast with the rugged outline of the nearby mountains.

ABOVE   Plans of ground-level columns, ground floor and first floor show the simplicity of the form and the careful disposition of windows and furnishings.

OPPOSITE

A south-facing balcony occupies the entire width of one end of the structure. Note how much more slowly the protected timber of the balcony is weathering.

DOWNLAND GRIDSHELL

Sussex, England | Edward Cullinan Architects | 2002

The three-humped whale of the Downland Gridshell at the Weald and Downland Museum at Singleton in Sussex is a building that could only have happened at the start of the twenty-first century. A magnificent structure, a romantic gesture and a tour de force of engineering and innovation, it has a mass of contradictory attributes that add to. rather than detract from, the overall achievement.

The Weald and Downland Museum is an open-air collection of rescued rural buildings from the south of England, which have been reconstructed and refurnished. Native plants grow in the gardens, and rural crafts and traditions have been revived and re-enacted. Representing a past that has virtually disappeared, even if some of its physical fabric remains, the museum straddles, almost entirely successfully, the gap between education and entertainment, between a history lesson and a theme park. This is the setting for the gridshell. The most eye-catching construction on the site, it is the only significant new building, yet it has a workaday purpose. Many grandstanding buildings serve grandiose functions; by contrast, the gridshell is a combination of store and workshop.

Another sort of building could have satisfied the museum's needs at a far lower cost. A simple pole barn, a building clad in corrugated metal, would have been more typical of today's rural vernacular – and, in a sense, a more appropriate continuation of the pragmatic rural tradition that most of the museum's other buildings represent. But the museum had greater ambitions, and was able to satisfy them because, for a brief period, money was available in a way that it hadn't been for a long time previously, and probably won't be again, thanks to the launch in the 1990s of the UK National Lottery. Like all national lotteries, it attracted an initial surge of enthusiasm from the public that sent ticket sales soaring. This was coupled with a willingness by the British government, before complications set in, to put much of the money into the construction of new buildings. During the short-lived honeymoon period, the Weald and Downland Museum won a substantial grant from one of the distributing bodies, the Heritage Lottery Fund, that allowed it to envisage a far more adventurous building than it would otherwise have been able to do.

Since the workshop would be largely dedicated to reconstructing and refurbishing the timber frames of other buildings on the site, the building needed a large, clear span. This was achieved by a development of the technology pioneered by Frei Otto and engineer Buro Happold at the Mannheim garden festival in the 1970s.

OPPOSITE
Walls of the Downland Gridshell are clad with
locally grown red cedar planks.

THIS PAGE
RIGHT   Despite using advanced technology, the building sits
         comfortably in its rustic setting.
BELOW   The form of the gridshell resembles two egg timers,
         laid end to end.

Buro Happold, also the engineer on the gridshell project, had in the meantime acquired considerable experience of working with green (unseasoned) oak, the main material used in the museum structure. Increasingly sophisticated computer technology allowed calculations of a complexity that would not have been possible earlier. Yet at the same time the building process was remarkably 'hands on'. Members of the architect's team, developing the ideas from an unbuilt structure that was intended to form part of an exhibition at Paris's Pompidou Centre, were frequent visitors to the site. They were not merely inspecting progress. They were carrying out practical experiments to see whether the calculated forms were achievable in practice.

Edward Cullinan, the founder of the practice, has described the gridshell as

'far beyond Hi-Tech, in that it prophesies a new and ecological use of materials in the light of the knowledge gained during the last, high-technology century.' It is remarkably unwasteful of materials and has a hard logic underlying the romanticism of its form. The three-humped shape is in no sense whimsical. It provides a greater stability and hence economy of materials than a simple tunnel shape would have done.

Although the visitor is immediately aware of the timber structure, it is in fact only the upper layer of the building. The lower layer, the store, is a concrete structure, dug into the chalk hillside, and so energy-efficient that it needs only a simple domestic boiler to heat it. Glulam posts and beams support the heavy roof of the store, made of 80 mm (3 inch) tongued-and-grooved British spruce, which acts as the structural

diaphragm floor for the gridshell. The gridshell itself is the first double-layer timber gridshell in the UK and only the fifth worldwide. It is a doubly curved shell made from oak laths, 50 mm (2 inches) wide and 35 mm (3⁄5 inches) thick, in four layers. Its double-hourglass shape, 48 metres (157 feet) long, is 16 metres (53 feet) across at its widest points and 11 metres (36 feet) wide at the waists. The internal height varies between 7 metres (23 feet) and 10 metres (33 feet).

Green oak was used because its high moisture content gives it great flexibility – essential during the forming process, since the laths in the gridshell were curved in both directions. Once the laths were in their final position, natural drying strengthened the structure. Oak, traditionally used for British boat building – and this structure in many

ways resembles an upturned boat – has the advantage of being twice as strong as equivalent sizes of other common timbers. This meant that smaller cross-sections could be used, and bent to the required radius, giving the whole structure a lighter look, and more than offsetting the fact that the oak required a higher bending force to achieve the desired curvature.

Ironically, given the underlying ethos that materials should as far as possible be sourced locally, the oak had to be imported. The UK had achieved its deforestation centuries before anybody started to worry about the depletion of the Amazon jungle, and there were no adequate supplies in the densely populated southeast of the country. British timber would have had to travel long distances by road. It was decided that it would be 'greener' to import the

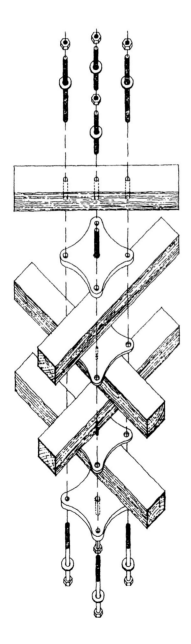

wood from Normandy, in France, just across the English Channel, where there are sustainable plantations.

It was essential that there were no defects in the oak laths. Natural defects in the wood were therefore cut out, and the remaining short lengths of timber were reconnected into 6 metre (20 foot) lengths using finger joints, which could be produced quickly and cheaply with minimum waste. The success of these finger joints relied in part on the latest in glue technology, which could accommodate the high moisture content of up to 40 per cent in the green oak. In contrast, the 6 metre (20 foot) lengths were then joined on site using traditional scarf joints to produce continuous laths – up to 37 metres (121 feet) long for the lattice laths and 50 metres (164 feet) long for the longitudinal rib laths. These scarf joints were similar to the techniques used on many of the historic buildings that make up the museum collection.

The gridshell was built flat on top of a special lightweight scaffold that stood on the timber floor and was then carefully pulled down into position and fixed. The architect and engineer worked closely with the contractor, the Green Oak Carpentry Company, which devised one of the most essential elements – a pinless connecting device for the laths, which gives them stability, but allows the

necessary degree of movement. The device consists of three plates. The centre plate has pins simply to locate the grid geometry of the middle lath layers, and the outer plates loosely hold the outer laths in place, allowing sliding during the formation of the shell. Two of the four bolts locating the plates were used to connect the diagonal bracing, bolted in place to provide shear stiffness after the shell had been formed. Green Oak has now patented this device.

In areas of high load, the laths were spaced 50 cm (20 inches) apart. Over the rest of the structure, this was increased to 1 metre (3 feet). Once the gridshell had been formed – the most exciting and nerve-racking part of the process – additional, diagonal bracing was added, consisting of longitudinal and transverse timber rib-laths, fixed to the nodes to provide shear resistance and to 'lock in' the shape. The bracing also supports the wall and roof cladding. Shear blocks were screwed into place between the layers of the shell. In this way, parallel lines of laths act compositely; the sizes and positions of the shear blocks were arranged to suit the forces determined by computer analysis. Only at this stage could the last of the temporary props be removed.

The success of the finger-jointing technique was demonstrated by the fact that, with some 10,000 finger-joints

THIS PAGE

BELOW   Clerestory glazing between the roof and the walls is
of polycarbonate, as glass would not have been able
to accommodate the movement.

RIGHT   The view into the building is light and welcoming.

OPPOSITE

LEFT   Plans show, from the bottom, the below-ground
store, the workshop and the central roof strip on
the double-egg-timer form of the gridshell.

RIGHT   Catenary arches of laminated green oak support
the awnings at either end of the building.

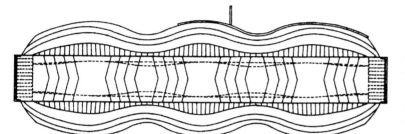

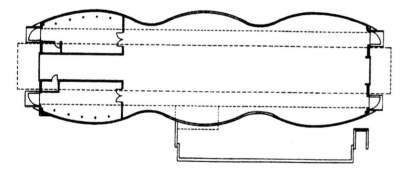

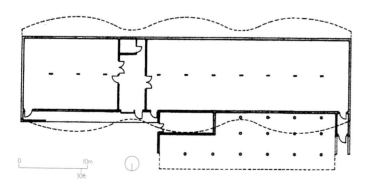

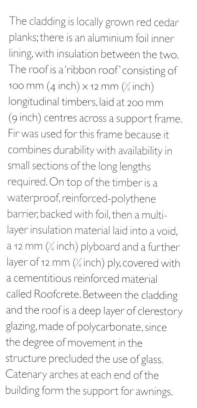

in the structure, there were only 145 breakages during shell-forming – even though the curvature of the structure was in some places only 6 metres (20 feet), the absolute minimum considered possible. Repairs consisted of introducing solid blocking between laths at the point of failure.

The cladding is locally grown red cedar planks; there is an aluminium foil inner lining, with insulation between the two. The roof is a 'ribbon roof' consisting of 100 mm (4 inch) × 12 mm (¼ inch) longitudinal timbers, laid at 200 mm (9 inch) centres across a support frame. Fir was used for this frame because it combines durability with availability in small sections of the long lengths required. On top of the timber is a waterproof, reinforced-polythene barrier, backed with foil, then a multi-layer insulation material laid into a void, a 12 mm (¼ inch) plyboard and a further layer of 12 mm (¼ inch) ply, covered with a cementitious reinforced material called Roofcrete. Between the cladding and the roof is a deep layer of clerestory glazing, made of polycarbonate, since the degree of movement in the structure precluded the use of glass. Catenary arches at each end of the building form the support for awnings.

These are made of laminated dry oak.

A well-thought-out environmental strategy minimizes the use of energy for heating and cooling the building. For the museum, the gridshell is much more than just a building fit for its purpose. It is also a reflection of the institution's aims, an icon and a draw in its own right. The museum has made the most of this, using the gridshell to hold weekend seminars on timber. On its website, at http://www.wealddown.co.uk/ downland-gridshell-construction-progress.htm, there is a detailed description, in words and pictures, of the construction process.

In 2002, the gridshell was a runner-up for the Stirling Prize, the UK's most prestigious award for architecture. It was narrowly beaten by an urban steel footbridge. The function and appearance of the two projects could not be more different, but each exemplifies the spirit of innovation in UK architecture at the start of the twenty-first century. That this innovative spirit can find expression in a material as redolent of history and tradition as timber is a tribute to the imagination and dedication of the team involved in the design and construction of the gridshell.

0   10m
    30ft

PAINTROCK CAMP

Hyattville, Wyoming, USA | Charles Rose Architects | 2000

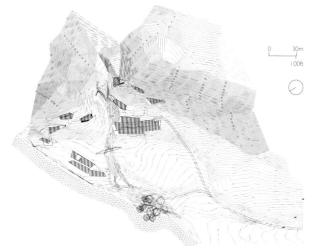

OPPOSITE

TOP   Paintrock Camp is set at the base of two converging
           canyons in the Big Horn Mountains.
BOTTOM   Buildings are carefully placed to take best advantage
           of views and to respond to the contours of the land.

THIS PAGE

ABOVE   With such magnificent scenery all around, the
           architect was keen to avoid creating a monumental
           structure.
RIGHT   Platforms and changes in level ensure that the
           visiting young people are in constant touch with
           nature and have plenty of places for informal social
           interaction.

The vast size and magnificent scenery of the USA mean that, if a charitable foundation wishes to give inner-city children a taste of country life, it can take them somewhere truly remote. Paintrock Camp at Hyattville, Wyoming, is such a project. Sponsored by the Alm Foundation, the private foundation of the chief executive of Coca-Cola Enterprises, it allows 76 young people from Los Angeles schools to enjoy horse riding, trekking and other outdoor activities during their summer vacation.

Set at the base of two converging canyons in the Big Horn mountains, Paintrock Camp is part of a 44,000 hectare (110,000 acre) ranch. With its substantial accommodation, plus facilities that include stables, a dining hall, a swimming pool and recreational areas, it could have required a very large building. But this would have been against the spirit of the place. Instead, Charles Rose Architects have created a number of separate structures. Oriented in a way that looks almost random, they are, in fact, carefully placed to take the best advantage of views and to respond to the contours of the land, so that, although the residents have a sense of enclosure and of access to other parts of the site on elevated walkways, they are constantly in touch with the outdoor world that they have come to Paintrock Camp to experience. By raising most of the buildings on stilts, the architect has reduced the impact

OPPOSITE

TOP LEFT   One side of the sleeping cabins can open up
entirely in good weather.

TOP RIGHT   Sections and plans of a typical sleeping cabin.

BOTTOM   Steel-framed and standing on steel platforms, the
sleeping cabins are clad with western red cedar.

THIS PAGE

ABOVE   The windows have wooden shutters.

RIGHT   Student accommodation and the dining hall make
up the main eastern camp. Housing for the camp
staff and visitors is placed to the west. To the
southwest are the stables and an offloading shelter
for arrivals and departures.

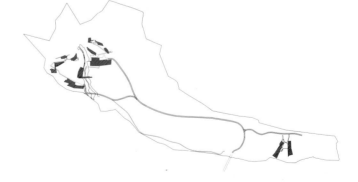

0   120m
400ft

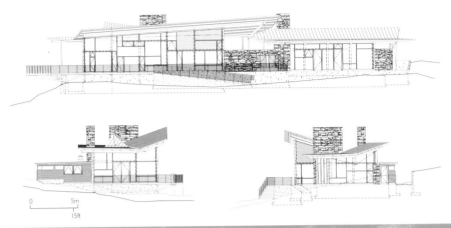

TOP LEFT   Recycled Douglas fir beams support the butterfly roof of the dining hall.

TOP RIGHT   Long section and cross sections through the dining hall, the largest building in the camp and its social centre.

BOTTOM   Timber frames the glazing of the dining-hall walls.

THIS PAGE

ABOVE   External sloping columns support the roof of the stable block.

TOP RIGHT   Inside the stable block, walls and stalls are clad in timber.

on the ground, and its use of platforms and walkways creates a series of spaces for informal interaction, or simply for standing and gazing, which make up a crucial part of the experience.

Most of the accommodation is contained in the main, eastern camp. Some way away, to the west, is housing for the camp director and for guests. And to the southwest, near the boundary of the site, are the stables and an offloading shelter to deal with arrivals and departures.

In the eastern camp, there are two clusters of student accommodation, connected by a bridge. Each cluster comprises three sleeping cabins for boys and three for girls, with a shower block for each cluster. Supported on steel platforms, the steel-framed cabins are roughly triangular in shape. They are compact, with bunk beds and storage all built in timber, and stand on a timber deck. The material used throughout is western red cedar. The cladding is of horizontal planks, and there are large rolling doors that can be opened completely in fine weather. On the opposite side are small timber shutters

that open out from the wall – with the guarantee of a dramatic view outside.

The dining hall forms the largest building in this grouping. Standing on a concrete foundation, and with some stone walls and a stone fireplace and chimney, it is nevertheless a largely timber-framed structure, with butterfly metal roofs supported on recycled Douglas fir beams and an almost Mondrian-like pattern of framing to the glazed walls. Recycled Douglas fir has also been used on the stables complex, where, externally, hockey-stick columns support the sloping roofs. There is a sheltered entrance between the two sets of stables, which are set at right angles to each other, and a combination of timber and corrugated metal cladding externally. Inside, the walls and stalls are clad in timber. The restricted palette of materials gives an architectural unity to the camp, in a manner that is far more pleasing and appropriate to the surroundings than any rigidly planned grid would have been. The camp engages continually with its environment, while creating a series of spaces and shelters that is both reassuring and intriguing.

REEVE RESIDENCE

Lopez Island, Washington, USA | Cutler Anderson Architects | 2002

OPPOSITE

TOP   Architect James Cutler sited and designed the
building to minimize its visual impact on the
surrounding landscape.

BOTTOM   A green roof, sloping at the angle of least resistance
to the prevailing wind, minimizes drag and visibility
from further up the slope.

THIS PAGE

ABOVE   Sketch showing how easily the house fits into the
landscape.

TOP RIGHT   A retaining wall of moss stone from Montana assists
with the camouflage.

There are generally two groups of
people who want houses on islands.
One group dreams of a tropical climate,
a beach and palm trees. The other wants
to be in touch with raw nature, to be
buffeted by the wind and stimulated by
ever-changing views. Lopez is the kind of
island that appeals to the second group.
The largest, flattest and most sparsely
populated of the three San Juan Islands
in Washington state, off the west coast
of the USA, it may have an organic food
store and good internet access, but it is
still a rugged spot, an ideal place for
watching orca whales and bald eagles.
With regular ferry services, Lopez is
a popular holiday and weekend
destination for people from Seattle,
who enjoy its sunnier climate, as well
as its other advantages.

When Sally and Tom Reeve, two Seattle
residents, decided to build a house on
Lopez, they chose to do so on the
southern tip, the most rugged part of
the island. With architect Jim Cutler, they
set about selecting a suitable spot. He
dissuaded them from their original
choice, on top of a ridge. As well as
being exposed to the winds, which often
reach 110 km (70 miles) per hour, such
a position would, says Cutler, 'have
desecrated the cliff'. Instead, he moved
them to a less obvious place, between
rocky outcrops, with a forest to the
north and the Pacific Ocean to the
south. The principal view from the coast

is of the house's retaining wall in a moss
stone brought in from Montana.
Surprisingly, the moss stone blends
better with the surrounding lichen-
covered rocks than does the local stone,
which is far too white. The other most
visible element is the roof, which covers
the entire property. Treetops here are
often sheared off by the wind, so the
architect gave the roof a low pitch, set
at a similar angle to that of the sheared
treetops. As well as making the roof less
visually intrusive, this also reduces its
wind resistance. Supported on a timber
deck, and highly insulated, the roof is
mostly covered in turf sod, planted with
local plants for increased camouflage.

Beneath the roof are three pavilions,
all angled slightly differently to take
advantage of the views. But all face
roughly south, to the sea. The central,
largest pavilion contains the kitchen and
a massive living room. To the east is the
'master suite' for the family, and to the
west is the bunkhouse, designed
compactly to house up to 12 visitors –
and visitors must be common in this
delightful location. The separate
pavilions means that, when seen from
the north, the house does not present
a single impenetrable wall, but allows
views through passages to the ocean.

The variation in the pavilions has been
made possible by placing the major
elements of the structure outside them,

so that the walls have to support only their own weight. Beams and columns are made of four pieces of fir, flitched with steel, chiefly at the connection points. On the northern side, where the roof is highest, the columns are cross-braced. In recognition of the composite nature of the construction, each X-shaped brace consists of one wooden element and one galvanized-steel plate, joined at the centre with a shear connector and a through bolt.

The building is clad with cedar shingles, although on the southern façade there is as much glazing as possible. This reaches its apogee in the main, central pavilion, where the timber-framed windows slide back to create an opening nearly 5 metres (16 feet) in width. Internally, timber is widely used, with white-pine wall cladding and built-in furniture contrasting with the darker wood of the underside of the roof. In an expression of hedonism, the building has a hot pool on the seaward side, and the stone terrace that links the pavilions is heated, so that those walking from one to another with bare feet will not suffer.

The house won an award from the Seattle chapter of the American Institute of Architects and a Wood Design Award. In their citation, the judges said, 'The home is a truly sustainable building because it starts with nature. It merges with the rock and flora of the landscape as if it grew there.'

Y HOUSE

Catskill Mountains, New York, USA | Steven Holl Architects | 1999

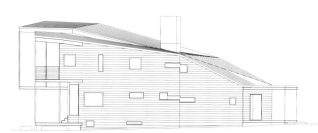

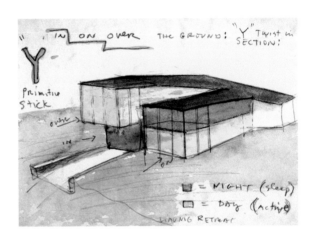

OPPOSITE

TOP   The twisted geometry of the Y House elevations allows it to take maximum advantage of views and redefines the conventional idea of how a house is used.

BOTTOM   Clad in red-painted cedar, and with its steel frame the colour of red oxide, the building makes reference to nearby stables.

THIS PAGE

ABOVE   In one of his characteristic watercolour sketches, Holl based his design on a Y-shaped stick, and defined zones for use during the day and at night.

RIGHT   The living-room balcony, on the west wing, oversails the partly buried children's bedrooms.

For an architect like Steven Holl, whose primary architectural interest is in designing to take advantage of changing light, the site for the Y House must have been nearly ideal. Located in the Catskill mountains, in New York State, it stands on a rising piece of ground and offers views of two different valleys. Holl designed a house that exploited the views and challenged ideas of how a house 'ought' to be, while accommodating the specific requirements of his clients.

In addition to the usual desire of weekenders for a great location within fairly easy reach of the city (New York City is a couple of hours' drive away), the owners wanted somewhere to store and display part of their art collection. Holl gave them a house of just over 325 square metres (3,500 square feet), with a double-height foyer space, ideal for showing art, which splits into two wings, distinguished from each other by their degree of privacy. In common with most of Holl's other projects, the design arose from a series of watercolour sketches. Holl likes to develop these over a period, sometimes several months. 'These sketches are indispensable for me to consider at the same time the conceptual idea, the texture, the colours and the light,' he says. 'At first, I am above all interested in defining where the light comes from and how it moves. In this way, I can already, from the first sketch, prefigure a certain

THIS PAGE
BELOW LEFT   The angle between the two wings allows each to
              have views along a different valley.
BELOW RIGHT   From bottom, ground-floor, first-floor and roof
              plans show the games that Holl has played with
              geometry and the customary disposition of spaces.
              1. living room, 2. master bedroom, 3. bedroom,
              4. dining room

OPPOSITE
The roofs run down towards the entrance, where
rainwater is collected in a cistern.

complexity of possible conditions. Sometimes I will work on numerous sketches and theories before convincing myself definitely to take a certain path.'

For this house in the Catskills, he developed an idea from what he called a 'primitive stick', resembling a catapult. He then defined zones of day and night, and twisted the axis of his 'Y', so that its stem was almost at right angles to its centre line. He was also very interested in the building's relationship with the ground: some parts are within the ground, others stand on it, and others fly over it. The east wing has the dining room at ground level, with the master bedroom above it. In the west wing, the children's bedrooms are set into the ground, with the living room above

them and oversailing them. By putting the two wings at an acute angle to each other, Holl allowed each to face directly towards one of the valleys, and also created a kind of chasm between them that brings in ever-changing light.

The building is steel-framed and steel-roofed, with the steel painted the colour of red oxide. It is clad in cedar, also painted red – a reference in material and colour to the stables of the surrounding farms. At the end of each wing, facing almost due south, are deep balconies, with roofs above them. These balconies are one above the other on the east wing, but on the west wing there is a balcony only on the upper level, projecting beyond the children's rooms below. As well as providing a

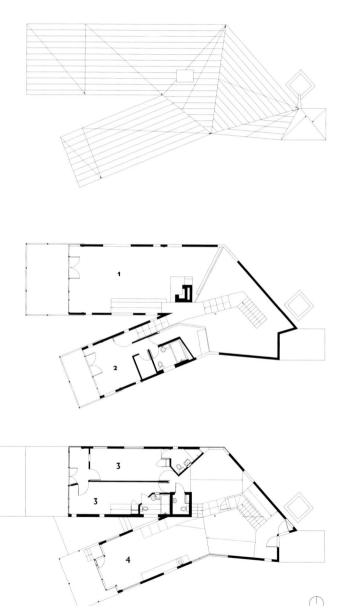

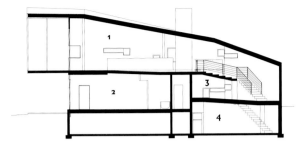

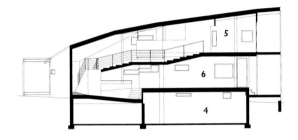

 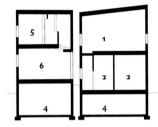

OPPOSITE
Ash is used for the floors and stair treads in the foyer.

THIS PAGE
ABOVE   Long sections through the west wing (top) and east wing (centre); cross sections through the foyers (above left) and the wings (above right).
RIGHT   Changes in level and orientation are most obvious at the point where the foyer bifurcates.

KEY   1. living room, 2. bedroom, 3. foyers, 4. basement, 5. master bedroom, 6. dining room

satisfactory termination to the forward and upward thrust of the house, the large overhangs allow in the winter sun, while providing shade in high summer. As the roofs run downward all the way to the entrance, rain is channelled back to there and collected in a cistern.

Internally, the house has ash floors and stair treads. Ceilings in the foyer are also of wood. The double-height entrance foyer is spatially complex, and it is there that the irregular disposition of windows, designed to make the most of the views, is most evident. Appropriately, perhaps, for a space intended for the display of art, there is a somewhat institutional feel, accentuated by the variation in levels.

Steven Holl has created a house that takes maximum advantage of its magnificent position, and which must interact with its ever-changing surroundings. It is an extravagant house, with a large external surface relative to its volume, costly in terms both of materials and energy consumption. Neither are its games with the levels of sleeping accommodation as revolutionary as one might first imagine. A cynic would simply see the children tucked away on the ground floor, and the service spaces pushed even lower into the basement, while the adults enjoyed all the best spaces. But as a luxury weekend retreat that responds to its site, it must be a great pleasure for Holl's clients to inhabit.

OPPOSITE
Carl-Viggo
Hølmebakk's summer
house at Risør,
Norway, reinterprets
the vernacular with
references to
Norway's classical
tradition and the use
of contemporary
construction
techniques.

Across great swathes of the world, timber was, for many centuries, the natural building material, the people's medium of choice. If wood was readily available, as it was in many temperate and tropical zones until deforestation, then it was used for all but the humblest of everyday buildings. The tools to work it were not sophisticated and would probably have been available to anybody accustomed to harvesting timber as fuel. Timber buildings were both durable and affordable.

As a result, a profusion of forms of vernacular timber construction have arisen. These have been affected by such diverse considerations as climate, the types of timber available and the randomness of historical development. In some places, the tradition of building with wood has virtually disappeared – London, for example, lost its timber buildings in the Great Fire of 1666 and issued ordinances to prevent their replacement. On the other hand, Helsinki, Finland's modern, forward-looking capital, carefully preserves an area of traditional log houses, sandwiched between busy urban highways. And in rural areas of many countries, the traditional forms remain, from the chalets of Switzerland to the coconut-wood shelters of Samoa. One of the pleasures of travel is to see how different cultures have used the same material in different ways.

How can those traditions be reinterpreted today? There are too many twee and unimaginative responses. In the USA, groups of enthusiasts are reviving cordwood masonry, building houses from what look like bundles of firewood. While their technical rediscoveries are fascinating, the architecture is, frankly, hideous and incongruous. In many other places, pastiche has created lifeless architecture, ill-suited to the twenty-first century.

But there is another way of approaching the vernacular. Architects can take elements of traditional architecture and incorporate them into buildings that are very much of the moment. In the countryside around Granada, Spain, Eduardo y Luis Javier Martín Martín have created a house that follows the form of traditional agricultural buildings and uses traditional materials, but is as far as it is possible to get from an unsympathetic barn conversion (see page 66). The architects have offered a compact and supremely rational way of living, within a restricted envelope.

In creating a Norwegian summer house, traditionally known as a 'hutte', Carl-Viggo Hølmebakk has used a 1,000-year-old technique of employing timber wedges to level the foundations, and has clad the building in Norwegian larch. But, by the adoption of prefabrication and the extremely precise detailing, he has given the building a modern twist, while the coherence of his plan draws strongly on the Norwegian classical tradition.

In Switzerland, Foster and Partners have used the traditional Swiss building material to create an apartment block that seems futuristic and is designed using the most advanced computer technology (see page 70). Yet it takes advantage of window details traditional to the Engadin valley and is clad in larch shingles, cut by traditionally trained craftsmen, as part of a strategy to make the building as environmentally sustainable as possible.

Swiss architect Peter Zumthor also used traditional techniques in his Hanover Expo pavilion (see page 56), although these were drawn more from cabinetmaking than from common building practice. And, rather than seeking durability, he designed a structure in which all the elements could be taken away and reused – presumably in more traditional building types.

At Chessy in France, Avant Travaux was concerned less with traditional techniques than with well-established building forms. It recreated the pitched roofs and freestanding nature of the *pavillons* beloved of the French bourgeoisie (see page 52). This may have been a slightly childlike interpretation, but why not? The architect was, after all, designing a centre for children.

Most intriguing, perhaps, is the Sami parliament building in the north of Norway (see page 60). The Sami have no architectural tradition, having lived a nomadic existence, so in commissioning a parliament building they were also looking to create a national architecture. Drawing on the forms of tents and reindeer stockades, the architects, Stein Halvorsen and Christian A. Sundby, have reinterpreted these familiar structures on a larger and more permanent scale. For the Sami, timber is not a traditional material to be superseded by more permanent and weighty materials, such as stone and brick. Instead, it represents a step towards a more settled existence than they knew before and an attempt to create an identifiable architectural language alongside a system of government and representation. As well as drawing on traditions, the architects have created what may be the vernacular of tomorrow.

Chessy, in the Ile de France, is only a stone's throw from Disneyland Paris at Marne la Vallée, but prides itself on being a world apart. Endowed with some historic buildings and plenty of trees and greenery, it promotes itself as a country town, despite the fact that it is rapidly being swallowed up by the Parisian agglomeration.

Mickey Mouse has no place here, but the children's leisure centre has been built in gardens that formerly belonged to the creator of a character equally famous among French children. Jean de Brunhoff was the creator of Babar the elephant. There is something of the enchanted dream of childhood about the leisure centre, but with none of Disney's brashness. At Chessy, children enjoy buildings that are like every child's drawing of a house, only more so. The three linked buildings, with tall, pitched roofs, have been positioned carefully among the existing trees, so that none needed to be cut down. The result is that birches, several gingko biloba and a magnificent cherry push right up against the roofs of the buildings, allowing the sensation from inside of being in a structure more akin to a tree house than a Wendy house. Two of the buildings are side by side, with one effectively cutting into the other. The third is positioned further back, and a discreet entrance is placed between the three, designed so that it maintains the sense of separate structures.

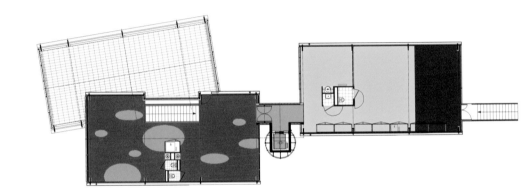

OPPOSITE

TOP   A common entrance (centre) provides access to
the three buildings.

LEFT   Glulam beams create the high pitches of the roofs.

RIGHT   Internally, bright colours help create a child-friendly
environment.

THIS PAGE

RIGHT   There are delicious views of the park that once
belonged to the creator of Babar the elephant.

BELOW   Section through one of the three buildings.

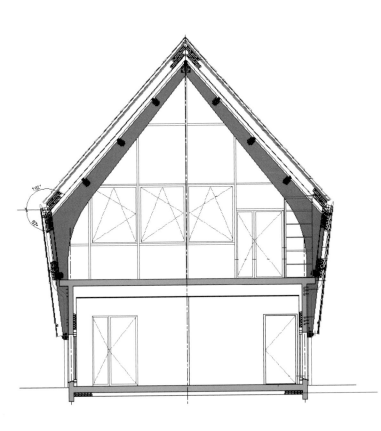

The buildings' roofs are not only taller than those on conventional houses, but also inhabitable, making them effectively upper storeys. Glulam beams were used to create the high pitch of the roofs, which are reminiscent of upturned boats. There are magnificent views out of the gable ends – glazed in the most enclosed of the structures and open in the others, with a railing against which to lean and gaze. While one of the three buildings has a conventional closed-in roof, the other two have strips of metal that provide some shade and changing light patterns for what is effectively an outdoor play area. Dreamy, pearlized colours are used on the roof coverings, and there are bright elements internally. For the 80 children who use the centre,

the experience should be life-enhancing and mind-stretching. Chessy is the kind of place inhabited by those French people who have embraced the dream of a *pavillon*, a stand-alone residence in a plot of land, away from the cheek-by-jowl existence of the city centre or the more contemporary soullessness of the HLMs (large housing schemes on the edges of cities). In their play and their interaction with nature, their children may be acquiring similar tastes. The *pavillon* is not the most sustainable form of urban planning, but this children's equivalent is enchanting. To create such charm without tipping over into sentimentality is the considerable achievement of these buildings, which evoke childhood stories without becoming over-explicit.

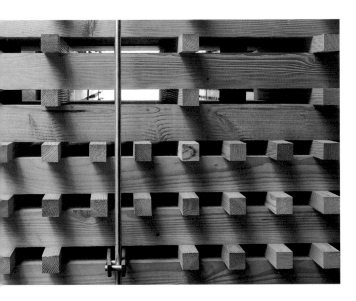

OPPOSITE
The maze consisted of stacks of timber, held together by tensioned stainless-steel rods, with a roof of galvanized gutters.

THIS PAGE
ABOVE   The longitudinal baulks of pine are separated by smaller larch elements, running crossways, allowing air to circulate freely.

RIGHT   For visitors, the Swiss pavilion provided a respite from some of the more strenuous aspects of the Expo.

If the term 'magic realism' could be applied to architecture as easily as it is to literature, then Swiss architect Peter Zumthor would be seen as its chief exponent. He uses materials austerely and with tremendous care to create buildings of apparent simplicity that have a special extra element, a spirit and a feeling to them. His concerns are not limited to materiality and light, but embrace sound and smell as well, to create an all-round experience that embeds itself in the visitor's memory. Perhaps his most famous work is the thermal baths at Vals, but he also has several wood buildings to his name and has worked with wood since his youth.

The son of a furniture-maker, Zumthor himself trained as a cabinetmaker, and brought this experience to bear most directly in his design of the Swiss pavilion at the Hanover Expo of 2000. Mindful of the white elephants that had remained in other cities decades after the end of their Expos, the Germans decided that – with the odd exception, such as Thomas Herzog's canopy (see pages 136) – all the buildings should be temporary.

As a country that has seriously embraced the idea of waste elimination and recycling, Germany decreed that the national pavilions should be not only demountable, but also reusable. One of the most literal interpretations of this concept, Zumthor's Swiss pavilion was

effectively a stack of construction timber that would season during the period of the Expo and could then be taken down and used wherever else it was needed. The building was anything but prosaic. Designed like a maze, the square pavilion was intended to entertain, rather than confuse. Blocks of parallel walls, some oriented north–south and some east–west, enclosed a series of openings, some of which were simple courtyards and others three-storey oval metal structures, allowing visitors to go up spiral stairs to platforms. The walls were made of freshly cut timber baulks, longitudinally of pine and crossways of smaller elements of larch. The lattices

that these created allowed air to circulate freely – an ideal environment for seasoning the timber. No nails, screws or glue were used to hold the elements of the lattice together, since these would have diminished the potential for reuse. Instead, Zumthor used stainless-steel rods in tension to give stability to the stacks. With every element sharply defined, the crisp springs at the top provided an industrial aesthetic reminiscent of electricity isolators.

Zumthor anticipated that, during the course of the Expo, the timber stacks would decrease in height by 120 cm (47 inches) from drying out and compression. As a result, the tension in the steel rods fell with time. The only other elements in the building were lengths of galvanized gutters forming the roof. With the aroma of the drying timber and the drumming on the guttering during storms, smell and sound had already been added to the visual aspects of the design. Zumthor augmented these sensations as part of his unusual approach to the function of the building. By the time that the Hanover Expo was being planned, after some rather prosaic shows, such as that in Lisbon, the organizers had grasped that these exhibitions had to be more than just glorified trade fairs or a chance for countries to show off their wares. But, despite the greater sophistication and simplicity of the Hanover Expo, it

was still a fairly exhausting experience for visitors. Zumthor's idea was that the Swiss pavilion should offer them some respite, a chance to recharge their batteries. Calling his pavilion a Sound Box, he described it as 'an event of the sensuous kind. Architecture, sounds, words, food, drink and dress blend together in a mix of theatre and excitement to create a complete happening.' Musicians played in some of the courtyards, actors staged events and cooks created dishes, all under the direction of the architect. 'We will offer a representation in real time for the relaxation of tired visitors,' Zumthor wrote before the Expo opened. 'When they have attentively visited the first 50 pavilions, they will be able to recharge themselves with us. Like this, after half an hour they will be ready for the next 50 pavilions.'

There was nothing obviously Swiss in Zumthor's pavilion, except for the timber itself, which came from Swiss trees. But, for many busy people, Switzerland has been a place of regeneration and recuperation, whether in the tuberculosis sanatoria of the past or in the ski slopes and walking trails of today. And the country's own successful Expo in 2002 showed that this is a country that can combine meticulous organization with a capacity to surprise and delight – again, very much in the spirit of Zumthor's pavilion.

THIS PAGE

LEFT   Zumthor calculated that, during the length of the Expo, drying out would cause the walls to decrease in height by 120 cm (47 inches).

BELOW   The quality of light filtering in was a vital part of the experience.

OPPOSITE
Three-storey oval metal structures in some of the courtyards of the pavilion offered a different experience.

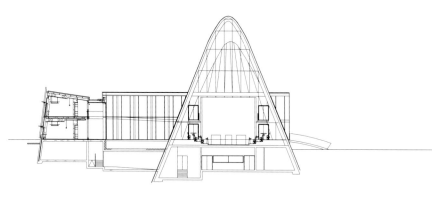

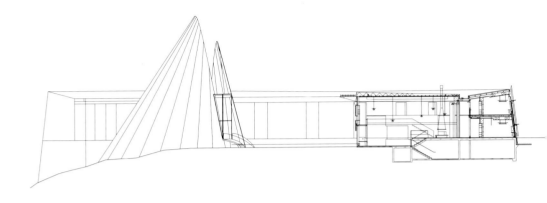

The Sami people of northern Norway and Finland (Lapland) are traditionally nomadic, so when they settle and want to erect buildings of civic significance, they have little tradition on which to draw. It is not surprising, therefore, that two representations of their nomadic life have informed the most significant building yet: the Sameting, or parliament building, in Karasjok, Norway, near the Finnish border. The traditional tepee-like tent, known as the *lavvo*, and the stockade used to enclose reindeer might seem too modest and workaday to fulfil a larger function. But the architects of the Sameting, who are based in Oslo, have adapted these forms into a building of great dignity and gravitas, which forms a vital focus in a town where most of the other development has been random and without architectural merit.

The dominant element is the debating chamber, which is based on a Sami tent. Its conical structure, clad in rough, untreated larch, looks rather like a rough-hewn, but carefully considered, woodstack. The cone is divided vertically into two unequal segments by a glass bridge. On the larger side is the debating chamber itself; on the smaller, an anteroom and access to the viewing gallery. The two vertical walls created by this slit are glazed, allowing light into the spaces, while maintaining the relatively closed exterior presence.

The rest of the building – based on a reindeer stockade – is in the form of a semicircle, with larger spaces for the Sami national library and a cafeteria contained within its embrace. The outer side of the semicircle has a battered, inward-sloping wall. Rough larch, the same as that on the debating chamber, is used to clad this part of the building, but here it runs horizontally instead of vertically. Since the building has only a simple timber canopy over the relatively modest entrance, reminiscent of the tent flap of the *lavvo* – or, in a more urban comparison, an up-and-over garage door – there is no strong sense of invitation. This is an inward-looking building in the style of traditional Sami dwellings. In one of the coldest places on earth, where there are months of darkness and winter temperatures fall regularly to below -50°C (-58°F), and which is infested by mosquitoes in summer, even the simplest structure acts as a refuge from the outside, rather than expressing a need to relate to it.

Internally, the story is very different. Offices and committee rooms occupy the outer, relatively blind face. In the centre of the building, facing onto a courtyard, is a two-level gallery used for informal meetings, strolling and unplanned interaction. It has some brightly coloured furniture, and the sense of colour is enhanced on important occasions when the Sami

wear national dress in red, yellow and green. Referring to the nomadic history of the Sami people, and described as a 'wander hall', it plays an important role in providing a sense of community. After all, for much of the year, there is unlikely to be much social interaction on the cold, dark streets of the town.

The largest spaces are the library and the café. Their roof is supported by glulam beams, up to 80 cm (32 inches) deep, with a span of up to 14 metres (46 feet). Steel columns support these beams, although where they meet the exterior glazing they are insulated and clad in wood to prevent the formation of cold bridges. Glulam is also used extensively in the debating chamber

building, with glulam arches and beams springing from a concrete ring beam.

Unfinished pine – though in a much more elegant and finished form than the external cladding – is used to clad the debating chamber interior. Thick fibreglass insulation keeps out the cold, as does the relatively low proportion of glazing. Cold air falling down from the surface of the glass is forced up again by heated air emerging from beneath the seats. Pine is also evident externally. In pressure-treated form, it is used as firings to support the rough larch. It provides enough ventilation behind the façade to allow the larch to dry out after rain. Since the larch is untreated, it will weather in time to a silver-grey colour.

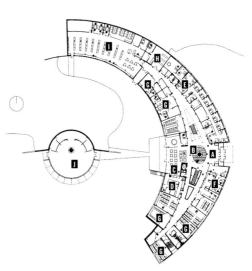

OPPOSITE

TOP LEFT   The interior of the library, which has deep glulam beams.

TOP RIGHT   Plan of the shared campus building of the Nicola Valley
             Institute of Technology and the University College of
             the Cariboo in British Columbia, Canada, showing an
             uncanny similarity to the plan of the Sami Parliament.

BOTTOM   Unfinished pine cladding gives a warm feeling to the
             impressive debating chamber, with focus towards
             the glazed wall.

THIS PAGE

BELOW   From bottom, ground- and first-floor plans of the
             parliament building show its stockade-like nature.

RIGHT   Bright furniture adds a touch of colour to the
             predominantly timber-coloured spaces.

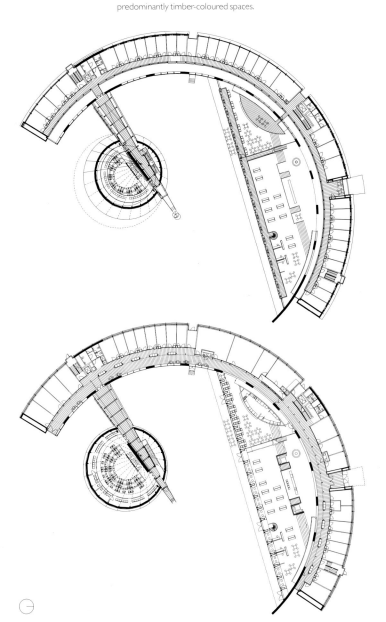

This magnificent building seems to be a solution to an issue unique in a modern western country: how to create an architectural identity for a people with little architectural history of their own. The Sami see their situation as similar to that of Native Americans, for example. In fact, the form of this building is not quite as unusual as it first appears. At a shared campus for the Nicola Valley Institute of Technology and the University College of the Cariboo in British Columbia, Canada, native and non-native students occupy the same space. The design, by architects Busby + Associates, was produced in consultation with local aboriginal elders. The resulting building also makes great use of wood. More surprising is the fact that, in plan, it is remarkably similar to the Sami parliament. It puts the main accommodation into a semicircular building, connected to a planned smaller circular space, which will be used as a ceremonial arbour. At Karasjok, of course, the parliamentary chamber – most certainly a ceremonial area – occupies the circular space. Could it be that a new architectural form is emerging in response to a set of sensitivities and requirements that previously were scarcely considered? If so, these two buildings, so far apart in Europe and America, will be seen as worthy pioneers.

COUNTRY HOUSE

Granada, Spain | Eduardo y Luis Javier Martín Martín | 2001

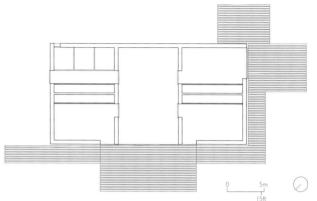

0   5m
15ft

TOP   The architects maximized the use of a restricted area by doing away with circulation space altogether.

BOTTOM   On the northern façade, glazing runs the full width of the central living room, and also extends to the bedrooms. Black poplar cladding recalls the vernacular of tobacco barns.

THIS PAGE

ABOVE   By making almost every opening a door, the architects maximized accessibility.

RIGHT   Decking, in a carefully considered configuration, effectively increases the footprint of the building.

The Moorish city of Granada in Andalucía, southern Spain, has a spectacular setting on the edge of the Sierra Nevada. Many visitors find that the backdrop of mountains enhances their experience of the Alhambra, but they may give less consideration to the fertile plains and meadows that stretch to the west of the city. In fact, this area, the Granada Vega, which occupies the basin of the Genil river, is also of historic importance, having long supported the city through its agriculture. And because of the special setting of Granada, the Vega is protected in a way that few such flat agricultural areas are.

Building a house there is not a straightforward business, and when the architects Eduardo y Luis Javier Martín Martín wanted to create a home for a client on the ruins of an agricultural building, they were limited to the volume of the building that they replaced. This gave them only 75 square metres (810 square feet) of space, on a single storey, which proved a wonderful discipline. Charged with producing a house with no wasted space, and no clutter, that would fit in with the agricultural vernacular, they produced an elegant, symmetrical building that sits comfortably in its surroundings, deals intelligently with the climate, and offers such a simple model for living that it makes many larger houses seem pokey and unnecessarily complex.

The first move was to eliminate all circulation space. Built to a very straightforward plan, the house is divided into three zones, running from north to south, each 4.25 metres (14 feet) wide. The living area occupies the whole of the central zone, and from it are openings into all the other rooms. On either side of the living room, there are, to the north, bedrooms. Behind them, on one side, are the shower and lavatory, and on the other is the kitchen/dining room. The building is as open as possible to the north (the shady side) and entirely closed on the south side. Along the south wall runs a continuous zone of cupboard space, and there are also walls of cupboards built into the spaces between the bedrooms and the auxiliary areas. By providing this ample storage, the architects have enabled the clients to keep the rest of the space free of clutter.

Exterior cladding is in black poplar, the same timber used for the tobacco-storage building that the house has replaced (agriculture in the Vega has shifted in the last century from linen, hemp and sugarbeet to asparagus, tobacco and vegetables). The only treatment this wood has received is against parasite attack. It should therefore age and fade naturally.

With the exception of one window, every opening in the outside of the building is a door, making it possible to enter any room directly. On the northern façade, sliding doors run the full width of the living space, and there are also doors opening into the bedrooms. At the back of the building, a skylight illuminates the servant spaces. The openings are framed in natural-coloured metal, and this palette of timber with a little metal extends to the interior, including the built-in furniture. The timber also extends beyond the boundary of the building into a series of decks at different levels on the northern and western sides, providing a link between the rectangular box of the structure and the nearby countryside. Tall, slender trees have been preserved in the building's surroundings, providing a pleasing vertical contrast with the low volume of the house.

Unlikely to distract the attention of visitors who approach with their eyes fixed on the more showy attractions of Granada itself, this little house is nevertheless an extremely satisfactory piece of architecture – an example of architects rising to the challenge of restrictions to create something so spare and rigorously considered that it is hard to see how it could be bettered.

THIS PAGE

TOP LEFT — The kitchen/dining room is on the more enclosed, southern side of the house.

ABOVE — The shower and lavatory are as uncluttered as every other element of the building.

OPPOSITE

TOP LEFT — Section showing the simplicity of the arrangement, and the careful consideration that has been employed.

TOP RIGHT — Timber has been used extensively, internally as well as externally.

BOTTOM — The central living room runs the full depth of the house.

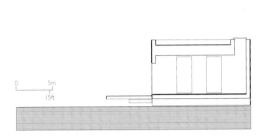

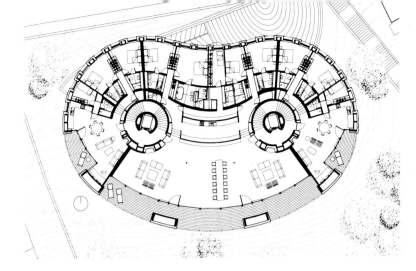

It looks rather as if a space ship has landed in the glitzy skiing resort of St Moritz – but a space ship that is not entirely alien, that in some indefinable way belongs. This is the effect of the Chesa Futura apartment building, designed by Foster and Partners, which combines an unusual rounded form that could only have been created by the latest computer technology with traditional construction techniques applied to locally sourced materials.

Foster and Partners established an international reputation as the epitome of high-tech style through a series of beautifully conceived and executed, if sometimes slightly mechanistic, buildings. This definition lost its sharpness as the architects embraced ubiquitous environmental issues and, at the same time, some of their straight lines and sharp angles softened into curves – but, nevertheless, timber is not a material immediately associated with the practice. Foster and Partners have used timber before, notably for a house in Corsica, completed in 1993, for the deputy mayor of Nîmes, Jean Bosquet, but the architect's signature materials remain steel and glass.

Indeed, it is on such steel-and-glass buildings that Foster have developed their use of parametric modelling, a 3D modelling process that allows the designer to specify or capture the geometric relationship between design features. The parameters that control those relationships can be modified to generate new versions of the design almost instantaneously. Having used this technique in the design of the Gateshead Music Centre in the north of England, the Swiss Re office building in London and the headquarters of the Greater London Authority, Foster have now also applied it to a relatively modest apartment building in Switzerland's Engadin valley. All the buildings mentioned here are non-orthogonal, and in all of them Foster have striven to achieve the greatest efficiency of form and performance. Lord Foster has written, 'The rapidity with which alterations can be made to a design generates a degree of creative freedom, allowing options to be worked up, assessed and improved upon in an organic fashion, providing important lessons along the way.' This approach also gives more authority to the architect, allowing a detailed dialogue between the architect, engineers and cost consultants, and at the same time drawing the contractor and the construction process into relatively early-stage discussions.

At St Moritz, the method has been used to create a flexible building of between six and 12 apartments that make the most of their orientation in terms both of views and environmental

performance, as well as exploiting the envelope determined by the planning regulations. The slightly alien nature of the structure seems appropriate in St Moritz, a resort whose native population is swollen tenfold at the height of the skiing season, mostly by rich foreigners. For some, the place is the epitome of glamour, but the more cynical *Rough Guide to Switzerland* says that it 'sticks out like a sore thumb. Seemingly plopped down unceremoniously amidst the quiet villages of the Engadin – although, of course, it was here long before they were, a spa as far back as the Bronze Age – St Moritz is a brassy, in-your-face reminder of the world beyond the high valley walls, the kind of place that gives money a bad name.'

St Moritz is densely built, and Foster's first concern was to create a building that could sit within the urban envelope, rather than sprawling out into the surrounding countryside – so the apartment block is lifted up on eight pilotis. This classic modernist move ensures that all the apartments have views and follows a Swiss tradition of protecting wooden buildings from prolonged contact with moisture from long-lying snow.

Planning required that at no point should the building be more than 15.5 metres (51 feet) above ground –

a complex constraint, given the sloping nature of the site, and one that the curved form exploits better than a rectilinear building would have done. Similarly, the curves reduce the apparent bulk, which is important since, by effectively eliminating the first two floors, the architect was obliged to make the three accommodation floors larger.

The accommodation consists of a frame of prefabricated glue-laminated beams and a skin of plywood sheets. The malleability of wood makes it easier to achieve the building's doubly curved shape. Since timber is a renewable material, it has good environmental credentials and, by sourcing the material locally, the architect was able to minimize the transport costs and fuel consumption. Compared with steel or concrete, the elements are relatively small and light, making them easier to bring in on narrow mountain roads.

Two concrete cores housing the lift shafts and stairwells provide further stability. This superstructure sits on a lightweight steel structure, supported on the eight sloping steel pilotis. The foundations consist of a sunken concrete box, which houses the plant rooms, car parking and storage spaces. Wherever possible, the architect has used prefabrication, since the winter holiday season restricted construction to eight months a year.

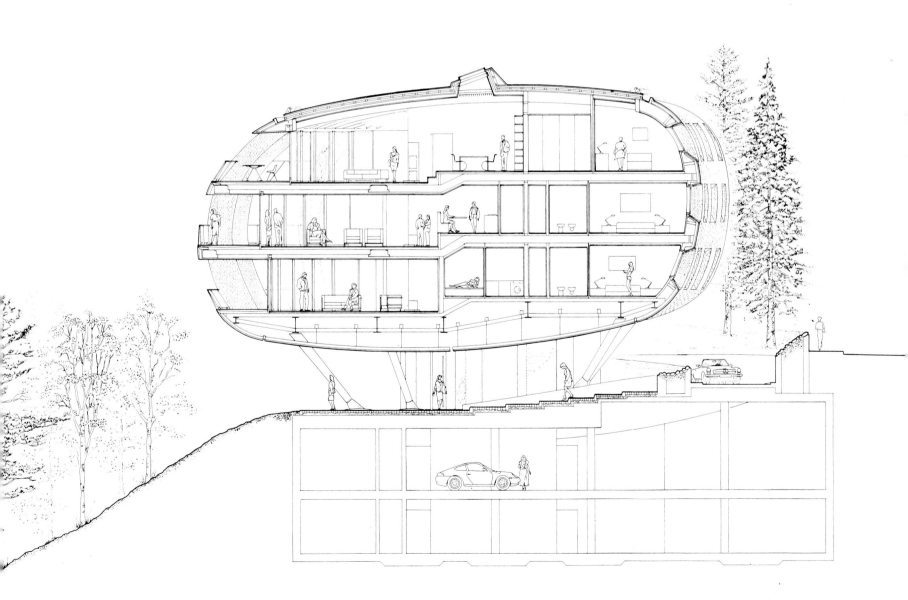

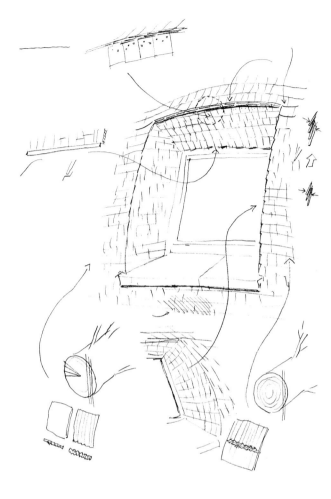

OPPOSITE
From the upper balconies, residents can look over
St Moritz to the mountains beyond.

THIS PAGE

ABOVE   Concept sketch by Norman Foster showing how
the larch shingles used for the cladding are cut in
different ways to offer different properties.

TOP RIGHT   Craftsmen cut the timber shingles in a traditional
manner.

By wrapping the windows around this curved form, the architect was also able to take maximum advantage of the panoramas of the town and the lake. The building stands to the north of the town, so that balconies are on the southern side, giving good views and letting in sunlight. On the north, which faces the mountains and the bleakest weather, the windows are small openings in the walls, which have an insulation-containing cavity 40 cm (16 inches) wide. Picking up a traditional Engadin design detail, the window surrounds are chamfered to allow in the maximum amount of light.

The building is clad in larch shingles, which weather and change colour with time. A traditional material, larch helps the building to blend in with its surroundings. A local family who has practised the craft for generations cut the shingles by hand. Cutting them both laterally and radially made the most efficient use of the material, so that only 80 trees were needed to provide the required 240 cubic metres (8,470 cubic feet) of shingles. The water-draining characteristics of one cut complement the structural strength of the other, and provide a variegated appearance.

By using trees that grow at the same altitude as the finished building, and cutting them in the winter when the wood is dry and contains no sap, it could be guaranteed that the shingles would not shrink. They were applied by hand, using nails, and have a life expectancy of 80 years. The roof is made from copper, another traditional local material. It is malleable enough to be formed on site, even in low winter temperatures.

The apartments have their bedrooms against the highly insulated northern façade, with living areas to the south, where they benefit from the sunlight and the views. Bathrooms and kitchens are in the middle section of the building, where there is less daylight. A building with walls that curve in two directions poses a challenge for the interior designer. There is no storage against the external walls, only on the internal partitions, which radiate from the cores.

The occupants of these apartments will doubtless find them delightful, and there is no denying that the architect has taken its environmental duty seriously. But whether this relatively small building will have much influence on construction in popular winter resorts is less certain, if only because such a distinctive form rarely bears repetition within a densely planned town.

OPPOSITE
At the Mason's Bend
Community Center
in Alabama, USA,
cypress planks were
not bought
readymade, but were
cut down and milled by
the students from
Rural Studio, who built
the project on a
very tight budget.

Architecture thrives on constraints – and the most difficult projects are often those where the architect has an entirely free hand and doesn't really know where to start. That is the problem of designing in a place with no cultural context. Think of the soulless housing that springs up on the outer edges of so many great cities or, worse, of the retail warehouses and out-of-town shopping centres that receive little public attention because they are not even considered to be places appropriate for 'architecture'.

It was in exactly this type of degraded environment that Spanish practice Roberto Ercilla y Miguel Ángel Campo designed a house on the outskirts of the Basque capital, Vitoria (see page 96). Rather than pretending to find inspiration in some ill-considered, recently built housing or opting for pastiche, they decided to draw on industrial design, since the house is in one of the most industrialized parts of Spain. From that decision, and from the client's specific requirements, the design ideas flowed freely to produce a building of beauty and practicality, although very unlike what might have been expected.

Another Spanish architectural practice, Nieto y Sobejano, faced a very different problem in Madrid, where, rather than too much freedom, there was hardly any freedom at all (see page 92). The form, the volume and the roof of an undistinguished house had to remain, although the client was eager for something modern and special. Cleverly, the architect achieved this by wrapping the house in a new coat to change the external appearance and, internally, by building a new spine to support the opening up of spaces that the client wanted. As with the Vitoria project, clever thinking was required to respond to such a distinctive site.

Quintáns Raya Crespo Architects faced the opposite problem when designing swimming pools for the province of La Coruña in Spain. The brief was to design a prototype that could then be built in a number of small towns, each in a different context. The flexibility and intelligence of the approach is demonstrated in the first built example, in the town of Puentedeume, where the architect managed to incorporate an old stone wall and a protected walnut tree, without compromising the integrity of its design.

Avoiding the imposition of a design signature on a project involves another kind of pragmatism. This was the approach of Peter Hübner in designing a school in Germany's Ruhr valley (see page 78). Since the ethos of the design was that students and parents should be involved as much as possible in the specification and construction of the buildings, the last thing Hübner wanted was to impose from above the dead hand of uniformity. So, although his work creates a strong skeleton for development, this relates to organization and materials, not to a recognizable aesthetic. Unlike those masterplanners who pay lip service to the concept of design diversity, yet make a plan so unyielding that their involvement will always dominate, Hübner is keen that, visually at least, his contribution should fade into the background. Timber played a dominant role in the construction of the school, partly for environmental reasons but also because it is a flexible material that can adapt to a range of uses and forms.

Such flexibility is demonstrated at Willoughby Barn in Missouri, USA, where the client, despairing of finding a new construction that suited her needs at a price she could afford, bought an old frame and had it transported and re-erected (see page 100). The architect, El Dorado, then created a modern building around the old frame, making extensive use of recycled timber.

Part of the driving force behind Willoughby Barn was the client's restricted budget, but this restriction was as nothing to that experienced by the Rural Studio in Alabama (see page 106). This university-based practice, founded by the late Samuel Mockbee, employs students to design houses and community buildings for some of the most deprived people in the western world. Given the minimal budgets, materials are not so much recycled as scavenged, resulting in buildings that are unconventional in appearance, but satisfy the needs of their clients. Whether some clients would prefer a more anonymous form of architecture has not been addressed amid the plaudits that the practice has received, but the close working relationship between the architecture students and their clients should ensure that the latter feel positive about the buildings by the time they are complete.

The concept of constructing buildings from waste materials was taken to extremes in the Tower of Babel, built by the Artists' Community of Ruigoord, Netherlands, to mark the new millennium. Intended to be as much a protest as a structure, this must have been one of the few inhabited buildings in which timber's propensity to burn was seen as an advantage: on the night of the millennium, its residents deliberately burnt it down.

SCHOOL

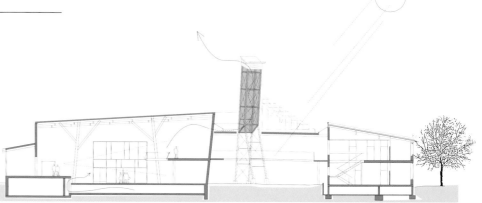

The new school at Gelsenkirchen-Bismarck, in Germany's Ruhr valley, does not, at first glance, have a strongly identifiable architectural style. But, if architect Peter Hübner had had his way, the agglomeration of buildings around a central street would have been even more diverse. Hübner wanted a different architect to design each building, and it was only his client's insistence that this would be too complex to administer that convinced Hübner to compromise. Instead, he achieved diversity by allocating individual buildings to different members of his associated practice, the enigmatically named plus +.

This was not because Hübner wanted to abdicate responsibility or because he lacked a strong agenda of his own. Indeed, a glance through his portfolio reveals a remarkable continuity. In schools and community buildings, he has concentrated on the use of timber, frequently in almost marquee-like structures. He has progressed from a somewhat over-expressive use of the material in such projects as a young people's centre in Stuttgart in 1984 to a more rationalist, though still creative, approach in a school sports hall in Oberhambach in 1994. Hübner's environmental approach is also reflected in the prevalence of planted roofs. Equally important is the involvement of clients and users in the process of design and construction.

At Gelsenkirchen-Bismarck, Hübner has produced his most radical vision yet. He won a competition to create an environmental school in a problem area with a solution that addressed not only the building fabric, but also the organization of the school itself and the learning that went on there. Hübner was keen that the strong architectural vision should not be seen as resulting in a deadening uniformity of appearance and approach – hence his desire to involve other architects wherever possible.

The need for something special was driven by the problems of the area and by the vision of Fritz Sundermeier, an educationalist associated with the local Protestant church organization. Gelsenkirchen-Bismarck is a former industrial suburb that grew up around a large coal mine at the end of the nineteenth century. In the 1960s, Turkish immigrant workers replaced the German workers. When the mine closed in the 1980s, the children of the Turkish workers were left stranded in an area of high unemployment. Since many of them did not read or write German, their prospects were bleak. Sundermeier and the Evangelische Kirche von Westfalen proposed the creation of a multicultural school for 1,100 students. Taking in large numbers of Turkish Muslim children and Catholics, it would carve its place in

THIS PAGE

ABOVE   Forming two-thirds of a circle, the workshop
            terminates the progression through the main set
            of buildings.

LEFT   Buildings are constructed with lightweight timber
            frames, and untreated Douglas fir is widely used
            for cladding.

OPPOSITE

TOP LEFT   Plan: 1. entrance, 2. cafeteria, 3. library, 4. theatre,
            5. central 'street', 6. classroom block, 7. workshop

TOP RIGHT   Colours may appear charmingly random, but, in
            fact, a colour consultant was used to draw up
            the palette.

BOTTOM   Green roofs, which are popular in Germany, form
            part of the environmental mix.

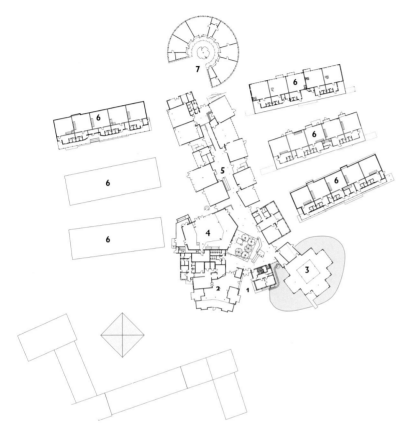

the community through its ecological initiatives, acting as a centre for the community, as well as for its pupils. The school was to be built on a former meadow near its predecessor, allowing a gradual transfer from the old building to the new. Also associated with the scheme was a considerable amount of self-build housing.

Hübner won the competition with the most holistic and radical of all the submissions. The main buildings cluster around an irregular internal street, a space whose form is determined by the shapes of the other buildings rather than carefully planned in itself. Just inside the entrance, there are a cafeteria on the left and the library, set within a pool, on

the right, with music rooms and a chapel above. Next come a theatre on the left and a more prosaic administration centre on the right, then a relatively small number of classrooms including four buildings for special classes, named 'chemist', 'cinema', 'laboratory' and 'studio'. Beyond the end of the street is a workshop in a building that occupies two-thirds of a circle. Farther away from the street is a sports hall.

Flanking the central row of buildings are the classroom blocks. These are being built gradually, to reflect the rate at which pupils are scheduled to move from the existing school – and the pupils have had a lot of involvement in the design. Along with their teachers and

OPPOSITE
The timber 'trees' in the theatre are rotated, relative to each other, in a deliberate attempt to increase the sense of randomness.

THIS PAGE
ABOVE   The library doubles as a church.
TOP LEFT   There are plenty of informal spaces within the school complex.
TOP RIGHT   Children have been involved in the development of the classrooms in which they study.

parents, they have worked with the architects on the planning, modelling and construction, being involved from the conceptual stages to the end, when they applied coats of paint. This means that the blocks vary, but are all based on a lightweight timber frame with a mixture of single- and double-storey accommodation providing a varied roofline. Natural daylight is used wherever possible and, with high levels of insulation and appropriate window sizes, natural ventilation is possible.

The three most complex structures are the workshop, the library/church and the theatre. All are made solely of timber. 'We try not to use steel if we build with wood,' says the architect. Douglas fir, grown in the Black Forest, is used widely, but in the theatre the architect turned to laminated timber because of the requirement to achieve 30 minutes' fire resistance.

The theatre space needs to be as flexible as possible, and, at first, gives a curiously random impression. It is irregular in shape and the roof is supported by timber 'trees', their trunks hexagonal and comprising six individual pieces of timber with a hollow core. 'Branches' emerge from these trunks at a variety of heights. In fact, the trees are identical, simply rotated at different angles – but this was against the architect's wishes, a uniformity imposed

by the engineer for ease of calculation. According to Hübner, 'The theatre is a game with geometrics. In a room with five angles, four trees support a roof of triangles, which create six angles and six branches.'

The large volumes of the theatre and the sports hall mean that natural ventilation would have been inadequate. The architect, working with its environmental consultant Transsolar, has used sophisticated passive ventilation techniques that, in the case of the theatre, also make use of the internal street. These include a combination of thermal chimneys and drawing in of air through underground pipes. Where Douglas fir is used for cladding, it is left untreated and in its natural colour, but elsewhere in the complex a wide range of colours is used. Creating a lively atmosphere, these are certainly not random. Indeed, the project used a dedicated colour consultant.

Hübner is convinced that this building – carefully thought out but leaving space for others to use their initiative – can create a positive environment for education in this deprived suburb. Experience has taught us to be cynical about the power of good buildings to bring about social regeneration, but few solutions have been as carefully considered as this one. It certainly deserves to succeed.

TOWER OF BABEL

Ruigoord, Netherlands | Artists' Community of Ruigoord | 1999

THIS PAGE

TOP RIGHT   Although, in fact, fairly near to the centre of Amsterdam, Ruigoord feels remote in its flat, wet setting.

RIGHT   The building was extraordinarily reminiscent of old paintings of the Tower of Babel.

BELOW   Occupants made an endless series of small accretions to the building, which gave it the stamp of individual personalities.

OPPOSITE

Using discarded wooden pallets for construction introduced a degree of uniformity.

Cultural Freeport Ruigoord and the Artists' Colony of Ruigoord are two of the names given to a community based on a boggy island in the Netherlands. Created in 1973, at the height of the hippy era, it has stubbornly continued to exist, its closest parallels perhaps being the houseboat dwellers of Amsterdam and the Danish enclave of Christiania – the type of 'alternative' living to which these conventional northern European societies seem uniquely tolerant.

The impression given by the romantic image of Ruigoord is that it is somewhere very remote. This is impossible, of course, in the Netherlands, Europe's most densely populated country. In fact, Ruigoord is a kind of suburb of Amsterdam, and easily accessible on the 82 bus from the city centre. A lot of the activity in Ruigoord is linked to performance, through the Amsterdam Balloon Company, and its manifesto says, 'Nation has made war on nation, but artists from different cultures have kept on cooperating and inspiring each other. It is for this spirit, on the interface between ancient traditions and avant-garde experiment, that Ruigoord stands open. Empower the Imagination, now especially, at the beginning of an as yet unformed new era.'

One of the most dramatic and complex events at Ruigoord was the decision, in 1999, to construct a Tower of Babel. This

was a reaction to the news that the enclave was under threat of redevelopment by international companies for industrial use.

Timber is a forgiving material, ideal for constructing a building without real plans. The community members sought donations from local firms, which, ironically, provided waste materials – in the form of hundreds of old timber pallets – from exactly the sort of industrial enterprise that the inhabitants of Ruigoord reject. Forming the basis of the structure, the pallets impose some material consistency that other seemingly random accretions could not obliterate. In fact, the finished structure, with its gently sloping, spiralling external ramp, bore a remarkable resemblance to the paintings of Pieter Breughel and Abel Grimmer. Most of the timber was unpainted, with decoration coming from the occasional coloured door, from banners and hanging mobiles, from a whimsical cantilevered birdcage, and from two items essential to modern communications: a postbox and a large sign bearing the address of the community's website.

Inside, the tower provided living and working quarters for the community's artists. This pragmatic, seemingly random, arrangement carried an echo, on a smaller and less threatening scale, of Hong Kong's former forbidden city, which, though confusing, also had its own logic and organization. In one way, the Tower of Babel resembled the most commercially rigorous of construction projects – it had a very tight and unmovable deadline for completion. It had to be finished by the last day of 1999, in time for an enormous party on the site, which culminated in the burning of the tower. Can this sort of dramatic gesture have any effect? Logic would say no, but four years after the tower was constructed, in reaction to the imminent threat to the community, Ruigoord is still there, still thriving and still organizing events.

MUNICIPAL SWIMMING POOL

Puentedeume, Spain | QRC Architects (Carlos Quintáns, Antonio Raya and Cristóbal Crespo) | 2002

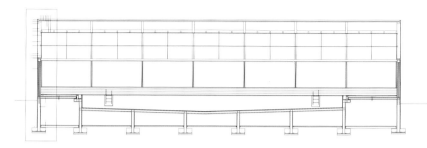

OPPOSITE

TOP The long section shows that the pool building is a simple, but well illuminated, box.

LEFT The architect has dealt with the horizontal nature of the building by adopting a stratified approach to materials: concrete up to a certain level, with timber above.

RIGHT There are large windows, with slender mullions, on the northern side of the swimming pool.

THIS PAGE

ABOVE The simple approach to design should allow similar pools to sit comfortably in a number of different locations.

RIGHT A long courtyard running between the back of the pool enclosure and the auxiliary building lets light come in to both.

Swimming pools are popular places, although they can be somewhat brutal, with noise echoing off hard surfaces to make a swim more of an ordeal than a refreshment. And, despite their popularity, pools rarely make money – operators of health clubs often treat them as 'loss leaders', attracting customers for more profitable activities.

For local authorities, therefore, pools can be a liability. The province of A Coruña in northwest Spain tackled the issue by trying to make it more controllable. Recognizing the need for a reproducible form of pool that could be adapted to local needs, it launched a competition to design just that. Local practice Quintáns Raya Crespo

Architects (since re-formed as VIER arquitectos) won the competition, and built its first example of such a pool in the town of Puentedeume.

This was a real design challenge, since the site was restricted and two elements had to be kept: a stone wall and a walnut tree. But the architects' concept – which consists of housing the pool itself in a rectangular box and putting all associated services in a building enclosed by a concrete wall – worked superbly. The pool building is oriented so that the windows that run the length of its north side look out past lawns and trees to the estuary of the River Eume. On the southern side, light can filter in through slender

THIS PAGE
LEFT   Grooved pine boards, treated for slip resistance, surround the pool at the level of the top of the water.
BELOW   Swimmers enjoy views towards the estuary of the River Eume.

OPPOSITE
RIGHT   Laminated timber panels cover the walls and ceiling above the pool.
LEFT   From bottom: basement and ground-floor plans show the supremely logical arrangement of spaces that still manages to encompass an existing walnut tree (bottom right).

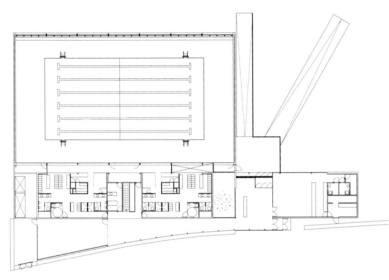

courts between the pool building and the service building. Another court, at right angles to the pool building, surrounds the walnut tree. The stone wall has been incorporated into the building's enclosure.

The architect has dealt with the horizontal nature of the building by adopting a stratified approach to materials. Concrete is used for the auxiliary building and the base of the pool building. Above this, on the pool building, is the glazing, with widely spaced and slender mullions. Above again is a timber superstructure with horizontal timber cladding. The superstructure is a post-and-beam construction of glulam elements, 1.4 metres (4 feet) deep, with the beams spanning 20 metres (66 feet) across the width of the pool hall. With some wind-bracing as well, these posts and beams form the skeleton onto which the cladding – consisting of pine, treated in an autoclave – is fixed. On the eastern face, the cladding is fixed to notches, cut into the superstructure, in a pattern that allows light to filter in.

This concept of enclosure within a timber box is continued internally, with walls and ceiling covered in laminated timber panels, treated to resist humid atmospheres. The modular nature of the structure made it possible to fix all the panels without any need for on-site cutting. Timber, again treated pine, is also used for the floor surrounding the pool. The boards, which are grooved for slip resistance, are laid with gaps between them, to accommodate any swelling and also to let water drain away between them; this is essential because they are level with the surface of the pool.

Puentedeume's pool hall is, therefore, a much gentler and more welcoming environment than many comparable buildings, and the careful consideration given by the architect to the use of materials means that the complex sits comfortably in Puentedeume, and should do so equally in other towns where the design is destined to be built.

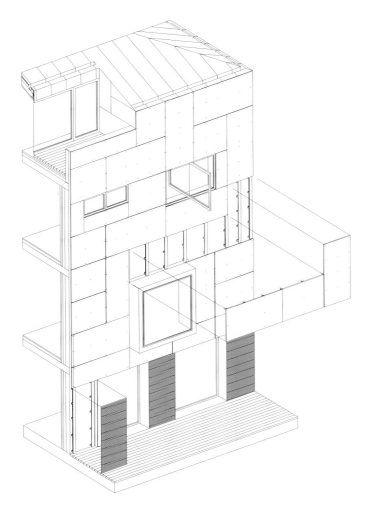

If some projects represent an architect's or a client's dream, a house in Chamartín, Madrid, came closer to being a nightmare. How does an architect give a client the modern and spacious house desired when byelaws appear to make it impossible? That was the problem faced by architects Fuensanta Nieto and Enrique Sobejano. Based in Madrid, the pair have an impressive portfolio of work in their home country and abroad, most of it on a larger scale than this house, and a great deal of it arts-based – but it is unlikely that they have faced a bigger challenge.

The client's house was in a pleasant street of buildings with little architectural distinction – to which their own house, constructed in the 1950s, was no exception. Since the house is set on a relatively generous plot, designing a replacement would have been a fairly straightforward challenge. However, byelaws prohibited demolition and required that the original building materials and volume be retained. Another requirement was that the house should keep a pitched roof, sloping in all four directions.

Even now, some civil servant may be toiling away to close a loophole in the byelaws that the architects exploited, since what they have done is to comply with the letter, but most definitely not with the spirit, of the law. They have kept the existing materials and walls of the house, but they have enclosed them in a new skin, so that the house looks completely different. Internally, they have built a spinal structure containing services and storage that has allowed them to create considerably more open floors than existed originally. In both cases, this new work makes extensive use of wood.

Aluminium battens fix the new façade to the existing exterior walls, leaving an air space between them so that the effect is of a ventilated façade. At the lowest level, this new façade is of undulating aluminium, but on the rest of the building it is of high-density bakelized boards. Bakelizing is a treatment with a resin that is cooked at about 150°C (302°F), to create a hard, impermeable surface. Commonly applied to specialist papers, it is also used on plywood, for applications such as formwork and advertising boards, where it is necessary to withstand repeated abrasion. On the house, the top layer is of elondo wood. Visually, the effect is attractive – a rich, reddish brown colour that varies slightly across the surface. The architect has carefully worked out the pattern of application, with some boards in portrait and others in landscape format, to fit in with the pleasing geometry of the windows. The effect, in a warm monotone, is almost like an abstract painting. The double

façade means that windows have attractively deep internal sills. Externally, the use of timber continues with a boarded strip to the front door. The pitched roof is in copper, to harmonize with the colour of the overcladding.

The new spine, built of boxwood, houses everything from ducting and a pantry to bookshelves and wardrobes, and runs across almost the entire width of the building. There is also an elegant timber staircase in ash, the timber also used for the floorboards. Both these allow the house to have the open, clear spaces that the owners wanted, but with the warmth of a natural material and without banishing personal possessions in an over-rigorous obsession with order.

This house was, in a sense, an unnecessary project, an example of the ridiculous hoops through which architects are sometimes forced to jump. Perhaps in the distant future an archaeologist will examine this house within a house, and wonder why it was wrapped up in this extraordinary way. It may be difficult to understand the planning foibles of the late twentieth and early twenty-first centuries, but they have certainly led the architect to produce a highly imaginative result.

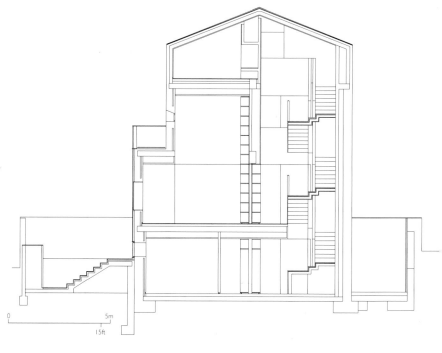

0   5m

15ft

HOUSE

San Prudencio Uleta, Vitoria, Spain | Roberto Ercilla y Miguel Ángel Campo | 2002

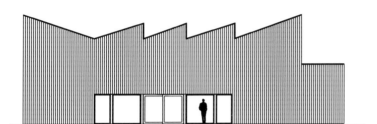
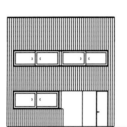
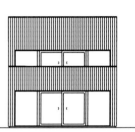

OPPOSITE

TOP   The stern façades offer few opportunities for
       inquisitive visitors to look in.
BOTTOM Locally sourced red cedar planks clad the building.

THIS PAGE

RIGHT   Sun loungers put outside on a sunny day provide
        one of the few hints at habitation.
BELOW   Model showing the sawtooth form of the roof that
        draws light into the heart of the building.

How can you create a sense of context in a region of ill-conceived and ill-controlled growth? This is a question that architects Roberto Ercilla y Miguel Ángel Campo tackled at San Prudencio Uleta, outside the Basque capital of Vitoria. Vitoria has one of the highest standards of living in Spain, resulting in a lot of poorly planned luxury housing on its outskirts. Dream homes have been created with little regard for their surroundings or the history of their location.

One home-owner who has thought about context is the owner of this house, not far from the city boundaries – and the chosen context is not what one would immediately have expected.

The building sits on the edge of an industrial park – the Basque country being, after all, one of the most heavily industrialized regions of Spain – and follows an industrial aesthetic that suits the client's particular needs. The architects, who come from Vitoria but have worked all over Spain, had previously used laminated timber in the construction of an uncompromising and uncompromisingly modern hotel building in a formless wasteland at the edge of Irun, on the French border. At San Prudencio Uleta, they have used such timber to create a building that provides exactly what the client needs, while making few concessions to traditional ideas of prettiness or domesticity.

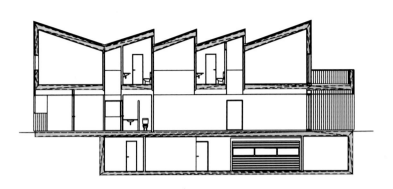
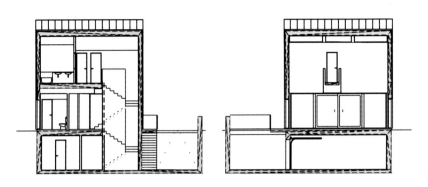

OPPOSITE

TOP   Sections through the house showing that both voids and the staircase help to draw light down to ground-floor level.

BOTTOM   The light wooden finishes and white-painted surfaces allow one to forget the relative paucity of direct, natural light.

THIS PAGE

BELOW   Plans show, from bottom, the basement-level parking, the main living area and the disposition of voids on the upper floor.

RIGHT   There is an almost ethereal quality to the light.

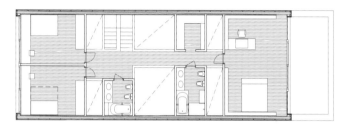

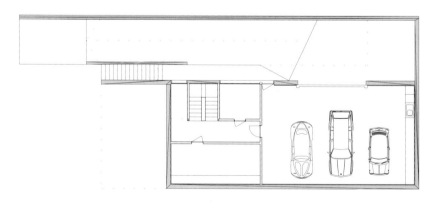

What the client wanted most in the house were light and privacy – apparently incompatible until you think of bringing in light from the roof. What better approach can there be than the sawtooth roof, beloved of those throwing up deep-plan and relatively inexpensive industrial space? At the house in San Prudencio Uleta, this guiding principle has resulted in a design with relatively few windows, and none leading into private areas. Instead, light streams in from the roof and is carried down into the central living space. The upper-floor bedrooms lead off a central corridor, but there are double-height spaces on either side that enable light to find routes down to the ground floor, where it is disseminated through the open-plan layout.

Structurally, the building is of reinforced concrete, but it is clad outside with local red cedar planks supported on pine battens. Both have received the same treatment to ensure durability. Timber used internally has been painted white to reflect the light. The effect on the visitor is uncompromising and faintly hostile – of a timber box that gives little away about the occupants or the purpose of the building. Only on sunny days, when sun loungers appear on the strictly rectilinear patio, is there a sense that this could be a building used for pleasure.

This is a house that recalls not only the industrial aesthetic, but also the artist's studio, the courtyard houses of southern Spain and the Arab world, and certain Japanese houses that turn their backs resolutely on the street to celebrate their own internal life. Beyond the toughness of the external timber planking there is a sense of luxury and privilege. And, even though this imaginative design seems to exclude the outside world, it also integrates better with its environment than most of its more extrovert neighbours.

WILLOUGHBY DESIGN BARN

Weston, Missouri, USA | El Dorado | 2001

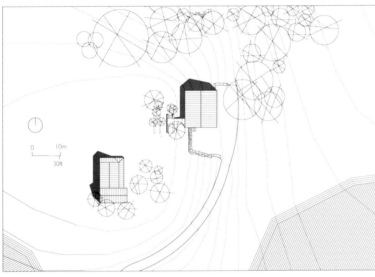

THIS PAGE

TOP RIGHT   The new barn sits on a sloping site.
ABOVE   Site plan.
BOTTOM RIGHT   The simple appearance, particularly as night falls, gives little indication of the sophistication in choice of materials or uses.

OPPOSITE

Cladding is in corrugated copper and corrugated fibreglass.

OVERLEAF

LEFT   By not heating the main spaces, the architect was able to leave the frame exposed, without fear of it warping.
RIGHT   The wood flooring was reclaimed from a former gymnasium.

Barn conversions are more common these days than the construction of entirely new barns, so it is exciting when a new barn is built – although this strikingly contemporary-looking building in Weston, Missouri, is not all it seems. The Willoughby Design Barn is not strictly a new building, nor is it entirely a barn. It is an ornament to its location and a testament to the ingenuity and dedication of a client and architect, who possessed more imagination than cash.

The client, graphic designer Ann Willoughby, wanted a 'barn' with an agricultural purpose that would also serve as an extension to her 1880s farmhouse, even though not physically connected to it. With her architect, Dan Maginn of El Dorado, based in Kansas City, Willoughby initially talked to local

carpenters and was disappointed by the conventional nature of the solutions they offered. But on a visit to Red Barn Farm, a nearby demonstration farm with traditional buildings, she saw a model of a timber-framed barn and was told that a similar, full-sized structure was about to be dismantled and turned into floor planks. Willoughby bought it and had it reconstructed at her farm on a new concrete foundation. The pine frame is a no-nonsense, elegantly proportioned, structure that the architect took as the starting point for the rest of the design and choice of materials.

Set on sloping ground (the house is at the high point, with commanding views), the new barn is largely workmanlike on the ground floor, with one side dedicated to farm storage and the other to car parking. A central timber staircase leads to the upper floor, which is designed as a place for Willoughby's clients and colleagues to visit – and stay, since she is a long way from a major urban centre. The space on one side of the stair is designated for entertainment and includes a dining table. Reflecting the changes in external level, there is an entrance on this side. The other side is described as a studio and has a sleeping loft above it, reached by a suitably agricultural-looking ladder. There is also a bathroom on the studio side, an enclosed room that is the only heated space; this may seem a spartan

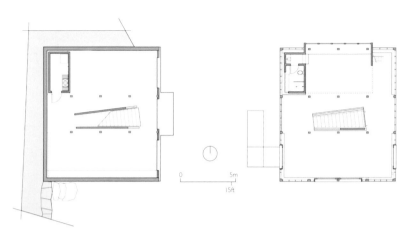

OPPOSITE

TOP   The plans (from left: ground and upper levels) were kept as simple as possible, with a single central staircase.

LEFT   An agricultural ladder leads up to the sleeping loft.

RIGHT   Panelling on the sides of the stairs and elsewhere is of reclaimed pine siding that was flood-damaged in 1933.

THIS PAGE

RIGHT   The enclosed bathroom is the only space with any heating.

BELOW   The sections show that, with the changes of level, it is possible to have entrances to both the ground and first floors.

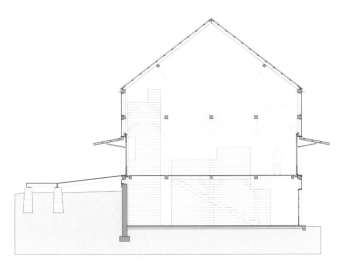

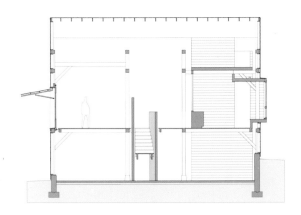

approach, but few visitors are likely to make the journey in a harsh Midwestern winter. The decision not to heat the rest of the space has made it possible for the frame to remain exposed, in conditions similar to those experienced in the first century of its life, thereby avoiding the drying-out problems that heating could cause. The architect has treated the frame with enormous respect, as an almost sculptural object. None of the new walls and openings – kept to a minimum – abuts the frame, so it maintains its independence.

Timber is used extensively on the interior. As with the frame, much of this is reclaimed. The wood flooring had its first life in a gymnasium, and the panelling is reclaimed pine siding that was flood-damaged in 1933. Cedar and glass were used for the sliding barn doors, and window openings are also framed in wood.

Although Willoughby Design Barn is a relatively small project, it is among the larger projects carried out by this particular architect. Founded in 1995, El Dorado is as much involved in the

process of making as in design, and has its own workshop. Indeed, a great deal of its early work concerned the making of furniture. On this project, it acted as general contractor and carried out the internal cabinetwork, also making the handles for the main doors, plus handrails and metal brackets.

If all this suggests a worthy, but rather retro, Arts and Crafts approach, that couldn't be farther from the truth. Internally, there is a clear, austere geometry, and the external appearance is both crisp and surprising. The architect originally planned to clad the building in galvanized iron, but then it learnt that it would cost only another $12,000 (around £8,000) for 16-ounce corrugated copper. This material has been used with the corrugations running horizontally and kept exposed on the inside of the building. Coupled with the use of corrugated fibreglass for glazing set in the walls and the roof, this creates a sense of restrained, appropriate warmth and luxury – all the more impressive when one learns that the entire project cost only $62.50 per square foot (around £400 a square metre).

MASON'S BEND COMMUNITY CENTER 2000

OPPOSITE

TOP   Old car windscreens form the glazing of the community centre.

BOTTOM   Students cut down cypress trees and milled them into planks to form the laminated beams.

THIS PAGE

ABOVE   Although relatively small, the community centre forms an important meeting place for the 150 people who live at Mason's Bend.

TOP RIGHT   Materials may be salvaged or very cheap, but the result is professional and impressive.

The latter half of the twentieth century demonstrated, through the ingenuity of engineering solutions, that timber can be a sophisticated and finely finished material. But there are circumstances in which its simple flexibility, its usability and reusability are much more important. Pre-eminent in this is the work of Rural Studio from Auburn University, Alabama.

There is no single distinct style to be found in this work since it is produced by successive groups of students. But there is an underlying ethos, established by the founders of Rural Studio, professors Dennis K. Ruth and Samuel Mockbee, the public face of the venture, who died at the end of 2001. Universities are always places of privilege, but rarely so glaringly as at Auburn, which is in Hale County, Alabama. This is the area immortalized by writer James Agee and photographer Walker Evans in the book *Let Us Now Praise Famous Men*, based on their stay there in 1936. Compassionate, engaged and angry, the book details a level of rural poverty that is almost unimaginable in the western world. Nearly 70 years later, improvements have been far fewer than might have been hoped. In 1999, average per capita income was only $12,661, and unemployment was about twice the national average.

In response to this situation, Mockbee and Ruth set up a programme for their architecture students that would involve them in designing and constructing houses and communal buildings for some of the most needy members of the community, working in both Hale County and adjacent Perry County. Described as 'context-based learning', it has a set of well-defined objectives:

- To give students of the School of Architecture the opportunity to learn the critical skills of planning, designing, and building in a concrete, practical, and socially responsible manner.
- To form leadership qualities in students by instilling the social ethics of professionalism, volunteerism, individual responsibility, and community service.
- To help communities, through partnerships with the state and local welfare agencies, provide suitable and dignified housing.
- To develop materials, methods, and technologies that will house the rural poor in dignity and mitigate the effects of poverty upon rural living conditions.

There are three distinct programmes of work, one for second-year students to build one-off houses, one for fifth-year students to create community buildings, and one for students from elsewhere. The results have been economic, practical and surprisingly architecturally imaginative. Obliged to scavenge, scrounge or buy materials cheaply, the

ANTIOCH BAPTIST CHURCH 2002

LEFT   Aluminium covers most of the exterior of the church.

BELOW LEFT   The opportunity to install a baptismal font (foreground) was one of the major reasons for having a new building.

BELOW   Although most of the materials were salvaged from the existing church, the effect is entirely contemporary.

BOTTOM   Joists, panelling and floorboards were among the salvaged materials.

HERO CHILDREN'S CENTER 1999

RIGHT   The barn-like buildings are clad in coloured timber
planking and have tin roofs.

BELOW   The covered passage joining the buildings doubles
as a waiting area.

students have built structures that vary from the sophisticated shack to works reminiscent of the early creations of Frank Gehry. Given the hot and humid climate, many of the structures are not entirely enclosed, since the provision of shelter is often enough. Much of the work takes place at a base in Newbern, where students have built their own accommodation and workspaces over time, using scavenged and experimental materials, including, most recently, cardboard clippings. Indeed, the students consider no material too humble, having even built one house almost entirely from recycled carpet tiles.

Even where materials appear relatively sophisticated, there has been significant input from the students. For example, a community centre at Mason's Bend has a central meeting space created from a sloping framework of laminated timber beams. But these were not bought readymade. Instead, the students cut down cypress trees and milled them into planks to create the laminated beams. Cypress was also used to create benches in the space. Neither of these elements is 'rough-hewn', but each has an elegant simplicity. The low, enclosing walls are of rammed earth, with a rusting metal drip on top. The sloping beams support a metal-framed canopy that is partly clad in aluminium sheet and partly in what seems an unusual, but attractive, glazing system – in fact, it is

made from the windscreens of 1980s General Motor Company sedan cars, salvaged from a Chicago scrapyard. Serving a multitude of functions, from a place of worship to a centre for childcare, this building also has an important symbolic one in a location where the 150 residents, comprising four extended families, live in trailer homes or very basic, decaying houses.

Whereas at Mason's Bend the salvaged material came from a distant city, the Antioch Baptist Church cannibalized an existing building. The previous church was no longer viable because of foundation problems, and lacked both a lavatory and a baptismal font. Having decided to construct a new building, the students salvaged everything they could from the old one – roof and floor joists, pine heartwood wall panelling, tongued-and-grooved boards and exterior corrugated metal. More than three-quarters of the materials in the new building come from its predecessor, but the building itself is entirely different in form. The interior is almost all of timber, with a long horizontal window offering views of the graveyard. However, this inner structure is protected from the elements by an outer wrapping, a cranked structure supported by handbuilt, composite wood-and-metal trusses and covered in aluminium. Only at one end does the inner timber box protrude.

SHILES HOUSE 2002

TOP LEFT   Oak shingles that clad the house were cut from old shipping pallets.

LEFT   Car tyres, clad in a cement render, support the timber staircase.

STUDENT HOUSING 1999

TOP RIGHT   Students design their own housing and studios, using them as an opportunity to experiment with materials.

ABOVE   Interiors are designed as much for practicality as for style.

PERRY LAKES PARK PAVILION 2002

RIGHT The pavilion roof is described as a 'dancing plane'.
BELOW Slender metal uprights support the cedar roof
    structure.

The student project at Greensboro in Hale County also fulfilled an urgent need. The HERO Children's Center is an addition to the Hale County Empowerment and Revitalization (HERO) Family Resource Center. It offers a place for various authorities to interview and counsel children who have been abused physically, mentally or sexually. Before it was built, these interviews had to take place in Tuscaloosa, one hour's journey away.

Serving a variety of complex functions, including the provision of an interview and observation room, with a one-way window disguised as a mirror, the construction is visually simple, consisting of a group of low, barn-like structures, clad in painted timber planking, with tin roofs. These buildings are joined by a loftier covered passage, which can serve as a waiting area. Supported by telegraph poles, with some rudimentary cross bracing, the passage has a pitched roof of paired timber members, covered in corrugated metal and perspex. This central walkway leads to a playground, with an enticing central play structure, again constructed from telegraph poles and smaller timber elements.

Second-year students also made use of telegraph poles for the Shiles House, lifting it above the wet ground. Much of the superstructure of this house is of timber, including a lattice supporting the roof, but where the building descends to ground level the structure consists of old car tyres. The tyres, clad in a cement render, also form the support for a central timber staircase. The exterior of the building is clad in oak shingles, cut from wooden shipping pallets.

Perhaps the most elegant of the students' projects is one of the simplest: a pavilion at Perry Lakes Park in Perry County. The park was closed to the public in 1974 when fish were stolen from the nearby US Southeastern Fish Cultural Laboratory. Following the closure of the laboratory in 1994, and with considerable local effort, the park reopened in 2001. The students designed the pavilion on the site of an old picnic spot at Barton's Beach, building it almost entirely from donated cedar, which they themselves cut down and took to be milled in Greensboro. Open on three sides, the pavilion has slender metal uprights supporting the cedar roof, which is described as 'a dancing plane' and rises slightly erratically from each set of supports until it reaches a height of 7 metres (23 feet) at the front. The cedar floor curves up to provide benches. Like all Rural Studio's buildings, the pavilion serves a multitude of functions: as a place for community gatherings, catfish fries, family reunions, and as an outdoor classroom for Judson College.

OPPOSITE
International superstar
Frank Gehry used
timber extensively in
the Maggie's Centre
for cancer treatment
in Dundee, Scotland,
which won the Royal
Fine Art Commission's
British Building of the
Year Award in 2004.

Buildings are increasingly used to define places. Since Frank Gehry designed the Guggenheim Museum in Bilbao, other cities have striven to achieve the 'Bilbao' effect, to create an architectural icon that can define and regenerate a whole city. Interestingly, Gehry, famed for his use of advanced computer technology and of space-age materials, such as the titanium cladding in Bilbao, turned to timber for the roof of his much smaller Maggie's Centre, the therapeutic healing facility that is his first building in the UK.

Whereas for Gehry timber only became an option when he was working on a more intimate scale, Santiago Calatrava – another architect whose buildings seem made to adorn postage stamps – used it on a far more extrovert scale on his wine production plant in the Rioja region of Spain (see page 126). More likely, in fact, to appear on a wine label than on a stamp, this building is an icon that will come to represent the wines of the producers and attract visitors from a distance.

The transport buildings shown here work much more directly in their own place. The Vancouver station is designed primarily to attract travellers to what, for them, is a new form of transport, but also, like the Aix-en-Provence TGV station, to form a core for the development of a new district (see pages 114 and 132). Both these buildings

are crucial parts of the planning jigsaw and set high standards of design that, it is hoped, will be followed in succeeding buildings. The reconstruction of Hounslow East station, in a much more mature setting, aims to restore some of the local pride originally engendered by the more iconic West London underground stations and, at the most basic level, make the station more accessible (see page 118). Its historical inspiration is expressed in a design that is forward-looking and inventive.

Thomas Herzog's building for the Hanover Expo site was also intended as a centrepiece for other developments, but in an interesting historical context (see page 136). It was one of the defining structures of the original Expo, and the only one to remain. Once the initial show had been dismantled, it had a new role to play in future development of the site – a building with more presence than any legacy building since, arguably, Paris's Eiffel Tower.

If Herzog's structure was a geographic centre for a set of temporary exhibition structures, then von Gerkan, Marg + Partner have created that crucial sense of place for their permanent sister, the normally soulless exhibition centre. Like all the buildings in this chapter, the Rimini Fiera (see page 140) uses timber in part to give a human scale and some warmth to buildings that, through their need to

be iconic, can too often seem cold and inhuman.

Smallest of the projects in this section, the View Silo House in the wide, open spaces of Montana (see page 122) shows that a landmark building is sometimes more in sympathy with a place than a building that tries unsuccessfully to disappear. Although it will be seen by far fewer people than the other buildings in this chapter, View Silo is a reminder that architects have to accept the responsibility of placemaking, even in remote locations.

BRENTWOOD SKYTRAIN STATION

Vancouver, Canada | Busby + Associates | 2002

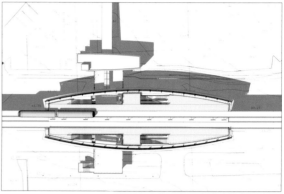

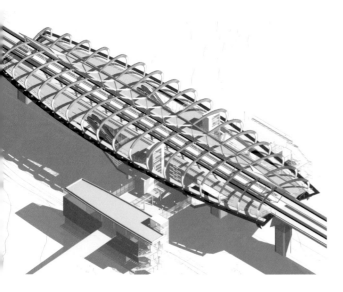

OPPOSITE

TOP   Brentwood station bulges in the middle, to accommodate lifts and stairs.

BELOW   Especially at night, the station is a tempting object, designed to attract commuters towards it.

THIS PAGE

ABOVE   The design is symmetrical about the rail tracks.

RIGHT   Skytrain carries travellers above the congested Lougheed Highway.

How to persuade commuters to abandon their cars is a problem facing most developed and developing cities. Everybody agrees that it is essential to reduce gridlock and improve urban air quality, but many car-drivers are resistant to change and reluctant to give up the freedom they associate with their vehicles. Some cities, such as London, which successfully introduced congestion charging at the start of 2003, favour the use of a stick. Others prefer the carrot. Vancouver, on Canada's west coast, is in the latter category.

With its great advantages of scenery and climate, Vancouver is determined to prevent uncontrolled suburban sprawl by concentrating development on a network of planned new centres. Although these centres are linked by a road, the Lougheed Highway, the city authority wants commuters to switch to more sustainable methods of transport than cars. It has therefore constructed the 13-station Millennium Line, the second line of its electrically powered Skytrain system, an elevated metro. Lifted above the road traffic, and providing panoramic views of city architecture and the surrounding mountains from front and rear carriages, Skytrain offers a pleasurable travelling experience. But how can passengers be persuaded to try it?

Part of the answer lies in station design. The Vancouver Rapid Transit Project

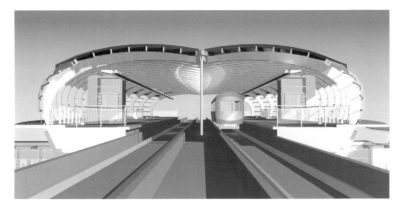

THIS PAGE
LEFT   Passengers arriving by train are drawn into the embrace of the glulam roof.
BELOW   Steel cross bracing across the elliptical central opening provides seismic stability.

OPPOSITE
TOP   The architect's design concept gives passengers access to light and views, lifting them above the pollution of the road beneath.
BOTTOM   The curved glulam beams are set into tall, white-painted steel shoes.

Office appointed local architect Busby + Associates to design two stations. The more dramatic is at Brentwood, which is intended to serve as the catalyst of a new town centre. The architect's solution – faintly reminiscent of a woodlouse, or some similar insect, squatting across the elevated railway lines – has translucent sides so that it can act as a beacon at night. Passengers approach the station by a new bridge straddling the main road, but running beneath the railway lines. From there, they take a lift or escalator to what turns out to be a simple, but elegant, shelter, made less formidable and more welcoming by the use of timber. The canopies over each platform are supported by curved, tapering glulam beams, set into tall, white-painted metal shoes, which rise from the concrete structure of the railway itself. This use of a curved glulam-and-timber roof is reminiscent of a station (albeit a temporary one) built in Rome by the architect Marco Tamino. Only 300 metres (980 feet) from St Peter's, the station was the terminus of a new metro line that provided access to celebrations to mark the new millennium.

The Brentwood station is wider in the centre than at the ends, to accommodate the lifts and escalators, so the canopies bulge out, curving in the horizontal, as well as in the vertical, plane. Flat glass in seven facets rises up around the concave curves of the sides, with all but its highest members covering the steel shoes. From outside, the dominant materials are steel and glass, but from the platforms the observer is far more aware of the glulam beams and of the curved-timber roof decks they support. This roof is made up from 50 x 100 mm (2 x 4 inch) softwood elements, nailed together in the centre. The two timber canopies meet only at the very ends, but there is steel cross bracing across the elliptical opening to provide seismic stability. Entirely practical in origin, the bracing adds the finishing touch to a drama that should surely tempt commuters away from their dreary cars on a road that offers neither views nor speed.

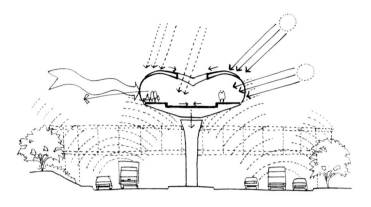

UNDERGROUND STATION

Hounslow East, London | Acanthus Lawrence & Wrightson | 2001

THIS PAGE

RIGHT   The timber-and-copper roof uses a combination
         of innovative engineering and top-quality detailing.
BELOW   Visualisation showing both phases of the project.

OPPOSITE
Oak struts support the diagrid roof.

Constructed in the early 1930s, the western section of London Underground's Piccadilly Line has some of the most architecturally distinguished stations on the network, including architect Charles Holden's designs for Southgate, Arnos Grove and Sudbury Town. The station at Hounslow East, however, had no such distinction. On a line that has become far more important since it was extended to Heathrow airport, Hounslow East is one of the last above-ground stations before the railway assumes its proper underground character on the way to central London.

The original station was nothing much to look at and access was difficult. It was set on an embankment, the continuation of a bridge over an adjacent cross road. Reaching the westbound platform after buying a ticket involved going out of the station, under a bridge and up the embankment. As well as being a less than satisfactory experience for passengers, the necessity for this procedure made it difficult to ensure that every passenger had bought a ticket before boarding a train.

When architects Acanthus Lawrence & Wrightson won a competitive tender to redesign the station, their answer sorted out the circulation with two new structures, one on either side of the embankment. The first to be built, and the more important, was on

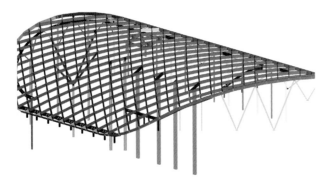

the west side and incorporated a ground-level ticket hall and ancillary accommodation. A stair and lift give access to the platform, which had to remain at its old, high level. A tunnel under the tracks leads to another set of stairs and a lift up to the other platform. The ticket office has a rounded shape that pays homage to the designs of Charles Holden. The station's palette of materials sets it apart from the mundane brick suburban houses nearby, but its swooping copper-covered roof brings it down on one side to the scale of the surrounding buildings. On the other side, at its full height, it provides shelter for passengers on the platform.

Timber has traditionally been used as a material for above-ground London tube stations, but at Hounslow East it is used on the roof in a novel and inventive manner. The architect worked with engineer Buro Happold and with Cowley Structural Timberworks, the UK's foremost expert in timber construction, to develop a two-way-spanning diagrid roof – a roof formed of relatively short elements, spanning in two directions. These make a grid of near squares, with a leg projecting from each corner of the square. An internal timber 'tree', consisting of oak struts emanating from a steel 'trunk', supports the roof in the centre. At the edges, it rests on the external walls.

This is not the classical environment in which lamella structures are used, since the approach is not used here, as is more common, to create an arched structure. Timber lamella structures were patented by Friedrich Zollinger in 1921, and were used commonly and successfully between the two world wars. Steel versions have been used – on the Houston Astrodome in 1965, for example – but have never proved widely popular. The two main advantages of lamellar construction are that regular elements can be used to form almost any shape, and that the relatively small elements are easy to handle on site.

At Hounslow, the roof is a constant barrel vault with a radius of 23.8 metres (78 feet). The design team selected a 1.25 metre (4 foot) grid, with each individual lamella being 2.5 metres (8 feet) long. The lamellas are identical, except that half are mirror images of the others, so there were no specials apart from wall plates and edge beams. The material used is a laminated veneer lumber (LVL), in this case made from Finnish softwood. It is dimensionally stable and can be supplied in large sizes, up to 40 x 1.8 metres (131 x 6 feet).

There are two main versions. In the one used for the lamella structure, the strands are oriented in the same direction. The other version, which has a cross-ply formation, was used for the roof decking, which was 27 mm (1 inch) thick, and tongued and grooved. This decking stiffens the structure and, to make proper contact with the lamella structure, the lamellas needed to twist slightly. Cowley Structural Timberworks, working with Buro Happold, developed a solution to make this possible. Each lamella is offset slightly at its junction with a cross member, and the edges of the lamellas are planed, so that the decking always lies flush to the top edge of the lamella and the wall plates on the bottom.

Since the timber structure was to be exposed internally, the connections had to be visually acceptable, as well as effective. Bolting is the traditional method of fixing, but this would have looked clumsy on such a relatively small structure. Instead, a special connector was used. This consists of a long, threaded bolt sleeve, glued into one end of a lamella. A threaded receiving tube went in the other end. On site, the bolts were threaded through a drilled hole in the cross member into the end of the next lamella in line, and the joint tightened up at an access hole in the side of the lamella. Crucial to the connection's success is a coupler embedded within the transverse beam. Normally, there are two of these bolts in each end but, where the shear forces demanded it, this number could be

increased to three. The offset angle between the lamellas, which was needed to allow the deck to lie flat, was achieved by offsetting slightly from the vertical the two connectors in the ends of the lamellas. Steel stanchions support the edges of the roof in most places, but where the edge of the roof cantilevers over the platform it is supported on triangulated timber struts that sit on specially shaped steel stanchion tops.

Above the deck of LVL, there is waterproofing, then a vapour barrier, insulation, battens, a ply skin and, finally, the handsome copper roof. Pre-patinated sheets, which had already acquired the characteristic green colour of an aged copper roof, were laid across the ply skin and joined by a combination of traditional and modern methods. The roof was rolled over the edge and a metre or so back under the eaves. Secret gutters, bull-nose eaves and soffits in copper contribute to the roof's continuous appearance. A translucent cladding material on the street elevation allows daylight to enter the ticket hall, ensuring that passengers can benefit fully from their last taste of natural light before plunging into the tunnels of the underground system.

VIEW SILO HOUSE

Livingston, Montana, USA | RoTo Architects | 2001

THIS PAGE
The tower draws on imagery of grain silos and, looking as if it is twisting in the wind, is a suitably tough structure for its rugged environment.

OPPOSITE

LEFT   Sections through the building show how the architect has made maximum use of its twisted geometry.

RIGHT   True vertical elements are clad vertically with cedar slats, reclaimed from pickling barrels.

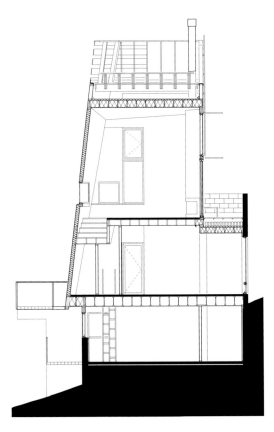

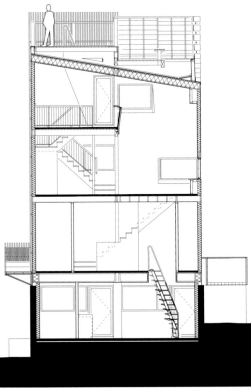

The mere name Paradise Valley is enough to make one yearn to live there – and for those who love wide-ranging views and a sense of space, Paradise Valley in Montana doesn't disappoint. Just north of Yellowstone Park, in what has become known as Marlboro Country, it offers warm summers and mild winters, dramatic storms and views of the Absaroka and Gallatin mountain ranges, made more spectacular by an absence of trees. Los Angeles-based RoTo Architects had a client, Ron Gompertz, who, after five years away from the valley, found the lure of the landscape too strong to resist and asked the practice to design him a house that, for the time being at least, would be a part-time residence.

The site, near Livingston at the north of the valley, covers 5.5 hectares (14 acres), with only one feature that offers any shelter at all – a bank of the Yellowstone River, which has now receded from this area. Gompertz was keen that part of the building should give him the opportunity to climb up and enjoy wide-ranging views. In response, the architect designed a house in two parts, only one of which has been built. Set into the ridge, the house consists of a tower (built) and a low-lying building that wraps around it (a future project). As well as satisfying the client's requirement for views, the rationale was that a dramatic tower would be more in sympathy with the qualities of the landscape than a medium-height,

blocky building. Drawing on the area's grain silos and elevators for inspiration, the tower is less obtrusive than a lower mass would have been. It has an asymmetric, twisted form, giving it a dynamic tension that suggests it is being distorted by the wind.

Entrance is at the southwest corner to the lowest level, which is partially set into the hillside. Two bedrooms and a bathroom occupy this space. A stair leads up to the next floor, where there is an office and living room. Over the living area is a mezzanine with space for cooking and eating. Above that, the staircase, now open to the elements, continues up to a viewing platform.

Although some masonry is used in the building, there is also extensive use of timber, both internally and externally. All of the window frames are timber, as is the construction of the mezzanine, which has a slatted timber railing – an effect repeated on the outside of the building. Whereas much of the cladding

is horizontal timber, on the south and west side there are elements projecting at true verticals from the main volume, which tapers upwards. These elements are clad with 2 x 2 cedar slats, reclaimed from barrels used for pickling. They are spaced between 1 and 6 cm (0.5 and 2 ½ inches) apart, and stand about 5 cm (2 inches) proud of a waterproofing layer. This layer consists of asphalt-roll roofing in bright red. The reclaimed timber is itself in a variety of colours, ranging from a silver-grey to a dark purple that is almost black. Set in front of the waterproofing, this creates a depth of colour and an effect that changes with the angle of view.

If constructed, the next phase of the building will be a low mass to the south of the silo. But the first phase is a fine structure in its own right, offering the client the views that he craves, while also acting as a handsome object in the landscape for other lovers of the region, whether they are observing from a distance or simply passing by.

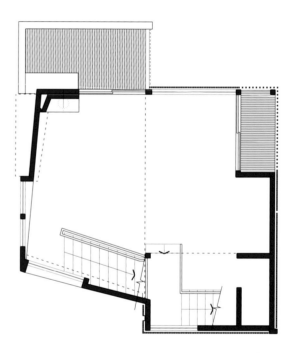

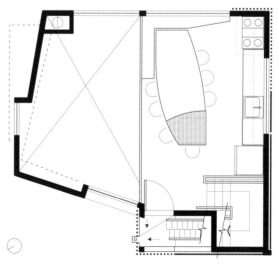

OPPOSITE
LEFT   Living room, with the mezzanine above.
RIGHT   Plans of the mezzanine floor (below), which
         contains the kitchen and dining areas, and of the top
         level, with its viewing platform.

THIS PAGE
LEFT   A certain amount of masonry mixes with the
         timber cladding.
BELOW   Entrance is at the lowest level, partially buried in the
         hillside, which also incorporates bedrooms and a
         bathroom.

BODEGAS YSIOS

Laguardia, Spain | Santiago Calatrava | 2001

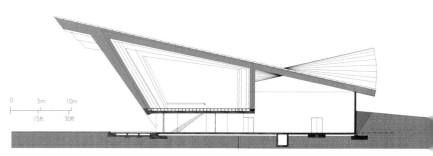

OPPOSITE

TOP   Section from front to back of the winery, showing that straight elements are used to create the undulating curves.

BOTTOM   The front façade is clad in strips of cedar, echoing the colour of the soil.

THIS PAGE

ABOVE   The short-side façades are treated in a much more workaday manner, clad with corrugated metal.

RIGHT   The central wave of seven rises to a higher peak.

Buildings can help to create a brand, and for a new winemaker in Spain, the Bodegas & Bebidas group, there seemed no better way to take advantage of this idea than to employ the country's most flamboyant architect-cum-engineer, Santiago Calatrava. With its new Bodegas Ysios in the Rioja region, the winemaker has ended up with a building that both serves its functional purposes and provides a great draw for visitors. Already featuring prominently on the company's website, the building is also sure to make an appearance on the wine labels.

Founded in 1998, Bodegas Ysios is near Laguardia, northwest of the Rioja capital of Logroño and at the foot of the Cantabrian mountain range. Climatically important for the quality of the wine, Rioja Alavesa, the mountains also provided inspiration for the form of the building. Eschewing his love affair with the zoomorphic form, Calatrava has instead designed a structure with a series of curving roofs. Faced with the constraints imposed by the building's function and a relatively limited budget, the architect continued his exploration of how to create curved structures using straight elements – a challenge that has fascinated him for years. Laminated beams of Scandinavian fir span nearly 26 metres (85 feet) from the front to the back of the building, rising and falling in a pattern that gives drama to the interior as well as the exterior.

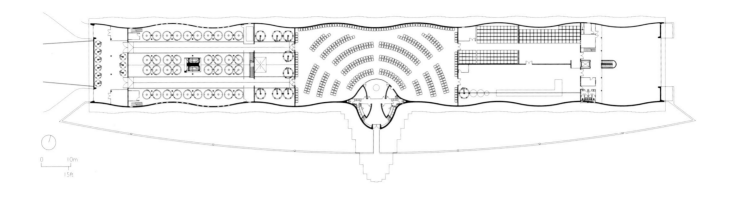

OPPOSITE
TOP Plan of the building, through which the wine-making process progresses from left to right.
BELOW The undulating roof echoes the form of the Cantabrian mountain range.

THIS PAGE
RIGHT The back of the building, visible from the vineyards, is in raw concrete.

THIS PAGE
RIGHT   The visitor entrance is at the dead centre.
BELOW   The extraordinary geometry also makes for
exciting spaces internally, with the laminated beams
of Scandinavian fir exposed.

OPPOSITE
Visitors are entertained on the upper floor at the
centre of the building, from where they can enjoy
magnificent views.

The building is 196 metres (643 feet) long, with a main entrance facing south towards the main road and the mountains set dramatically behind it. It has load-bearing concrete walls that undulate on both the front and back façades. This undulation adds stability to the long walls. Keeping the curves synchronized on the two walls means that the distance between them remains constant. Where the walls bulge outwards most, they are at their highest; where they are most concave, they are at their lowest. Since the most concave element on the front will be matched by the most convex at the rear, and vice versa, this means that the beams spanning from front to back are at constantly changing angles. Hence, with one simple geometric concept, the architecture has generated drama in three dimensions.

For a working winemaker, this would probably be enough, but the building also has a grand axial entrance for visitors. The central element, the middle wave of seven, pokes up 10 metres (32 feet) higher than the rest to accommodate a dining and seating area for visitors, with a glazed wall offering dramatic views over the plains to the hilltop town of Laguardia. By projecting forward, this gesture also allows the insertion of a visitors' lobby, without interfering with the linear production processes. These are straightforward.

Grapes (up to 1 million kilograms/ 2.2 million pounds per harvest) come in at the western end, and move through production and barrel-cleaning areas to storage in the central area, and then on to bottling, bottle storage and shipping. Finished bottles of wine leave the building on the eastern side.

The small east and western façades are simply clad in corrugated metal, as if this were almost any industrial shed. In contrast, the 'public' south façade is clad in strips of cedar, in a conscious echo of the colour of the soil. The back of the building, visible to those working in the vineyards, is raw concrete, and the roof is clad in reflective aluminium. 'The effect of sunlight on the roof creates a wave-like movement, like the changes in tonalities of the surrounding vineyards,' says Calatrava. It is fortunate that these vineyards are at too low a level for the people working in them to be dazzled by reflections from the roof on days of remorseless sunshine.

Bodegas Ysios is growing tempranillo grapes on 65 hectares (176 acres) and takes its winemaking as seriously as its architectural commission. Repeat business will depend on the quality and price of the product. But for those visiting the region, Calatrava's building will doubtless prove an attraction. There can be few better three-dimensional advertisements.

RAILWAY STATION

Aix-en-Provence, France | AREP | 2001

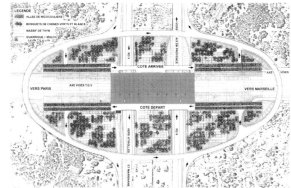

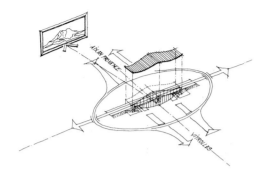

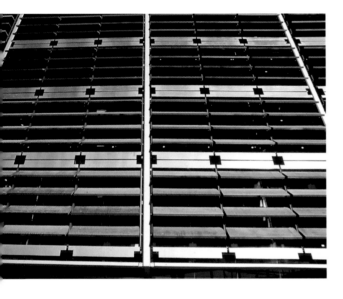

OPPOSITE

TOP   The TGV station is intended to form the
           centrepiece for new development.

BOTTOM   The screen of untreated cedar louvres on the
           western façade is controlled automatically to
           provide shade from the sun.

THIS PAGE

TOP LEFT   Views of Mont Sainte-Victoire were central to the
           architect's concept for the station.

TOP RIGHT The station, seen from the west, with Mont Sainte-
           Victoire behind it.

ABOVE   The cedar louvres are arranged in rows of four, with
           strips of solid aluminium between them.

France has been a world leader in fast, efficient rail travel, with its network of TGV trains setting the standard. The first of these lines was built between Paris and Lyons, in the centre of the country, and this line has now been extended to Marseilles, the country's second city. Journey times between Paris and Marseilles can now be as short as three hours. This line, like other TGV lines, was built relatively easily for three reasons. The French countryside is fairly lightly populated; the development that accompanies a TGV line is seen as beneficial; and the planning system is able to force through new projects. As a result, TGV trains run on new, straight tracks, and this means that they frequently need new stations outside the city centres.

The extension of the TGV line south has led to the construction of three such stations, at Valence, Avignon and Aix-en-Provence. This last station, the most southerly of the three, is actually midway between Aix and Marseilles, and within easy reach of Marseilles airport. All three stations were built under the direction of Jean-Marie Duthilleul of AREP (Aménagement Recherche Pôles d'échanges), the station development arm of SNCF, the national railway.

Working with landscape architect Desvigne & Dalnoky, Duthilleul decided that each station should respond to its context, rather than following a unified pattern. This may seem common sense, but it represented a departure from established practice. Previously, each line had its own identity, echoed in the unified design of its stations. A precedent had, however, been set by Santiago Calatrava's show-stopping station at Lyons-Satolas airport. A similar change in approach was also evident in London, where the newest underground line, the Jubilee Line extension, designed in the 1990s, broke away from a previously unified approach by employing a different architect to design each station. On the TGV line, the decision was also influenced by the fact that the stations were set in 'nowhere places', where they might well form the catalyst for new development, and therefore it was essential that they should have distinct local identities.

Some unifying characteristics are evident, however. The hot summers in the south of France make it necessary to provide protection from the heat and dazzle of the sun. At the Aix-en-Provence station, the station structure provides shade to the ground-level track and platforms. It has a curving roof that peaks in the middle, supported on chunky glulam columns of untreated iroko, spaced to echo the carriage lengths of the trains, from which rise angled steel struts. The eastern side of the station is open, offering views to

Mont Sainte-Victoire, the local landmark, made famous in the paintings of Paul Cézanne, a native of Aix.

Mont Sainte-Victoire has a distinctive, symmetric, almost Japanese form, and this is eerily echoed in the appearance of the western façade of the station. Blocking one's view of the mountain, the station seems almost to replace it. The western side is enclosed by a double glass wall that curves outwards, and in front of it is a screen of untreated red cedar louvres coming down to a few metres above ground level. These smallish louvres, crisply detailed in bands four deep, on an aluminium framework and with solid aluminium set between them, respond to the sunlight, shutting to keep out the hot afternoon sun and, with the glass façade, providing shelter from the biting Mistral wind.

The station is surrounded by a sea of car parks, above which the wooden 'mountain' appears to float. Once some trees have grown, it will seem more directly set in a landscape. Reflecting the French tradition of formal planting, their position and significance have been calculated precisely by the landscape architect; in this case, each station has been addressed in a similar manner, although with some local differences. At Aix, plantings of thyme, myrtle and santolina will provide some of the evocative smell of the garrigue.

Twin avenues of planes trees, the traditional companions of French country roads, will delineate the approaches to the stations. Rows of cypress trees will provide station windbreaks, and the parking areas will be both shaded and softened by 'orchards' of local trees (in Valence and Avignon) and by holm and white oaks in Aix-en-Provence. These are also being used to mark the outline of future development – something that is likely to happen, given the enthusiasm with which the French embrace their new stations, and the care and consideration that has been given to their design and situation.

THIS PAGE

TOP CENTRE   Chunky glulam columns of untreated iroko, spaced at the length of train carriages, support the roof.

BOTTOM LEFT   Timber floors help give an air of calm to the upper waiting area.

ABOVE   Angled steel struts emerge from the top of the columns.

OPPOSITE

TOP LEFT   The form of the roof is reminiscent of Mont Sainte-Victoire.

TOP RIGHT   Trains serve two island platforms.

BOTTOM   Users of the station are protected from the elements, but not isolated from their surroundings.

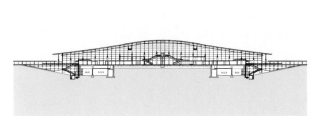

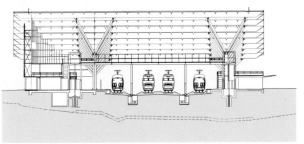

EXPO CANOPY

Hanover, Germany | Herzog + Partner | 2000

0   500m
1500ft

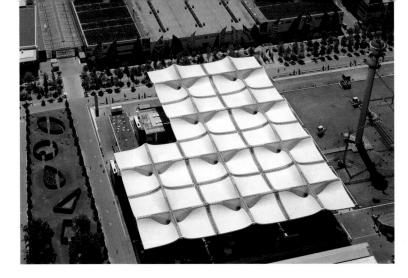

OPPOSITE

TOP   Although its size is impressive, the canopy occupied a fairly small proportion of the Expo site.

BOTTOM   Even to the casual observer, this collection of timber umbrellas is obviously something special.

THIS PAGE

ABOVE   Plan and elevation of one of the ten units that make up the canopy.

TOP RIGHT   Above the timber is a synthetic, impermeable translucent skin.

BELOW   Concept sketch showing the scheme as a provider of shelter.

Professor Thomas Herzog, who is based in Munich, is an architect who has always treated architecture as a form of research. Every one of his relatively modest number of built works pushes technology and thinking forward in a new way. But this is not a cold-hearted embrace of technology. Deeply involved with environmental issues, Herzog is concerned to create true sustainability in his buildings, in terms both of materials and the way that buildings relate to their surroundings.

This is reflected in his canopy at the Hanover Expo, held in 2000. As part of a project that, by its nature, involves a great deal of 'throwaway' structures, Herzog's canopy was the one item always intended to have a permanent life. In a place where the weather is decidedly unpredictable, this massive structure provides protection from both rain and sun – in a manner that allows even the casual observer to understand that timber is being shown off and exploited in entirely new ways.

The L-shaped canopy consists of ten elements (arranged as if two are missing from a four by three rectangular array). Each measures roughly 40 x 40 metres (130 x 130 feet) and is more than 20 metres (66 feet) high. In each, a weighty, square-sided mast supports a wavy square umbrella of latticework on glulam beams. These ten panels are joined together and provide each other with structural stability, minimizing deformation under snow and wind loads.

The analogy with trees is obvious, with the beams curving up out of the trunks at a 'branchlike' angle, and the canopy of smaller elements taking the role of small branches and twigs. But the romanticism of this concept, with the traditional role of trees as shelter, does not tip over into feyness, and it is evident that every element of this ornate structure is working hard and playing a vital role.

The supporting columns comprise four whole, sawn-timber sloping uprights, connected by glulam panels and steel elements, with steel feet anchoring them to a below-ground concrete ring beam. Each of the timber columns is 16 metres (53 feet) long and comes from a single silver fir tree up to 200 years old from the southern Black Forest. Cut in half longitudinally to speed drying, and then joined together again, these trees were selected very carefully, by a combination of observation and ultrasonic testing. The columns may be metaphorical 'trunks', but the real tree trunks are used, counter-intuitively, upside down, since the stresses are highest at the tops of the columns and so the larger diameter was needed there. The diameter is 95 to 110 cm (37 to 43 inches) at the top and 68 to 74 cm (27 to 29 inches) at the bottom.

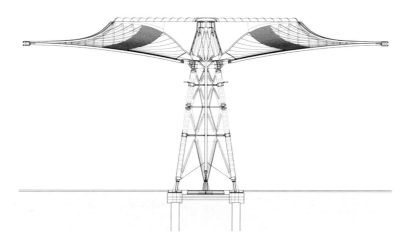

The large size of these trees meant that there was no option but to dry them naturally. After seven months of drying, it was found when construction started that they had not dried as much as expected. This meant that the engineer, IEZ Natterer, had to scale down some predictions of their behaviour.

Triangular frames of glulam, covered with laminated veneer lumber (LVL), 33 mm (1⅓ inches) thick, stabilize the towers. Some steel elements are also included, to prevent horizontal distortion. All the steel connectors that join the timber elements had to be designed specially to take into account the distortion caused by the continued drying of the timber. Bolts had to be accessible for further tightening. A central steel structure at the top of each tower passes down the loads from the cantilevered beams and the umbrella itself – technically termed a 'ripped shell'. These doubled curved shells, and the cantilevered glulam beams that support them, have a very 'natural' form that could only be obtained by the use of the most sophisticated mathematical modelling.

The cantilever beams are 19 metres (62 feet) long and consist of two elements – a lower curved beam that follows the curvature of the ripped shell and an upper straight element. As they move towards the ends of the beams, these two elements come together into a single straight beam. The shells themselves each weigh 36 tonnes (35 tons) – a lot of weight to support on cantilevers above the ground. The form of the ribs that make up the shell is dictated by the design forces, and they are joined by layers of LVL boarding. Most connections are by screws, but gluing was used in the areas under highest stress.

Additional layers of boards above the ribs provide further stability. The umbrella is covered with a synthetic impermeable translucent skin, 5 mm (⅕ inch) above the timber structure, supported by a series of cables. It provides considerable protection from rain, while the gap permits ventilation. The fixing of the membrane is more complex than the simple infill observed from below. In fact, this is the only sleight of hand seen in this building – what looks awe-inspiring but very natural from below has some of its workings concealed above the canopy itself, where few will ever see them.

When dealing with structures of this level of innovation, architects and engineers cannot rely on their gut instincts. Wind and snow loads were simulated in a wind tunnel and produced unexpected results. With most canopy roofs, the wind load is directed upwards. It turned out that at Hanover this load was directed downwards, providing an additional load on the structure. Design details were worked out carefully, not only to provide structural performance, but also to ensure that the structure was protected from rotting. Construction was also complex, with large prefabricated elements being transported to site as outsize loads, and a considerable amount of pre-assembly taking place under cover, adjacent to the site.

The result is a structure that, while not appearing effortless to construct, looks supremely logical, as well as elegant. It is a virtuoso piece of work, made possible by the fact that it had only to provide one simple function – that of shelter. Serving its purpose admirably at Expo, and still in place, it should have sent some of its visitors away with a renewed appreciation of the potential of timber. As Herzog moves on to his next discovery, we should expect to see some of the lessons learnt at Hanover filtering through into other buildings.

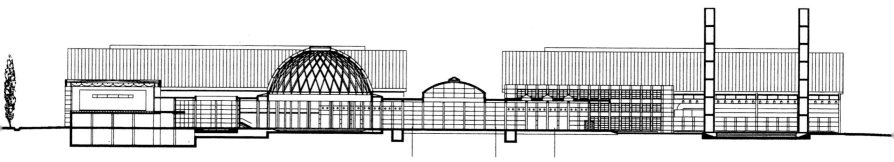

OPPOSITE

TOP   A strongly axial design leads visitors through the centre of the exhibition complex.

BOTTOM   A circular dome, 30 metres (100 feet) across, is one of the main features.

THIS PAGE

ABOVE   The repetitive nature of the halls offered savings in the construction process.

RIGHT   There are echoes of classical architecture, appropriate to Rimini's history.

The thought of going to an exhibition centre makes most people's hearts sink. Important meeting places, where there is a chance to see new things and meet new people, exhibition centres are usually huge and, far too often, soulless, confusing and depressing. They give the impression of being cut off from the normal world, in an artificial and ugly environment of bad ventilation and worse design.

How refreshing, then, to see a new exhibition centre that bucks all these trends by being logically laid out, airy, full of natural light – and which makes generous use of timber, a material not commonly associated with such locations. That is the case at the Nuova Fiera di Rimini, which uses warmth and natural materials to make the exhibition centre a welcoming place to visit and to do business in.

Although Italy hosts the Milan furniture fair and important exhibitions in Bologna, Germany has certainly been the leader in massive fair halls for the last few decades. Now Rimini has decided to catch up. Appropriately, it appointed Hamburg-based architects von Gerkan, Marg + Partner, who had, as part of their distinguished portfolio, designed exhibition halls in cities including Hanover and Düsseldorf.

Rimini is a town that, although commonly associated with cheap-and-

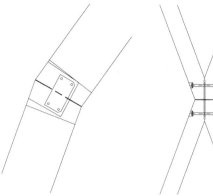
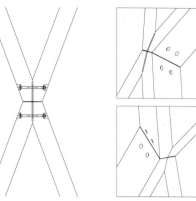

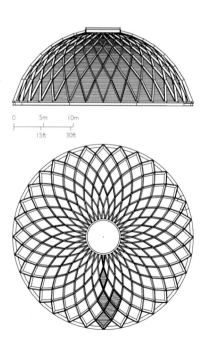

cheerful seaside holidays, has a long and illustrious architectural history. This is reflected in the approach to the fair buildings, which the architect has described as 'orientated around the Emilia-Romagna tradition, which has characterized European architectural history since the ancient world and the Renaissance'. The design is strongly axial but, within the clear geometry, classical elements are interpreted in a contemporary manner.

With an exhibition area of 80,000 square metres (860,000 square feet) and a service area of 50,000 square metres (538,000 square feet), the Fiera consists of 12 exhibition halls, congress and conference rooms, event areas, restaurants, shops, administration offices and auxiliary and storage rooms – quite enough, if the organization were bad, to make it disorientating and overwhelming. To avoid this, the architect created an entrance forecourt with a fountain and four tall, square light towers to signal its presence from a distance. The main entrance has a portico and a circular domed meeting space. The single-storey exhibitor halls are arranged along colonnaded walkways. They are modular to allow maximum flexibility of use, despite the formal layout.

Both the dome roof and the roofs of the exhibition halls are in timber.

The exhibition hall roofs span a 60 x 100 metre (197 x 320 foot) column-free space, a feature that is popular with exhibition organizers. Along their central apex, glazing rises above the timber structure, again bringing light into the space. The architect has used a lamellar structure for the roofs, developed in the 1920s by Friedrich Zollinger. However, by using modern techniques of laminated timber construction, it has managed to produce spans far larger than those Zollinger could achieve. The use of identical, relatively small elements, which are easy to transport, is ideally suited to projects like the Rimini Fiera, where all 12 exhibition halls have the same form.

Each roof consists of lamellar wood, laid in a framework made up of a regular mesh of diagonally placed, rhomboid-shaped elements, measuring 3 x 6 metres (10 x 20 feet). Each beam in this module has the same section, 16 x 70 cm (6 x 28 inches), wherever it is within the roof. At the ends of the roof are arches, also in lamellar wood, with a rectangular section of 50 x 70 cm (20 x 28 inches). Where four beams join, the joints connecting them have to be able to transfer both the bending-axial stress and the shear stress, to ensure that the structure is truly continuous. Specially developed for the project, these joints consist of a steel plate, connected to the four beams by pins

OPPOSITE
Rhomboid-shaped elements make up the
rectangular roofs.

THIS PAGE
The halls offer exhibitors plenty of column-free
space.

and a central X-shaped element bolted
to the plate. The cavities between the
steel and the wood are filled with a
special grout. Once the assembly is
complete, these steel elements are
completely invisible.

At 3 metre (10 foot) intervals, the roof
is connected via hinges to a steel
perimeter box beam, which is
supported, in turn, on a concrete
substructure. Most of the stability of the
roof comes from its vaulted shape and
from the way it is restrained at the
edges, but there is also a contribution
from the planking that clads it, which is
fixed from the top by nails. The circular
dome uses similar technology. It has a
30 metre (100 foot) diameter and is
22 metres (76 feet) high at the crown.
A central oculus brings in light both to
the space and to illuminate the dome
itself, which has a latticework of
structural members coming together
towards the apex. Timber boarding
behind runs circumferentially.

The architect used Scandinavian timber
and took care to ensure it was all from
renewable resources. At the opening
ceremony, architect Volkwin Marg said
that the building was intended to
connect the 'past to the future, with
references to antique architecture and
with a dedication to the culture of this
region and of this country, which has
survived all the ages. It is an architecture

oriented towards an ecological future
that distinguishes the greatest and most
modern European wood-covered
building, a portrait of beauty, technology
and respect for the environment.' He
described the success of the project in
this way: 'The Italians have organized a
competition and we have worked with
them to create a unique experience: as
disciplined as Prussians, hardworking as
Swabians, and always punctual, they cost
half as much as Germans and are timely
in making decisions.'

Another way of looking at this synergy
of cultures is that a distinguished
German architect has embraced an
Italian architectural tradition to build
a complex that, by making wise and
imaginative use of Scandinavian timber,
transcends the usual bland international
experience of visiting exhibitions.

In the early 1990s, organizations such as DEGW, the London-based architect and space planner, put a lot of effort into analyzing the timescale in which change took place in buildings. DEGW argued that change happened at one speed (the slowest) to the fabric, at another to the services and at a third (the most rapid) to the fixtures and fittings. Since then, the prominence of such thinking has declined. Some of the arguments seem self-evident, while others are too mechanistic for the less formalized approaches to work that we have today.

But it holds true that the interiors of buildings are generally less long-lasting than the exteriors, since fashion, use and ways of working dictate relatively rapid change. This is often reflected in the design: the interior architect can create something specifically for the current needs of the user, rather than having to think too much about flexibility over a long period. Moreover, an interior has to cope only with the depredations of human traffic and not with those of the weather. The interior is therefore an excellent place for experimentation. Designing interiors is also often a way for architectural practices to establish themselves, in that clients seem more willing to trust new practices with the inside of a shop, a bar or even an office than with the structural and planning complexities of a whole building.

Wood has always been a vital part of interiors. It is a traditional material for furniture and floors, and in recent years it has won back a large part of the market from carpet. Tactile and warm both to the touch and the eye, wood makes environments more welcoming. But it can do much more than that.

At the Bally store in Berlin, Craig Bassam has used a high-quality flooring product to cover all surfaces in a sophisticated manner that makes the Swiss company's concerns with quality and detail evident to all customers (see page 150). Since the shop has large windows, Bassam's near-obsessive use of timber manages to avoid the relentless, and rather claustrophobic, impression created in, for example, some traditional Finnish timber houses.

At Renzo Piano's Parco della Musica in Rome, there is an equal level of technical control, but it serves a different end (see page 154). Here timber – American cherry – helps to create the carefully calculated acoustic. The application of timber for its acoustic properties is not restricted to concert halls. For example, at Temple Quay House, a large government office building in Bristol, southwest England, local architect Stride Treglown used slotted timber panels with an acoustic quilt behind them to reduce reverberation within the atrium.

The relative ease of shaping timber, particularly in its engineered forms, makes it suitable for the creation of 'pods', or insertions into large, open spaces, which are becoming popular in offices eager to foster less formal ways of working. Timber has been used in that way in the office of TBWA\Chiat\Day in San Francisco, and also in the UK headquarters of the Danish pharmaceuticals company Lundbeck (see pages 166 and 168). Nor is this idea of inserting a timber object into a large, open space restricted to office buildings. At the Peckham Library in South London, which won its creator, Will Alsop, the UK's most prestigious architectural award in 2000, ellipsoid timber pods are used to house an interview area and an early-learning centre. Working with the superlative craftsman Gordon Cowley, Alsop created these from timber ribs, covered with oriented strand board and with ply-clad triangular elements. Even more expressively, the Russian practice Savinkin/Kuzmin has used plywood in an extraordinarily sensuous way to create the Moscow nightclub Cocon Club. The initiative even extends to a lavatory pod, with a door that opens like a submarine door, suspended high above the rest of the club. The most extraordinary aspect of this project is that, although plywood seems an integral part of the design, it was not the architect's first choice. Originally the architect wanted to clad

the pods with titanium and abandoned the idea only because the necessary technical expertise was not available.

All the projects mentioned above are new constructions, but timber also has an important part to play in heritage buildings. A sympathetic material in that it was generally available at the time when the old buildings were constructed, timber can also be treated as relatively temporary. If you build in concrete, your decision is probably irrevocable; and, while stone and brick masonry constructions may be removable, they are heavy and have an air of permanence. The latest thinking about the conservation and restoration of historic buildings is that all interventions should be reversible. There is an awareness that some of the 'correct' thinking of the past, in terms of restoration, was flawed and that we may be making similar errors today. The only way to avoid this is to ensure that everything that is done can be undone. At the castle of Peñaranda in northern Spain, the architect has created a new use for a hollow stone tower by inserting a timber structure that can simply be taken away again when no longer required (see page 162). In other chapters of this book, we see how timber can be designed for durability and long life; at Peñaranda, it is timber's potential impermanence that makes it such an attractive choice.

BALLY STORE
Berlin, Germany | Craig Bassam Studio | 2001

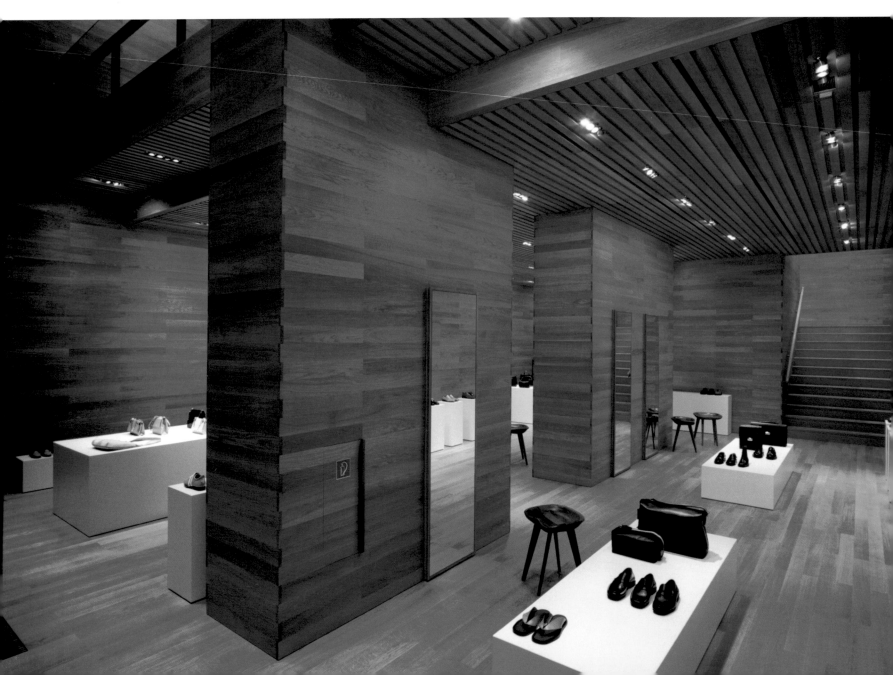

OPPOSITE

TOP LEFT   Bally's store on Berlin's elegant Kurfürstendamm
            was chosen as the flagship for the new design
            approach.
TOP RIGHT   The warmth and simplicity of the wooden interior
            act as a magnet for shoppers.
BOTTOM   Oiled light oak was used for floors, walls and
            ceilings. The stools, based on a Swiss tractor seat,
            were carved in the Bally factory with the same
            technology it uses to make its shoe lasts.

THIS PAGE

ABOVE   Clothes are displayed like art objects against the
            simple background.
RIGHT   White-lacquered cubes form one of the few other
            elements in the design.

Given that fashion is such a fickle business, it was a great achievement for Swiss shoe company Bally to have stayed at the top of the tree since its foundation in the Swiss hamlet of Schönenwerd in 1851 by Carl Franz Bally. Not surprisingly, when the Texas Pacific Group bought the company in 1999, it decided that the time had come for an updating. The company brought in Scott Fellows as its creative director, who set about brightening the brand, while sticking to some of its traditional strengths. And he appointed Australian architect Craig Bassam (now of Bassam Fellows) to design a new concept retail store in Berlin, as well as a headquarters building in Switzerland.

This was Bassam's first retail project and it shows – not through any incompetence, but in the freshness of his thinking. He turned the 300 square metre (3,230 square foot) store at 219 Kurfürstendamm, Berlin's prime shopping street, into an extraordinarily homogeneous, understated, yet rich, space. Designed to show off the latest lines not only in shoes, but also in accessories and clothes, the store is utterly contemporary, but nevertheless reflects the Swiss craftsmanship and obsession with detail and quality that have always characterized Bally.

Bassam achieved this by using a Swiss-made standard flooring product of fine-grained, oiled light oak throughout – on

THIS PAGE

LEFT   Red-lacquer trays in the cupboards add a touch of
        colour.

RIGHT   Construction by Swiss cabinetmakers ensured a
        beautiful level of finishes.

OPPOSITE

TOP LEFT   Handrails to the stairs, deliberately designed to be
            comfortable to hold, are supported by oxidized
            bronze.

TOP RIGHT   Slightly offsetting the oak planks one above the
            other adds a degree of visual dynamism.

LEFT   There are 300 square metres (3,230 square feet) of
            selling space, on two levels.

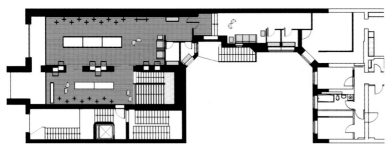

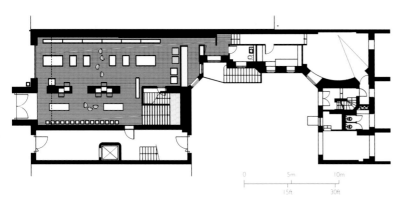

the floors, naturally, but also on walls, ceilings and stairs. The feeling in this two-storey building is of being enclosed in an intricately worked box. The success of the project derives from its superb execution, which, in this case, was carried out not by German construction workers, but by Swiss cabinetmakers. They have achieved a perfect grid of materials, with, for example, alternate elements offset horizontally by a very small amount at the ends of walls, imparting a minute sense of movement to this serene environment. Even the design of supports for the handrails was carefully considered, with a hand-friendly oval section supported on oxidized bronze.

'Switzerland suggests a certain connection with nature and also the precise character associated with its people,' Bassam has said, referring to photographic research of old Swiss chalets with all-wood interiors. But whereas those chalets showed off traditional, but relatively rough, craftsmanship, his Berlin store is so smooth and finely finished that you want to stroke its walls. The timber throughout is European oak, used as boards with a section of 10 x 125 cm (4 x 61 inches) on the walls and floors. Bassam has used slats on the ceilings, and boards with black rubber insets on the stairs.

He has added only two elements to this harmonious interior. Shoes are displayed on simple white-lacquered cubes that sit on rubber feet. Intended to be used for a museum-style display, these cubes can be moved around the store easily, with their changing configuration generating changes of atmosphere. The other element is the furniture. Stools, benches, stacked storage trays and even hangers are all solid walnut. Red-lacquer trays in the cupboards add the only touch of colour. The stools, adapted from a Swiss tractor seat, were carved in the Bally factory in Switzerland, using the same technology as shoe lasts.

Bassam also used oak for a roof extension to the Bally headquarters building in Switzerland, and rolled out ideas from the German store to other key locations. In 2003, a new top management team was installed at Bally, who will doubtless introduce new ideas. As stores increasingly become places of entertainment, Bassam's approach looks even more restrained and tasteful, while reflecting the Zeitgeist of shop-as-museum. Swiss craftsmanship makes products that are built to last and, just as few Bally customers would discard their purchases at the end of one season, Bassam's store design should be allowed to enjoy the longevity to which it is suited.

PARCO DELLA MUSICA

Rome, Italy | Renzo Piano Building Workshop | 2002

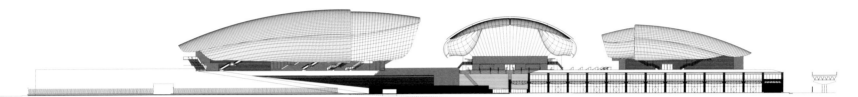

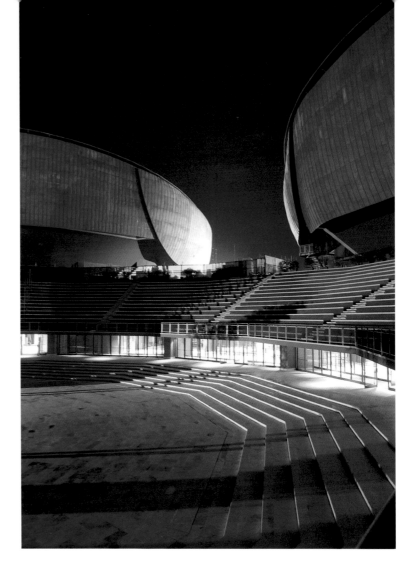

OPPOSITE

TOP By designing each hall as an independent structure, Piano was able to respond to their individual acoustic needs, without having to shoehorn them into an all-embracing structure.

BOTTOM The bulbous lead roofs are supported on arched glulam beams.

THIS PAGE

ABOVE A common concourse underlies and links the three halls.

BELOW Vegetation fills the spaces between the three halls.

RIGHT The amphitheatre defined by the halls is a magnificent space for outdoor performances.

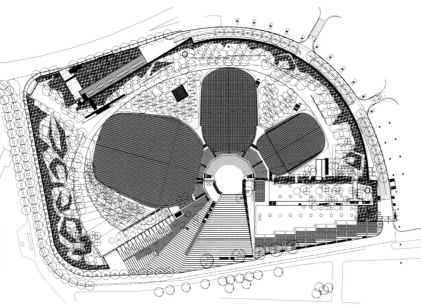

A contender for the title of the world's greatest living architect, Renzo Piano is a lover of materials. He received the first Spirit of Nature Wood Award, an international prize administered from Finland, for his work with timber, of which the most prominent example is the Tjibaou Cultural Centre, in Noumea, New Caledonia. It is one of the best-known yet least visited of contemporary buildings, because of its remote position. Far more accessible to most people is Piano's Parco della Musica in Rome. This building, fulfilling a long-felt need, is not, externally, a celebration of wood. The three linked concert halls are in brick, topped by wonderfully expressive bulbous lead roofs, albeit lined in pine and supported on arched glulam beams.

But if you reach the concert halls themselves, you will find yourself surrounded by timber, used in a sensuous and – given the precision of today's acoustic design – immensely practical manner.

Piano's masterstroke in the design of the Parco della Musica was not to try to shoehorn all the functions into one massive structure, but to put each of the three concert halls, of different sizes and functions, into a separate building. This makes sense both architecturally and acoustically. The three are linked and define between them a fourth space, an amphitheatre for open-air performances. This strategy paid dividends when, unsurprisingly for such

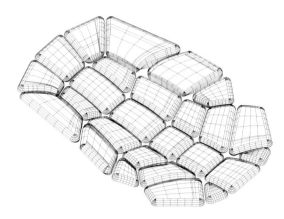

an archaeologically rich city, the site turned out to be perched on the ruins of a sixth-century Roman villa, which then had to be incorporated in the complex – a much simpler task with an agglomeration of buildings than with a monolith. As described by Piano, 'In our project the three halls, each set in a container resembling a giant sound box, were arranged symmetrically around an empty space, which became the fourth auditorium, not included in the initial programme: an open-air amphitheatre. Amid these constructions, luxuriant vegetation established a connection with the nearby park of Villa Glori.'

Each hall has a different function and, as well as being designed to give optimum quality for live performances, offers high-quality recording opportunities. Acoustic consultant Helmut Muller went through a lengthy process to ensure that the acoustics were right. First, he built models with reflective surfaces and used lasers to trace the route followed by reflections. From this he created diagrams of acoustic response, and fed the data into a computer to simulate the reflections of the soundwaves. The final stage entailed analogical tests – that is, using real sound – on large-scale models.

The smallest hall, with a capacity of 700, is intended for concert operas, baroque and chamber music, and theatrical plays. Based on some of the

solutions that Piano first used at the IRCAM experimental music centre in Paris 20 years ago, the space is extremely flexible, with movable walls and floors. This flexibility makes it possible for the hall to host symphonic performances, for which the stage arch opens up and thereby redefines the layout of the stage.

The 1,200-seat hall is also flexible in terms both of acoustics and layout, with a movable stage and an adjustable ceiling. It has space for a large orchestra and choir, but can also be used for ballet and contemporary music.

The largest hall, with 2,756 seats, is for symphony concerts with large orchestras and choir. It is about the largest feasible size for a concert hall – any bigger and there would be unacceptable reflections from the back. As well as straightforward acoustic considerations, audiences tend to feel uncomfortable in a larger space. The hall is of the vineyard type, meaning that the audience is divided into small, asymmetric rising blocks of steps. The vertical surfaces of the divisions create early lateral reflections into all sections of the audience. The best-known representative of the type is still Scharoun's Berlin Philharmonie, completed in 1963. At Parco della Musica, the orchestra is placed midway between the centre and the rear of the hall, with some seating behind it.

THIS PAGE

TOP LEFT A total of 26 curved caissons form the roof of the large symphony hall.

ABOVE In the symphony hall, some of the audience is seated behind the orchestra.

OPPOSITE

TOP The vineyard design, used in the symphony hall, places the audience in small, asymmetric groups at different levels.

BOTTOM Using cherry for the caissons gives a wonderfully warm feeling to the symphony hall.

OVERLEAF

However good the design, a performance space never reaches its full potential until it is filled with people.

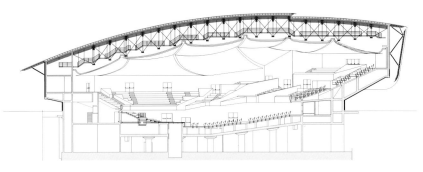

The medium-sized hall seats 1,200. It has an
adjustable ceiling and a movable stage for maximum
flexibility of use.

Seating 700, the smallest hall is intended for more
experimental works.

The timber used for all the hall interiors
is American cherry. A rich and warm-
looking timber, with colour variations
that add depth to its appearance, it also
performs superbly acoustically. Many
acoustic designers, especially in more
northerly latitudes, favour larch for
concert-hall interiors, but, although it
can be beautiful, it lacks the sensuality
of cherry. At Parco della Musica, the
cherry is most dramatic in appearance
in the largest hall, where it is used in the
26 curved caissons that descend from
the ceiling.

Even such matters as the choice of
fire retardants needed to be considered
carefully. Piano commissioned test
panels from manufacturers, and

eventually selected a retardant
known as Saverlack, on the basis of its
transparency and its adhesion.

The combination of its complex local
politics, funding problems and over-
abundance of archaeological riches
means that Rome has taken a long time
to achieve the musical centre it needs.
But Piano's solution was well worth the
wait. And despite the centre's use of the
most up-to-date technology and bold
contemporary architectural forms, it is
an eminently comfortable setting for the
music of the last three centuries – not
least because of the use of timber and
its association with traditional musical
instruments.

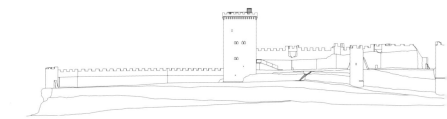

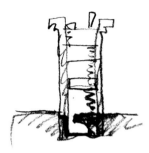

OPPOSITE

TOP   The castle at Peñaranda is in a commanding position, set on a ridge above the town.

LEFT   The design makes maximum use of views outwards.

RIGHT   There is a clear distinction between the new, timber interventions and the existing stone fabric.

THIS PAGE

TOP   The architect saw the tower as 'a hollow tube of stone emerging from the rocks' and designed accordingly.

ABOVE   The first of many stairs lead visitors to the entrance.

RIGHT   The simple square form and the crenellations on top give the tower a satisfyingly story-book aspect.

Castile, in northern Spain, is scarcely short of castles, but the castle at Peñaranda is one of the most dramatically sited. Peñaranda, 92 km (57 miles) from Burgos, is on the River Duero, which formed an important border in the long battles between the Moors and the Christians. The castle is set on a ridge above the town, with its battlements running dramatically along the contour. Most prominent of all is the Torre del Homenaje, a square tower with satisfying crenellations, looking exactly as a castle tower should look.

Although magnificent from the outside, the castle had lost its roof and internal floors. The city council stabilized and restored the structure, and then decided that its best use would be as a combined visitor centre and viewing point. In keeping with the best of heritage thinking, the council did not wish to make an irreversible intervention and therefore commissioned architects Carazo, Grijalba & Ruiz to design an internal structure that was, as far as possible, self-supporting.

The architects designed a timber structure, described as 'like a piece of furniture', which guides visitors up through the tower on eight levels. The architects saw the tower as resembling 'a hollow tube of stone emerging from the rocks', and wanted to maintain that impression. They felt that they could do this more successfully by suspending the

THIS PAGE

ABOVE Visitors emerge at the top within a glass enclosure.
LEFT   Stairs are made from old railway sleepers, treated
       with creosote.

OPPOSITE

LEFT   Slatted screens of oiled oak help to define the
       route.
RIGHT The walkways have a top surface of oiled oak, on
       birch ply.

new structure from a few key points, rather than by supporting it from the ground. All the staircases, except the last one, consist of a single timber flight. Only at the very top do visitors use a small spiral staircase, contained within a glass enclosure, just visible like a periscope above the battlements.

The timber platforms are constructed of 18 x 40 cm (7 x 16 inch) laminated wood boards, with an oiled-oak top surface on birch ply. These boards are supported on sawed wooden joists, 8 x 16 cm (3 x 5 inches) in section. The steps of the stairs have been created from old railway sleepers, treated with creosote. Climbing all those stairs could be dispiriting, but the architect has ensured that at every level the experience is different. Making use of the few existing openings in the original structure, it has positioned windows on some of the platforms. Elsewhere, the visitor is both drawn on and intrigued by oiled-oak slatted screens, which both define the path and offer views. And the view from the top is magnificent, making it all worthwhile. Peñaranda has found an excellent use for its dominating tower, which should prove highly popular. But if the city decides one day that this use is no longer appropriate or desirable, it will be able to remove the architect's clever intervention without damaging the structure of a monument that has already stood for centuries and should remain for centuries to come.

OFFICE FOR LUNDBECK PHARMACEUTICALS

Milton Keynes, England | Artillery Architecture & Interior Design | 2001

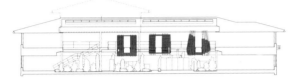

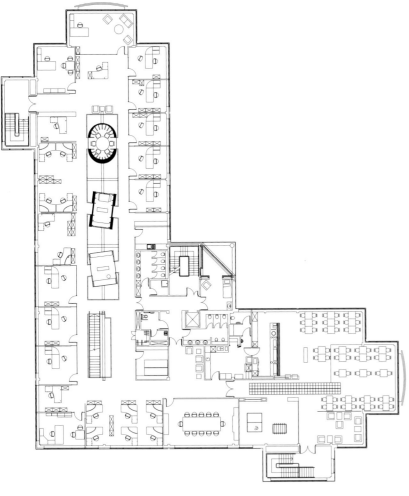

OPPOSITE

TOP LEFT   The timber pods formed the centrepiece of a design aimed at creating an interesting space within an anonymous Milton Keynes office building.

TOP RIGHT   The glulam uprights and plywood cladding extend below the floors of the pods, creating an intriguing impression from underneath.

BOTTOM LEFT   A steel frame supports the glulam-and-plywood structures.

CENTRE BOTTOM   Workers can look down past the pods to the floor below.

BOTTOM RIGHT   The effect is as if the pods are hanging in space.

THIS PAGE

ABOVE   First-floor plan, showing the key role played by the three pods.

TOP LEFT   The ground floor is designed as a 'garden', a recreational space.

TOP RIGHT   An elegant staircase joins the two levels.

New ways of working in offices may have been pioneered by the 'creative' industries, but, by the end of the 1990s, the realization that communication was as important as doing mechanical tasks was spreading more widely. This is reflected in the UK headquarters of Danish company Lundbeck Pharmaceuticals, created by Artillery Architecture. The office is in Milton Keynes, a somewhat soulless town, but with good communications. Lundbeck took space in one of a number of mundane, speculative red-brick buildings on a business park, but was open to Artillery's relatively radical proposals for creating a home for its 67 employees. Offices are on two floors, and the architect cut a hole in the centre to create an atrium that acts as a hub, surrounded by working spaces. Natural light from the rooflight comes right down into the heart of the building.

Artillery treated the ground floor as a 'garden', a recreational space with plenty of plants, and with informal meeting spaces within it, plus some functional areas housing office equipment and storage. But the most dramatic intervention is at upper level, where the architect has created three 'pavilions'. Each a different shape – a cube, a trapezoid and an oval – they are all structurally the same. A steel framework is fixed to beams crossing the space, but there are glulam uprights and horizontal plywood cladding. The height of the cladding varies at different points on the perimeter, and there are spaces between the individual elements, so that occupants can see out, although from a distance the pods will be relatively opaque. The pavilions are used for informal meetings, to house a reference library and for photocopying equipment. Beams and cladding extend beneath the floors of the pavilions, so that the view from below is intriguing. Overall, the effect is of contemporary basketwork writ large.

Sales teams and their support services are the major users of this building, which means that many of them are only present infrequently. It was therefore thought important to make it a pleasant place for them to work, and one where they could catch up with information and office gossip. Other facilities include a gym, a canteen and a games room, but none is as visually distinctive as the three pavilions. As well as being a clever way of providing some facilities within what could otherwise be the dead space of the upper level of the atrium, the pavilions give out an important signal about Lundbeck. It is clearly a company that has thought about the working environment of its staff and is open to new ideas – both of which should help with recruitment in an increasingly competitive market.

TBWA\CHIAT\DAY OFFICES

San Francisco, USA | Marmol Radziner & Associates | 2002

THIS PAGE

BELOW   In contrast to the rectilinear nature of the existing
building, meeting rooms have sinuous profiles.

TOP RIGHT   The architect has cut through the existing
floorplates to insert the meeting rooms.

BOTTOM RIGHT   Detail showing edge-on timber, used with
polycarbonate glazing.

OPPOSITE
The contrast between the retained structure and
the new insertions can be seen clearly at reception.

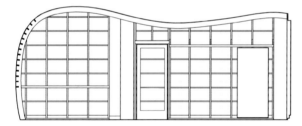

Advertising agencies often have the most playful of office buildings. Working their staff very hard, but with an ethos that work should be 'fun', and needing to impress their clients with their imagination and visual pizzazz, agencies can end up with offices that are not far removed from playpens. In those terms, Marmol Radziner & Associates' offices for TBWA\Chiat\Day in San Francisco could be seen as a positive paragon of sobriety. There are no primary colours, and there is a connection with the building's own history. But the architects have created an environment that is unusual without being wilful, playful without being perverse, and curvaceous without being disorientating.

Housed in an old brick warehouse, across the street from the headquarters of new client Levi Strauss, the office's design makes reference to the hulks of ships that were once abandoned there by those optimists taking part in the Gold Rush. It has exposed and exploited the great solid structure of the warehouse, while making insertions that set up a dynamic tension with the rectilinear grid.

Floorplates are cut away to create a double-height space that holds meeting rooms with sinuous profiles, clad in Douglas fir planks. In places, this cladding is solid, in others it is spaced to allow light to enter through polycarbonate glazing, also used on the ends of

the pods. By making all the cladding horizontal, the architect has strengthened the analogy with ships, although, in places, it is deliberately applied very loosely in a manner that is certainly not shipshape – more reminiscent of some rough-hewn crates. Inside the pods, the tightly planked and curved timber ceilings convey a feeling of captains' cabins in old sailing ships.

Workstations are ranged around the walls on the upper two levels, and there a sterner geometric is applied, with their enclosure in square forms reminiscent of crates, albeit rather refined ones. Made of apple plywood, they have wheatboard tops. Rice-paper lanterns

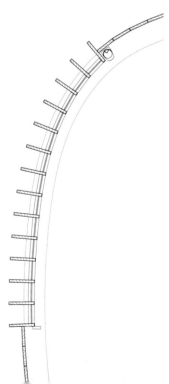

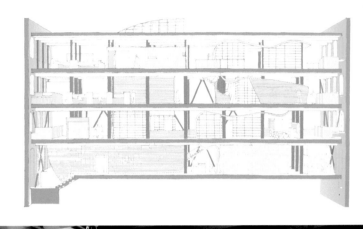

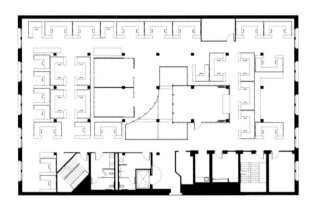

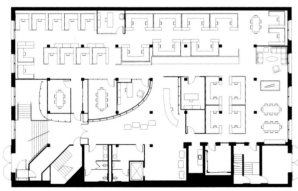

above the workstations provide a far less formal effect than does conventional task lighting. Other sturdy, but well-made, timber furniture looks as if it too would be at home on a high-class ship.

The floor in the reception area is of cork, but elsewhere is of Douglas fir ply. The odd touch of bright colour, as in the red soft furnishings and carpet of the waiting area, prevents there being an unrelieved brown palette.

Cleverly, the architect has banished those functions, such as editing suites, that require true isolation to the basement, creating out of the rest of the space a sense of community and communication, but with some privacy attached.

Chiat\Day has a history of commissioning imaginative buildings, from architects including Frank Gehry and Rem Koolhaas. If its San Francisco office is more restrained than some of its previous efforts, then it reflects the fact that advertising itself has grown up. Marmol Radziner has created an environment where it should be possible to work hard and to have fun, without the relentless pressure to be jolly and zany that some of the more wacky advertising offices impose, and which could, ultimately, be both wearying and depressing.

OPPOSITE
Shigeru Ban, who is
always interested in
using materials in new
ways, employed
laminated veneer
lumber (LVL) when
he designed the Imai
Hospital Daycare
Centre in Odate,
Japan.

Timber is a solid, reliable material that will perform well if you don't ask too much of it; it cannot be used to achieve terrifically long spans and, compared with more modern materials, it is likely to be rather chunky. Such an assessment is right only up to a point. These were the known attributes of timber in the days when using it in construction meant cutting down trees of variable properties, doing a bit of seasoning and knocking up a building within the restrictions that this imposed. But, even in the Middle Ages, designers were using their ingenuity to achieve longer spans in halls and churches. Now – with engineered timber solutions, computer-based calculations and a knowledge of how to enhance the behaviour of timber by using it in conjunction with other materials – there is almost nothing that cannot be achieved.

Nearly every project in this book demonstrates the new understanding that architects and engineers have of timber and of the uses to which it can be put. What distinguishes the projects in this chapter is that the design teams responsible for them have taken that understanding to an extreme, achieving something that may previously have been considered impossible. And what they have done is clear for all to see. These are buildings likely to prompt even the casual observer to say, 'I didn't know you could do that with wood.'

At the Bodegas Vina Perez Cruz in Chile, José Cruz Ovalle has used glulam, a material that has become almost common currency, in one of the most exciting and sensual ways possible (see page 174). The curves he achieves, the imaginative use of spaces that in the hands of a less-talented architect would otherwise be wasted, and the timber interior that gives one the impression of being inside a wine barrel, are combined with a profound understanding of the building's intended purpose.

Flitched timbers – timber elements combined with steel plates – have a longer pedigree than glulam, since their development was not dependent on the introduction of modern adhesive technology. But, again, exploitation of their potential has advanced rapidly in the last few years, and in the enclosure for the Greenport Carousel in New York, Sharples Holden Pasquarelli have taken this to a new level, using an exceptionally hard timber that, in combination with other steel elements, allowed the members to be more slender than might previously have been thought possible (see page 196).

Shigeru Ban also used steel elements in combination with timber in his design of the Atsushi Imai memorial gymnasium (see page 180). Ban is an architect whose work is dedicated almost entirely to achieving things not previously thought possible. Fascinated by the lightweight and by minimizing the use of materials, he has used laminated strand lumber to create a dome that not only spans 28 metres (92 feet), but also has to support heavy snow loads.

Concrete is used in conjunction with timber at the timber-engineering school in Biel-Bienne, Switzerland, where architect Meili & Peter set out to challenge preconceptions about timber buildings (see page 190). The school's four-storey building, with large windows, looks too big and too open either to function structurally or to satisfy fire requirements. Partly by concealing a stabilizing timber structure, and partly through the use of a concrete spine, the architect found a way around both these problems to create a building that adds to the ways timber can be used.

If the Biel-Bienne building is likely to be emulated more for its technical solutions than for its appearance, so, in a very different way, is Shin Takamatsu's Myokenzan worship hall in Japan (see page 184). This is such a bravura building – doing extraordinary things not only with timber, but also with steel and glass – that there is unlikely to be a place in the world for any imitators. But the astonishing nature of the achievement should prove to doubters that the potential for materials, and particularly for timber, is virtually limitless.

BODEGAS VINA PEREZ CRUZ

Paine, Chile | José Cruz Ovalle | 2001

OPPOSITE

TOP   With three buildings beneath a common roof,
       covered spaces are created for deliveries or for
       workers to rest away from the heat of the sun.
BOTTOM The Mediterranean-type climate is ideal for
       winemaking, and the scenery is also reminiscent
       of that area.

THIS PAGE

ABOVE   Site plan of the wine estate.
TOP RIGHT The roof slopes up gently from the centreline of
          the complex.
BELOW   The three buildings are set at slight angles to
          each other.

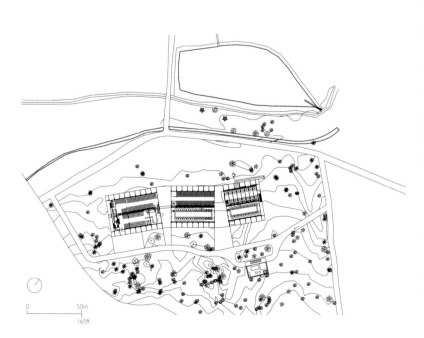

Chile's wine industry is growing fast, and the Maipo valley, only about 50 km (30 miles) from Santiago, has ideal growing conditions. Its quasi-Mediterranean climate includes many days without rain. But families such as the Perez Cruz, who have planted grapes on 530 hectares (1,310 acres), see their estates simply as working places. They have not developed the culture of day trips and longer tourist visits that lies behind the design of such extraordinary structures as Bodegas Ysios in Spain, designed by Santiago Calatrava (see page 126).

This makes it all the more remarkable that the Perez Cruz have commissioned a truly extraordinary building to serve what are, effectively, purely industrial purposes. The design, by expatriate Catalan José Cruz Ovalle, is sensuous and expressive – he has a secondary career as a sculptor – yet fulfils all the specific needs of wine production and storage. Above all, it is a magnificent expression of the use of wood. Apart from some concrete columns and walls of stone infill used up to shoulder height, it is an undiluted celebration of the material. Externally, sinuously curved glulam columns support the overhanging roof. Internally, the impression is of the kind of coopering technology used to create the barrels that, to a winemaker, are second in importance only to the grapes. The

material used is radiata pine, the main plantation timber in Chile. Although in the past the material has been seen as of relatively poor quality, this is changing as plantations become better managed.

The roof unifies three buildings that run roughly from southwest to northeast, but with a change in direction of about 15° between each to allow the buildings to follow the topography. Intended as a sign of respect for the land, this also reduces the mass and formality of the structure, so that the pairs of curved glulam beams do not march along a single line, diminishing to a vanishing point, but curve in plan in a manner reminiscent of their curves in the vertical plane. Informal covered courtyards, created between each pair of buildings, can be used for delivery of grapes or collection of bottles. At other times, they serve as informal shady spaces for relaxation.

The oversailing roof, which slopes up gently from the centreline of the building complex, has a visible gridded timber structure on its underside. This grid tightens up at the openings between the buildings in response to the greater spans required. Seen from either of its long sides, the complex is dominated by the view of the roof overhang and the glulam beams. The extraordinary structure that the roof protects is only really evident from the

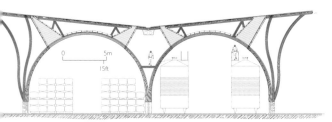

	OPPOSITE
TOP LEFT	At the ends of the 'hidden passages', set between the tops of the two arches, there are views out.
TOP RIGHT	The form of the passages is determined by the curves of the arches, and by the clerestory glazing.
BOTTOM	A cellar, used for secondary fermentation, is also a dramatic space.
	THIS PAGE
TOP LEFT	Section through the complex, showing the twin arched forms under a common roof.
ABOVE	The clerestory glazing brings a magical quality of light to the timber-lined spaces.
TOP RIGHT	Fermentation vats are a reminder that, however beautiful it may be, this is very much a working building.

other ends. Each building consists of two barrel-arched forms, running side by side. Entirely of wood, these are clad in relatively small timber elements. The consideration given to this cladding is symptomatic of the care that has gone into the whole design. Whereas the arching cladding runs directly along the curves of the arches, at the ends the cladding runs at 45° to the vertical, rather than simply vertically or horizontally. This is a much more lively experience for the eye, drawing it up and across the surface.

The functions of the buildings are arranged logically, with the entire eastern building and half of the central one dedicated to fermentation. The other half of the central building is for barrel storage. In the western building are the bottling plant, an area for bottle storage and an area for receiving materials. In some places there is a single space within the vault, in others there is a mezzanine floor. There is also a cellar beneath part of the building complex, used for secondary fermentation. With temperature control crucial, and also a requirement for some natural light, there is a clerestory on both sides at about the point where the walls would meet the roof, if this were not a continuously curved structure. Used primarily for ventilation, the clerestory also allows light to filter down into the building, imparting a wonderful glow to

the timber interior, as well as exposing the curved glulam beams at the top of the structure.

The point at which two barrel-vaulted structures meet is potentially awkward or, at best, a waste of space. But here the architect has made his most brilliant move. He has put in a floor, turning this into a continuous corridor of almost hexagonal shape. The lower two 'diagonals' curve, being the outer sides of the barrel vaults. Sometimes these open up to allow inspection of production processes. Between the buildings there are open views into the courtyard. Laminated timber elements are here in a range of sizes, sometimes solid and sometimes with gaps, unified by their colour. It resembles a showcase of the material in all its variety, while also giving the enticing impression of being in a secret place, in the sort of corridor that is sometimes created from the lost space beneath a castle's walls.

Despite being fiercely utilitarian, the Bodegas Vina Perez Cruz also has a magical quality, tempting the observer to speak in the superlatives that too often appear on winetasting notes. Winemaking may, after all, be an industrial process, but it is also one that has a special mystique and magic – and it deserves a building to match.

ATSUSHI IMAI MEMORIAL GYMNASIUM

Odate, Japan | Shigeru Ban | 2002

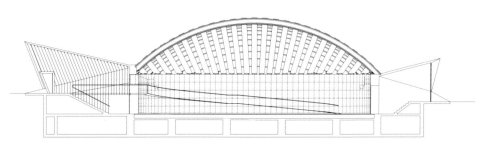
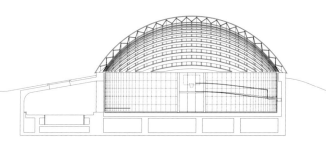

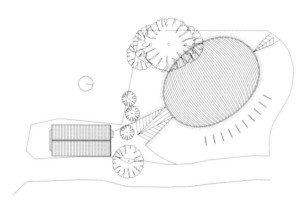

Japanese architect Shigeru Ban loves to experiment with technology, particularly with concepts of lightness, and is probably best known for his experiments in using cardboard as a structural material, most notably in the Millennium Dome in London and at Expo 2000 in Hanover. So, at first sight, he might not seem the obvious choice to design a structure in a place subject to notoriously heavy snow loads – a condition that normally calls for a very heavy structure to provide support. But Ban is an architect who embraces a challenge, and his design for the Atsushi Imai memorial gymnasium at Odate, in the north of Japan, is an indication of what can be achieved by a combination of imagination and rigorous analysis. Faced with having to construct a dome measuring 20 x 28 metres (66 x 92 feet), as the most efficient way to cover the gymnasium, Ban chose to use laminated strand lumber (LSL), a lightweight material and not one that would appear particularly suitable for dealing with snow loads.

The gymnasium is attached to a hospital, at which Ban had already built a smaller structure, a children's day-care centre, also using lightweight timber elements. But that building, although both innovative and beautiful, was considerably smaller and used laminated veneer lumber (LVL), in combination with a stabilizing steel outer frame. In

the gymnasium, Ban was far more dependent on the timber itself to fulfil the structural requirements.

There is a very different precedent for the use of timber in Odate: a much larger dome designed by Toyo Ito, which uses massive cedar elements, in combination with steel trusses – a solution that Ban would shun both for the wastefully large volume of materials used and because it contravenes all his aesthetic concerns. Ban looked briefly at creating a moulded dome in LVL, but the material was not technically suitable. Moving on to the concept of using arches, he had to discover a way to make the structure strong enough to support snow loads. He chose LSL, a material similar to LVL but incorporating lower-grade logs, such as aspen, that are unsuitable for conventional timber products. Crucially, it is possible to make LSL in lengths of up to 28 metres (92 feet) and widths of up to 2.5 metres (8⅕ feet).

Ban's solution was to create a lattice of Vierendeel arches parallel to the longer axis, crossed by trussed arches parallel to the shorter axis. 'Just as paper could surprisingly be a structural member depending on how it is used, thin plywood can also be a structural element, which spans a distance beyond expectations,' Ban explains. 'With this solution, much less wood is used than

in an ordinary dome structure using other wood lamination.'

The trussed arches consist of LSL chord members and steel diagonal members. Since it is impossible to shape LSL 60 mm (2⅜ inches) thick into a curve, the material was divided into three lamina, each 20 mm (⅞ inch) thick. These were curved individually and then relaminated. The Vierendeel truss, which uses elements at right angles for stiffening, with no diagonal bracing members, consists of steel-pipe horizontal members, steel vertical members and LSL shear panels. The plywood is left entirely exposed internally, and the zigzag connections create a fascinating three-dimensional effect. The timber echoes the colour of the sports floor.

A circular concrete ring beam forms the base of the structure. Since surrounding buildings are low, Ban has set his building deep into the ground, so that it is already semi-submerged before being affected by snow. Only the dome itself and the two entrances are above ground level. The dome is clad with translucent polycarbonate that has

strips of stainless steel lying above it. This combination allows a considerable amount of light into the gymnasium, giving it an appealing sense of connection with the outside world. A swimming pool, set to one side of the dome, is entirely underground, and also benefits from strips of light penetrating its roof.

Ban was more confident in his innovation than the structural engineers with whom he worked, who persuaded him to incorporate more steel elements than he believed were necessary, either before or after construction. Nevertheless, the elements of the building were subjected to a very severe testing regime, with the behaviour of every component rigorously examined. As an architect who develops his ideas from one project to the next, Ban is likely to continue flying ever more freely from the constraints of conservative engineering. Having shown just what can be done with the cardboard tube, he is now demonstrating that, even under the most demanding technical conditions, timber can be used as a lightweight material in a manner more commonly associated with steel.

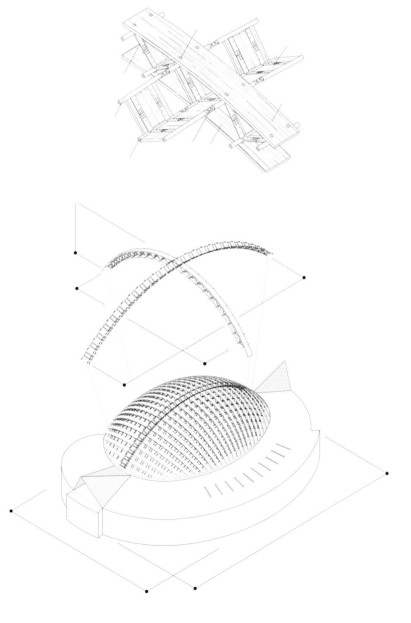

MYOKENZAN WORSHIP HALL

Kawanishi, Japan | Shin Takamatsu Architect and Associates | 1998

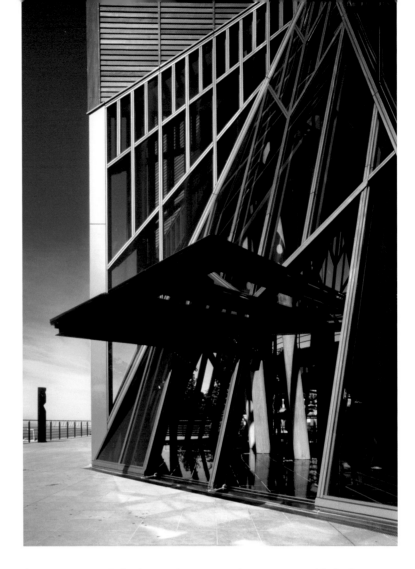

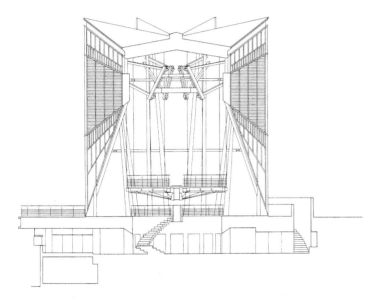

OPPOSITE

TOP   Although known as an iconoclastic architect, Shin
Takamatsu responded to the special nature of the
building's setting.

BOTTOM   The worship hall is set amongst a dense forest of
cedars, with traditional Japanese buildings nearby.

THIS PAGE

ABOVE   Both the geometry and the materials change as one
travels up the building.

RIGHT   Locally harvested cedar was used on the worship
hall.

An extrovert complexity characterizes the work of Shin Takamatsu, an architect with many admirers and many detractors. Often colliding with the modern world and contemporary culture, he does not seem the obvious person to design a Buddhist place of worship. But this is exactly what he did at the Myokenzan worship hall, working for the first time with timber.

There is no established iconography for Buddhist places of worship, so Takamatsu had more freedom and a more open brief than if he had been designing a church or a mosque. What he did have was an important site, a place of great beauty, where the Buddhist holy man Myoken Bosatsu used to come to worship. Set in a wood of dense cedars, the worship hall is near a group of traditional Japanese buildings. Takamatsu himself has said about the building and about the use of timber, 'It is reasonable to hope that wood might provide some new way of deciphering beauty and structuring our culture; viz. by "structuring our mind". Since ancient times, Japan has used wood to construct the foundations of our "mind". People usually say that wood ought to be used on the spot where it is cut. This very simple way of thinking also applies to the "structure of the mind".' The timber, which was harvested locally, is cedar, so the constructed form stands amid the natural material. But nobody could

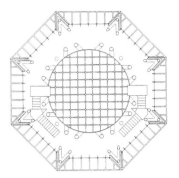

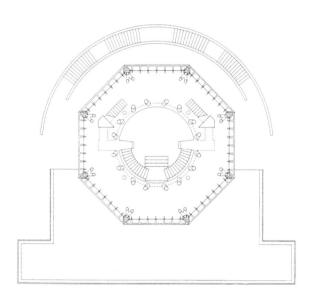

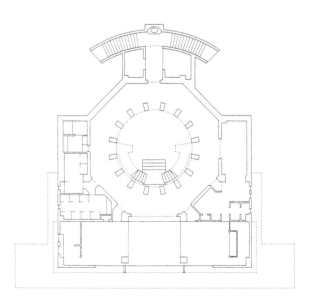

OPPOSITE
At night, with light shining out, the building looks like a magnificent mutant crystal.

THIS PAGE

LEFT   Plans showing (from the bottom) the first basement, ground floor and second floor give a sense of the changing geometry.

ABOVE   Transparent at ground level and at the top, the building has an opaque mid-section of cedar cladding.

mistake this for a 'natural' building, or for one that in any way evolved from its surroundings. Instead, it is a bravura tour de force, a symmetrical structure with a number of superimposed symmetries.

It has soaring form, whose plan transmutes from a circle to an octagon to an eight-pointed star. The materials change as one moves upwards, from concrete below ground level to steel to timber, and finally to a top level of glass, with an all-glass floor that allows the visitor to look down and be dazzled by the extraordinary geometry. The construction is not a simple layering of one material above another. Rather, the angular steel structure comes up from the circular concrete of the

underground structure to frame the prayer hall itself, and from within the prayer hall the timber structure rises up, clad externally in horizontal timber slats, a near-solid layer sandwiched between the filigree glazing above and below. Impressive enough in the daytime, the building is stunning at night, when light shines out of a form that seems like some kind of mutant crystal.

Buddhism is a famously non-prescriptive religion, so there is no one to dictate what the thoughts should be of those inside the worship hall. But it is a good bet that they will include at least a measure of amazement and awe at the achievement, both imaginative and structural, of this remarkable building.

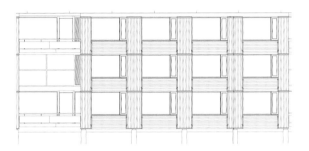

Assumptions about what a timber building should be, how it behaves and what is possible in timber construction have been turned on their head at a timber-engineering school in Switzerland. Architect Meili & Peter, appointed in competition to design the new teaching block, took the opportunity to challenge received ideas. It has succeeded in creating a building that is elegant and stern. Although, at first sight, the building hides its unconventional nature, its sheer scale and the size of the window openings are indications that something odd is going on – that this is, in fact, a revolutionary building. Almost every aspect is different from a traditional Swiss timber building. Aesthetically, there is none of the homely look of the conventional house, or indeed of the sheds that surround the new building. This is a more abstract architecture, concerned with defining spaces and bringing in light in a manner more usually attempted with the contemporary palette of steel and concrete.

At four storeys, the building was the tallest timber structure ever built in Switzerland, and there had to be considerable work done on the way that it would behave in fire. Other innovations include the extensive use of prefabrication, a floor design taken from first principles and an approach to acoustics that, although not entirely successful, provides some interesting

data. Given the identity of the client, it seems appropriate that the architect treated the building in some senses as a laboratory of ideas for timber construction. And, despite the fact that funding problems led to a gestation period of nearly ten years, the architect was sufficiently forward-thinking for its ideas to still seem radical when the building opened.

At 18 × 20 × 100 metres (59 × 66 × 328 feet), this is a substantial building. But it is not a single, continuous structure. Instead, it is made up of a series of self-supporting timber boxes, themselves built from storey-high prefabricated timber frames. The spaces between the boxes are used for circulation and to create balconies. Along the centreline of the building, where the main corridor runs, are four vertical concrete shafts, housing the lift, the staircases and the wc blocks. Each shaft is offset from its immediate neighbours to one side of the corridor or the other.

The corridor floors themselves are of heavily post-tensioned concrete. This concrete spine braces the building both longitudinally and transversely, and plays a key role in fire safety. Clad in steel, the corridors offer guaranteed escape routes. The balcony areas also act as fire havens. The spine does not extend the full length of the building. At the northwestern end, where a more

conventional, freestanding beam-and-post structure was adopted, are a library, lecture hall and exhibition space, occupying the full width of the building.

The rest of the building consists of five 'cells' made from prefabricated elements and capable of being divided into a varying number of classrooms. Six prefabricated timber wall elements, each measuring 2.7 × 9.7 metres (9 × 32 feet), make up a single storey of each cell. The floor/ceiling structure was also prefabricated. These floors, spanning between the walls, are hanging structures. The boxes are entirely independent, with no load transferring to the cores of the building. The fourth floor is a continuous structure, set behind a balcony. It holds together the 'cells' of the lower three floors. The roof, a timber structure supported on glulam beams, cantilevers above this upper storey. It is almost flat, draining down to the centreline of the building.

The palette of materials had a strong industrial influence: as-cast concrete, concrete floor screeds, steel and factory-finish timber. The use of solid panels provides the necessary stiffness to allow the large window openings. Façade elements, fixed from outside to allow replacement if necessary, are solid oak. The boards are simply planed and untreated, so quality control in their selection was crucial. Slightly inclined

THIS PAGE

TOP LEFT   The northwest façade has no openings, providing stability to the open structure that lies behind it.

ABOVE   The palette of materials is entirely contemporary.

OPPOSITE

LEFT   The top-level structure runs continuously from one end of the building to the other, helping to tie together the disparate elements lower down. Because it is relatively sheltered, cladding can be in less durable larch.

RIGHT   Cross sections through classrooms (top) and the foyer and exhibition space (bottom).

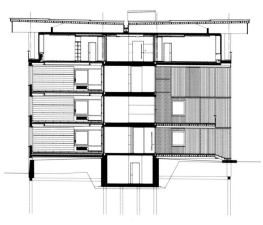

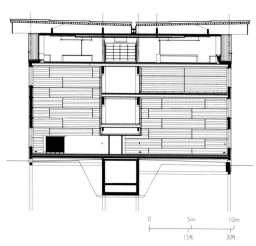

OPPOSITE

TOP   First- and second-floor plans, showing the concrete
core elements in the protected central corridors.

BOTTOM   Internal cladding throughout the building is of solid
pine.

THIS PAGE

ABOVE RIGHT   The steel-lined corridor, with concrete cores rising
through it.

BELOW   Exits from the cores are specially designed to
preserve fire integrity.

surfaces ensure that water will run off. The façade is ventilated, and panels are connected from the outside by special steel connections. There are no openings in the cladding on the northwest façade, providing further bracing to the open structure behind. Internal cladding throughout the building is of solid pine. Since the top floor is set back beneath the roof, and therefore receives considerable protection, it was possible to use larch, a less durable timber, for the cladding.

The timber floors formed the most complex element of the building because they had to deal with large loads and spans. In recognition of this, a separate competition was held for their design. The winning solution consists of a series of triple-cell elements, 32 cm (13 inches) high, spanning 8.5 metres (28 feet) as a single-span beam. These are made up of horizontal and vertical solid boards, joined by adhesives. To improve the acoustic performance, sections of the lower elements about 1.2 metres (4 feet) long were cut out to increase the sound-absorbing surfaces. However, this also had implications for fire safety, so calculations and full-scale tests were carried out to prove that the floor would still be able to provide 30 minutes' fire resistance.

To reduce the transmission of impact sound, it was necessary to increase the weight of the floor. In this case, a traditional method was used. First, 60 mm (2½ inches) of sand were laid on top of the floor structure. Above this went a floating particleboard floor, with a wearing surface above it of lightweight concrete. Between each trio of box elements there is a gap of 120 mm (5 inches), in which a sprinkler system was inserted – another part of the fire strategy.

A study produced at the school found that not all aspects of the building were entirely successful. Although acoustic isolation is good between floors and between the cells and the corridor, there is an unacceptable level of transmission between classes. The study notes that 'the floor system is considered as a rather expensive solution'. It also finds that the overhang of the roof does not provide as much solar protection to the top floor as was hoped, and that it can become very hot in summer. It is probably unavoidable that there should be some elements of failure in such an experimental building. What is equally important is that the school, which had undergone expansion and been upgraded to a university of applied science, has a building that both reflects its new role and demonstrates the possibilities of development in timber construction.

CAROUSEL HOUSE

New York, USA | ShoP/Sharples Holden Pasquarelli | 2001

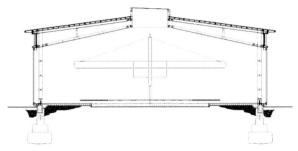

Old-fashioned carousels are romantic objects, evoking instant nostalgia, even in those people who have never seen one. So it is not surprising that the installation of an historic carousel in Mitchell Park, Greenport Village, on New York's Long Island, is playing a key role in the park's development. Built as a carnival ride in North Tonawanda, New York, in the 1920s, the 12 metre (40 foot) diameter carousel was donated to Greenport in 1995 by the Northrop-Grumman foundation. It is a traditional 'brass ring' carousel, with wooden horses and brass rings hanging from the edge, most of them tantalizingly out of reach. But fortunate or skilful children can grasp a ring at the height of their horseback journey and, even today, win a free ride.

The 1.4 hectare (4 acre) park is part of Governor George E. Pataki's programme to improve access to waterfronts across the state of New York. It links key elements in Greenport Village, including the ferry, Commercial Wharf, the main street and the marina. This former brownfield site is being given an identity by the New York architects SHoP/Sharples Holden Pasquarelli, working with the New York office of engineer Buro Happold. Key to the project is the use of timber, with a timber boardwalk linking important points, and the same timber used for shady arbours and the carousel house. This last item formed the major element of the first phase of the project.

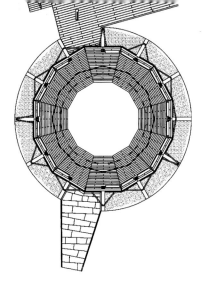

Neither the carousel itself nor the public are as hardy as they used to be. It was therefore essential to provide shelter from the elements, but without either disguising the structure within a bland box or cutting off contact with the outside world, which is part of the pleasure of the ride on a fine day. The solution was a 12-sided timber enclosure that allows the carousel to be used all year round, responding to the changing weather with either partial or complete opening. In the worst weather, just one side opens partially. It echoes the form of the carousel and allows wonderful views, but is obviously a piece of advanced twenty-first-century engineering, rather than a romantic throwback. The effect is reminiscent of a beautifully constructed contemporary hatbox, enclosing a cherished old hat. It is an additional layer, both contrasting with and enhancing the charm of the original object.

Since the designers wanted the structure to be as slender as possible, they used flitched timber elements, which combine the best qualities of timber and steel in a composite, sandwich construction; the steel provides strength and stiffness, and the enclosing timber offers lateral stability, plus protection from fire and corrosion. Flitched beams are not a new form of construction – they were commonly used in industrial warehouses – but

newly available materials have made it possible to update techniques.

The carousel house uses high-strength steel plates, 1.9 cm (¾ inch) thick, sandwiched between ipe, an extremely hard and durable hardwood from Brazil that is naturally fire-resistant and resistant to rot. Flush-mounted stainless-steel hardware has also been used. The beams span about 8 metres (26 feet) from the exterior columns to a central compression ring that supports the stressed-skin plywood roof panels. The columns, constructed from double 10 × 30.5 cm (4 × 12 inch) ipe members, are spaced apart to provide slits of light between these deep members. The resulting increase in stiffness allowed the columns to be used as cantilevers, eliminating the need for cross bracing, which would have obstructed the doors. Shear key connectors were used at a number of locations to transfer the forces between the steel and timber members at the critical joints. These incorporated specially machined stainless-steel 'sex bolts' (the engineer's terminology), up to 30 cm (12 inches) long to pass through all members.

An independent system of structural steel members, propped off the main timber frame, supports the doors. These members house the worm-drive system that activates the doors. Each pair of doors is about 5 × 5 metres

(16 × 16 feet). The use of the worm drive allows the doors to be opened in winds of up to 64 kph (40 mph). Each door operates independently and can be opened or closed in less than 60 seconds. The engineers worked out the pattern of the doors by studying wave patterns common to both the movements of the carousel ride and its location on Peconic Bay. In this way, they provided both the greatest weather protection and the most atmospheric ride.

OPPOSITE
The Carter/Tucker
house, in Breamlea,
Australia, was the first
on which Sean Godsell
exploited the idea of
an outer cladding of
slats. The appearance
of the building changes
when flaps over the
door and various
windows are opened.

Not all buildings are immutable. Some have moving elements, and the appearance of others changes when seen from different angles, whether through reflection or obstruction of view. One of the most interesting ways in which this change can happen is through the use of timber screens and shutters.

As environmental awareness has increased, building designers in temperate zones have come to see the importance of shade – a quality long appreciated by the inhabitants of hotter countries. The *brise-soleil* – a horizontal metal grid, set above windows to keep out the hot summer sun – became a cliché of late-twentieth-century corporate buildings. Geographically, it was dominant in the region between the two zones where the wooden shutter is commonplace: the cold climates where the shutter provides insulation at night and the hot countries where loss of light is secondary to keeping out the heat.

Some contemporary architects have embraced the concept of the timber shutter, an object that may once have seemed irrevocably traditional. Aires Mateus, one of the most resolutely non-decorative of practices, has used shutters to animate the timber façade on its student accommodation in Coimbra, in Portugal (see page 222). The

appearance of the façade changes as individual students open and close their shutters. Herzog & de Meuron adapts an even more traditional-seeming element in its Paris apartment building (see page 208). It takes the technology of the rolltop desk and turns it into Oregon pine blinds for the informal courtyard façade of its building, giving residents both shade and privacy, while still allowing them to peep out.

If shutters are an integral part of much European architecture, then screens are similarly important in Japan. Kengo Kuma, who successfully blends the traditional with the contemporary, and feels a particular affinity with wood, makes excellent use of cedar slats in his Hiroshige Ando Museum (see page 218). The need for shade is only part of the story; the slats articulate a building with an almost simplistic design by providing a constantly changing pattern of light and shadow. Tadao Ando has created a shifting pattern of strips of light and shade with more substantial timber elements in his Buddhist temple on the island of Shikoku. He uses the elements for an outer, permeable wall, with interior circulation space.

Sean Godsell has taken a similar approach in Australia, with an outer casing of timber slats to his Peninsula House near Melbourne (see page 202). He takes advantage of the relative

abundance and extreme strength of the native jarrah timber to use it in slender and sharply detailed slats. The benefit of being able to see out, while being, at least partially, protected from prying eyes, is also a quality of an innovative old people's home near Paris, where the architect Maast has created a latticework of red cedar on the edge of a communal balcony (see page 214). This is part of a menu of timber, including wooden venetian blinds, that gives warmth to a building intended to be very much part of its local community, while still maintaining a distinctive identity.

Baumschlager and Eberle have married the sensation of changing patterns through timber slats with the 'moving parts' approach of blinds. The architects used this technique on a bank and office building in Wolfurt, Austria. Lattice screens slide back and forward, both changing the view of the building from outside and creating an enticing striped-light effect for those who choose to keep the screens across their windows.

All the effects described above are most likely to be enjoyed by those either inside the building or outside it, but relatively nearby. But changing patterns can also work on a larger scale. When Bucholz McEvoy designed Limerick County Hall in Ireland, it knew that part of its task was to create a new civic identity (see page 226). The brilliantly engineered timber screen on the main façade acts as a *brise-soleil* for the building's occupants, but is also designed on a large enough scale to provide a constantly changing and intriguing view for drivers on the nearby road. In this case, a device normally considered primarily as a way of maintaining privacy has mutated into a form of advertising. Who said that wood wasn't a versatile material?

PENINSULA HOUSE

Melbourne, Australia | Sean Godsell | 2002

THIS PAGE

ABOVE The house is set into a hillside.

BELOW Seen from above, the house appears to have
burrowed into the landscape.

RIGHT The jarrah cladding is very slender and beautifully
detailed.

OPPOSITE
Glazing to the lower level of the sitting room can
open up completely.

Jarrah is one of the densest and most durable of timbers. A native of south-western Australia, it has a high oil content that makes it resistant to insect attack. Melbourne architect Sean Godsell has used it to create a semi-transparent enclosure for a weekend house near his native city.

Put in the simplest terms, the house is a box within a box. The inner box is made of glass and pre-oxidized steel, the outer of recycled jarrah slats, fixed by galvanized steel elements to another oxidized steel frame. Staircases on either side, contained between the two frames, lead up to the carport and to the bedroom. Since jarrah is such a tough material, the slats can be very

slender – they are only 10 x 35 mm ($\frac{2}{5}$ x 1$\frac{2}{5}$ inches) in section. On the west side of the building, the secondary enclosure goes down to ground level; on the east side, it comes down to the top of the ground floor and then bends out at 90 degrees to shade a deck of radiata pine. There is a minimal south elevation, since the house is dug into a hillside. The northern façade is open, allowing sun worship and views of the sea.

Godsell has invented a clever way of shading his house from heat and glare. But the sharply defined slats do much more than that; they provide privacy and a changing appearance as one moves around the exterior. Seen straight on, they are relatively

ABOVE   Transitions between inside and outside are clearly defined, but not forbidding.

LEFT   The cladding angles out at 90 degrees to shade a deck of radiata pine.

OPPOSITE

Viewed straight on, the timber cladding is relatively transparent, but it becomes more opaque when seen from an angle.

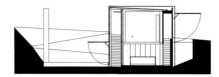

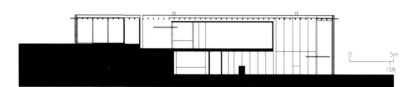

OPPOSITE

TOP LEFT Ground-floor plan, showing the open nature of the front of the house, and the much more enclosed areas behind.

TOP RIGHT First-floor plan, with the bedroom cantilevering over the living room.

BOTTOM The bathroom leads into a private courtyard, with overhead shading.

THIS PAGE

ABOVE The cladding permits views out from the building, but filters the quality of light that enters.

TOP Short and long sections through the building, showing its intimate relationship with the surrounding landscape.

transparent, but become more opaque when viewed obliquely. For the house's inhabitants, they offer a pattern of shade that changes with the position of the sun, giving an indication of both the time of day and the time of year.

This second enclosure is impeccably detailed, subtly articulating the transition from ground floor to first floor, and from first floor to roof. Godsell has given similar care to the inner enclosure. The entire house is designed to give the inhabitants maximum pleasure in both the building itself and its surroundings. Visitors arrive at the upper level, at the carport on the south side of the building. From there, they go down the stairs to the double-height living room. This has fixed glazing on its upper half, but the lower half opens out entirely onto a veranda. Tucked back behind the living room are a kitchen and laundry with, behind them, a library. In contrast to the openness of the living room, this is a much more enclosed space, a pleasant retreat on a stormy day. The second staircase leads up to the bedroom, which cantilevers into part of the living space. Its end wall is also glazed, providing views out through the living room. Behind its bathroom, this bedroom has its own private courtyard, shaded by the jarrah slat roof.

This is not the first time that Godsell has used an enclosure of jarrah slats for a house by the sea. He first did it on the Carter/Tucker house, in Breamlea, Victoria, an even more enclosed-looking building that can open up in a number of intriguing ways (see page 200). Taking some of his ideas from the concept of the traditional Japanese house, Godsell has written that both these houses explore 'notions of inner room (*moya*) and enclosed veranda (*hisashi*). The main difference between the Carter/Tucker house and the Peninsula house is that, whereas the former treats the three main spaces similarly, in the latter the three primary spaces are very different in dimension, volume and quality of light.'

Godsell has received plaudits internationally for his beautifully considered work, but is less popular locally, where some decry his style and others his uncompromising criticism of much contemporary architecture. A tough, resilient substance such as jarrah seems an appropriate choice of material for this other native of south-western Australia.

APARTMENT BLOCKS

Paris, France | Herzog & de Meuron | 2001

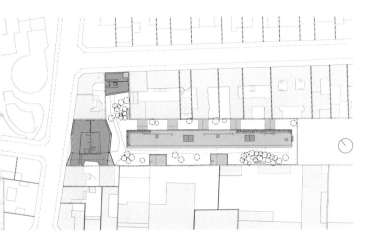

ABOVE   The development consists of two blocks with street frontages, and one set into the more private backlands.

RIGHT   The long three-storey block has a green roof.

BELOW   The two blocks with street frontages, seen to either side of the white corner block, maintain the scale of Parisian streets, although using more contemporary materials.

OPPOSITE

At night, the façades are enlivened by changing patterns of open and closed shutters, of light and of darkness.

Anybody who has ever had access to a rolltop desk, especially as a child, will know the sensual, auditory and imaginative pleasure of opening and shutting the top – of making something solid curve away and disappear. Swiss architects Herzog & de Meuron have used this approach on a social housing development in the 14th arondissement of Paris, near the Gare de Montparnasse, creating sensuously curved timber shutters that use a rolling mechanism.

Named the Rue des Suisses after one of the streets onto which it faces – an appropriate name, given the provenance of the architects – the development was carried out for a forward-looking public-housing agency, Régie Immobilière de la Ville de Paris (RIVP). Building new social housing in the city centre was an admirable initiative since all but the wealthiest are increasingly being driven out to the soulless, and sometimes violent, areas that ring the city. By holding a competition and then appointing an internationally known and extremely innovative practice, RIVP was making a bold move, and one that paid off.

When the two founders of the practice, Jacques Herzog and Pierre de Meuron, won the Pritzker Prize, probably architecture's most prestigious award, in 2000, the jurors made the following comments: 'They refine the traditions

of modernism to elemental simplicity, while transforming materials and surfaces through the exploration of new treatments and techniques';'One of the most compelling aspects of work by Herzog and de Meuron is their capacity to astonish';'All of their work maintains throughout the stable qualities that have always been associated with the best Swiss architecture: conceptual precision, formal clarity, economy of means and pristine detailing and craftsmanship.' All these virtues were demonstrated in the Rue des Suisses project.

To the visitor, Paris is an immensely pleasing city – not only on account of its numerous architectural set-pieces, ranged mainly, but not exclusively, along the Seine, but also on account of the rhythm of its everyday buildings. Housing, typically between four and eight storeys high, with shops or other commercial activity at ground level, is rarely architecturally distinguished, although sometimes there are intriguing or imaginative details. More important are both the harmony of the individual buildings and the way they behave with each other, creating a coherent appearance that is not uniform enough to be dull.

This was part of what Herzog & de Meuron had to address, with two of its buildings breaking into the urban grid, on rue des Suisses and rue Jonquoy,

which runs at near right angles. Both these buildings have façades that match their neighbours in scale, although not at all in treatment. But this was not the whole of the commission. The project, which provides a total of 57 dwellings, includes not only the two seven-storey blocks with street frontages, but also a long three-storey block set within the city block, with no street frontage.

These backlands are another aspect of Paris, often not apparent to the visitor. Some blocks may be so small that they contain little more than a lightwell, but if you walk through certain neighbourhoods you quickly realize that the paucity of side streets means that the roads enclose an area much larger than could be accounted for by the footprint of the perimeter buildings. What is this area like? It is a much less ordered, formal world, sometimes described, confusingly, as a *cité*. Often closed off by a security gate, or just generally unwelcoming to the outsider, it typically contains a random assemblage of buildings that consitute a secret world. This is a place where, with little or no traffic, children can spill outside to play, a quiet world apart without the theatre of a city street – representing, in short, the introduction of an almost suburban way of life into the heart of the city.

Herzog & de Meuron have addressed this condition by introducing a three-

THIS PAGE

ABOVE LEFT   Small maisonettes are set opposite the three-storey block.

ABOVE   A sharp bend in the path that leads away from the block enhances the sense of privacy.

OPPOSITE

TOP   The plan is a series of 'T' shapes, enclosing private gardens behind for ground-floor residents.

BOTTOM   Exposed concrete at the ends of the block has wires on it, up which plants are being trained.

OPPOSITE
The timber shutters enclose spaces that are
generous enough to sit in.

THIS PAGE
RIGHT   With the shutters open, residents can supervise
children playing in the space below.
BELOW   Shutters have a different profile on each floor, but all
use the mechanism of the rolltop desk.

storey block that is both formal enough
to impose some rhythm on the site and
relaxed enough to reflect the virtues of
its more easygoing environment.
Entrances are on the long southern
façade, and it is here that the architect
has deployed the roller-shutter device.
Made of Oregon pine, on aluminium
guide rails, the shutters are used on all
three storeys. They cant out from a
dramatically overhanging roof on the
top floor, with a similar, though less
extreme, slope on the middle floor, and
follow an almost vertical line on the
ground floor, where they come down
to overlap the small, raised concrete
platform, on which the apartments have
been sited to keep them apart from the
public passageway. At all levels, the living
rooms and bedrooms are at the front,
with service spaces set behind.

On the ground and middle floors, there
are open terraces with floors of broad
timber planks behind the shutters,
whereas on the top floor there is simply
a small balcony railing in front of the
full-height sliding glazing. Set out with
garden furniture, these are places for

both relaxing and supervising children.
The architect has acknowledged that, at
this temperate latitude, there are times
when the sun is welcome and others
when it needs to be kept out. The
shutters serve to exclude the sun, as
well as providing privacy and security.

The ground-floor apartments are
ⲧ-shaped, enclosing small, private, north-
facing gardens at their rear. This is a truly
suburban touch for a city-centre site,
but the south façade, with its blend of
individual and communal relationships, is
more successful. It changes constantly, as
individuals alter the position of their
blinds – a facility also employed on the
street-facing blocks with perforated
aluminium shutters. The roller shutters
have their own natural perforation
between the slats, so that, even when
the shutters are closed, the occupants
of the apartments can gain some sense
of the exterior world. Herzog & de
Meuron have shown that its original
approach to materials and their
deployment can be made to work
within the restricted funds associated
with a social-housing project.

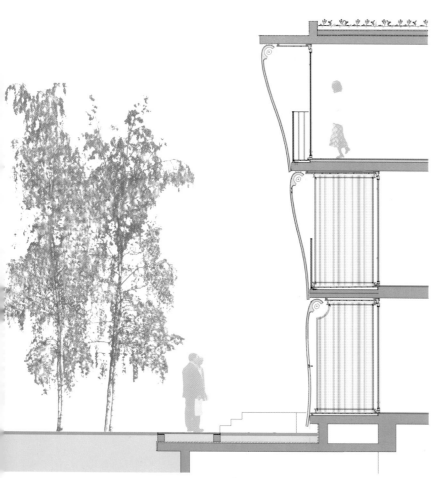

HOUSING FOR FRAIL ELDERLY PEOPLE (MAPAD)

Tremblay-en-France, France | Maast | 2002

TOP Horizontal timber cladding is used widely
throughout the project.

BOTTOM There are private, enclosed spaces, with plenty of
planting.

THIS PAGE

ABOVE In contrast to the widespread use of timber, the
reception building is distinguished as a stark
white block.

RIGHT Set on a triangular site, the development both
engages with the outside world and provides some
sheltered spaces for the disorientated.

France is carrying out a programme to replace conventional, asylum-like old people's homes with accommodation more suited to the needs of today's elderly people. New types of home are being built and evaluated; they vary according to the needs of the residents and the locations. Among these are MAPADS (maisons d'accueil pour personnes âgées dépendants), which are designed to house between 40 and 80 elderly people with physical or mental problems. One of the first to be built was at Tremblay-en-France, a small town near Paris's Roissy airport. Isabelle Manescau and François Marzelle of architect Maast won a competition to design the building on a triangle of land beside a main road.

Their challenge was to create a building that related visibly to the town and allowed residents to see their surroundings, yet would not seem hopelessly out of scale with the single-family homes and small blocks of flats among which it was set. They achieved this by dividing the project into a three-storey accommodation block and a large multi-use hall, linked by the reception, medical and administration building. With generous use of gardens and planting, plus the widespread use of wood for cladding, blinds and screens, they have created an environment that is domestic, without being twee, and which both engages with the town and offers places of refuge from it.

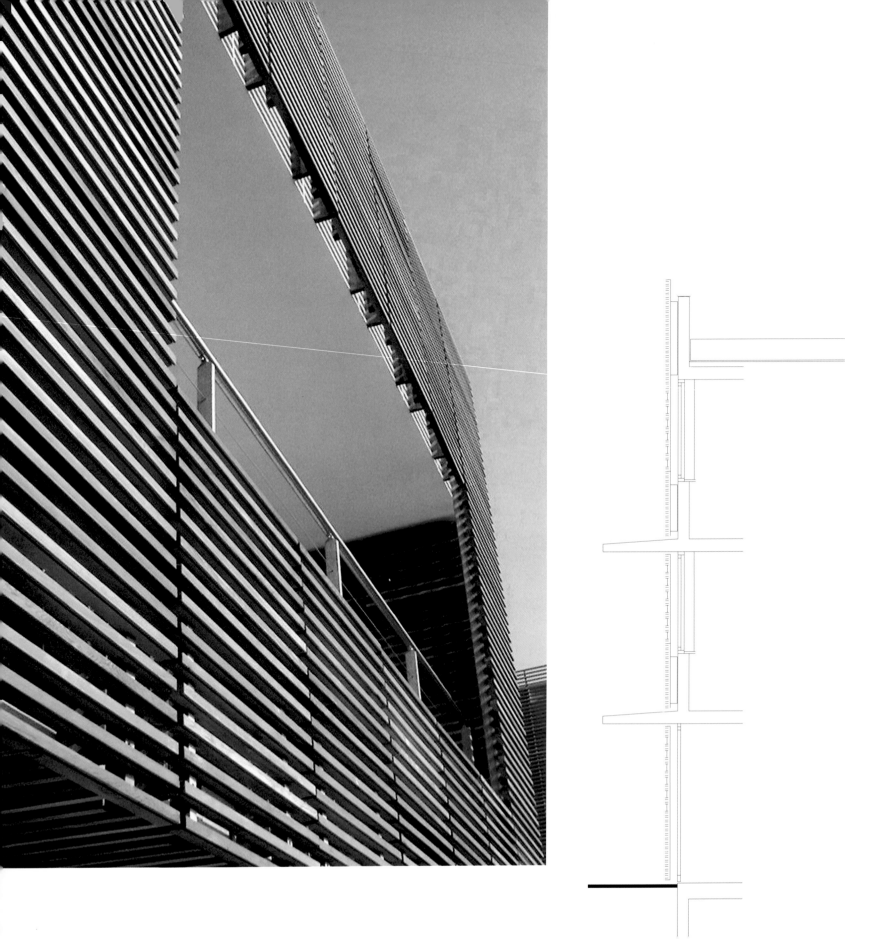

OPPOSITE

LEFT   High-quality cladding, with horizontal cedar slats, helps give a domestic scale to this relatively large development.

RIGHT   There is a slab-edge concrete sunbreak, with a timber screen set in front of it.

THIS PAGE

ABOVE   Sitting areas have views within the development.

TOP RIGHT   On the outer face, upper rooms have balconies and lower rooms private gardens. Cladding is with Navirex plywood panels, and rooms have timber venetian blinds.

The accommodation building is set on a curve, following the line of the road, from which it is separated by rows of fruit trees. On its inner side are two gardens, one on either side of the reception building, with the more contained of the two dedicated to the use of those residents who are disoriented. Internal streets run along the garden side of the accommodation building, widening into seating areas that allow social interaction. This design places the bedrooms on the street side. Although, even with the barrier of trees, this will make the bedrooms less quiet than they would otherwise be, research shows that most residents of old people's homes and sheltered accommodation are eager to look out on everyday life and people. On the ground floor, each room opens onto a small enclosed garden. On the upper level, these gardens are replaced by a continuous balcony.

Rooms have walls clad in timber panels, with timber venetian blinds to the full-height windows that open onto the balcony. Externally as well, timber cladding is used – Navirex plywood panels on the outer façade and red cedar in a latticework effect on the courtyard side. This cladding is echoed on the hall, a large rectangular building whose different form does not stand out because of the similarity of external treatment.

The residents, some of whom may go out only infrequently, can look in one direction from their balconies and in the other direction from the seating area. By opening and closing the blinds, they change the look of the building's outer façade, bringing home clearly to the townspeople that this is an inhabited building and very much a part of their environment.

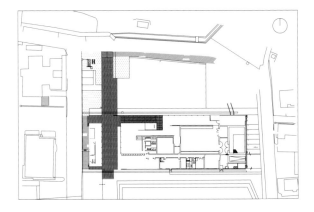

OPPOSITE

TOP   The form of the building could not be simpler
        – a large, rectangular enclosure.
BOTTOM Both during the day and at night, the timber slats
        of the enclosure play tricks with the eyes.

THIS PAGE

TOP   Site plan, showing the museum running parallel to
        a road and at right angles to a right of way.
ABOVE   The public right of way running through the
        western end of the gallery acts as a divider between
        two types of accommodation.
RIGHT   Six rows of cedar slats make up the roof.

'Wood sits at the centre of Japanese culture. I use chopsticks every day,' said Kengo Kuma, when he won the second Spirit of Nature Wood Architecture Award in September 2002. 'My body feels uncomfortable in a concrete structure,' he added. 'I don't like the smell, I don't like the feeling of concrete.' He also talked about the importance of continuing to build with native Japanese cedar if the country's traditions are to be maintained. Although Japanese cedar is often more expensive than imported materials, only by continuing to use it will Japan be able to maintain and regenerate the forests that cover about 60 per cent of the landmass and are intricately bound up with the country's traditions and culture.

At the woodblock print museum in Bato, about an hour's drive north of Tokyo, Kuma has used local Yamizo cedar on the roof and to create screens. The museum houses the work of Hiroshige Ando, one of the country's most renowned artists, and the building might have turned out to be an exercise in traditional Japanese architecture. But when Kuma won the award, the judges cited the way 'he has successfully combined new and traditional elements to produce architecture that is completely modern while still being sensitively and carefully adjusted to its existing surroundings' – and the Bato museum is an excellent example of that. It is crisply engineered and uncompromising in a manner that brings

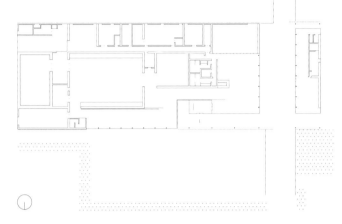

right up to date the traditions for which Kuma has shown so much respect.

Covering an area of 5,587 square metres (60,138 square feet), the building could not be simpler in its overall form: a single-storey rectangle, with some mundane elements tucked away in a basement. It has a shallow-pitched roof and, within its barn-like dimensions, contains two exhibition spaces, an open gallery and all the usual administrative and educational functions of a gallery. A public right of way runs through the western side of the gallery, and beyond this are a restaurant and shop.

The building has a simple steel frame. Its roof, made of six rows of cedar slats on each of the two slopes, rests above glazing, allowing the slats to cast shadows into the building. On the long, north side of the building, the most public aspect, there is a screen of cedar slats running almost the entire length, again in front of glass. This combination of timber slats and glass allows the timber to be seen from both inside and out. It also allows bright sunlight during the day and interior lighting at night to set up patterns of constantly changing reflections, which make it difficult to be sure what is structure and what is reflection, and to identify the boundary between inside and out.

There is no such confusion when one comes to the exhibition areas themselves. These are entirely enclosed, so that the lighting can be controlled. Gentle fibre-optics provide an acceptable degree of illumination for the delicate paper of Ando's prints. But the timber theme continues in suspended cedar ceilings, with dark voids behind them. Combined with the dark blue of the walls, this creates a sufficiently subdued environment to allow visitors to focus on the work.

Other materials used within the building are also local and traditional: Karasuyama *washi* handcrafted paper for the walls and dark Ashino-ishi stone for the floors. Kuma has described his concept of the museum as a place that 'expresses the artistry and tradition of Hiroshige by means of a traditional, yet subdued exterior'. A guide to the museum explains that, for those interested simply in the architecture, it is not necessary to visit the exhibition spaces. Indeed, despite the importance of Hiroshige Ando's work, there are likely to be many visitors who come simply to see how well Kengo Kuma has realized his concept.

THIS PAGE

TOP LEFT   The building contains two exhibition spaces, open gallery space and, to the far side of the public passage, a restaurant and shop.

TOP RIGHT   Light coming through the layers of timber and glass has an almost tactile quality.

ABOVE   Exhibition spaces are entirely enclosed, to protect the delicate works on paper.

OPPOSITE

TOP   Traditional handmade paper, called *washi*, is used for the inner walls of the circulation areas.

BOTTOM   A simple steel frame supports the cedar cladding.

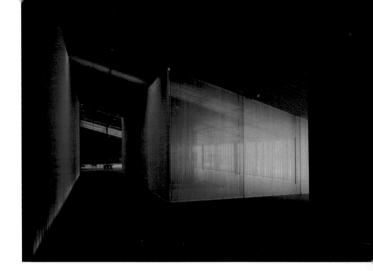

STUDENT HOUSING

Coimbra, Portugal | Aires Mateus e Associados | 1999

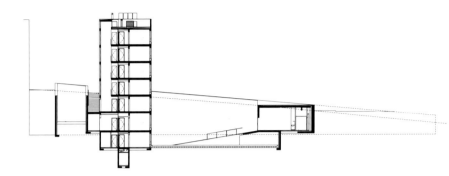

OPPOSITE

TOP   The architect chose to put almost all the accommodation into an eight-storey block at one end of the triangular site.

BOTTOM   Façades are clad in phenolic-resin-bonded plywood.

THIS PAGE

ABOVE   The concrete façades are nearly entirely blank, except for slot windows at one corner.

RIGHT   Rooms in the eight-storey tower face into an enclosed courtyard.

Founded in the thirteenth century, the University of Coimbra is the second oldest in Europe and still a dominant presence in this small Portuguese town, midway between Lisbon and Oporto. Its latest accommodation, squeezed onto a small triangular site, has been designed by Aires Mateus e Associados, a practice that has been described as 'Portugal's rising star'. Although the practice is more accustomed to working with concrete, in Coimbra it uses timber to animate one of the flush and severe façades that are becoming a trademark of the practice's work.

The complete lack of adornment on the façades somehow makes the window openings seem smaller, so that one is

surprised to discover that there is, in fact, plenty of natural light inside. The architect has chosen to put most of the accommodation at one end of the site, where it rises to the maximum permitted level of eight storeys. This block runs north–south, with, at right angles to it, a two-storey accommodation building. A common room, tucked into the apex of the triangle, helps create a central, enclosed courtyard, into which the bedsitting rooms of the tall building face.

In total, there are 52 bedsitting rooms, with a toilet and shower shared by each pair of rooms. The rooms in the tall building face east, across the courtyard, while those on the southern block look

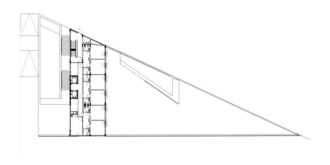

THIS PAGE
LEFT   Rooms are offset from each other, to increase the animation of the façade.
BELOW   Plans of the lower floor, ground floor and a standard floor show the disposition of student rooms around the courtyard and common room.

OPPOSITE
There are three different heights of cladding panel, and the timber shutters are the same height as a medium-sized panel.

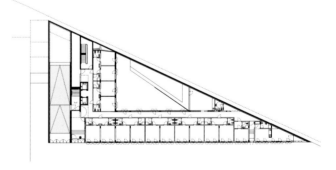

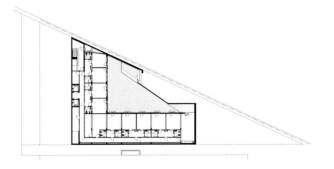

outward, to the south. It is these façades that are animated with timber.

At Coimbra, the architect has clad the façades in 8 mm (⅓ inch) phenolic-resin-bonded plywood in front of a ventilated cavity, insulation and blockwork. These are 80 cm (31 inches) wide, and come in three different heights. The window openings to the bedsitting rooms are the same height as a medium-height façade panel and are two panels wide. These windows have 20 mm (¾ inch) thick pivoting plywood shutters that the residents open and close as they wish. With windows on each floor slightly offset from those below, this creates an attractively randomized pattern. At night, if all residents choose

to shut out the light, the building closes down completely. This game with patterning is similar to one the practice used on a university building in Lisbon, although there the irregular patterning was permanent.

The timber façades contrast with the others, which are of exposed, white concrete blocks, with a rough surface that can reflect the sunlight. The only openings on the tall block on these façades are on the south-west corner, where horizontal slots allow light to penetrate into small common rooms. Otherwise, these are forbidding, if beautifully realized, faces that give no sign of the playfulness of the timber façade that lies around the corner.

LIMERICK COUNTY HALL

Limerick, Ireland | Bucholz McEvoy | 2003

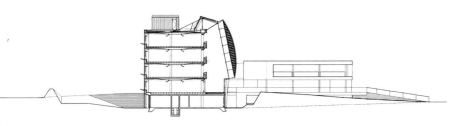

THIS PAGE

TOP   The *brise-soleil* shades the 75 metre (246 feet) long western façade of the office building, from which the council chamber juts out.

ABOVE LEFT   The façade is designed to make a good impression on motorists driving past.

ABOVE RIGHT   By canting the glazing and using the *brise-soleil*, the architect was able to give staff views of the constantly changing skies of western Ireland, without overheating the building-length atrium.

OPPOSITE
Vertical bowstring trusses, joined by timber members, make up the screen.

In a resurgence of municipal pride, numerous civic buildings have been constructed or planned in Ireland during the past decade. Designed by distinguished architects from within the country and from overseas, they tend to emphasize openness, communication and an environmental agenda.

Bucholz McEvoy made its mark when it won an open international competition in March 1996 to design the county hall for Fingal County Council in Dublin. Based in Dublin, the principals of the practice are US-trained Merritt Bucholz and Karen McEvoy, who, although trained in Dublin, has worked overseas with practices including Michael Graves

& Associates and Emilio Ambasz & Associates. The pair followed up the acclaimed Fingal building with the Limerick County Hall at Dooradoyle, which was the Irish entry at the 2002 Venice Biennale, while still under construction.

Although largely a concrete structure, Limerick County Hall has a delicately engineered timber *brise-soleil* along its long western façade that forms an integral part of the environmental strategy of the building, as well as creating its public face. For people driving past on a newly created road, the county hall provides a continuously changing face, a signal that this is a building of some importance, with

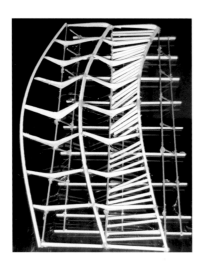

enough glimpses inside to reflect the
concept of 'open government'.

The siting of the building is unusual. It
forms a gateway to semi-natural open
space along the Ballynaclough River, but
stands at the intersection of this open
space and a large shopping centre,
surrounded by a parking area. The
offices for 260 council staff are
contained within a north–south
oriented rectangle, with an atrium
running the full length of the western
façade. This is intended to provide
pedestrian links with both existing and
future developments, and the architect
has embraced the concept of the
adjacent shopping mall to offer a similar
level of accessibility within its building.
Rather than being lost within this large
building, the council chamber juts out
at right angles from the western façade,
so that the timber screen forms a
backdrop to it.

At 75 metres (246 feet) long and 15
metres (49 feet) high, the timber screen
was engineered by Paris-based RFR.
Not merely an add-on, it forms an
engineering whole with the canted glass
wall behind it. The architect was keen
that those inside the building should be
able to enjoy the rapidly changing skies
of the west of Ireland, and tipping the
glazing made this possible. The screen
itself consists of a number of vertical
bowstring trusses, of which the lower

chord is of galvanized steel and the
catenary arch and all connecting
members are of timber. The lower
chords provide stiffening and
anchorages for the glass wall.

Horizontal timber members join
the trusses to each other, running from
the bow of one to the bottom chord
of the next. Combined with the cross
members in the trusses themselves,
which are more numerous than pure
structural considerations would require,
they provide shading from the sun when
it is in all orientations from south to
west. The members linking the trusses
are triangular in profile, reflecting the
fact that they have to support their own
self-weight across a span of 3 metres
(10 feet) and also deal with snow loads.

A variety of materials can be used for
brises-soleil, and, given the widespread
use of concrete in the rest of the
structure, timber might not have
seemed the obvious choice. But, as far
as the architect was concerned, it was; it
symbolized the green credentials of the
building, provided a low embodied
energy and helped to provide a link with
the natural landscape. The building has
a design life of 120 years, and the timber
had therefore to be robust. It is an
inherently tough material, but detailing,
design and processes all had to be
considered carefully to ensure that it
did not suffer.

The timber is Scots pine (Pinus sylvestris) from the Czech Republic, which was shipped to Austria in 12 metre (39 foot) lengths for cutting into strips, gluing, forming and finishing. The common, although increasingly controversial, copper chrome arsenic (CCA) treatment for timber has been banned in Austria for a number of years, and instead a copper chrome boron (CCB) treatment was used. Establishing the correct gluing regime required some rather nerve-racking testing.

Gluing is an exact science, and there is an interaction between the preservative treatment and the glue. In glulam, the type of glue required changes with the laminate thickness. Whereas the requirements are well documented for timber that has had a CCA treatment, there is little information available for CCB-treated timber. The glulam contractor, Weihag, advised increasing the thickness of the laminate strips from 25 to 33 mm (1 to 1⅓ inches). It made up a sample and put this in a weathering oven. When the results turned out well, it adopted this strategy on the building. Another potential difficulty was the

connection between the timber and the galvanized steel chord. To isolate the two incompatible materials from each other, all fixings from the steel to the timber were done in stainless steel.

It was essential that water did not gather in any of the timber joints, so the sections were always designed with a sharp upper surface to ensure that water would fall away from the joints. Birds can be as much of a hazard as rain, since their droppings are not only unsightly, but also a source of chemical attack. The lateral members were spaced to ensure that they were too close together for pigeons to perch. On the top parts of the trusses, where this ploy would not work, anti-bird spikes were used.

Such attention to detail is unlikely to be noticed by the motorists speeding past, but they should certainly appreciate the interest of the façade. And, if they are driving some distance, they may be able to see how the building compares with others in Ireland's distinguished roster of civic offices.

THIS PAGE

LEFT   The council offices are designed to be as welcoming and accessible as a nearby retail centre.

BELOW   An atrium runs the whole length of the western façade.

OPPOSITE
The combination of glazing and the timber screen allows views out, while keeping out the worst of the sun.

APARTMENT BLOCKS
Rue des Suisses, Paris
CLIENT Régie Immobilière de la Ville de Paris
ARCHITECTS Herzog & de Meuron
PROJECT TEAM Béla Berec, Andrea Bernhard,
Christine Binswanger, Jacques Herzog,
Robert Hösl, Sacha Marchal, Mario Meier,
Pierre de Meuron
COMPETITION 1996 Herzog & de Meuron
PROJECT TEAM Béla Berec, Christine
Binswanger, Jacques Herzog, Susanne Kleinlein,
Mario Meier, Pierre de Meuron, Reto Oechslin,
Stephan Wolff
CONSTRUCTION SUPERVISION Cabinet
A.S. Mizrahi
GENERAL CONTRACTOR Bouygues SA

ATSUSHI IMAI MEMORIAL GYMNASIUM
Odate, Japan
PROJECT TEAM Shigeru Ban, Nobutaka Hiraga,
Soichiro Hiyoshi, Keita Sugai
STRUCTURAL ENGINEERS TIS & Partners –
Norihide Imagawa, Yuuki Ozawa
MECHANICAL ENGINEERING ES Associates
GENERAL CONTRACTORS Obayashi Gumi

BALLY STORE
Berlin, Germany
ARCHITECT Craig Bassam, Bassam Fellows
(formerly Craig Bassam Studio)
GENERAL CONTRACTOR Blumer Schreinerei
LIGHTING DESIGNER Dinnebier Licht Berlin

BODEGAS VIÑA PEREZ CRUZ
Chile
ARCHITECT José Cruz Ovalle
ASSOCIATE ARCHITECTS Hernán Cruz S,
Ana Turell S-C
COLLABORATOR M. Ramirez
STRUCTURAL ENGINEER R.G. Ingenieros y
Mario Wagner
TECHNICAL COORDINATION Ramón Goldsack

BODEGAS YSIOS
Laguardia, Spain
ARCHITECT Santiago Calatrava
ENGINEER Santiago Calatrava

BRENTWOOD SKYTRAIN STATION
Vancouver, Canada
CLIENT Rapid Transit Project Office
DESIGN TEAM Busby + Associates
ARCHITECTS B. Billingsley, M. Bonaventura,
P. Busby, S. Edwards, T. Mullock, M. Nielsen,
R. Peck, A. Slawinski
STRUCTURAL ENGINEER Fast & Epp Partners
MECHANICAL ENGINEER Klohn Crippen
ELECTRICAL ENGINEER Agra Simons
LANDSCAPE ARCHITECT Durante Kreuk
PUBLIC ARTIST Jill Anholt

CAROUSEL HOUSE
Greenport, New York, USA
CLIENT Village of Greenport, New York
ARCHITECTS SHoP/Sharples Holden Pasquarelli
STRUCTURAL ENGINEER Buro Happold
Mechanical Engineer: Laszlo Bodak Engineer, PC
ELECTRICAL ENGINEERS Leonard J Strandberg
& Associates/ Laszlo Bodak Engineer PC
CIVIL ENGINEER Leonard J Strandberg &
Associates
LANDSCAPE CONSULTANT Quennell
Rothschild & Partners, LLP
LIGHTING CONSULTANT Universe Lighting
GENERAL CONTRACTOR Carriage Hill
Associates, Inc.
ELECTRICAL CONTRACTOR Johnson Electric
Construction Corporation
DOOR FABRICATOR L. D. Flecken, Inc.
CARPENTER J.E. O'Donnell Construction
Co, Inc.

CHESA FUTURA APARTMENT BUILDING
St Moritz, Switzerland
CLIENT SISA Immobilien AG
ARCHITECTS Foster and Partners –
Norman Foster, Graham Phillips, Stefan Behling,
Matteo Fantoni, Sven Ollmann, Kate Carter,
Jooryung Kim, Judit Kimpian, Tillman Lenz,
Cristiana Paoletti, Stefan Robanus,
Carolin Schaal, Horacio Schmidt,
Thomas Spranger, Anna Sutor, Michele Tarroni,
Huw Whitehead, Francis Aish
Küchel Architects – Arnd Küchel, Vic Cajacob,
Martin Hauri, Georg Spachtholz,
Francesco Baldini, Thomas Henz,
Thomas Kaufmann, Richart Kevic
STRUCTURAL ENGINEERS Edy Toscano AG,
Ivo Diethelm GmbH, Arup
MECHANICAL & ELECTRICAL ENGINEERS
EN/ES/TE AG, R & B Engineering GmbH
ACOUSTIC ENGINEERS Edy Toscano AG
QUANTITY SURVEYOR Davis Langdon &
Everest
CLADDING CONSULTANT Emmer Pfenninger
Partner AG
LIGHTING Reflexion AG
CONCRETE O Christoffel AG
TIMBER STRUCTURE Holzbau Amann
GLASS LOBBIES Buehlmann AG
SHINGLES Patrick Staeger
ROOF Dachtechnik AG
WINDOWS HFF Fenster und Fassaden AG
COMPUTERS Siemens Switzerland
DOORS Lualdi
STONE Vogt
PLASTERBOARD Palombo
KITCHENS Bulthaup
TIMBER FLOORS Hagetra
METALWORK Pfister

CHILDREN'S LEISURE CENTRE
Chessy, France
CLIENT SAN du Val d'Europe, Ville de Chessy
ARCHITECTS Philippe Lankry Avant Travaux
Architects
ENGINEERING/CONTRACTOR Alto Ingénierie
SA
SHELL SPE
WOOD STRUCTURE Paris Charpentes
ROOFING Répisol
EXTERIOR JOINERY SHMI
INTERIOR JOINERY Fériaud
FLOORING EFI
PLUMBING SEED
ELECTRICS Thevenet
LIFT Kone

COUNTRY HOUSE
Granada, Spain
Eduardo y Luis Javier Martín Martín

DOWNLAND GRIDSHELL
Sussex, England
CLIENT Weald & Downland Open Air Museum
ARCHITECTS Edward Cullinan Architects –
Edward Cullinan, Robin Nicholson, John Romer,
PROJECT ARCHITECT Steve Johnson
STRUCTURAL ENGINEER Buro Happold –
Michael Dickson, Richard Harris, Chris Williams,
James Rowe, Oliver Kelly, Shane Dagger
SYSTEMS ENGINEER Buro Happold –
Doug King, Simon Wright
QUANTITY SURVEYORS Boxall Sayer –
Clive Sayer, David Foster, Paul Comins
PLANNING SUPERVISOR Boxall Sayer –
Clive Sayer
MAIN CONTRACTOR E.A. Chiverton –
Mike Wigmore, Chris Silverson
CARPENTERS The Green Oak Carpentry
Company – Andrew Holloway, Stephen Corbett
SCAFFOLDING Peri UK Ltd./Peri GmbH –
Howard Ball, Jurgen Kurth

EXPO CANOPY (EXPO-DACH)
Hanover, Germany
CLIENT Deutsche Messe AG
REPRESENTATIVE OF THE MANAGING BOARD
Sepp D. Heckmann
DIRECTOR FOR CENTRAL TECHNICAL OFFICE
Dr.-Ing. Rainar Herbertz
ARCHITECTS Herzog + Partner BDA –
Prof. Thomas Herzog, Hanns Jörg Schrade
PROJECT ARCHITECT Roland Schneider
Assistants: Jan Bunje, Peter Gotsch,
Moritz Korn, Thomas Rampp, Stefan Sinning
REALIZATION BKSP Projektpartner GmbH,
Hannover
PROJECT SUPERVISOR Ingo Brosch
Assistants: Wilfried Peters, Hans-Joachim Kaub
STRUCTURAL ENGINEERS IEZ Natterer GmbH,
Wiesenfelden – Prof. Julius Natterer, Dr.-Ing.
Norbert Burger, Assistants: Andreas Behnke,
Alan Müller, Johannes Natterer, Volker Schmidt;
Ingenieurbüro Bertsche, Prackenbach –
Peter Bertsche, Assistant: Peter Fitz;
Ingenieurbüro kgs, Hildesheim –
Prof. Dr.-Ing. Martin H. Kessel, Dirk Gnutzmann,
Assistants: Klaus Winkelmann, Georg Klauke
VIBRATION REPORT Technische Universität
München, Institut für Tragwerksbau –
Prof. Dr.-Ing. Heinrich Kreuzinger
PROOF ENGIENEERS FOR STRUCTURAL
ANALYSIS Ingenieurbüro Speich-Hinkes-
Lindemann, Hannover – Prof. Dr.-Ing. Martin
Speich, Dipl.-Ing. Josef Lindemann
COLOUR DESIGN Prof. Rainer Wittenborn,
München
LIGHTING DESIGN Ulrike Brandi Licht, Hamburg
PROJECT SUPERVISORS Marana Müller-Wiefel,
Oliver Ost
MEMBRANE PLANNING & ENGINEERING
IF Jörg Tritthart, Dr.-Ing. Hartmut Ayrle,
Reichenau/Konstanz, Engineers and architects
for lightweight structures
PROJECT MANAGEMENT Assmann Beraten
und Planen GmbH, Hamburg –
Dr.-Ing. Wolfgang Henning

SOIL REPORT Dr.-Ing. Maihorst & Partner,
Hannover
FIRE PROTECTION Hosser, Hass & Partner,
Braunschweig
FOUNDATIONS Renk Horstmann Renk,
Hannover
SURVEYING SERVICES Descoll v. Berckefeldt,
Hannover
EXTERNAL WORKS Dieter Kienast, Vogt Partner,
Zürich

HIROSHIGE ANDO MUSEUM
Bato, Japan
ARCHITECTS Kengo Kuma & Associates
COOPERATIVE ARCHITECTS Ando Architecture
Design Office
STRUCTURAL ENGINEERS Aoki Structural
Engineers
MECHANICAL ENGINEERS P.T. Morimura
& Associates
GENERAL CONTRACTORS Obayashi
corporation

**HOUNSLOW EAST UNDERGROUND
STATION**
London, England
CLIENT & PROJECT MANAGER London
Underground (Infraco JNP Limited)
ARCHITECT Acanthus Lawrence & Wrightson
QUANTITY SURVEYOR Dearle & Henderson
STRUCTURAL ENGINEER (roof &
superstructure) Buro Happold
CIVIL ENGINEER/M&E ENGINEER Infraco JNP
Limited
CLADDING CONSULTANT Buro Happold
CONTRACTOR Gleeson MC Limited
TIMBER FABRICATOR Cowleys Structural
Timber
STEEL FABRICATOR S H Structures
TIMBER TESTING University of Bath

HOUSE
San Prudencio Uleta, Vitoria, Spain
CLIENT Melquiades Pérez de Eulate
ARCHITECTS Roberto Ercilla and
Miguel Angel Campo

HOUSING AND STUDIOS
Newbern, Alabama, USA
Rural Studio, Auburn University

**HOUSING FOR FRAIL ELDERLY PEOPLE
(MAPAD)**
Tremblay-en-France, France
ARCHITECTS maast – François Marzelle and
Isabelle Manescau
CLIENT SAGE
ENGINEER GEC Ingénierie

LIMERICK COUNTY HALL
Ireland
ARCHITECT Bucholz McEvoy Architects, Dublin,
Ireland
CLIENT Limerick County Council
STRUCTURAL ENGINEER Michael Punch and
Partners
SERVICES ENGINEER Buro Happold
FAÇADE ENGINEER RFR Paris
QUANTITY SURVEYOR Boyd Creed Sweett
CONTRACTOR John Sisk and Co Ltd

MUNICIPAL SWIMMING POOL
Puentedeume, Spain
CLIENT Diputación Provincial de A Coruña
ARCHITECTS QRC architects – Antonio Raya,
Carlos Quintáns, Cristóbal Crespo
COLLABORATORS Santiago Sánchez, Enrique
Antelo, architects
CONSTRUCTION MANAGERS Antonio Raya,
Carlos Quintáns, Cristóbal Crespo, Diputación
de A Coruña Architectural Services – Marian
Juárez, Javier Fafián
CONTRACTOR Construcciones Mouzo y
Souto S.L.
MECHANICAL ENGINEERING INSELT, S.L.
CARPENTRY & WOOD Carpintería Hijos de
Romay, S.L.
AIR CONDITIONING ALTAIR, S.L.
ELECTRICITY ALTAIR, S.L.
WATER PURIFICATION AQUANOR, S.L.
STAINLESS STEELWORK Industrias Caamaño
S.L.
DAMPPROOFING Firestone Giscosa S.A.,
INSTALLATOR TRADISCO
PLUMBING Instalaciones RAYPA S.L.

MYOKENZAN WORSHIP HALL
Kawanishi, Japan
DESIGN TEAM Shin Takamatsu, Mitsuo Manno,
Masafumi Sato
STRUCTURAL ENGINEERS Toda Corporation
Structural Engineering Dept.
MECHANICAL ENGINEERS Architectural
Environmental Laboratory
GENERAL CONTRACTOR Toda Corporation

NUOVA FIERA EXHIBITION CENTRE
Rimini, Italy
CLIENT Ente Autonomo Fiera di Rimini,
Presidente Lorenzo Cagnoni
ARCHITECTS gmp – von Gerkan, Marg +
Partner
DESIGN Prof. Volkwin Marg
PROJECT MANAGER Stephanie Joebsch
TEAM Yasemin Erkan, Hauke Huusmann, Thomas
Dammann, Wolfgang Schmidt, Regine Glaser,
Helene van gen Hassend, Mariachiara Breda,
Susanne Bern, Carsten Plog, Marco Vivori,
Eduard Mijic, Arne Starke, Dieter Rösinger,
Olaf Bey, Uschi Köper, Beate Kling,
Elisabeth Menne, Dagmar Weber, Ina Hartig
LOCAL PARTNER ARCHITECT Dr. Clemens
Kusch
STRUCTURAL ENGINEERS Favero & Milan
CONSULTANTS Schlaich Bergermann und
Partner
TECHNICAL & ELECTRICAL ENGINEERS
Studio T.I.
CONSULTANT Uli Behr
LANDSCAPE DESIGN Studio Land, Milan

OFFICE FOR LUNDBECK
PHARMACEUTICALS
Milton Keynes, England
PROJECT CREDIT Artillery Architecture &
Interior Design

PAINTROCK CAMP
Hyattville, Wyoming, USA
ARCHITECT Charles Rose Architects Inc.
PRINCIPAL-IN-CHARGE Charles Rose
PROJECT TEAM Charles Rose, Eric Robinson,
David Gabriel, David Martin, Franco Ghiraldi,
Lori Sang, Takashi Yanai, Maryann Thompson,
Marios Christodoulides, Patricia Chen,
Heidi Beebe

PARCO DELLA MUSICA
Rome, Italy
CLIENT City of Rome
ARCHITECTS Renzo Piano Building Workshop
DESIGN TEAM, COMPETITION 1994 K. Fraser
(architect in charge), S. Ishida (senior partner)
with C. Hussey, J. Fujita and G.G. Bianchi, L. Lin,
M. Palmore, E. Piazze, A. Recagno, R. Sala,
C. Sapper, R.V. Truffelli (partner), L. Viti;
G. Langasco (CAD operator)
CONSULTANTS, COMPETITION 1994 Ove
Arup & Partners (structure and services);
Müller Bbm (acoustics); Davis Langdon &
Everest (cost control); F. Zagari, E. Trabella
(landscaping); Tecnocamere (fire prevention)
DESIGN TEAM, DESIGN DEVELOPMENT
1994–1998 S. Scarabicchi (partner in charge),
D. Hart (partner), M. Varratta with S. Ishida,
M. Carroll (senior partners) and M. Alvisi,
W. Boley, C. Brizzolara, F. Caccavale, A. Calafati,
G. Cohen, I. Cuppone, A. De Luca, M. Howard,
G. Giordano, E. Suarez-Lugo, S. Tagliacarne,
A. Valente, H. Yamaguchi; S. D'Atri, D. Guerrisi,
L. Massone, M. Ottonello, D. Simonetti (CAD
operators); D. Cavagna, S. Rossi (models)
CONSULTANTS, DESIGN DEVELOPMENT
1994–1998 Studio Vitone & Associati
(structure); Manens Intertecnica (services);
Müller Bbm (acoustics); T. Gatehouse,
Austin Italia (cost control); F. Zagari, E. Trabella
(landscaping); Tecnocons (fire prevention);
P.L. Cerri (graphic design)
DESIGN TEAM, CONSTRUCTION PHASE
1997–2002 S. Scarabicchi (partner in charge)
with M. Alvisi, D. Hart (partner) and P. Colonna,
E. Guazzone, A. Spiezia
CONSULTANTS, CONSTRUCTION PHASE
1997–2002 Studio Vitone & Associati
(structure); Manens Intertecnica (services);
Müller Bbm (acoustics); Techint/Drees &
Sommer (site supervision)

PARLIAMENT BUILDING
Karasjok, Norway
CLIENT Statsbygg
ARCHITECTS Stein Halvorsen AS Sivilarkitekter
MNAL and Christian A. Sundby Sivilarkitekter
MNAL

PENINSULA HOUSE
Melbourne, Australia
ARCHITECT Sean Godsell
PRINCIPAL Sean Godsell
PROJECT TEAM Sean Godsell, Hayley Franklin
STRUCTURAL ENGINEER Felicetti Pty Ltd
LANDSCAPE ARCHITECTS Sean Godsell with
Sam Cox
GENERAL CONTRACTOR Kane Constructions
(Vic) Pty Ltd

RAILWAY STATION
Aix-en-Provence, France
CONTRACTING AUTHORITIES RFF, SNCF
(Station Development Office)
DELEGATED CONTRACTORSHIP LN5 (SNCF)
PROJECT MANAGEMENT AND SITE
SUPERVISION Station Design Office (SNCF),
AREP
ARCHITECTS AREP – Jean-Marie Duthilleul,
Etienne Tricaud, Marcel Bajard, Eric Dussiot,
Gérard Planchenault (site manager)
LANDSCAPE ARCHITECT DPLG Desvigne
et Dalnoky
STRUCTURAL ENGINEER ARCORA
TECHNICAL ENGINEER Trouvin/BETEREM
GENERAL ENGINEERING COORDINATION
OTH
SCHEDULING, SITE MANAGEMENT
& PROGRAMMING COPIBAT

REEVE RESIDENCE
Lopez Island, Washington, USA
ARCHITECTS Cutler Anderson
PROJECT ARCHITECT Janet Longnecker
PROJECT DESIGNER Jim Cutler
PROJECT TEAM Julie Cripe, Bruce Anderson
CONTRACTOR Alford Homes (Lowell Alford)
MASON Tony Rothiger
STRUCTURAL ENGINEER Coffman Engineers –
Craig Lee, Deann Arnholtz
ROOFING CONSULTANT Ray Wetherhold

REMODELLING OF A HOUSE
Chamartín, Madrid, Spain
ARCHITECTS Nieto y Sobejano –
Fuensanta Nieto, Enrique Sobejano
COLLABORATORS Carlos Ballesteros,
Mauro Herrero, Juan Carlos Redondo
SITE SUPERVISION Fuensanta Nieto, Enrique
Sobejano, Arquitectos Miguel Mesas Izquierdo,
Aparejador
STRUCTURE N.B.35 S.L., Eduardo Gimeno
MECHANICAL ENGINEER Aguilera Ingenieros
S.A., Pedro Aguilera
MODELS Estudio Nieto-Sobejano

SCHOOL
Gelsenkirchen-Bismark, Germany
PROJECT MANAGERS Peter Hübner,
Martin Müller, Martin Busch
PROJECT TEAM FOR MAIN BUILDING
Peter Hübner, Filip Hübner (workshop);
Christoph Forster, Ulrike Engelhardt (studio);
Martin Müller (laboratory); Martin Busch
(pharmacy); Mathias Gulde (cinema);
Olaf Hübner (theatre); Thomas Strähle
(town hall); Peter Hübner, Reiner Wurst
(library); Peter Hübner (chapel); Olaf Hübner,
Akiko Shirota (bar and music); Thomas Strähle
(public meeting hall); Olaf Hübner (market
place); Bärbel Hübner (interior architecture);
Martin Busch (pyramid)

SHELTER
Campo de Vallemaggia, Switzerland
ARCHITECT Architect dipl. ETH Roberto
Briccola

STUDENT HOUSING
Campus II, Coimbra, Portugal
ARCHITECTS Manuel Aires Mateus, Francisco
Aires Mateus
COLABORATORS Henrique Rodrigues da Silva,
Filipe Nassauer Mónica, Gabriela Gonçalves,
Nuno Marques
STRUCTURAL ENGINEER Planear,
José Carvalheira
ELECTRIC ENGINEER Ruben Sobral
MECHANICAL ENGINEER Galvão Teles
PROJECT MANAGEMENT Rui Prata Ribeiro
CLIENT Serviços da Associação Social,
Universidade de Coimbra

**SWISS ENGINEERING AND TECHNICAL
SCHOOL FOR THE WOOD INDUSTRY**
Biel-Bienne, Switzerland
CLIENT Bau-, Verkehrs- und Energiedirektion
des Kantons Bern
ARCHITECTS Marcel Meili, Markus Peter
with Zeno Vogel
ARCHITECTS' COLLABORATORS
Andreas Schmidt and Thomas Schnabel,
Othmar Villiger, Thomas Kühne,
Urs Schönenberger (Project), Marc Loeliger
(Competition)
STRUCTURAL ENGINEER Jürg Conzett –
Conzett, Bronzini, Gartmann AG
STRUCTURAL ENGINEER'S COLLABORATORS
Reto Tobler (Concrete), Rolf Bachofner (Wood
construction)
ART Jean Pfaff
SITE MANAGEMENT Bauleitungsgemeinschaft
Hofmann + Huggler

SWISS PAVILION
Hanover Expo, Germany
Architect büro Peter Zumthor

TBWA\CHIAT\DAY
San Francisco
ARCHITECT Marmol Radziner and Associates
MANAGING PARTNER Leo Marmol, AIA
DESIGN PARTNER Ron Radziner, AIA
PROJECT MANAGER Anna Hill
PROJECT ARCHITECTS John Kim, Su Kim,
Brendan O'Grady
PROJECT TEAM Paul Benigno, Juli Brode, Patrick
McHugh, Chris McCullough, Daniel Monti,
Bobby Rees, Renee Wilson, Annette Wu
FURNITURE COORDINATOR Michael Holte

TOWER OF BABEL
Artists' Community of Ruigoord

VEGETAL BUILDINGS
Sanfte Strukturen – Marcel Kalberer, Bernadette
Mercx, Anna Kalberer, Peedy Evacic u.a

VIEW SILO HOUSE
Livingston, Montana, USA (Phase One – Silo)
CLIENT Ron Gompertz
ARCHITECTS RoTo Architects, Inc. –
Clark Stevens, AIA, Principal
COLLABORATORS Ben Ives, Dave Kitazaki,
Kirby Smith
TEAM Carrie DiFiore, Eric Meglassen
CONSULTANTS MT Structural –
John Schlegelmilch, Principal

VISITOR CENTRE
Peñaranda, Spain
ARCHITECTS Eduardo Carazo Lefort,
Julio Grijalba Bengoetxea, Victor J. Ruiz Mendez
PROJECT TEAM Detet Renner, Carlos Ruiz
QUANTITY SURVEYOR A. Grijalba Grijalba
CLIENT Ayuntamiento de Peñaranda de Duero
(Burgos)
CONTRACTOR Ortega S.A., Yofra (wood)

WILLOUGHBY DESIGN BARN
Weston, Missouri, USA
CLIENT Ann Willoughby, Willoughby Design
Group
DESIGN TEAM El Dorado, Inc.
PROJECT ARCHITECTS Dan Maginn (AIA),
Josh Shelton
STAFF Doug Hurt, Brady Neely, Chris Burk

Y HOUSE
Catskill Mountains, New York, USA
OWNER Herbert Liaunig
ARCHITECT Steven Holl Architects
PROJECT ARCHITECT Erik F. Langdalen
PROJECT TEAM Annette Goderbauer,
Brad Kelley, Justin Korhammer, Yoh Hanaoka,
Jennifer Lee, Chris McVoy
SITE ARCHITECT Peter Liaunig
STRUCTURAL ENGINEERS Robert Silman
Associates P.C.
LIGHTING CONSULTANT L'Observatoire
International
CONTRACTOR Dick Dougherty
CUSTOM MADE FURNITURE Face Design,
Chris Otterbein